VOLUME FORTY EIGHT

ANNUAL REPORTS IN
MEDICINAL CHEMISTRY

VOLUME FORTY EIGHT

ANNUAL REPORTS IN
MEDICINAL CHEMISTRY

Editor-in-Chief

MANOJ C. DESAI

Gilead Sciences, Inc.
Foster City, CA, USA

Section Editors

ROBICHAUD • DOW • WEINSTEIN • McALPINE • PRIMEAU •
LOWE • BERNSTEIN • BRONSON

AMSTERDAM • BOSTON • HEIDELBERG • LONDON
NEW YORK • OXFORD • PARIS • SAN DIEGO
SAN FRANCISCO • SINGAPORE • SYDNEY • TOKYO

Academic Press is an imprint of Elsevier

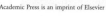

Academic Press is an imprint of Elsevier
525 B Street, Suite 1800, San Diego, CA 92101-4495, USA
225 Wyman Street, Waltham, MA 02451, USA
The Boulevard, Langford Lane, Kidlington, Oxford, OX5 1GB, UK
32, Jamestown Road, London NW1 7BY, UK
Radarweg 29, PO Box 211, 1000 AE Amsterdam, The Netherlands

First edition 2013

ISBN: 978-0-12-417150-3
ISSN: 0065-7743

For information on all Academic Press publications
visit our website at www.store.elsevier.com

Printed and bound in USA
13   14   15   16   12   11   10   9   8   7   6   5   4   3   2   1

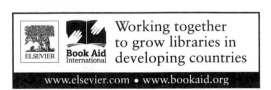

CONTENTS

5. Beyond Secretases: Kinase Inhibitors for the Treatment of Alzheimer's Disease   57

Federico Medda, Breland Smith, Vijay Gokhale, Arthur Y. Shaw, Travis Dunckley, and Christopher Hulme

6. Orexin Receptor Antagonists in Development for Insomnia and CNS Disorders   73

Scott D. Kuduk, Christopher J. Winrow, and Paul J. Coleman

Section 2
Cardiovascular and Metabolic Diseases
Section Editor: Robert L. Dow, Pfizer R&D, Cambridge, Massachusetts

7. Discovery and Development of Prolylcarboxypeptidase Inhibitors for Cardiometabolic Disorders   91

Sarah Chajkowski Scarry and John M. Rimoldi

Section 3
Inflammation Pulmonary GI Diseases

Section Editor: David S. Weinstein, Bristol-Myers Squibb R&D, Princeton,
New Jersey

Section 4
Oncology

Section Editor: Shelli R. McAlpine, School of Chemistry, University of New South Wales, Sydney, Australia

Section 5
Infectious Diseases

Section Editor: John Primeau, Hancock, Maine

Section 6
Topics in Biology

Section Editor: John Lowe, JL3Pharma LLC, Stonington, Connecticut

Section 7
Topics in Drug Design and Discovery

Section Editor: Peter R. Bernstein, PhaRmaB LLC, Rose Valley, Pennsylvania

Section 8
Case Histories and NCEs

Section Editor: Joanne Bronson, Bristol-Myers Squibb, Wallingford, Connecticut

28. To Market, To Market—2012   471

Joanne Bronson, Amelia Black, T. G. Murali Dhar, Bruce A. Ellsworth,
and J. Robert Merritt

CONTRIBUTORS

PREFACE

The *Annual Report in Medicinal Chemistry* is dedicated to furthering genuine interest in learning, chronicling, and sharing information about the discovery of compounds and the methods that lead to new therapeutic advances. ARMC's tradition is to provide disease-based reviews and highlight emerging technologies of interest to medicinal chemists.

A distinguishing feature of *ARMC* is its knowledgeable section editors in the field who evaluate invited reviews for scientific rigor. The current volume contains 28 chapters in 8 sections. The first five sections have a therapeutic focus. Their topics include CNS diseases (edited by Albert J. Robichaud); cardiovascular and metabolic diseases (edited by Robert L. Dow); inflammatory, pulmonary, and gastrointestinal diseases (edited by David S. Weinstein); oncology (edited by Shelli R. McAlpine); and infectious diseases (edited by John Primeau). Sections VI and VII review important topics in biology (edited by John A. Lowe) and new technologies for drug optimization (edited by Peter R. Bernstein). The last section deals with case histories and drugs approved by the FDA in the previous year (edited by Joanne Bronson).

The format for Volume 48 follows our previous issue. We continue with the personal essays, written by MEDI hall of famers, which express the personal stories and scientific careers of "drug hunters," as well as a chapter listing new chemical entities that have entered Phase III. The volume opens with essays by Ashit K. Ganguly, Christopher A. Lipinski, and Malcolm MacCoss. In section VII, a chapter whose subject matter was jointly developed by IUPAC and MEDI, we collate and define many terms that are considered to be essential components of the medicinal chemist's expanded repertoire. In the current volume, the last section has been consolidated with new title, "Case histories and NCE," which includes case histories for the two recently approved cancer drugs, crizotinib and vemurafenib.

This would not have been possible without our panel of section editors, to whom I am indeed very grateful. I would like to thank the authors of this volume for their hard work, patience, dedication, and scholarship in the lengthy process of writing, editing, and making last-minute edits and revisions to their contributions. Further, I extend my sincere thanks to the following reviewers who have provided independent edits to the manuscripts: Michael Bishop, George Chang, Margaret Chu-Moyer, Michael Clarke, Andrew Combs,

Chris Cox, Kevin Currie, Michael Dillon, Gene Dubowchik, Carolyn Dzierba, Gary Flynn, William Greenlee, Randall Halcomb, Nicholas Lodge, Mary Mader, Paul Orenstein, Anandan Palani, Paul Renhowe, Paul Roethle, Greg Roth, Joachim Rudolf, David Sperandio, James Taylor, Bingwei Yang, and Will Watkins. Additionally, I would like to thank Rachel Sumi for editorial help with personal essays.

Finally, I would like to thank Andy Stamford, who was the section editor for the cardiovascular and metabolic diseases for Volumes 40–47, for his dedication to identifying the most relevant topics of interest to the medicinal chemists and his editorial expertise.

I hope the material provided in this volume will serve as a precious resource on important aspects of medicinal chemistry, and, in this way, maintain the tradition of excellence that *ARMC* has brought to us for more than four decades. I am excited about our new initiatives; I look forward to hearing from you about them and welcome your suggestions for future content.

MANOJ C. DESAI, Ph.D.
Gilead Sciences, Inc.
Foster City, CA

Personal Essays

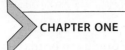

CHAPTER ONE

Challenges in Drug Discovery at Schering–Plough Research Institute: A Personal Reflection

A.K. Ganguly
Stevens Institute of Technology, Hoboken, New Jersey, USA

Contents

Much has happened since I joined Schering Corporation in 1968, including its name change to Schering–Plough Research Institute (SPRI), and, a few years after my retirement, the disappearance of that name altogether after the merger with Merck. While I believe this was a natural partnership and will bear fruitful results, for the purposes of this article, I will refer to the company I worked for as SPRI. After leaving Schering, I have joined academic life at the Stevens Institute of Technology. In this article, I will avoid cataloging every event that occurred during my tenure at SPRI. Instead, I will concentrate on a few key trends in medicinal chemistry drug discovery research[1] in the pharmaceutical industry in general and SPRI in particular.

When I joined SPRI, its reputation for the discovery of novel steroids was already well established. It was in 1950, for example, when Noble, working alone in the laboratory of Charney, discovered that cortisone (used for the treatment of rheumatoid arthritis) could be converted to prednisone, a much more potent compound, using microbiological oxidation. He also converted hydrocortisone to prednisolone. These drugs are still widely used. It should be noted that this study showed very early on that natural products could be converted into more potent drugs using chemical and microbiological modification. In fact, there are several drugs used in human medicine

Annual Reports in Medicinal Chemistry, Volume 48
ISSN 0065-7743
http://dx.doi.org/10.1016/B978-0-12-417150-3.00001-6

which are either obtained from natural sources or whose origin can be traced to compounds obtained from nature, such as β-lactam antibiotics, taxotere, etc.

Steroid research at SPRI began and prospered under the guidance of Hershberg and led to the discovery of betamethasone, valisone, and mometasone furoate. His leadership extended to all areas of research and was ably supported by Herzog, Oliveto, Sperber, Shapiro, and Topliss. Indeed, Topliss guided the discovery of diazoxide, quazepam, and flutamide; he also developed the concept of "Topliss Tree," which is now a must-read for all medicinal chemists. During this period of time, Villani discovered claritin, the first example of a widely used nonsedative antihistamine. It was my good luck in the beginning of my career to be surrounded by all these medicinal chemists who had taken different approaches to drug discovery and succeeded. I learned a lot from all of them.

When I joined SPRI in 1968, Weinstein[2] had already launched a very successful program in infectious diseases with the discovery of gentamicin, an antibiotic complex produced by *Micromonospora* spp., which were introduced to SPRI by Ludeman. Although it was not a well-studied microorganism compared with *Streptomyces* spp., it proved extremely valuable for the subsequent discoveries of sisomicin, rosaramicin, everninomicins, halomicins, and other antibiotics. Under the strong leadership of Daniels,[3a] supported by Wright, Nagabhushan, Mallams, and McCombie, an intensive program of research toward chemical modification of aminoglycosides was undertaken, which led to the discoveries of netilimicin[3b] and isepamicin.[3c] It should be mentioned that proper understanding of the enzymatic mechanisms by which bacteria develop resistance led to the discoveries of these antibiotics, in collaboration with the efforts of the microbiologists Miller and Waitz.

I was assigned to work on nonaminoglycoside antibiotics: specifically, to determine their chemical structures followed by chemical modification to obtain more potent antibacterials. I enjoyed these scientific challenges immensely. With the help of my colleagues, Girijavallabhan, Saksena, Liu, Sarre, and Pramanik, I was able to solve the very complex structures of oligosaccharide antibiotics, including ziracin[4] (**1**), along with the macrolide rosaramicin[5] (**2**), and an ansamycin called "halomicin"[6] (**3**).

Two of these, ziracin and rosaramicin, were advanced to the clinic for Phase III trials. Although they were found to be safe and efficacious in the clinic, they were not advanced further for different reasons. For example, ziracin is a potent antibiotic that is active against gram-positive bacteria, including those which are resistant to methicillin and vancomycin. However, it could not be formulated reproducibly to be used as an intravenous drug. As for rosaramicin, it is a macrolide antibiotic that has a similar activity profile to erythromycin, with additional activity against gram-negative organisms. It was deemed safe and efficacious by the clinicians, but the management at SPRI was not quite convinced of its commercial prospects, and it was dropped from further development. Although these events were disappointing—and not rare in the industry—we learned a great deal from these experiences by bringing in new science and technologies to the organization, which helped in our future endeavors. All the antibacterials mentioned above were obtained from natural sources, and I have been deeply interested in this area of research; however, it became obvious that without being able to discover commercially successful drugs from this source in a timely manner, it would be impossible to pursue this line of investigation. It will, therefore, have to wait for another day when newer technologies become available for collecting, screening, and creating large libraries of natural products, using combinatorial biosynthetic approaches.

At this point, Merck announced the discovery of thienamycin[7] (4), another natural product with a novel structure of a β-lactam antibiotic. The structure of 4, besides containing a β-lactam ring, was totally different than those represented by the well-investigated structures of penicillins and cephalosporins. Thus, it provided us with a novel opportunity to enter the field of β-lactam antibiotics, an area of immense importance to the infectious disease community. Our work began when the general structure of thienamycin was revealed without its stereochemical detail and following the announcement of Woodward's synthesis of penem antibiotics. As penems synthesized by Woodward[8] had 6-acylamino side chains similar to those found in penicillins and cephalosporins, and possessed weak antibacterial activity, we decided to synthesize penems with substituents at C-6, similar to the one present in thienamycin. Since Merck had not disclosed the absolute stereochemistry of thienamycin, we decided to synthesize all the possible stereoisomers of our compounds. After extensive investigation and the development of new synthetic methodologies, we progressed Sch-34343[9] (5) to the clinic. Sch-34343 is a broad-spectrum antibiotic similar in spectrum and potency to imipenem[10] (6), a chemically modified derivative of thienamycin. Our compound, unlike imipenem, did not possess activity against *Pseudomonas* spp. In general, the penems were chemically and enzymatically more stable than imipenem. Sch-34343 was found to be safe and efficacious in the clinic and was progressed to the Phase III stage in the trial. However, its further development was abandoned when it showed suggestions of long-term toxicity in animal studies. Developing novel synthetic routes for synthesis of penems would not have been possible without the creativity and diligence of Afonso, Girijavallabhan, and McCombie.

4 R = H
6 R = —HC=NH

5

I wish to recall now the details that led to the inventions of ezetimibe,[11] posaconazole,[12] and boceprevir,[13] which are presently used in human medicine; and lonafarnib,[14] which was initially being developed for treatment of cancer and has found use in treating the devastating disease progeria. The paths to the discovery of these drugs were very different and illustrate the importance of keeping all options open in drug discovery. Newer technologies, including combinatorial chemistry, X-ray crystallography to determine the structures of ligands complexed with proteins, refinement of fragment-based drug discovery using NMR, and high-throughput assays using mass spectrometry, among others, became available and were used with greater frequency. However, the expectations for using these modern technologies to shorten the time required to discover drugs turned out to be unrealistically high and led to disappointment in some quarters. Nevertheless, time and experience have taught us, medicinal chemists, how and when to use these technologies successfully.

The importance of lowering low-density lipoprotein in the treatment and prevention of coronary heart disease has been established both in the clinic and in epidemiological studies. Statins, which inhibit 3-hydroxy-3-methylglutaryl coenzyme A, thus lowering the biosynthesis of cholesterol in the liver, have been successfully used in the clinic. These drugs include lovastatin, simvastatin, and atorvastatin. Of the two sources of cholesterol in our bodies, approximately 70% is derived from biosynthesis in the liver; the remaining 30% comes from the food we consume. In our institute, we discovered ezetimibe, which is a potent inhibitor of absorption of cholesterol from food. Subsequently, in collaboration with Merck, we also introduced Vytorin, which is a combination of ezetimibe and simvastatin. In the case of ezetimibe, we started working toward the discovery of acyl coenzyme A cholesterol acyltransferase (ACAT) inhibitors, which were implicated in lowering cholesterol levels in hamsters when they were fed a high-cholesterol diet. However, their effect on nonrodents was not known. Starting with amides represented by structure **7**, known in the literature as ACAT inhibitors that lowered the levels of cholesterol ester in hamster liver in the above model without lowering serum cholesterol level, we synthesized several analogs of **7**, including those with restricted conformations.

7

8 R_1 = OCH$_3$, R_2 = OCH$_3$, R_3 = R_4 = H
9 R_1 = OCH$_3$, R_2 = OH, R_3 = R_4 = H
10 R_1 = F, R_2 = OH, R_3 = OH, R_4 = F

After extensive studies, we concluded that monocyclic β-lactams had the best promise, not as inhibitors of ACAT, but as agents that lowered serum cholesterol in various species of animals, including monkeys and dogs, when they were fed a high–cholesterol diet. Indeed, the most potent cholesterol absorption inhibitors were very poor ACAT inhibitors, and it was therefore decided, at this juncture, to test every new compound *in vivo*. Several β-lactams were synthesized by Clader, Burnett, Dugar, and Rosenblum, to name a few. When tested to understand structure–activity relationships, it was concluded that Sch-48461 (**8**) was the most potent analog for lowering cholesterol absorption when administered orally in hamsters, rats, monkeys, and dogs. *In vivo* experiments suggested that **8** underwent extensive metabolic degradation. Van Heek,[15] in the group of Davies, demonstrated that one of the major active metabolites of Sch-48461 was the glucuronide of compound **9**. Based on this information, the decision was made to further explore the structure–activity relationship of Sch-48461, keeping in mind to block possible sites of its metabolism while including sites that could be premetabolized. Based on these studies, it was clear that ezetimibe (**10**) was the most potent of all analogs synthesized. Indeed, in hamsters, rats, monkeys, and dogs, ezetimibe showed dramatic improvement *in vivo* when administered orally, compared to **8**. Subsequently, it was demonstrated, in collaboration with Merck, that the combination of ezetimibe with simvastatin, called Vytorin, produced a significant synergy in lowering serum cholesterol in animal models, described above, as well as in humans. Ezetimibe is also used as monotherapy. It should be noted that the mechanism of action of ezetimibe was elucidated[16] only after FDA approval was received for its commercialization. It was discovered that ezetimibe blocked the activity of the dietary cholesterol transporter NPC1L1, which is expressed at the apical surface of enterocytes.

A commercial synthesis of ezetimibe was developed by Thiruvengadam in the development group led by Walker. Throughout my career, I have recognized that a close working relationship between research and development groups is essential for the success of any project, and I found such a partner in Walker.

The discovery of the antifungal posaconazole (**11**) followed a different pathway. We wished to discover an azole antifungal that was active against *Aspergillus*, *Cyptococcus*, and *Histoplasma* spp., as well as active against fluconazole-sensitive and resistant *Candida* infections. It was well known that azoles worked by inhibiting lanosterol 14α-demethylase, responsible for the biosynthesis of ergosterol, which, in turn, is an essential component of the fungal cell membrane. Itraconazole (**12**) was known to possess activity against *Aspergillus* spp. and attracted our attention. We argued that it would be of interest to synthesize azole antifungals, such as itraconazole, without possessing the 1,3–dioxolane ring system because it will be expected to be unstable under acidic conditions in the stomach. Thus, we synthesized all the possible stereoisomers of the corresponding tetrahydrofurans and, after considerable effort, we identified Sch–51048 (**13**) as the most potent antifungal with broad–spectrum activity, including activity against *Aspergillus* spp.

11 R = (S)(S)-CH(Et)(CH(CH₃)(OH))
13 R = CH(Et)₂
12

Having 5-(R) absolute stereochemistry in **13** for antifungal activity was essential. It is worth remembering that without synthesizing all the possible stereoisomers, it would not have been possible to discover Sch–51048 and, eventually, posaconazole. It appears to me that there is always one more issue to address before declaring victory, and this case is no different. When Sch–51048 was administered in animals, it was metabolized to a more potent antifungal. Although the structure of the metabolite was not known, it was clear from mass spectral studies that the parent compound had undergone monohydroxylated metabolic oxidation in the side chain, probably at the primary (creating three asymmetric centers) or secondary (creating four asymmetric centers) site. It was a daunting task, but Saksena synthesized all the possible stereoisomers, and it was clear that posaconazole possessed the most potent antifungal activity in addition to the desired pharmacokinetic properties. Based on its *in vitro* antifungal activity, which was superior to all then clinically used antifungals (particularly against serious infections caused by *Aspergillus* and other species of fungi, including *Candida*), it was advanced to clinical studies and subsequently approved by the FDA for human use.

The discovery of boceprevir (**14**) was achieved through substrate–based drug design. By this time, we had established a relationship with Pharmacopeia to do combinatorial chemistry as well as an in–house facility to solve X–ray crystal structures of proteins. Hepatitis C virus infection is found around the world, and approximately 200 million people suffer from it. Before the discovery of boceprevir, the standard therapy for the disease was pegylated α–interferon, alone or in combination with ribavarin. Although 80% of genotype-2-infected patients respond well, only 40% of genotype-1 patients show improvement. This lack of effectiveness in the latter group of patients, along with recurrence of infection in interferon-treated patients, prompted several groups to take a different approach, which led to the discovery of orally active Hepatitis C viral protease inhibitors. Boceprevir from our group and telaprevir[17] from Vertex were approved by the FDA in quick succession.

Hepatitis C virus has a positive-strand RNA and encodes a polypeptide that is posttranslationally modified with the help of a serine protease, NS3, to produce mature virions. Inhibition of NS3 protease was therefore deemed an excellent target for discovering a new cure for this disease. We screened a very large number of compounds obtained from the collections of compounds in our files, along with those we generated in our collaborative work with Pharmacopeia, but found no significant lead to pursue. At this point, we started working on the problem using substrate-based drug design. Our starting point was the ketoamide **15** containing the sequence of the viral polypeptide, which covered P6–P5′. It showed excellent potency in inhibiting the NS3 protease. The X-ray crystal structure of the protease was available in our lab and helped guide the discovery of boceprevir. Extensive modification of **15** was required to impart acceptable drug-like characteristics and to provide selectivity against the related human neutrophil elastase. Our approach was to truncate the structure of **15** both on the P and P′ sides with a hope of identifying molecules with smaller molecular weights that still possessed good activity. Extensive synthetic work by Njoroge and others in Girijavallabhan's group, accompanied by X-ray crystal structures of the inhibitors complexed with the protease determined by Strickland, led finally to the discovery of boceprevir.

The initial lead for the discovery of lonafarnib (**16**), a farnesyl transferase inhibitor of mutated Ras protein, was identified by Bishop and his colleagues from our library of compounds, which contained a large number of loratadine analogs. It was not totally surprising that one of them showed activity. I suspect other companies' libraries of compounds will contain analogs of their own successful drugs, which might also serve as initial leads in drug discovery. Pharmacopeia scientists made libraries of loratadine analogs using combinatorial chemistry; in parallel, Doll, Girijavallabhan, and Weber used chemical synthesis and X-ray crystallography to identify several active compounds. Based on all the data, including pharmacokinetic properties, etc., we progressed lonafarnib to the clinic to be tested on lung carcinoma. Although its clinical efficacy against lung carcinoma did not entirely meet our expectation, its effectiveness in progeria patients was extremely gratifying. Progeria is a devastating disease in children who have a genetic mutation that leads to the production of the protein progerin. Like Ras protein, progerin also gets farnesylated for its function, and, therefore, Gordon[18] of Boston Children's Hospital argued for lonafarnib's use in the treatment of progeria. Supplied by Merck, lonafarnib is currently being used for the treatment of children with this disease around the world. The discovery of commercially successful drugs is vitally important to providing the great sums of money necessary to develop the next generation of drugs for the future. As I have noted in this article, many of our attempts toward this end are not always successful. However, in addition to the discovery of important drugs that are widely used and commercially successful, if one can be associated with the discovery of a drug—even by chance—that can be used for only a small number of patients suffering from such a devastating disease, then all of our effort in drug discovery is worthwhile.

There are, of course, other drugs discovered at SPRI that are under development at Merck, including vorapaxar[19] (**17**), the first member of a new class of antiplatelet agents known as thrombin receptor antagonists. The inspiration for this work was derived from the structure of the alkaloid himbacine (**18**), which was synthesized by Chackalamannil. Vorapaxar, whose absolute stereochemistry was opposite to that of **18**, was discovered after extensive synthetic work and is now undergoing extensive clinical trial.

I find academic life challenging, yet rewarding in a different sense, as when you are able to get young students excited about pursuing careers in science. My interest in natural products led us to synthesize himgaline[20a] (**19**), in collaboration with Chackalamannil at SPRI and to synthesize[20b] compound **20** as a novel pharmacophore based on the structure of *Erythrina* alkaloids, as well as compound **21** as a potent inhibitor[20c] of HIV-1 protease. These are examples of some of the ongoing projects in our laboratory.

In the beginning of this article, I mentioned the merger of SPRI and Merck. Similar megamergers have taken place recently in our industry as it passes through a difficult period. These changes are, I am sure, necessary for the future health of the industry; however, to see so many of our bright scientists made redundant in the process is disheartening. We need to make sure that the infrastructure of scientists built so well and patiently by every member of our industry is not lost for good. As the new model of business is practiced, it will be important to remember that a strong research program will always need creative scientists on the premises.

ACKNOWLEDGMENTS

It is impossible in a brief article to acknowledge the valuable contributions made by each member of the various teams involved in drug discovery at SPRI over several decades. I have identified several senior investigators without the intention of ignoring others. In addition, I would like to acknowledge the contributions of Greenlee, Piwinski, Strader, Bullock, and Pickett.

REFERENCES

1. Ganguly, A. K. Taylor, J. B., Triggle, D. J., Eds.; Comprehensive Medicinal Chemistry II, Vol. 8; Elsevier Ltd: Oxford, 2007, pp 29–51.
2. Wagman, G.; Weinstein, M. J. *Annu. Rev. Microbiol.* **1980**, *34*, 537.
3. (a) Daniels, P. J. Kirk-Othmer Encyclopedia of Chemical Technology, Vol. 2; 3rd ed.; John Wiley & Sons, Inc; New York, 1978, p 819. (b) Wright, J. J. *J. Chem. Soc., Chem.*

Commun. **1976**, 207. (c) Nagabhushan, T. L.; Cooper, A. B.; Tsai, T.; Daniels, P. J. L.; Miller, G. H. *J. Antibiot.* **1978**, *21*, 681.

4. Ganguly, A. K.; Pramanik, B.; Chan, T. M.; Sarre, O.; Liu, Y. T.; Morton, J.; Girijavallabhan, V. *Heterocycles* **1989**, *28*, 83. Ganguly, A. K. *J. Antibiot.* **2000**, *53*, 1038.

5. Ganguly, A. K.; Liu, Y.-T.; Sarre, O.; Jaret, R. S.; McPhail, A. T.; Onan, K. K. *Tetrahedron Lett.* **1980**, *21*, 4699–4702.

6. Ganguly, A. K.; Szmulewicz, S.; Sarre, O. Z.; Greeves, D.; Morton, J.; McGlotten, J. *J. Chem. Soc., Chem. Commun.* **1974**, 395–396.

7. Albers-Schonberg, G.; Arison, B. H.; Hensens, O. D.; Hirshfield, J.; Hoogsteen, K.; Kaczka, E. A.; Rhodes, R. E.; Kahan, J. S.; Kanhan, F. M.; Ratcliffe, R. W.; Walton, E.; Ruswinkle, L. J.; Morin, R. B.; Christensen, B. G. *J. Am. Chem. Soc.* **1978**, *100*, 6491–6499.

8. Ernest, I.; Gosteli, J.; Woodward, R. B. *J. Am. Chem. Soc.* **1979**, *101*, 6301–6305.

9. Ganguly, A. K.; Afonso, A.; Girijavallabhan, V. M.; McCombie, S. *J. Antimicrob. Chemother.* **1985**, (Suppl. C), 1.

10. Leama, W. J.; Wildonger, K. J.; Miller, T. W.; Christensen, B. G. *J. Med. Chem.* **1979**, *22*, 1435–1436.

11. Clader, J. *J. Med. Chem.* **2004**, *47*, 1. Clader, J. Taylor, J.B., Triggle, D.J., Eds.; Comprehensive Medicinal Chemistry II; Vol. 8; Elsevier Ltd: Oxford, UK, 2007; p 65. Rosenblum, S. B.; Huynh, T.; Afonso, A.; Davis, H. R.; Yumibe, N.; Clader, J.; Burnett, D. A. *J. Med. Chem.* **1998**, *41*, 973–980 and references cited in all the above articles.

12. Saksena, A. K.; Girijavallabhan, V. G.; Lovey, R. G.; Pike, R. E.; Wang, H.; Liu, Y. T.; Pinto, P.; Bennett, F.; Jao, E.; Patel, N.; Desai, J. A.; Rane, D.; Cooper, A. B.; Ganguly, A. K. In *Anti-infectives: Recent Advances in Chemistry and Structure Activity Relationships*; Bentley, P.H., O'Hanlon, P.J., Eds.; Royal Society of Chemistry: Cambridge, 1997; p 180.

13. Venkatraman, S.; Bogen, S. L.; Arasappan, A.; Bennett, F.; Kevin Chen, K.; Jao, E.; Liu, Y.-T.; Lovey, R.; Hendrata, S.; Huang, Y.; Pan, W.; Parekh, T.; Pinto, P.; Popov, V.; Pike, R.; Ruan, S.; Santhanam, B.; Vibulbhan, B.; Wu, W.; Yang, W.; Kong, J.; Liang, X.; Wong, J.; Liu, R.; Butkiewicz, N.; Chase, R.; Hart, A.; Agrawal, S.; Ingravallo, P.; Pichardo, J.; Kong, R.; Baroudy, B.; Malcolm, B.; Guo, Z.; Prongay, A.; Madison, V.; Broske, L.; Cui, X.; Cheng, K.-C.; Hsieh, Y.; Brisson, J.-M.; Prelusky, D.; Korfmacher, W.; White, R.; Bogdanowich-Knipp, S.; Pavlovsky, A.; Bradley, P.; Saksena, A. K.; Ganguly, A. K.; Piwinski, J.; Girijavallabhan, V.; Njoroge, F. G. *J. Med. Chem.* **2006**, *49*, 6074–6086.

14. Strickland, C. L.; Weber, P. C.; Windsor, W. T.; Wu, Z.; Le, H. V.; Albanese, M. M.; Alvarez, C. S.; Cesarz, D.; Rosario, J.; Deskus, J.; Mallams, A. K.; Njoroge, F. G.; Piwinski, J. J.; Remiszewski, S.; Rossman, R. R.; Taveras, A. G.; Vibulbhan, B.; Doll, R. J.; Girijavallabhan, V. M.; Ganguly, A. K. *J. Med. Chem.* **1999**, *42*(12), 2125–2135. Ganguly, A. K.; Doll, R. J.; Girijavallabhan, V. G. *Curr. Med. Chem.* **2001**, *8*, 1419.

15. Van Heek, M.; France, C. F.; Compton, D. S.; McLeod, R. L.; Yumibe, N. P.; Alton, K. B.; Sybertz, E. J.; Davis, H. R., Jr. *J. Pharmacol. Exp. Ther.* **1997**, *283*, 157.

16. Altman, S. W.; Davis, H. R., Jr.; Zhu, Li-ji; Yao, X.; Hoos, L. M.; Tetzloff, G.; Iyer, S. P. N.; Maguire, M.; Golovko, A'.; Zeng, M.; Wang, L.; Murgolo, N.; Graziano, M. P. *Science* **2004**, *303*, 1201.

17. Revill, P.; Serradell, N.; Bolos, J.; Rosa, E. *Drugs Future* **2007**, *32*(9), 788.

18. Gordon, L. B.; Kleinman, M. E.; Miller, D. T.; Neuberg, D. S.; Giobbie-Harder, A.; Gerhard-Herman, M.; Smoot, L. B.; Gordon, C. M.; Cleveland, R.; Snyder, B. D.; Brian Fligor, B. W.; Robert Bishop, W. R.; Statkevich, P.; Regen, A.; Sonis, A.; Riley, S.; Ploski, C.; Correia, A.; Quinn, N.; Ullrich, N. J.; Nazarian, A.;

Liang, M. G.; Huh, S. Y.; Schwartzman, A.; Kieran, M. W. *Proc. Natl. Acad. Sci. U.S.A.* **2012**, *109*(41), 16666–16671.

19. Chackalamannil, S.; Wang, Y.; Greenlee, W. J.; Hu, Z.; Xia, Y.; Ahn, H.-S.; Boykow, G.; Hsieh, Y.; Palamanda, J.; Agans-Fantuzzi, J.; Kurowski, S.; Graziano, M.; Chintala, M. *J. Med. Chem.* **2008**, *51*(11), 3061–3064.

20. (a) Shah, U.; Chackalamannil, S.; Ganguly, A. K.; Chelliah, M.; Kolotuchin, S.; Buevich, A.; McPhail, A. *J. Am. Chem. Soc.* **2006**, *128*, 12654–12655. (b) Ganguly, A. K.; Wang, C. H.; Biswas, D.; Misiaszek, J.; Micula, A. *Tetrahedron Lett.* **2006**, *47*, 5539–5542. (c) Ganguly, A. K.; Alluri, S. S.; Caroccia, D.; Biswas, D.; Wang, C. H.; Kang, E.; Zhang, Y.; McPhail, A. T.; Carroll, S. S.; Burlein, C.; Munshi, V.; Orth, P.; Strickland, C. *J. Med. Chem.* **2011**, *54*, 7176–7183.

CHAPTER TWO

My Perspective on Time, Managers—and Scientific Fun

Christopher A. Lipinski
Waterford, Connecticut, USA

Contents

I am a scientist. Over the course of my career, I have learned that success as a scientist requires effective time management, not just to get things done but to protect and nurture the scientific creativity so crucial to innovation. (Warning to the reader: some of the time management techniques described below may border on the just-barely acceptable!) But individual careers do not progress in a vacuum; they require active partnerships with management to move forward. Let me be clear: I am a scientist, not a manager. In my 32 years at Pfizer I never had more than three reports, never had anything to do with budgets or high-level strategy, and almost never attended a meeting of senior managers. (I will discuss a notable exception to that rule later.) Nevertheless, I have the greatest respect for good management. In fact, a good manager can be a scientist's best friend, as I learned during the career change that led me to develop the "Rule of 5."

1. CONTROLLING YOUR SCIENTIFIC LIFE: TIME MANAGEMENT

I have never had much patience for all the interruptions that can plague the scientist, and in my last 5 years at Pfizer I had an active plan to preserve me from all the worthless meetings and interruptions that even by 1997 were beginning to become increasingly annoying. My time

Annual Reports in Medicinal Chemistry, Volume 48
ISSN 0065-7743
http://dx.doi.org/10.1016/B978-0-12-417150-3.00002-8

management solution took root during a 1997 visit to the Pfizer Sandwich UK site. I noticed that the message button on my host's office phone was flashing. When I asked him about it, my host replied that he was fed up with all the time-wasting phone calls, so he just let the voice mail accumulate; after 25 messages nobody could leave him any more voice mails. A light bulb went off in my head and, when I returned to Groton that winter of 1997, I copied his behavior. Unfortunately, that meant that I was difficult to reach, something my supervisor diplomatically reminded me of during my annual evaluation. I hit on the idea of a secret number—one not published in the corporate directory, but one I made available to my wife, our administrative assistant, the immediate members of our technology group, and selected others. While I was not easily reached by phone, I always did respond to a note on my office chair or computer monitor. I figured that if someone were motivated enough to walk to my office, it was probably worthwhile talking to them. The strategy worked wonders in helping me regain control over my time—so well, in fact, I decided to tackle the issue of endless meetings. If I were asked about my availability for a meeting I suspected was worthless, I would reply positively but then cancel at the last minute. Anything important could be relayed to me later. Why did I take such drastic measures? Simply put, a scientist has to have time to think. Constant interruptions kill thinking. You need periods of 90 min of uninterrupted thinking to have any chance of being creative.

While such steps may seem a form of self-preservation, ensuring the creative process actually benefits the organization. In the mid 1990s at Pfizer Groton, for example, we had something called a scientific committee that issued a report detailing the medicinal chemists' concerns. This was the prelude to the very good formal scientific ladder at the Groton site,[1] which made it possible for someone like me with no interest whatsoever in management to retire at the scientific equivalent of the department director position. As part of this scientific committee report, I interviewed every lab head in medicinal chemistry as well as the two management department directors. The medicinal chemistry lab heads were pretty uniformly concerned about straying very far from the formal written project goals. Ironically, it was the two directors who were adamant about medicinal chemistry lab heads being free to spend 15% of their time on their own ideas; in fact, there was already a history of off-the-books skunk works evolving into fully approved projects.[2] The message here is that no written plan can capture all the possibilities for success. Making sure you have the time to think allows you to explore the kind of extracurricular ideas that can prove valuable to the entire organization. Remember: your time is YOUR time and, you should protect it with all the energy you can muster (even if you have to game the system a bit).

2. FOSTER INNOVATION WITH SCIENTIFIC FUN

We have established that time to think is essential. What else fosters innovation? A playful attitude—I call it scientific fun. Skunk works and informal collaborations can be extremely rewarding, but note: you still have to get your official job done. Rewards can only be expected if something of real demonstrable value accrues to the organization as a result of your extra-curricular efforts. Being happy, feeling good about yourself and the esteem of others are great rewards. But there is a risk in this, and one should keep it in mind. Be creative, but keep your eyes on the prize.

From early on at Pfizer I got myself into all kinds of off-the-books activity. I met Frank Clarke (former Ciba research head) at an American Chemical Society (ACS) poster session. I peppered him with questions about his Apple IIe program to control a Mettler titrator and to measure acid/base pK_a. Thus began a long friendship that provided me a lot of help—and computer code. Early on, I latched onto the importance of understanding your molecule because it might help explain its biology or help in activity optimization. In Groton, we had a very active summer intern program in medicinal chemistry. All the chemists competed for the few chemistry undergraduates. I quickly learned how wonderful engineering students were as summer interns. It did not matter what type of engineer—electrical, mechanical, or chemical—they all did great work. There was something in their smart pragmatism that was perfect for my growing interest in physicochemical measurements. Expanding beyond summer interns, I worked with young women lab assistants who became pregnant and did not want to work with lab chemicals, as well as people recovering from lab accidents or from serious illness who were willing to do analytical work with minimal chemical exposure. I learned how to scrounge extra lab space. I also learned that the natural ally for my skunk works was the pharmaceutical sciences department, an effort that led to off-the-books cross-departmental collaborations[3] and cross-project publications[4] and was one of the early factors in my thinking on the Rule of 5. Why did I do all this? Because I enjoyed it—it was scientific fun.

3. A GOOD MANAGER IS A SCIENTIST'S BEST FRIEND

In 1988, the aldose reductase medicinal chemistry program in which I was working was completely stalled because of serious clinical issues with Sorbinil, our Phase III clinical candidate.[5] I was incredibly frustrated and took an evening course on retrosynthetic analysis at the University of Connecticut in Storrs just so I would not go brain dead. I decided to complain to my department director. To my surprise, he responded by saying, "How can I help you, Chris?" I had to think fast. "Well, I would really like to do more of this physicochemical measurements work that I have been doing part time," I answered. He offered that he could find a $20,000 budget for equipment, but I would have to find a person to do the work, find lab space and meet with the chemistry managers to get their signoff. I was ecstatic, since I knew none of this would be a problem. From despair to elation—all thanks to a supportive manager.

My next breakthrough also came from good management. I was invited to a senior management meeting between medicinal chemistry and pharmaceutical sciences managers. This was effectively a 3-h gripe session about what was being thrown over the discovery/development wall and the resulting pharmaceutical science horror stories about the properties of the newly nominated discovery clinical candidates. I was the only scientist present. It turned out that solubility was a key problem, and I determined to do something about it. The meeting completely changed my physicochemical lab operation. From Gene Fiese (later to be president of the American Association of Pharmaceutical Sciences (AAPS)) who ran our general pharmaceutics profiling lab, I learned about some very preliminary unpublished Pfizer work measuring solubility by precipitation. We had an HP diode array in our lab that I knew was quite sensitive to absorbance due to light scatter from precipitation. I later met up with Roger McIntosh, a software designer with expertise in pump systems, with whom I had previously worked at Pfizer. Over beer and hot dogs, I sketched out an idea for an automated solubility assay to measure precipitation on our UV diode array. Roger delivered the pumps and software, and we were off and running on what I believe was the industry's first discovery zone automated solubility assay. Primitive by today's standards, we ran 1000 solubilities our first year.

This new solubility assay probably violated every text-book rule about the qualities of a proper solubility assay, but we had a problem in discovery, and I was convinced the new assay—as rudimentary as it was—would be part of the solution. As it turned out, the pharmaceutical sciences department proved to be one of my strongest allies. They were fed up with the incredibly insoluble compounds they were being stuck with, and they figured that anyone who could improve the solubility of discovery compounds was their friend.

Here again, good management came into the picture. After the solubility assay had run for a year, I was asked to give a short 15-min presentation at our annual project review. The budget was tight, and nearly everybody put in a plug for more manpower. I gave a straightforward science presentation with no pitch for extra personnel. After about 5 min, I noticed that John LaMattina, Pfizer's future head of research, was getting agitated. He peppered me with questions about the solubility assay. How many compounds did you run in a year? 1000. Could you run 10,000 in a year? Yes, but not with the current assay. The meeting broke up and I returned to the lab. Six weeks later, my department director dropped by. "Chris, the chemistry manager's authorized a new hire for you. We want you to run 10,000 solubility assays per year." While I had not requested the additional manpower, management recognized that a crucial problem could be addressed, and it was up to me to get the job done.

Now I was really having fun. I knew I had to hire the right person for the solubility assay, and the official Pfizer hiring process was dreadfully slow. So I did the "old boy network" bit. Professor Mike Smith at the University of Connecticut said he had the perfect person for me. Paul Feeney was a very talented materials science major with programming skills. He had already been admitted to graduate school but could not attend because of family issues. I phoned Paul, and in the first 5 min I knew he was my person. In fact, when he interviewed at Pfizer, Paul had the hire offer from me before he could leave Groton that day.

Now it was my turn to be the good manager. I knew the pressures at Pfizer, and I knew I had to protect Paul's time. His job was to design a solubility measurement robot, build it and write the computer code to control it. I promised that nothing would detract from this—no screening, no stupid meetings, nothing. It took seven months, and Paul delivered splendidly.

The lab got transformed rapidly from what originally looked like an analytical lab into what we would now recognize as a high-tech lab with computers, robots, industrial-grade overhead wiring, heavy-duty backup systems, etc. We got a large former chemistry lab custom revamped for our needs. We were unique. There was no such lab anywhere in the Pfizer world, or anywhere else for that matter. I loved it. For the first time in my career, I hired people whose work was way outside of medicinal chemistry: an electrical engineer from the Navy sound lab, a biochemist, and a molecular biologist. I used to ask jokingly, "What went wrong in the lab today?" I had to protect my people's freedom to try something new—to have scientific fun. We were a high-tech new lab. Of course things would go wrong; that was to be expected.

In the spring of 1992, I was asked to make a decision: run both a med-chem and a technology lab, or choose to run a single lab. I made my decision by tracking my time, a very good decision method. In my experience, people inevitably choose to spend more time on the fun stuff. I was putting in a huge number of hours on the new lab and loving every minute of it, so the decision to run a technology lab in Pfizer's new technology zone was a no-brainer. But I did have a concern. I was one step below the top of the scientific ladder. If I left medicinal chemistry, I would leave behind my support structure and risk losing that last promotion. There was no precedent in the industry for what I was doing, nor was there a clear path for international recognition, which was required for the top position in our scientific ladder. I went to see John LaMattina. He acknowledged the risk but gave me his support for my career direction change. John came through for me again when a few years later, in early 1997, I previewed for him what the "Rule of 5" paper might look like. I wanted to know the chances of approval by the Pfizer Groton publication committee. I still remember the conversation. John did not answer me at first. He looked up at the ceiling—thinking. After several minutes he said, "Go for it, Chris."

Being part of a technology zone led me to acquire a lot of new skills. For one, I had a lot more direct contact with computational chemistry. My immediate boss was Beryl Dominy, a computational chemist, later to become a coauthor on the Rule of 5 publication.[6] I became a member of a small upstart group called precandidate technology. Our combination of solubility, permeability and metabolic stability assays on every compound made by traditional chemistry proved the value of early drug-like properties in compound optimization. Eventually, we grew into a really industrial multi-Pfizer site operation and expanded to include toxicity assays. I soaked up the pharmaceutical sciences literature, and I learned a great deal about commercial databases and how to combine literature and internal data.

In early summer 1995, I heard about JMP, a statistical graphing program from SAS.[7] It was incredibly graphic (great for a chemist) and dead easy to use. It also introduced me to Kaplan–Meier survival plots. I mentioned earlier that unstructured time is critical to thinking and innovation. Summer for me was the best time for scientific fun. Lots of people were on vacation, and there were fewer of those dreaded meetings. JMP was a perfect scientific-fun fooling-around tool. I realized that Kaplan–Meier survival plots could be used for looking at the relationship of properties to compound quality. I knew that size (molecular weight), lipophilicity and hydrogen bonding were all important to oral absorption, and I knew from the Derwent World Drug Index that United States Adopted Name (USAN) and International Non Proprietary Name (INN) were fields in the World Drug Index (WDI) and that these names were given to compounds at entry to Phase II clinical. The first Rule of 5 plots were created on a pleasant June afternoon, and I immediately noticed that for the four parameters in the Rule of 5, the cutoff value at the 90% level was 5 or (2 times 5). This activity had absolutely nothing to do with any formal project work, and it would never have happened if I had a full Outlook calendar.

The "Rule of 5" became operational at Pfizer because of a casual conversation.[8] Beryl Dominy (another of those good managers) had stopped by my office to chat, and I was describing what I had done with JMP and what I called the "Rule of 5" observation. I blurted out, "Too bad we couldn't make it part of the internal Pfizer compound registration system." Beryl looked at me and said, "Let's do it." So, we did it. Beryl controlled the registration process, and he did the hard computational lifting. We never asked for permission. We just did it. Our medicinal chemistry managers knew there was a problem with poor solubility, but they did not know who the culprits were. When they discovered a registration method existed to flag problematic compounds, they could ask the hard questions at quarterly meetings. The Rule of 5 illustrates the synergy possible between scientists and managers. The chemistry managers became Rule of 5 supporters because it solved a problem for them; for the organization, it improved the overall chemistry physicochemical compound quality.[9]

The Rule of 5 is a very simple rule-based computational solution to improve the probability of improving oral absorption. Are there other lingering problems awaiting solution? There are many. Here is one of my favorites: biologically active compounds are not uniformly distributed in chemistry space but rather are tightly clustered.[10] As a result, screening truly diverse libraries is the worst way to discover a drug.[11] Nevertheless, evolution has probed the limits of ligand druggability through natural products.[12] The problem, however, is that we cannot currently analyze natural products in the same way we categorize synthetics. We do not have a handle on natural product shapes. The scaffold approach, which works so well for synthetics, does not work for natural products. The complex and often ring containing structures and intramolecular H-bonding possibilities mean that we do not know shape. There is no high-throughput method for calculating intramolecular hydrogen bonding energy. The recent work on cyclosporine illustrates the issue.[13] Cyclosporine is orally available and a clear Rule of 5 outlier. It is also a molecular chameleon. It changes shape depending on the environment. In a lipid environment, the backbone N-methyls stick outside. In polar media, the N-methyls are buried in the middle, and the polar functionality presents to the outside. I think a lot of natural products have this chemical molecular chameleon property. If we could calculate intramolecular H-bonding energies, we could determine shape and, therefore, better analyze the natural products' hundreds-of-millions-of-years evolution experiment on the limits of druggability. If only I had some time to think.

REFERENCES

1. http://www.nature.com/drugdisc/nj/articles/nrd1623.html.
2. Melvin, L. S.; Johnson, M. R. *NIDA Res. Monogr.* **1987**, 1046-9516, *79*, 31–47.
3. Lipinski, C. A.; Fiese, E. F.; Korst, R. J. *Quant. Struct.-Act. Relat.* **1991**, *10*(2), 109–117.
4. Desai, M. C.; Thadeio, P. F.; Lipinski, C. A.; Liston, D. R.; Spencer, R. W.; Williams, I. H. *Bioorg. Med. Chem. Lett.* **1991**, *1*(8), 411–414.
5. Maggs, J. L.; Park, B. K. *Biochem. Pharmacol.* **1988**, *37*(4), 743–748.
6. Lipinski, C. A.; Lombardo, F.; Dominy, B. W.; Feeney, P. J. *Adv. Drug Deliv. Rev.* **1997**, *23*(1–3), 3–25.
7. Much newer versions of JMP software are available www.jmp.com/software/.
8. http://www.biosolveit.de/video/lipinski/index.html?ct=1.
9. http://www.beilstein-institut.de/bozen2002/proceedings/Lipinski/Lipinski.pdf.
10. Hert, J.; Irwin, J. J.; Laggner, C.; Keiser, M. J.; Shoichet, B. K. *Nat. Chem. Biol.* **2009**, *5*(7), 479–483.
11. Lipinski, C. A. *J. Pharmacol. Toxicol. Methods* **2000**, *44*, 235–249.
12. Lachance, H.; Wetzel, S.; Kumar, K.; Waldmann, J. *Med. Chem.* **2012**, *55*(13), 5989–6001.
13. Alex, A.; Millan, D. S.; Perez, M.; Wakenhut, F.; Whitlock, G. A. *MedChemComm* **2011**, *2*(7), 669–674.

CHAPTER THREE

A Career in Medicinal Chemistry—A Journey in Drug Discovery

Malcolm MacCoss

Bohicket Pharma Consulting LLC, Seabrook Island, South Carolina, USA

Contents

I first picked up an interest in chemistry as a grammar school student in the UK in the mid-1960s, when I became fascinated with the design of experiments designed to better understand the molecular world in which we live. I went on to study chemistry at the University of Birmingham, where I received a B.Sc. in 1968 and a Ph.D. in 1971.

1. EARLY DAYS IN NUCLEOSIDE, NUCLEOTIDE, AND NUCLEIC ACID CHEMISTRY

As a graduate student at Birmingham, I was interested in the nucleic acid research which was being carried out in the laboratories of Professors A.S. Jones and R.T. Walker. This was the decade after the double helix had been solved and the molecular underpinnings of cellular biology, that are taken for granted today, were being understood in depth by the application of strong synthetic chemistry to the fundamental understandings of biology. In particular, I was fascinated by Professor Khorana's work in

Annual Reports in Medicinal Chemistry, Volume 48
ISSN 0065-7743
http://dx.doi.org/10.1016/B978-0-12-417150-3.00003-X

23

which he completed the first total synthesis of a single gene, as well as Professor Ikehara's efforts on a total synthesis of a tRNA. In my own doctoral work, I was drawn to the application of difficult synthetic chemistry in solving biological problems and I prepared oligodeoxynucleotide analogues which lacked the phosphate linking group,[1] but which had similar interbase spacings. It was fascinating to observe the results obtained in a cell free biological system with molecules that I had made in the chemistry laboratory.[2] It is a fascination that was to shape my life and still does today.

In late 1971, I emigrated to Canada and continued my training as a postdoctoral fellow in the laboratory of Professor Morris Robins at the University of Alberta where I delved deeper into the synthetic complexities of nucleoside and nucleotide chemistry.[3–6] The passion for a complete understanding of the underlying detail in organic chemistry, a constant in the Robins lab, stood me in very good stead as my career developed, and it also sparked my subsequent fascination with the coupling of structural information with reactivity and, later, with intermolecular interactions—another important element of drug discovery.

In 1976, I moved to the United States to take up a position at Argonne National Laboratory in Illinois. There, I worked in the Division of Biological and Medical Research with Dr Steve Danyluk to solve structural problems in nucleic acids and their analogues using NMR methods,[7–9] including ^{17}O NMR.[10–12] In addition, I started work on a novel series of phospholipid prodrugs of anticancer agents[13,14] (and characterized their macromolecular aggregation characteristics[15]), which took me directly into the world of drug discovery.

My journey into that world was completed in 1982, when I joined Merck Research Laboratories in Rahway, NJ, which was one of the strongest synthetic laboratories in the industry. I arrived there at a glorious time for Merck Research—Merck was to launch over the next decade, norfloxacin, enalapril, imipenem, ivermectin, lovastatin, lisinopril, simvastatin, and finasteride. They were "America's Most Admired Company" for 7 years in a row, and the science excellence and enthusiasm around the laboratories was palpable. I stayed there for 26 years, until 2008 when I left to move a few miles up the Garden State Parkway to join Schering–Plough. In 2010, I left "big pharma" to start my own consulting company, Bohicket Pharma Consulting LLC, that now serves a number of clients, from small biotechs and start-ups to large multinational pharmaceutical firms. The underlying driver throughout my career is a passion to take our chemistry skills and use them to design and create new medicines for human patients. It is a noble cause and the task is demanding—but extremely rewarding.

2. ANTIVIRALS

When I joined Merck, there was already an effort underway to prepare antiherpes drugs which were based on the acycloguanosine format that had been initiated by Burroughs–Wellcome. Prior to this effort, nucleoside chemistry at Merck had been restricted to efforts in the anticancer area and it had been difficult to identify active molecules that did not also show the attendant toxicities seen with antimetabolite approaches. Acyclovir demonstrated that polymerase inhibitors could be prepared that had little or no such associated toxicology. Working with Wally Ashton, Dick Tolman, and the group at Rahway, along with the virologists at Merck West Point, I quickly got involved with other acycloguanine analogues,[16–18] including ganciclovir and ganciclovir cyclic phosphate[19,20] and I became associated with the biochemical and antiviral understanding of this class of compounds.[20–22] It was an excellent introduction to the industry because, I quickly learned how chemists, biochemists, and biologists (and in a highly competitive field, lawyers) have to work together as a close-knit team in order to succeed—something which I suspect is rarely experienced in academia.

Acyclovir   Ganciclovir   Ganciclovir 1',3'- cyclic phosphate

In the late 1990s, with the continued interest in anti–HIV drugs and the identification of the CCR5 receptor as a HIV virus cell entry coreceptor, I worked with virology colleagues at Merck West Point (Emilio Emini and Daria Hazuda), on CCR5 antagonists, with the immunology being done at Rahway (Marty Springer). Although several compounds almost made it, none progressed into the clinic, although one was tested successfully as a topical agent in primate models using a vaginal application. The concept was finally realized by Pfizer[23] who were able to commercialize SELZENTRY® (maraviroc) as a CCR5 receptor antagonist for the treatment of HIV.

Returning to the nucleoside area, my work switched to focus on inhibitors of hepatitis C RNA-dependent RNA polymerase, again with virology colleagues at Merck West Point (Daria Hazuda, Dave Olsen, and Steve Carroll) and collaborators at Isis Pharmaceuticals. The group came up with 2'-C-methyltubercidin which is an extremely potent inhibitor (as the triphosphate) of the target polymerase. The compound was moving forward,[24] until we came upon a tissue retention issue in preclinical studies that we were not able to overcome. This was quite a disappointment because it really looked like we had made a potent, highly selective polymerase inhibitor without the toxicological issues associated with so many other nucleoside analogues. As is so often the case, the issues only became apparent late in the preclinical program.

2'-C-methyltubercidin

When I moved to Schering–Plough in 2008, it was a pleasure to pick up on the HepC projects there. The groups there were just completing the work on the NS3 protease inhibitor VICTRELIS® (boceprevir) as well as the backup, narlaprevir, and they were working hard on a number of additional HepC targets.

It is a tribute to all the antiviral researchers around the world to observe how the global threat of HIV, in such a relatively short time period, has been addressed with effective new medicines. Now, the scourge of HepC (the leading cause of liver transplants), until recently considered mostly incurable, seems to be on the verge of a total cure, based on recent clinical trial data conducted by Gilead. Exciting times indeed.

3. ELASTASE INHIBITORS

The project at Merck to design inhibitors of PMN elastase was one of the longest running projects at Rahway when I joined the team and it was one in which the level of scientific understanding of the mechanism of inhibition was of the very highest order. This led to a drug design initiative which was scientifically complex but very rewarding. Because of the large amounts of elastase that needed to be inhibited in order to block matrix destruction by this serine protease, we decided to tackle the project using suicide inhibitors. Of course, we went forward with some trepidation on this front, due to concerns about the potential for the generation of non-specific binding to off-target proteins that might lead to idiosyncratic immune-based toxicity. The electrophile that was selected to react with the active site serine was a β-lactam, but we wanted to ensure that this was not such a reactive entity as used, for example, in antibiotics. This led to the design of a mono cyclic β-lactam scaffold and a gem diethyl substitution, α to the β-lactam carbonyl, which fitted nicely into the P1-binding site; this afforded selectivity, as well as reducing the reactivity of the β-lactam, except when it was located on the active site of elastase.[25] After much optimization of the rest of the molecule, the team designed compounds such as those shown below which had second-order rate constants of ~2,000,000 $M^{-1} s^{-1}$, but which were stable in hot water and to external nucleophiles such as N-acetylcysteine.

The detailed mechanistic evaluation of how these molecules interacted with the enzyme active site, and the slow regeneration of active enzyme, was a superb piece of mechanistic enzymology carried out by Blaine Knight and his colleagues.[26–28]

Pharmacologically, this was also a terrific program, championed by Phil Davies, Rick Mumford, and Euan MacIntyre, and led to some of the very best biomarkers of elastase enzyme activity in blood, based on antibodies generated to elastase-derived neoepitopes,[29] at a time when biomarkers were certainly not as prevalent in drug discovery programs as they are today. Another key element to this program was the decision to progress forward with an inhibitor that was basic and cell penetrant (and which we believed inhibited the elastase *inside* the azurophilic granule before the elastase exited the cell).[30] Such a molecule performed markedly better in pharmacological models of disease than the earlier acidic analogues which were much less cell penetrant and which inhibited elastase after it had been released into the blood. Ultimately this program led to a clinical candidate that was evaluated by DuPont–Merck in clinical trials against rheumatoid arthritis and cystic fibrosis after clear demonstration of bio-chemical efficacy using biomarkers of enzyme–inhibitor complex forma-tion and catalytic activity. Scientifically this was a truly exciting time for me and it demonstrated once again the absolute necessity of close multi-disciplinary teams in drug discovery, where world class experts from dif-ferent disciplines tackle the problems as a single team, with single goals. One still ponders whether a drug is possible from this approach to elastase inhibition, particularly in the areas of α1-antitrypsin deficiency and selected leukemias and myeloid dyscrasias.

4. SUBSTANCE P ANTAGONISTS AND THE DISCOVERY OF APREPITANT AND FOSAPREPITANT

Many companies and academic groups had been working on tac-hykinin receptors since the original discovery in 1931 by von Euler and Gaddum of a "…depressor substance in certain tissue extracts…," which they called Substance P. Many peptide like analogues had been prepared as antagonists, but small molecule nonpeptide antagonists proved elusive until 1991, when Snider et al.[31] from Pfizer described, CP-96,345, a potent, subnanomolar antagonist that was a simple, disubstituted quinuclidine. This molecule was a breakthrough in the field, but had only modest oral bioavail-ability and in animal models showed cardiovascular side effects that we and others were able to attribute to off-target blockade ($IC_{50} \sim 240$ nM) of the L-type Ca^{++} channel. This compound was followed a couple of years later[32] with CP-99,994 which was also subnanomolar, but was an even simpler disubstituted piperidine; in addition, it had minimal cardiovascular off-target effects, due to a much reduced binding to the L-type Ca^{++} channel, which we attributed to the lower basicity of the molecule in this particular case.

The excitement that these discoveries brought to the tachykinin field was huge, and Merck, like many other companies, invested in a big team to con-struct better Substance P antagonists. At the time, the biology indicated that such molecules would have utility in a number of diseases including pain, asthma, arthritis, emesis, migraine, and even psychiatric disorders such as depression. I was asked to lead the medicinal chemistry efforts in Rahway on this target, in collaboration with a similar group in our Terlings Park Centre for Neuroscience in the United Kingdom. This made for a huge team of talented chemists, pharmacologists, biochemists, and neuroscientists that, once again, proved the value of a closely knit group of scientists working together toward a common goal.

We began by seeking other ways to append the pharmacophores onto the central scaffold, and in addition, to lower the basicity of the core scaffold ring (to reduce the Ca^{++} channel liability, *vide supra*). We were pleased to be able to replace the benzylic amine used in CP-96,345 and CP-99,994 with a benzylic oxygen, despite previous beliefs that the –NH– was necessary. However, it became clear that the optimal substitution pattern on the benzylic ether phenyl ring was the CF_3 electron withdrawing group located at the 3- and 5-positions of the phenyl ring. This was in marked contrast to the optimal substitution for a benzylic amine which was shown to be the electron-donating CH_3O group at the 2-position of the phenyl ring. In order to reduce the basicity, we chose to replace the piperidine with a morpholine group which necessitated the installation of a morpholine acetal from which to append the benzylic ether mentioned above. This required some clever chemistry, driven by Jeff Hale and Sandy Mills, in order to get the desired axial and *cis* orientation (relative to the phenyl ring at the 3-position). Further elaboration of potential substituents on the morpholine nitrogen led to the methyl triazolinone which became our first generation preclinical candidate. We decided to seek more metabolic stability, so we added a benzylic methyl group on the benzylic ether (introduced *via* the ester by using the Tebbe reagent); fluorination on the 3-phenyl ring at the *para* position was also introduced to ensure maximal coverage *in vivo*.[33]

This was advanced into preclinical development and became EMEND® (aprepitant) which was approved by the FDA and launched in 2003; it is still the only approved Substance P antagonist and is used to treat chemotherapy-induced nausea and vomiting (CINV) when dosed with other antiemetics. Alongside our development of EMEND®, we decided that it would also be appropriate to have an i.v. formulation in our armamentarium for CINV in order to give physicians flexibility in treatment options. However, we knew this was going to be difficult because compounds in the series had high molecular weight and were quite insoluble (in fact, a novel formulation technology was used at the time to ensure good oral bioavailability). We felt that we needed to install a charge on the molecule(s) in order to get them soluble enough for i.v. delivery. Thus, while our colleagues at Terlings Park set about investigating new analogues with cationic character, my group in Rahway decided to look at phosphate prodrugs of EMEND® itself. As it turned out, we were not able to advance the excellent cationic molecules synthesized at Terlings Park and so our phosphate prodrugs were extensively investigated. These needed to be quite unstable in the body so that they could be rapidly degraded to the parent EMEND®, and the only reasonable site for attachment of the phosphate was on the triazolinone ring. This led us to prepare IVEMEND® (fosaprepitant),[34] which satisfied all our requirements. It was approved by the FDA and launched in 2008.

CP-96,345   CP-99,994   R = –H, aprepitant
                                      R = –PO₃H₂ ; fosaprepitant

This project was a true milestone in my career. Not only was it the largest Merck team in terms of resources, but also the scope of the chemistry and the biology involved, all carried out under the most intense competitive pressure, made it quite an experience. To follow a drug literally from a design on a blackboard, through synthesis in my group, and on to the marketplace was immensely rewarding. It is one of the most difficult things to do in biomedical science, and I will always be indebted to the team who made it possible.

5. HYPOGLYCEMIC AGENTS AND THE DISCOVERY OF SITAGLIPTIN

While I had spent a lot of time at Merck working on antiviral agents as well as on G-protein coupled receptors (GPCRs), I had also spent quite a lot of time looking at potential treatments for type 2 diabetes mellitus. I got my introduction to this fascinating area of biomedicine early in my career, working with Eve Slater, Cathy Strader, Dick Saperstein, and Peggy Cascieri, on a licensed-in compound. We did not know the mechanism of action, but when dosed orally it had a profound effect lowering blood glucose in ob/ob mice. From a chemistry perspective, we tackled this the "old fashioned way" using the *in vivo* biology in ob/ob mice to guide our structure–activity relationships. We were hoping to decipher the mechanism as we went along, but even after we had an optimized molecule ready for advancement into toxicology, we still had not been able to definitively identify the mechanism, other than it was an insulin secretagogue, acting on β-cells in the pancreas. As a result, the compound was not advanced, but I maintained my interest in this disease, which is proving to be such a health care burden around the world. One approach that I pursued was in the area of glucagon antagonists, and we made some interesting molecules which were potent antagonists.

Yet another approach was initiated by Nancy Thornberry who was looking for a new enzyme target in the diabetes field. She selected dipeptidyl peptidase IV (DPP4), which is an enzyme that degrades the incretin glucagon-like peptide-1 (GLP-1). GLP-1 is produced after a meal and stimulates the pancreatic β-cells to secrete insulin, but only when glucose levels are elevated, thus removing the concern of hypoglycemic overshoot which is seen with other insulin secretagogues. GLP-1 agonists were contemplated, but such molecules proved hard to identify in screening and so Nancy chose to target DPP4. The idea was to prolong the normally short half-life of GLP-1 by inhibiting its breakdown by DPP4.

Again, this proved to be a very competitive area, with several major pharmas already ahead of us. In addition, a small European company had already completed some clinical trials with their lead molecules. These leads were licensed in and though they proved to be ultimately flawed from a side effect profile, they were invaluable in identifying the cause of the side effects that were being observed as resulting from off-target activity on closely related family members of the enzyme class. Ann Weber and her group in my department set about tackling the medicinal chemistry on this project and did a fabulous job bringing together very clean molecules with excellent PK and toxicology profiles. This was done under immense time constraints, and not a single issue was left to chance. It was a marvelous effort—again reemphasizing to me the absolute importance of closely integrated chemistry, biology, and pharmacology teams all rapidly advancing toward a single goal. This endeavor allowed us to design clean, selective molecules without the side effects discussed above and ultimately led us to JANUVIA® (sitagliptin).[35] The speed with which this project was carried out was electrifying. At the time, the typical timeline for major pharma from entry into toxicology until the first pivotal dose in Phase III was 4.12 years; for sitagliptin, the timeline was 2.25 years!![36] JANUVIA® was launched in 2006, and JANUMET®, a fixed dose combination of sitagliptin and metformin, was launched in 2007. These two products are now major frontline therapies in the world wide fight against diabetes, treating millions of patients.

Sitagliptin

6. CONCLUSIONS

My career in pharma could not have happened without all of my collaborators over the years. They have all been great colleagues and friends and they have contributed to my ongoing education in drug discovery, an endeavor that my colleague Paul Anderson rightly calls a lifelong learning process.[37] They have taken me down a path that I would not have thought possible when I left graduate school to start a career in chemistry. I have learned that there can be few feelings that surpass those that come with being associated with a drug that treats human disease. What we do is a noble profession, applying our skills and knowledge to address age old problems that affect humankind. I believe that true innovation occurs only at the interfaces between disciplines, such as chemistry and biology, and we must continue to foster such innovation as we go forward in tackling the devastating diseases of the twenty-first century.

REFERENCES

1. Edge, M. D.; Hodgson, A.; Jones, A. S.; MacCoss, M.; Walker, R. T. *J. Chem. Soc., Perkin Trans. I* **1973**, *3*, 290.
2. Jones, A. S.; MacCoss, M.; Walker, R. T. *Biochim. Biophys. Acta* **1973**, *294*, 365.
3. Robins, M. J.; Ramani, G.; MacCoss, M. *Can. J. Chem.* **1975**, *53*, 1302.
4. Robins, M. J.; MacCoss, M.; Naik, S. R.; Ramani, G. *J. Am. Chem. Soc.* **1976**, *98*, 7381.
5. Robins, M. J.; MacCoss, M. *J. Am. Chem. Soc.* **1977**, *99*, 4654.
6. Robins, M. J.; MacCoss, M.; Wilson, J. S. *J. Am. Chem. Soc.* **1977**, *99*, 4660.
7. MacCoss, M.; Ezra, F. S.; Robins, M. J.; Danyluk, S. S. *J. Am. Chem. Soc.* **1977**, *99*, 7495.
8. MacCoss, M.; Ezra, F. S.; Robins, M. J.; Danyluk, S. S. *Carbohydr. Res.* **1978**, *62*, 203.
9. MacCoss, M.; Ainsworth, C. F.; Leo, G.; Ezra, F. S.; Danyluk, S. S. *J. Am. Chem. Soc.* **1980**, *102*, 7353.
10. Schwartz, H. M.; MacCoss, M.; Danyluk, S. S. *J. Am. Chem. Soc.* **1983**, *105*, 5901–5911.
11. Schwartz, H. M.; MacCoss, M.; Danyluk, S. S. *Magn. Reson. Chem.* **1985**, *23*, 885.
12. Schwartz, H. M.; MacCoss, M.; Danyluk, S. S. *Tetrahedron Lett.* **1980**, *21*, 3837.
13. MacCoss, M.; Ryu, E. K.; Matsushita, T. *Biochem. Biophys. Res. Commun.* **1978**, *85*, 714.

14. Ryu, E. K.; Ross, R. J.; Matsushita, T.; MacCoss, M.; Hong, C. I.; West, C. R. *J. Med. Chem.* **1982**, *25*, 1322.
15. MacCoss, M.; Edwards, J.; Seed, T. M.; Spragg, P. *Biochim. Biophys. Acta* **1982**, *719*, 544.
16. MacCoss, M.; Chen, A.; Tolman, R. L. *Tetrahedron Lett.* **1985**, *26*, 1815.
17. MacCoss, M.; Chen, A.; Tolman, R. L. *Tetrahedron Lett.* **1985**, *26*, 4287.
18. MacCoss, M.; Tolman, R. L.; Ashton, W. T.; Wagner, A. F.; Hannah, J.; Field, A. K.; Karkas, J. D.; Germershausen, J. I. *Chem. Scr.* **1986**, *23*, 113.
19. Field, A. K.; Davies, M. E.; DeWitt, C. M.; Perry, H. C.; Scholfield, T. I.; Karkas, J. D.; Germershausen, J. I.; Wagner, A. F.; Cantone, C. L.; MacCoss, M.; Tolman, R. L. *Antiviral Res.* **1986**, *6*, 329.
20. Germershausen, J. I.; Bostedor, R.; Liou, R.; Field, A. K.; Wagner, A. F.; MacCoss, M.; Tolman, R. L.; Karkas, J. D. *Antimicrob. Agents Chemother.* **1986**, *29*, 1025.
21. Karkas, J. D.; Germershausen, J. I.; Tolman, R. L.; MacCoss, M.; Wagner, A. F.; Liou, R.; Bostedor, R. *Biochim. Biophys. Acta* **1987**, *911*, 27.
22. MacCoss, M.; Wagner, A. F.; Cantone, C. L.; Strelitz, R. A.; Chen, A.; Ashton, W. T.; Hannah, J.; Tolman, R. L.; Bostedor, R.; Germershausen, J. I.; Karkas, J. D.; Perry, H. C.; Field, A. K. *Nucleosides Nucleotides* **1989**, *8*, 1155.
23. Price, D. A.; Gayton, S.; Selby, M. D.; Ahman, J.; Haycock-Lewandowski, S.; Stammen, B. L.; Warren, A. *Tetrahedron Lett.* **2005**, *46*, 5005.
24. Olsen, D. B.; Eldrup, A. B.; Bartholomew, L.; Bhat, B.; Bosserman, M. R.; Ceccacci, A.; Colwell, L. F.; Fay, J. F.; Flores, O. A.; Getty, K. L.; Grobler, J. A.; LaFemina, R. L.; Markel, E. J.; Migliaccio, G.; Prhavc, M.; Stahlhut, M. W.; Tomassini, J. E.; MacCoss, M.; Hazuda, D. J.; Carroll, S. S. *Antimicrob. Agents Chemother.* **2004**, *48*, 3944.
25. Knight, W. B.; Green, B. G.; Chabin, R. M.; Gale, P.; Maycock, A. L.; Weston, H.; Kuo, D. W.; Westler, W. M.; Dorn, C. P.; Finke, P. E.; Hagmann, W. K.; Hale, J. J.; Liesch, J.; MacCoss, M.; Navia, M. A.; Shah, S. K.; Underwood, D.; Doherty, J. B. *Biochemistry* **1992**, *31*, 8160.
26. Chabin, R. M.; Green, B. G.; Gale, P.; Maycock, A. L.; Weston, H.; Dorn, C. P.; Finke, P. E.; Hagmann, W. K.; Hale, J. J.; MacCoss, M.; Shah, S. K.; Underwood, D.; Doherty, J. B.; Knight, W. B. *Biochemistry* **1993**, *32*, 8970.
27. Green, B. G.; Chabin, R.; Mills, S.; Underwood, D. J.; Shah, S. K.; Kuo, D.; Gale, P.; Maycock, A. L.; Liesch, J.; Burgey, C. S.; Doherty, J. B.; Dorn, C. P.; Finke, P. E.; Hagmann, W. K.; Hale, J. J.; MacCoss, M.; Westler, W. M.; Knight, W. B. *Biochemistry* **1995**, *34*, 14331.
28. Underwood, D. J.; Green, B. G.; Chabin, R.; Mills, S.; Doherty, J. B.; Finke, P. E.; MacCoss, M.; Shah, S. K.; Burgey, C. S.; Dickinson, T. A.; Griffin, P. R.; Lee, T. E.; Swiderek, K. M.; Covey, T.; Westler, W. M.; Knight, W. B. *Biochemistry* **1995**, *34*, 14344.
29. Carter, R. I.; Mumford, R. A.; Treonze, K. M.; Finke, P. E.; Davies, P.; Si, Q.; Humes, J. L.; Dirksen, A.; Piitulainen, E.; Ahmad, A.; Stockley, R. A. *Thorax* **2011**, *66*, 686.
30. Mumford, R. A.; Chabin, R.; Chiu, S.; Davies, P.; Doherty, J. B.; Finke, P. E.; Fletcher, D.; Green, B.; Griffen, P.; Kissinger, A.; Knight, W. B.; Kostura, M.; Klatt, T.; MacCoss, M.; Meurer, R.; Miller, D.; Pacholok, S.; Poe, M.; Shah, S.; Vincent, S.; Williams, H.; Humes, J. *Am. J. Respir. Crit. Care Med.* **1995**, *151*, A532.
31. Snider, R. M.; Constantine, J. W.; Lowe, J. A.; Longo, K. P.; Lebel, W. S.; Woody, H. A.; Drozda, S. E.; Desai, M. C.; Vinick, F. J.; Spencer, R. W. *Science* **1991**, *251*, 435.

32. Desai, M. C.; Lefkowitz, S. L.; Thadeio, P. F.; Longo, K. P.; Snider, R. M. *J. Med. Chem.* **1992**, *35*, 4911.
33. Hale, J. J.; Mills, S. G.; MacCoss, M.; Finke, P. E.; Cascieri, M. A.; Sadowski, S.; Ber, E.; Chicchi, G. G.; Kurtz, M.; Metzger, J.; Eiermann, G.; Tattersall, F. D.; Rupniak, N.; Williams, A.; Rycroft, W.; Hargreaves, R.; MacIntyre, D. E. J. *Med. Chem.* **1998**, *41*, 4607.
34. Hale, J. J.; Mills, S. G.; MacCoss, M.; Dorn, C. P.; Finke, P. E.; Budhu, R. J.; Reamer, R. A.; Huskey, S. W.; Luffer-Atlas, D.; Dean, B. J.; McGowan, E. M.; Feeney, W. P.; Chiu, S. H. L.; Cascieri, M. A.; Chicchi, G. G.; Kurtz, M. M.; Sadowski, S.; Ber, E.; Tattersall, F. D.; Rupniak, N. M. J.; Williams, A. R.; Rycroft, W.; Hargreaves, R.; Metzger, J. M.; MacIntyre, D. E. J. *Med. Chem.* **2000**, *43*, 1234.
35. Kim, D.; Wang, L.; Beconi, M.; Eiermann, G.; Fisher, M. H.; He, H.; Hickey, G. J.; Kowalchick, J. E.; Leiting, B.; Lyons, K.; Marsilio, F.; McCann, M. E.; Patel, R. A.; Petrov, A.; Scapin, G.; Patel, S. B.; Sinha Roy, R.; Wu, J. K.; Thornberry, N. A.; Weber, A. E. J. *Med. Chem.* **2005**, *48*, 141.
36. CMR International Performance Metrics (Major Company Comparison), 2001 and 2002.
37. Anderson, P. S. *Annu. Rep. Med. Chem.* **2012**, *47*, 3.

Central Nervous System Diseases

Section Editor: Albert J. Robichaud
Sage Therapeutics, Inc. Cambridge, Massachusetts

Selective Inhibitors of PDE2, PDE9, and PDE10: Modulators of Activity of the Central Nervous System

Morten Jørgensen, Jan Kehler, Morten Langgård, Niels Svenstrup, Lena Tagmose
Discovery Chemistry & DMPK, H. Lundbeck A/S, Valby, Denmark

Contents

1. INTRODUCTION

Phosphodiesterases (PDEs) play a central role in controlling the levels of cyclic nucleotides in the brain and in peripheral tissue. The cyclic nucleotides cAMP and cGMP are second messengers of vital importance for the transmission of extracellular signals from hormones or neurotransmitters into intracellular signals. The cyclic nucleosides are formed by either adenylate or guanylate cyclase, enzymes located on the cytoplasmic side of the cell membrane. Adenylate cyclases catalyze the conversion of ATP into cAMP and guanylate cyclases catalyze the conversion of GTP into cGMP; this takes place as part of a signaling cascade in which the signal generally originates from a G protein-coupled receptor in an interaction mediated

Annual Reports in Medicinal Chemistry, Volume 48
ISSN 0065-7743
http://dx.doi.org/10.1016/B978-0-12-417150-3.00004-1

by a G protein. The PDE superfamily comprises 11 different families of PDEs (21 different gene products) that differ in their substrate specificity and their enzyme kinetic properties (K_m and V_{max}) as well as in their expression profile in human tissue.

Compared to the kinases, the PDEs are attractive drug targets owing to a relatively low physiological concentration of the natural substrate. Furthermore, it is often more practical to set up a counterscreening panel against 10 other PDE families as opposed to the hundreds of related kinases necessary in a detailed counterscreening effort in a kinase inhibitor project, not to mention the thousands of other enzymes having ATP as a substrate. However, due to the similarity of the substrates of PDEs and kinases (and as a consequence, the similarity of the chemical matter available to address the two target classes), they provide challenges in terms of CNS druggability and overall physicochemical properties, an issue that will be discussed for the three major CNS-relevant PDEs in this chapter.

PDEs of essentially every family are highly expressed in the brain and central nervous system, with each specific PDE subtype expressed in distinct brain regions and cellular subtypes. The vast majority of neurons express one or more types of PDE isoforms, making it possible, at least in theory, to address specific areas of the brain with the appropriate isoform-selective PDE inhibitor.

Several R&D programs are exploring the use of inhibitors of PDE2, 9, and/or 10 for a range of CNS diseases, such as Alzheimer's disease (AD), schizophrenia, Huntington´s disease, and various mood disorders. Compounds targeting AD and schizophrenia have advanced into the clinic in recent years. The scope of this chapter is to focus on the three currently most promising PDEs as drug targets for CNS indications, on the recent progress to identify small molecule inhibitors, and on the challenges to develop compounds for these targets.

2. PHOSPHODIESTERASE (PDE) STRUCTURES

The inhibitory effect of the PDE-isosteric ligands is caused by their competitive binding for the nucleotide (cAMP or cGMP) substrate binding site in the catalytic domain of the various PDE enzymes. The binding sites of the subfamilies have been described in the literature.[1] All crystal structures from the PDE family can easily be aligned and the variance of the binding sites can be mapped out as shown in Fig. 4.1. It is common

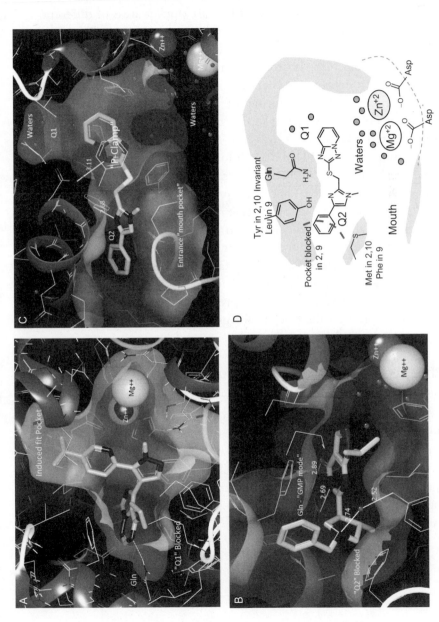

Figure 4.1 (A) Side view of PDE2 with docked pose of the clinical candidate **4** (see Table 4.2). The trifluoromethyl–pyridyl side chain occupies the large hydrophobic pocket formed by induced fit upon ligand binding to the PDE2 enzyme. (B) Side view of PDE9 "3JWS" complex showing the GMP mode binding motif to the invariant Gln. (C) Top view of PDE10 complexed with an analogue of Lu AE90074 showing the open "Q2" selectivity pocket. (D) Schematic view of the PDE10 layout from (C) with annotation of some of the differences between PDE2, 9, and 10.

to use the invariant glutamine (Gln), to which the purine bases of cAMP and cGMP make hydrogen bonds, as a reference point for the pocket description. On both sides of the Gln, there are pocket areas of different size and character dependent on the PDE subtype. The small pocket pointing inwards is referred to as Q1, and the pocket pointing out toward the "mouth" opening, Q2. In front of the Gln there is a hydrophobic region between a conserved phenylalanine ("roof") and isoleucine ("floor") where the aromatic purine of cAMP/cGMP is "clamped." The area in front of the water/metal-ions complex is where the hydrolysis of the cyclic esters takes place.

The various isoforms of PDEs have different capabilities of hydrolyzing cAMP and cGMP; PDE2 acts on both substrates (K_m's of similar size), PDE9 is cGMP specific, and PDE10 is cAMP preferring (K_m difference >20-fold).[2] A "glutamine-switch" (180° rotation around the bond connecting the amide group to the rest of the side chain) hypothesis has been proposed[2] as a mechanistic explanation for the substrate preference, but many later studies have indicated that the situation is far more complex.[1] Despite the controversy about the "glutamine-switch," it is clear that the substrate preference "motif" can be translated into the preferred binding motif found for the discovered ligands for the three isoforms in this review.

2.1. Phosphodiesterase 9 (PDE9) structure

From the landscape of known PDE9 inhibitors, it seems evident that a GMP-like motif is needed. The structure of a 1-isopropyl analogue of **1** (PF-4447943, IC$_{50}$ = 66 nM)[3] is an example of this where the pyrimidinone core interacts with the donor–acceptor pair of the invariant Gln-453 in the GMP mode. The GMP mode of action will ensure selectivity against the AMP-preferring PDEs. By addressing the small Q2 pocket and by forming a hydrogen bond to the PDE9 unique Tyr-424, it has been possible to create PDE9 selective inhibitors.

1 (PF-4447943)
IC$_{50}$ = 8 nM
Selectivity vs PDE1 > 150-fold

2.2. Phosphodiesterase 10 (PDE10) structure

A recent PDE10 inhibitor review[4] comprehensively describes the PDE10 landscape. The layout and description of the binding site used in this review is very similar to the one we are using although different annotations are used. The most important selectivity features to note for the PDE10 site are the AMP-binding mode of the invariant Gln (no known exceptions but a few inhibitors bind without interaction with Gln); and binding to the Q2 pocket involving Tyr-683 that ensures selectivity toward all other PDEs, in contrast to the mixed selectivity profiles obtained for the non-Q2 pocket inhibitors.

2.3. Phosphodiesterase 2 (PDE2) structure

There are currently only two X-ray structures publicly available for PDE2 in complex with a ligand situated in the catalytic site; an inhibitor complex with **2** (IBMX, 3-isobutyl-1-methylxanthine, $IC_{50} = 40 \ \mu M$)[5] and a complex with **3** (EHNA, erythro-9-(2-hydroxy-3-nonyl)adenine, ($IC_{50} = 1 \ \mu M$)).[6] Both ligands bind via hydrogen bonds (bidentate motif) to the invariant Gln-859 which adopts the GMP mode conformation when **2** binds and the AMP mode in the case of **3**. In that way, the two structures nicely illustrate the dual substrate nature of the enzyme. As described for PDE9, the Q2 pocket in PDE2 is also small (as opposed to the large hydrophobic Q2 pocket in PDE10). However, a deep hydrophobic pocket is generated by induced fit when EHNA binds to the PDE2A enzyme. The side chain of EHNA pushes Leu-770 out to the solvent exposed surface of the protein and thereby opens up a deep hydrophobic pocket. This pocket is closed/blocked by Leu-770 in the absence of ligand substituent protruding into the pocket (e.g. binding of **2**) or just in absence of ligand (APO structure - PDB 1Z1)[7]. Docking experiments of **4** strongly suggest that the 2-(2-methyl-2H-pyrazol-3-yl)-5-trifluoromethyl-pyridine group, by which full selectivity of this compound series toward other PDEs is gained, occupies this hydrophobic pocket.[8,9] In PDE10 and PDE9, this pocket is blocked by Leu625 and Met365, respectively. There are only few differences between PDE2 and PDE10 in the amino acid residues lining this "induced" hydrophobic pocket, but between PDE2 and PDE9 there are several differences. Induction and occupancy by ligand of a similar hydrophobic pocket to the one occupied by EHNA when it binds to PDE2, has not been observed in any published X-ray structures of either PDE10 or PDE9. It could therefore indicate that this pocket constitutes a selectivity pocket for PDE2.

3. PDE2 INHIBITORS AND DUAL PDE2 + PDE10 INHIBITORS

One of the first PDE2 inhibitor programs culminated in the identification of a very potent and selective compound **5** (Bay 60-7750).[10] In general, the compounds from this series are characterized by poor CNS drug-like properties as illustrated by the data in Table 4.1.[11] The combination of a relatively high molecular weight and topological polar surface area and the presence of a hydrogen bond donor are probably responsible for the *p*-glycoprotein liabilities associated with the lead compound.[12] In 2004 and 2007, a series of benzodiazepinones as PDE inhibitors were reported.[13] Despite good calculated properties, the best compound, **6** (ND70010), is a Pgp substrate.[12] Nevertheless, it was efficacious in animal assays of anxiety[14] and was advanced to phase I clinical trials in 2005. One can speculate that the two hydrogen bond donors in the carboxamide are responsible for the poor CNS penetration. This moiety is present in 23% of the exemplified compounds, while 36% contain a nitrile in that position, suggesting that it is possible to reduce the HBD count by two and still maintain the topological polar surface area in a favorable range. A recent disclosure explored **9** (tofisopam)[15] and related compounds.[16] These compounds have potential

Table 4.1 Calculated data for the PDE2 inhibitors[11]

Compound	$MW_{example}$ MW ± SD	$HBD_{example}$ HBD ± SD	$TPSA_{example}$ TPSA ± SD	$A \log P_{example}$ $A \log P$ ± SD	$MPO_{example}$ MPO ± SD
5	477 452 ± 42	2 1.6 ± 0.6	98 92 ± 11	4.2 4.3 ± 0.5	2.8 2.6 ± 0.6
6	399 410 ± 57	2 0.8 ± 1.0	85 76 ± 19	3.0 3.6 ± 1.2	4.8 4.3 ± 0.9
7	430 452 ± 33	2 2.2 ± 0.4	81 92 ± 10	5.0 4.6 ± 0.6	2.1 2.3 ± 1.0
8	314 339 ± 20	1 1.1 ± 0.3	88 84 ± 9	0.6 1.4 ± 0.6	5.8 5.7 ± 0.3
9	383 385 ± 26	0 0.9 ± 0.9	62 74 ± 13	4.3 4.3 ± 0.8	4.3 4.0 ± 0.5
10	397 414 ± 29	2 1.8 ± 0.7	81 82 ± 14	3.5 3.9 ± 0.7	3.8 3.7 ± 0.6
11	424 402 ± 48	2 1.3 ± 0.5	68 72 ± 14	3.4 4.3 ± 1.2	4.2 3.8 ± 0.7

issues for CNS applications; in particular, two or more phenolic hydrogen bond donors, leading to concerns with respect to brain penetration and phase II metabolism. Additionally, two structurally distinct chemotypes: pyridopyrimidines **7**, with poor CNS drug-like properties, and oxindoles **8** that display promising calculated profiles have been reported.[17] Additional work may have been discontinued on these structural classes,[18] perhaps due to the lack of rotatable bonds, the secondary amide linker, and the tricyclic core suggesting that the solubility may have been an issue. More recently, two different (but structurally somewhat related) chemotypes have been disclosed as PDE2 inhibitors, namely 8-aminoquinolines (**10**) and dihydro-quinolin-2-ones (**11**).[19,20] Compounds from this series contain a (4-amino-pyridin-3-yl)-methanol substructure; an advantage being that an intramolecular hydrogen bond may mask one of the two hydrogen bond donors. Additionally, some of these compounds feature an ester group instead of the primary alcohol, raising concerns with respect to plasma stability and potential BBB penetration issues for the corresponding carboxylic acids. Both of these scaffolds appear to have relatively poor CNS drug-like characteristics. Furthermore, a majority of compounds contain the mono-substituted thiophene, a structural alert.[21] Nevertheless, compound **11** is a potent PDE2 inhibitor (IC$_{50}$ = 1.8 nM).

5   6   7   8

9   10   11

12   13

PDE10 IC$_{50}$ 7.16 nM   PDE10 IC$_{50}$ 1.9 nM
PDE2 IC$_{50}$ ~1000 nM   PDE2 IC$_{50}$ 11 nM

Recent disclosures include tetraaza-cyclopenta[*a*]naphthalenes like **12** and **13** as PDE10 inhibitors, with a number of these compounds being dual inhibitors of both PDE10 and PDE2.[22] However, structurally related **14** (benzo[*e*]imidazo-triazines) and **15** (pentaaza-cyclopenta[*a*]naphthalenes) were found to be primarily PDE10 inhibitors, or compounds showing additional significant inhibitory activity versus PDE2.[23] In 2012, compounds like **16** (triazolo-quinoxalines) were identified to be dual PDE2 and PDE10 inhibitors.[24] In 2013, a patent application appeared on the same scaffold and again dual inhibitors were identified.[25] The application additionally contains several radioligands and claims their use as diagnostics. The exemplified compound **17** was highlighted in a publication that discussed the SAR and *in vivo* potency as a dual inhibitor (PDE2 $IC_{50} = 2.8$ nM; PDE10 $IC_{50} = 35$ nM).[26] From the extensive data in the patent applications, it is evident that the 3-phenoxy-propan-1-ol moiety is responsible for the exquisite PDE2 selectivity for the highlighted compound **16** (PDE2 $IC_{50} = 14$ nM; PDE10 $IC_{50} > 1000$ nM). A recent publication includes selective PDE2 inhibitors **18** and **4**, with the latter compound owing its selectivity to the 5-trifluoromethyl-pyridin-2-yl group.[8,9] This compound has relatively poor human PK properties, but reportedly was possible to achieve full target coverage by using a modified release formulation. The compounds in Table 4.2 represent an improvement in terms of CNS drug-likeness over the compounds in Table 4.1.

4   **14**   **15**   **16**   **17**   **18**

4. PDE9 INHIBITORS

PDE9A is a high-affinity, cGMP-specific enzyme encoded by a single gene. It is subject to a complex pattern of regulation, yielding approximately 20 human splice variants.[27,28] In central areas where it is expressed, PDE9A is thought to be the major regulator of cGMP levels, and it has the lowest K_m among the PDEs for this nucleotide.[29,30]

Table 4.2 Calculated data for the PDE2 and/or PDE10 inhibitors

Compound	$MW_{example}$ $MW \pm SD$	$HBD_{example}$ $HBD \pm SD$	$TPSA_{example}$ $TPSA \pm SD$	$A \log P_{example}$ $A \log P \pm SD$	$MPO_{example}$ $MPO \pm SD$
4	414	0	77	3.0	4.8
	402 ± 36	0.3 ± 0.5	78 ± 12	3.2 ± 0.6	3.2 ± 0.7
14	325	0	52	3.8	4.7
	355 ± 49	0.1 ± 0.3	60 ± 12	3.7 ± 1.2	4.8 ± 0.8
15	360	0	65	3.5	4.9
	334 ± 41	0.1 ± 0.4	75 ± 11	3.0 ± 1.0	5.2 ± 0.6
16	369	1	73	3.3	5.0
	375 ± 49	0.1 ± 0.3	54 ± 10	4.6 ± 0.9	4.2 ± 0.6
17	394	0	56	3.6	5.5
	407 ± 46	0.5 ± 0.5	6813	3.8 ± 1.1	4.4 ± 0.9
18	474	0	92	3.8	3.9
	434 ± 41	0.2 ± 0.4	78 ± 15	4.0 ± 0.5	4.0 ± 0.5

The search for selective PDE9A inhibitors has yielded a number of interesting compounds from various classes. It appears that PDE9A has a very pronounced preference for compounds displaying variations of the purinone scaffold, that is, flat, aromatic heterobicyclic compounds capable of forming the characteristic double hydrogen bond to the active site Gln, as observed in structures of many other PDE inhibitors such as sildenafil and vardenafil.[31] These structural characteristics are also recognizable in the chemical classes that have resulted from the four major discovery efforts disclosed so far (Table 4.3).

The first published selective PDE9A inhibitor, **19** (BAY 73-6691), belongs to the pyrazolopyrimidinones.[32,33] This compound selectively inhibits human PDE9A with an *in vitro* IC_{50} of 55 nM and greater than 25-fold selectivity versus other PDEs.

The currently most advanced PDE9A program is aimed at identifying an inhibitor for the treatment of cognitive deficits in AD and other neuropsychiatric disorders.[34,35] The *in vitro* and *in vivo* profile of the development compound **1** (PF-4447943) was recently disclosed,[3,36,37] and it appears to penetrate the CNS in several animal species and humans, causing an elevation in the levels of cGMP in the CSF *in vivo*.[37] Additional compounds of interest in the pyrazolopyrimidinone scaffold include **20**.

Table 4.3 Calculated data for the PDE9 inhibitors

Compound	$MW_{example}$ $MW \pm SD$	$HBD_{example}$ $HBD \pm SD$	$TPSA_{example}$ $TPSA \pm SD$	$A \log P_{example}$ $A \log P \pm SD$	$MPO_{example}$ $MPO \pm SD$
1	395	1	98	-0.2	4.3
	416 ± 26	1.0 ± 0.1	77 ± 12	2.8 ± 1.1	4.3 ± 0.5
19	357	1	59	3.6	4.8
	328 ± 23	1.1 ± 0.4	65 ± 8	3.4 ± 0.7	4.9 ± 0.5
20	365	1	88	1.2	4.8
	416 ± 32	1.0 ± 0.1	85 ± 13	2.1 ± 1.1	4.9 ± 0.4
21	409	1	69	3.1	5.0
	407 ± 53	1.2 ± 0.5	75 ± 14	2.8 ± 0.8	4.8 ± 0.8
22	405	1	99	0.7	5.2
	421 ± 26	1 ± 0	87 ± 11	1.6 ± 1.1	5.2 ± 0.2
23	386	1	85	1.3	5.6
	387 ± 23	1.1 ± 0.4	87 ± 9	1.5 ± 0.7	5.3 ± 0.5
24	444	1	85	2.5	4.2
	422 ± 29	1.1 ± 0.2	84 ± 16	2.0 ± 1.3	4.1 ± 0.5
25	399 ± 33	1.0 ± 0.2	67 ± 12	2.0 ± 1.2	5.0 ± 0.7

19 (Bay 73-6691)
IC_{50} = 55 nM
Selectivity vs PDE1 > 25-fold

Another program employing the pyrazolopyrimidinone scaffold to discover selective inhibitors of PDE9A resulted in **21–23**.[38] Although no detailed biological data have been disclosed, this focused compound class seems to be quite potent and selective versus PDE1.

24 (imidazotriazinones)[39] and **25** (imidazopyrazinones)[61] are newer scaffolds to display activity for the inhibition of PDEs, with both compound classes being highly potent and selective.

Looking at the CNS-targeted PDE9 inhibitors described above, an interesting pattern emerges. Taken as a group, the PDE9 inhibitors seem to be more homogeneous and significantly more polar than the PDE2 and PDE10 inhibitors, indicating possibly that the PDE9 active site is more discriminating in the types of pharmacophores it accepts. All compound classes can be described as guanine bioisosteres, having presumably similar binding modes (double hydrogen bond to glutamine-453). Overall, it would appear that PDE9 inhibitors are more polar (dogD values going into the negative region) than PDE2 inhibitors.

5. PDE10 INHIBITORS

The discovery of PDE10A was simultaneously reported in 1999 by three independent groups.[40–42] PDE10A has the most restricted distribution of all the 11 known PDE families with the PDE10A mRNA highly expressed only in the brain and testes.[40,43,44] In the brain, mRNA and protein are highly enriched in the striatum[40,45,46] which may indicate a potential use of PDE10A inhibitors for treating neurological (e.g., Huntington's disease) and psychiatric disorders.[47] However, PDE10A inhibitors have also been claimed to be useful as treatment for cancer,[48] diabetes,[49] and obesity.[50] The presence of PDE10A mRNA (but not protein) in testes has prompted a patent application claiming the use of PDE10A inhibitors in kits for evaluating the semen quality and compositions for preserving a sperm sample, especially during cryo-preservation.[51] Recently, PDE10A was also found in the retina and pineal gland and the PDE10A levels were found to fluctuate with circadian rhythm, perhaps indicating a potential use for PDE10A inhibitors for the treatment of retinal diseases.[52] In the CNS, PDE10A plays an essential role in regulating cAMP/PKA and cGMP/PKG signaling cascades by controlling the magnitude, duration, and cellular location of cAMP/cGMP elevation. Biochemical and behavioral data indicate that PDE10A inhibition activates cAMP/PKA signaling in the basal ganglia, leading to the potentiation of dopamine D_1 receptor signaling, and concomitant inhibition of dopamine D_2 receptor signaling. Preclinical evidence in a range of animal models suggests that a PDE10A inhibitor could provide efficacy on positive, cognitive, and negative symptoms of schizophrenia, which is also the primary indication followed by the multitude of companies that have PDE10A inhibitors in development (Table 4.4).

In a recent analysis of the landscape of PDE10A inhibition,[4] it was observed that there is a move toward more drug-like characteristics in compounds as measured by CNS MPO scores of all published PDE10A inhibitors.

Table 4.4 Calculated data for representative PDE10 inhibitors

Compound	$MW_{example}$ $MW \pm SD$	$HBD_{example}$ $HBD \pm SD$	$TPSA_{example}$ $TPSA \pm SD$	$A \log P_{example}$ $A \log P \pm SD$	$MPO_{example}$
26	392	0	53	4.8	4.5
	387 ± 42	0.21 ± 0.41	57 ± 10	4.6 ± 0.8	
29	488	0	90	4.1	4.8
	417 ± 37	1.3 ± 0.7	77 ± 16	5.0 ± 0.8	
33	465	0	95	2.5	4.8
	423 ± 52	0.8 ± 0.8	88 ± 14	2.7 ± 0.9	
34	332	0	61	2.9	5.9
	406 ± 39	0.23 ± 0.5	57 ± 13	3.7 ± 0.8	
35	487	1	87	3.6	3.7
	419 ± 59	1.2 ± 0.5	80 ± 13	3.5 ± 1.0	
36	401	1	97	1.4	5.3
	418 ± 44	1.6 ± 0.8	98 ± 19	1.8 ± 0.8	
37	394	0	61	4.2	4.8
	421 ± 26	1 ± 0	87 ± 11	1.6 ± 1.1	
38	406	0	71	4.3	4.4
	384 ± 53	0.5 ± 0.7	70 ± 16	4.0 ± 1.0	

The highly selective compound **26** (MP-10) was the first compound to enter into clinical trials. In spite of having molecular properties outside the golden triangle, MP-10 demonstrated good oral bioavailability, high CNS exposure, and good efficacy in a number of animal models. Notwithstanding this promising preclinical profile, the compound did not demonstrate antipsychotic efficacy in a phase 2 study, and its development has been terminated. The discovery of **26** has been extensively reviewed elsewhere.[53,54]

Previously, 4-(pyridin-3-yl)cinnolines[55] were reported as selective inhibitors of PDE10, but more recently the same group reported a series of biaryl ethers like e.g. **29** as novel PDE10A inhibitors.[56] The biaryl ethers seem to follow the general trend of having calculated properties mainly outside the golden triangle. The main issue (besides affinity) in the optimization of the biaryl ethers was related to problems with ADME parameters, particularly permeability and Pgp-mediated efflux (Fig. 4.2). In this case, the solution was to change the heterocycle from benzimidazole to benzothiazole, thereby reducing both HBD and PSA.

The results of an optimization from a novel HTS-derived tetrahydropyridopyrimidine scaffold **30**, represented by structure **31**,[57] (Fig. 4.3) was recently published. Initial optimization of the R1 and R2 substituents revealed a flat SAR for PDE10A. The main challenges during the optimization (besides potency and selectivity) were ADME related, namely Pgp-liability and plasma clearance. The molecular properties of the majority of

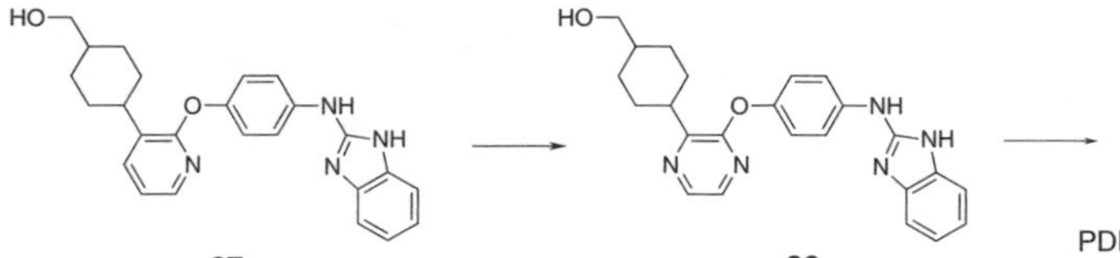

Figure 4.2 Hit-to-lead optimization of a series of biaryl ethers.

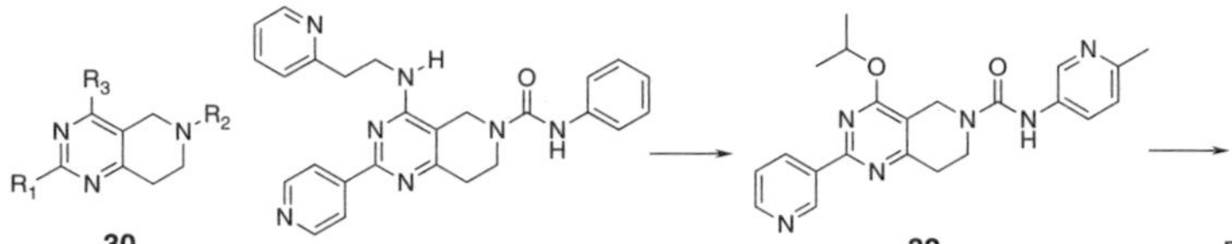

27

PDE10A IC_{50} = 422 nM
PDE3A IC_{50} = >30,000 nM
P_{app} = 23.4 x 10^{-6} cm/s
Efflux Ratio = 45.6

28

PDE10A IC_{50} = 0.092 nM
PDE3A IC_{50} = >30,000 nM
P_{app} = 41.6 x 10^{-6} cm/s
Efflux Ratio = 76.7

29

PDE10A IC_{50} = 0.199 nM
PDE3A IC_{50} = >30,000 nM
P_{app} = 8 x 10^{-6} cm/s
Efflux Ratio = 1.1

Figure 4.3 Hit-to-lead optimization of a series of tetrahydropyridopyrimidines.

30

31

PDE10A IC_{50} = 94 nM
Efflux Ratio > 3.5

32

PDE10A IC_{50} = 7.9 nM
Efflux Ratio < 1.5
Solubility = Very low

33

PDE10A IC_{50} = 1 nM
PDE5 IC_{50} = 116 nM
PDE6 IC_{50} = 44 nM
P_{app} = 32 × 10^{-6} cm/s
Efflux Ratio = 0.75

the compounds fall outside the golden triangle. Pgp–mediated efflux problems were solved by exchanging the amino substituent at R3 for an alkoxy substituent in **32**. The resulting optimized compound, **33**, exhibits nanomolar potency, excellent pharmacokinetic properties, and a clean off–target profile. It displays *in vivo* target engagement as measured by increased rat striatal cGMP levels upon oral dosing, as well as dose–dependent efficacy in a key pharmacodynamic assay predictive of antipsychotic activity. It was also reported to improve cognition in rat and rhesus monkey.[58]

Compound **34** (Lu AE90074 Fig. 4.4) was recently published as a result of a lead optimization effort on a series of phenylimidazole–based PDE10 inhibitors. **34** demonstrated good efficacy in preclinical models of

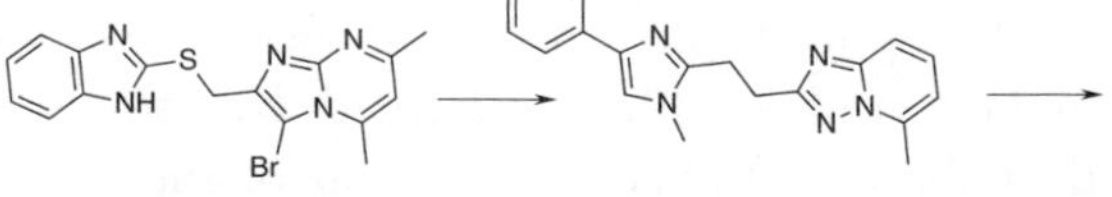

PDE10A IC_{50} = 410 nM
PDE Cross = >20
Clint HLM (l/h) = 2.0

PDE10A IC_{50} = 14 nM
PDE Cross = >100
Clint HLM (l/h) = 3.5

34

PDE10A IC_{50} = 12 nM
PDE Cross = >100
Clint HLM (l/h) = 0.6

Figure 4.4 Hit-to-lead optimization of a series of phenyl imidazoles.

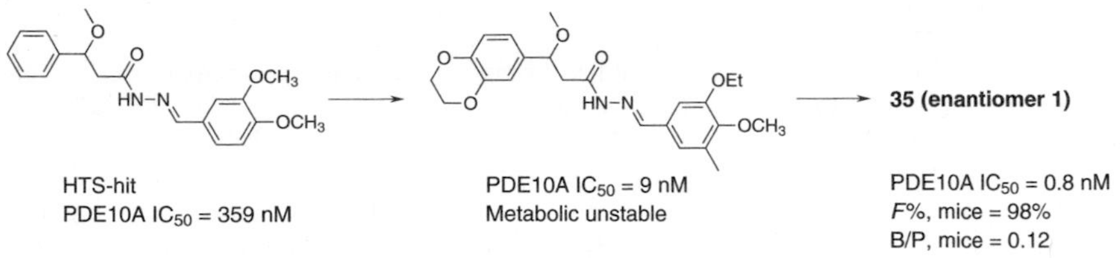

HTS-hit
PDE10A IC$_{50}$ = 359 nM

PDE10A IC$_{50}$ = 9 nM
Metabolic unstable

35 (enantiomer 1)

PDE10A IC$_{50}$ = 0.8 nM
F%, mice = 98%
B/P, mice = 0.12

Figure 4.5 Hit-to-lead optimization of a series of N-acylhydrazones.

antipsychotic activity, and the optimization program focused on improving metabolic stability and brain penetration while removing HERG liability. This compound's properties fall within the golden triangle, although the vast majority of the compounds in the series did not.

A series of HTS-derived N-acylhydrazones, typified by the compounds shown in Fig. 4.5, were the focus of a recent disclosure.[59,60] Again, the minority of the compounds fell inside the golden triangle—however, the resulting optimized lead compound **35** (enantiomer-1) demonstrated subnanomolar inhibition of PDE10A *in vitro*—was reported to have excellent oral bioavailability and to be active in the CAR mouse model. It was also active in a hyperactivity assay and a memory model, despite poor brain penetration.

Additionally, compounds from three other scaffolds were reported to be PDE10 inhibitors, but the data have not yet been published; hence, knowledge of these chemotypes and series can only be surmised from their published patent applications.

6. CONCLUSIONS

Although superficially similar, the analyses above demonstrate that much of the currently pursued inhibitors of PDE2, 9, and 10 differ in a number of ways. PDE9 appears from the published patent literature to be very restrictive in its binding motif requirements. In essence, only one chemotype which recognizes the GMP bidentate-binding motif represents the PDE9 patent chemical space. A much larger diversity of compounds is present in the PDE2 patent literature. It appears as if the less stringent requirements for binding to the invariant Gln in combination with the induced hydrophobic (selectivity) pocket offer a much broader variety of chemotypes capable of inhibiting this enzyme. PDE10 also has high diversity in the reported patent literature chemical space when compared to PDE9. The PDE10 unique Q2 pocket provides an additional opportunity for diversity in compounds targeting this enzyme. In special cases like **26**, high affinities can be obtained without even interacting with the central Gln.

REFERENCES

1. Ke, H.; Wang, H.; Ye, M. Structural insight into the substrate specificity of phosphodi-esterases. Phosphodiesterases as drug targets. *Handb. Exp. Pharmacol.* **2011**, *204*, 121–134.
2. Zhang, K. Y.; Card, G. L.; Suzuki, Y. *Mol. Cell* **2004**, *15*, 279–286.
3. Verhoest, P. R.; Proulx-Lafrance, C.; Corman, M.; Chenard, L.; Helal, C. J.; Hou, X.; Kleiman, R.; Liu, S.; Marr, E.; Menniti, F. S.; Schmidt, C. J.; Vanase-Frawley, M.; Schmidt, A. W.; Williams, R. D.; Nelson, F. R.; Fonseca, K. R.; Liras, S. *J. Med. Chem.* **2009**, *52*, 7946–7949.
4. Chappie, T. A.; Helal, C. J.; Hou, X. *J. Med. Chem.* **2012**, *55*, 7299–7331.
5. Pandit, J.; Forman, M. D.; Fennell, K. F.; Dillman, K. S.; Menniti, F. S. *Proc. Natl. Acad. Sci.* **2009**, *106*, 18225–18230.
6. PDB Code: 4C1I.
7. Iffland, A.; Kohls, D.; Low, S.; Luan, J.; Zhang, Y.; Kothe, M.; Cao, Q.; Kamath, A. V.; Ding, Y. H.; Ellenberger, T. *Biochemistry* **2005**, *44*, 8312–8325.
8. Helal, C. J.; Chappie, T. A.; Humphrey, J. M. Patent Application WO 2012168817 A1 20121213, 2012. Helal, C. J.; Chappie, T. A.; Humphrey, J. M.; Verhoest, P. R.; Yang, E. Patent Application US 20120214791 A1 20120823, 2012.
9. The SAR around the two recent Pfizer PDE2 scaffolds was presented at the 244th National ACS Meeting in Philadelphia by C.J. Helal in a talk entitled 'Identification of a brain penetrant, highly selective phosphodiesterase 2A inhibitor for the treatment of cognitive impairment associated with schizophrenia (CIAS)'.
10. Niewoehner, U.; Schauss, D.; Hendrix, M.; König, G.; Böß, F. G.; Van der Staay, F. -J.; Schreiber, R.; Schlemmer, K. -H.; Toshiya, T. Patent Appilication DE 10108752 A1 20020905, 2002. Niewoehner, U.; Schauss, D.; Hendrix, M.; König, G.; Böß, F. G.; Van der Staay, F. -J.; Schreiber, R.; Schlemmer, K. -H.; Toshiya, T.; Grosser, R. Patent Application WO 2002050078 A1 20020627, 2002. Böß, F. -G.; Hendrix, M.; König, G.; Niewohner, U.; Schlemmer, K. -H.; Schreiber, R.; Van Der Staay, F. -J.; Schauss, D. Patent Application WO 2002009713 A2 20020207, 2002.
11. General procedures for the patent analysis: The compounds exemplified in CNS-relevant patent applications were extracted from Scifinder as an SD file. An in-house Pipeline Pilot protocol was used to read the SD file and a MOL of the Markush structure with up to five R-groups to perform an analysis that generated: (1) Average physico-chemical properties; (2)R1–R5 group analysis by pie charts, 5 top R1–R5 groups exem-plified; (3) A full R group table and the structures of all exemplified compounds; (4) Compounds not matching the probe. The Pipeline Pilot patent tool also generated an Excel-readable file with calculated physico-chemical properties for all compounds, and the file also contained MPO and the Golden Triangle analyses. All hydrogen atoms bound to N or O were counted as hydrogen bond donors. Sulfur atoms were not included in the calculation of the topological polar surface area (TPSA).
12. In-house profiling of these compound in the MDCK permeability assay showed them to be strong Pgp substrates. Bay 60-7750 was further found to be rapidly cleared in human liver microsomal preparations.
13. Bourguignon, J. -J.; Lugnier, C.; Abarghaz, M.; Lagouge, Y.; Wagner, P.; Mondadori, C.; Macher, J. -P.; Schultz, D.; Raboisson, P. Patent Application WO 2004041258 A2 20040521, 2004. Abarghaz, M.; Biondi, S.; Duranton, J.; Limanton, E.; Mondadori, C.; Wagner, P. Patent Applications EP 1548011 A1 20050629, 2005 and EP 1749824 A1 20070207, 2007.
14. See for exampleMasood, A.; Huang, Y.; Hajjhussein, H.; Xiao, L.; Li, H.; Wang, W.; Hamza, A.; Zhan, C.-G.; O'Donnell, J. M. *J. Pharmacol. Exp. Ther.* **2009**, *331*, 690.
15. The psychotropic profile of tofisopam has been evaluated clinically: Bond, A.; Lader, M. *Eur. J. Clin. Pharmacol.* **1982**, *22*, 137. The compound has further been claimed in a

patent application as a PDE10 inhibitor: Nielsen, E. B.; Kehler, J.; Nielsen, J.; Brøsen, P. Patent Application WO 2007082546 A1 20070726, 2007. Tofisopam has been demonstrated to be a PDE inhibitor with activity in a mouse model of negative symptoms of scizophrenia: Runfeldt, C.; Socała, K.; Wlaź, P. *J. Neural. Transm.* **2010**, *117*, 1319.

16. Bernard, T. Patent Applications US 20070161628 A1 20070712, 2007 and FR 2870539 A1 20051125, 2005.

17. Beyer, T. A.; Chambers, R. J.; Lam, K. T.; Li, M.; Morrell, A. I.; Thompson, D. D. Patent Application WO 2005061497 A1 20050707, 2005. Chambers, R. J.; Lam, K. T. Patent Application WO 2005041957 A1 20050512, 2005.

18. Chambers, R. J.; Abrams, K.; Garceau, N. Y.; Kamath, A. V.; Manley, C. M.; Lilley, S. C.; Otte, D. A.; Scott, D. O.; Sheils, A. L.; Tess, D. A.; Vellekoop, A. S.; Zhang, Y.; Lam, K. T. *Bioorg. Med. Chem. Lett.* **2006**, *16*, 307.

19. Feng, Y.; Arancio, O.; Deng, S.; Landry, D. W. Patent Applications WO 2010074783 A1 20100701, 2010 and US 20120076732 A1 20120329, 2012.

20. De Leon, P.; Egbertson, M.; Hills, I. D.; Johnson, A. W.; Machacek, M. Patent Application WO 2011011312 A1 20110127, 2011.

21. For a general discussion, see for example Stepan, A. F.; Walker, D. P.; Bauman, J.; Price, D. A.; Baillie, T. A.; Kalgutkar, A. S.; Aleo, M. D. *Chem. Res. Toxicol.* **2011**, *24*, 1345.

22. Hoefgen, N.; Stange, H.; Langen, B.; Egerland, U.; Schindler, R.; Pfeifer, T.; Rundfeldt, C. Patent Application WO 2007137820, 2007. The SAR with respect to PDE10 and PDE2 inhibition has been discussed in the primary literature: Malamas, M. S.; Ni, Y.; Erdei, J.; Stange, H.; Schindler, R.; Lankau, H. -J.; Grunwald, C.; Fan, K. Y.; Parris, K.; Langen, B.; Egerland, U.; Hage, T.; Marquis, K. L.; Grauer, S.; Brennan, J.; Navarra, R.; Graf, R.; Harrison, B. L.; Robichaud, A.; Kronbach, T.; Pangalos, M. N.; Hoefgen, N.; Brandon, N. J. *J. Med. Chem.* **2011**, *54*, 7621.

23. Stange, H.; Langen, B.; Egerland, U.; Höfgen, N.; Priebs, M.; Malamas, M. S.; Erdei, J. J.; Ni, Y. Patent Applications WO 2010054260 A1 20100514, 2010 and WO 2010054253 A1 20100514, 2010.

24. Lankau, H. -J.; Langen, B.; Grunwald, C.; Höfgen, N.; Stange, H.; Dost, R.; Ugerland, U. Patent Application WO 2012104293 A1 20120809, 2012.

25. Andres-Gil, J. I.; Rombouts, F. J. R.; Trabanco-Suarez, A. A.; Vanhoof, G. C. P.; De Angelis, M.; Buijnsters, P. J. J. A.; Guillemont, J. E. G.; Bormans, G. M. R.; Celen, S. J. L.; Vliegen, M. Patent Application WO 2013000924 A1 20130103, 2013.

26. Andres, J.-I.; Buijnsters, P.; De Angelis, M.; Langlois, X.; Rombouts, F.; Trabanco, A. A.; Vanhoof, G. *Bioorg. Med. Chem. Lett.* **2013**, *23*, 785.

27. Guipponi, M.; Scott, H. S.; Kudoh, J.; Kawasaki, K.; Shibuya, K.; Shintani, A.; Asakawa, S.; Chen, H.; Lalioti, M. D.; Rossier, C.; Minoshima, S.; Shimizu, N.; Antonarakis, S. E. *Hum. Genet.* **1998**, *103*, 386–392.

28. Rentero, C.; Monfort, A.; Puigdomenech, P. *Biochem. Biophys. Res. Commun.* **2003**, *301*, 686–692.

29. Fisher, D. A.; Smith, J. F.; Pillar, J. S.; St Denis, S. H.; Cheng, J. B. *J. Biol. Chem.* **1998**, *273*, 15559–15564.

30. Soderling, S. H.; Bayuga, S. J.; Beavo, J. A. *J. Biol. Chem.* **1998**, *273*, 15553–15558.

31. Wang, H.; Ye, M.; Robinson, H.; Francis, S. H.; Ke, H. *Mol. Pharmacol.* **2008**, *73*, 104–110.

32. Wunder, F. *Mol. Pharmacol.* **2008**, *68*, 1775–1781.

33. (a) Hendrix, M.; Baerfacker, L.; Erb, C.; Hafner, F. -T.; Heckroth, H.; Schauss, D.; Tersteegen, A.; Van Der Staay, F. -J.; Van Kampen, M. Patent Application WO 2004099211 A1 20041118, 2004.

34. Menniti, F. S.; Kleiman, R.; Schmidt, C. *Schizophr. Res.* **2008**, *102*, 38–39.

35. Schmidt, C. J.; Harms, J. F.; Tingley, F. D.; Schmidt, K.; Adamowicz, W. O.; Romegialli, A.; Kleiman, R. J.; Barry, C. J.; Coskran, T. M.; O'Neill, S. M.; Stephenson, D. T.; Menniti, F. S. *Alzheimers Dement.* **2009**, *5*, P331.
36. Verhoest, P. R.; Fonseca, K. R.; Hou, X.; Proulx-Lafrance, C.; Corman, M.; Helal, C. J.; Claffey, M. M.; Tuttle, J. B.; Coffman, K. J.; Liu, S.; Nelson, F.; Kleiman, R. J.; Menniti, F. S.; Schmidt, C. J.; Vanase-Frawley, M.; Liras, S. *J. Med. Chem.* **2012**, *55*, 9045–9054.
37. Kleiman, R. J.; Chapin, D. S.; Christoffersen, C.; Freeman, J.; Fonseca, K. R.; Geoghegan, K. F.; Grimwood, S.; Guanowsky, V.; Hajos, M.; Harms, J. F.; Helal, C. J.; Hoffmann, W. E.; Kocan, G. P.; Majchrzak, M. J.; McGinnis, D.; McLean, S.; Menniti, F. S.; Nelson, F.; Roof, R.; Schmidt, A. W.; Seymour, P. A.; Stephenson, D. T.; Tingley, F. D.; Vanase-Frawley, M.; Verhoest, P. R.; Schmidt, C. J. *J. Pharmacol. Exp. Ther.* **2012**, *341*, 396–409.
38. Eickmeier, C.; Doerner-Ciossek, C.; Fiegen, D.; Fox, T.; Fuchs, K.; Giovannini, R.; Heine, N.; Hendrix, M.; Rosenbrock, H.; Schaenzle, G. Patent Application WO 2009068617 A1 20090604, 2009. Giovannini, R.; Doerner-Ciossek, C.; Eickmeier, C.; Fiegen, D.; Fox, T.; Fuchs, K.; Heine, N.; Rosenbrock, H.; Schaenzle, G. Patent Application WO 2009121919 A1 20091008, 2009.
39. Ripka, A.; Shapiro, G.; McRiner, A. Patent Application WO 2012040230 A1 20120329, 2012.
40. Fujishige, K.; Kotera, J.; Michibata, H.; Yuasa, K.; Takebayashi, S. I.; Okumura, K.; Omori, K. *J. Biol. Chem.* **1999**, *274*, 18438–18445.
41. Loughney, K.; Snyder, P. B.; Uher, L.; Rosman, G. J.; Ferguson, K.; Florio, V. A. *Gene* **1999**, *234*, 109–117.
42. Soderling, S. H.; Bayuga, S. J.; Beavo, J. A. *Proc. Natl. Acad. Sci. U.S.A.* **1999**, *96*, 7071–7076.
43. Fujishige, K.; Kotera, J.; Omori, K. *Eur. J. Biochem.* **1999**, *266*, 1118–1127.
44. Coskran, T. M.; Morton, D.; Menniti, F. S.; Adamowicz, W. O.; Kleiman, R. J.; Ryan, A. M.; Strick, C. A.; Schmidt, C. J.; Stephenson, D. T. *J. Histochem. Cytochem.* **2006**, *54*, 1205–1213.
45. Seeger, T. F.; Bartlett, B.; Coskran, T. M.; et al. *Brain Res.* **2003**, *985*, 113–126.
46. Kotera, J.; Sasaki, T.; Kobayashi, T.; Fujishige, K.; Yamashita, Y.; Omori, K. *J. Biol. Chem.* **2004**, *279*, 4366–4375.
47. Kehler, J.; Nielsen, J. *Curr. Pharm. Des.* **2011**, *17*, 137–150.
48. Niewoehner, U.; Bauser, M.; Ergueden, J. -K.; Flubacher, D.; Naab, P.; Repp, T. -O.; Stoltefuss, J.; Burkhardt, N.; Sewing, A.; Schauer, M. Patent Application WO 2002048144 A1 20020620, 2002.
49. Sweet, L. Patent Application WO 2005012485 A2 20050210, 2005.
50. Black, S. C.; Gibbs, E. M. Patent Application WO 2005120514 A1 20051222, 2005.
51. Richard, F.; Guillemette, C.; Aragon, J. P.; Hebert, A.; Leclerc, P.; Blondin, P. Patent Application CA 2766540 A1 20120809, 2012.
52. Donello, J. E.; Yang, R.; Viswanath, V.; Leblond, B.; Beausoleil, E.; Pando, M. P.; Desire, L. J. R.; Casagrande, A. -S. Patent Application WO 2012112918 A1 20120823, 2012.
53. Verhoest, P. R.; Chapin, D. S.; Corman, M.; Fonseca, K.; Harms, J. F.; Hou, X.; Marr, E. S.; Menniti, F. S.; Nelson, F.; O'Connor, R.; Pandit, J.; Proulx-LaFrance, C.; Schmidt, A. W.; Schmidt, C. J.; Suiciak, J. A.; Liras, S. *J. Med. Chem.* **2009**, *52*, 5188.
54. Hoefgen, N.; Grunwald, C.; Langen, B. *Drugs Fut.* **2012**, *37*, 577.
55. Hu, E.; Kunz, R. K.; Rumfelt, S.; Andrews, K. L.; Li, C.; Hitchcock, S. A.; Lindstrom, M.; Treanor, J. *Bioorg. Med. Chem. Lett.* **2012**, *22*, 5903.
56. Rzasa, R. M.; Hu, E.; Rumfelt, S.; Chen, N.; Andrews, K. L.; Chmait, S.; Falsey, J. R.; Zhong, W.; Jones, A. D.; Porter, A.; Louie, S. W.; Zhao, X.; Treanor, J. J. S.; Allen, J. R. *Bioorg. Med. Chem. Lett.* **2012**, *22*, 7371.

57. Raheem, I. T.; Breslin, M. J.; Fandozzi, C.; Fuerst, J.; Hill, N.; Huszar, S.; Kandebo, M.; Kim, S. H.; Mac, B.; McGaughey, G.; Renger, J. J.; Schreier, J. D.; Sharma, S.; Smith, S.; Uslaner, J.; Yan, Y.; Coleman, P. J.; Cox, C. D. *Bioorg. Med. Chem. Lett.* **2012**, *22*, 5903.

58. Smith, S. M.; Uslaner, J. M.; Cox, C. D.; Huszar, S. L.; Cannon, C. E.; Vardigan, J. D.; Eddins, D.; Toolan, D. M.; Kandebo, M.; Yao, L.; Raheem, I. T.; Schreier, J. D.; Breslin, M. J.; Coleman, P. J.; Renger, J. J. *Neuropharmacology* **2013**, *64*, 215–223.

59. Cutshall, N. S.; Onrust, R.; Rohde, A.; Gragerov, S.; Hamilton, L.; Harbol, K.; Shen, H.-R.; McKee, S.; Zuta, C.; Gragerova, G.; Florio, V.; Wheeler, T. N.; Gage, J. L. *Bioorg. Med. Chem. Lett.* **2012**, *22*, 5595.

60. Gage, J. L.; Onrust, R.; Johnston, D.; Osnowski, A.; Macdonald, W.; Mitchell, L.; Urogdi, L.; Rohde, A.; Harbol, K.; Gragerov, S.; Dorman, G.; Wheeler, T.; Florio, V.; Cutshall, N. S. *Bioorg. Med. Chem. Lett.* **2011**, *21*, 4155.

61. Svenstrup, N.; Simonsen, K. B.; Rasmussen, L. K.; Juhl, K.; Langgaard, M.; Wen, K.; Wang, Y. Patent Application WO 2013110768 A1 20130801, 2013.

CHAPTER FIVE

Beyond Secretases: Kinase Inhibitors for the Treatment of Alzheimer's Disease

Federico Medda[*,†], **Breland Smith**[*,‡], **Vijay Gokhale**[*,†], **Arthur Y. Shaw**[*,†], **Travis Dunckley**[§], **Christopher Hulme**[*,†,‡]

[*]BIO5 Oro Valley, University of Arizona, Oro Valley, Arizona, USA
[†]Department of Pharmacology and Toxicology, College of Pharmacy, University of Arizona, Tucson, Arizona, USA
[‡]Department of Chemistry and Biochemistry, University of Arizona, Tucson, Arizona, USA
[§]Neurogenomics Division, Translational Genomics Research Institute, Phoenix, Arizona, USA

Contents

1. INTRODUCTION

Alzheimer's disease (AD) is a neurodegenerative condition whose most evident symptom is a progressive decline in cognitive functions. It accounts for 70% of all cases of dementia in elderly people, and estimates indicate that it will affect roughly 40 million individuals worldwide by 2020. Currently, treatment options for AD are limited and represent a major unmet therapeutic need.[1,2]

Annual Reports in Medicinal Chemistry, Volume 48
ISSN 0065-7743
http://dx.doi.org/10.1016/B978-0-12-417150-3.00005-3

In addition to the overwhelmingly prominent β-amyloid hypothesis being evaluated in a multitude of clinical trials through small molecule modulation of γ- and β-secretase, aberrant phosphorylation of the *tau* protein is believed to significantly contribute to the development of AD. *Tau* is a cytoplasmic protein involved in the stabilization of microtubules under normal conditions. In AD, neuronal *tau* has been found to be excessively phosphorylated, with subsequent generation of aggregates of phosphorylated *tau* protein, known as "neurofibrillary tangles" (NFTs). NFTs and amyloid plaques are considered the most common hallmarks of AD and are correlated with neurofibrillary degeneration, neuronal death, and dementia.[3,4]

Interestingly, several protein kinases have been implicated in neuronal development and, in particular, their overexpression and aberrant activation have been shown to play a significant role in the development of AD via *tau* phosphorylation, which over the last 10 years has led to increased efforts to discover small molecule partners that modulate their functional activity.[5–7]

This chapter focuses on compiling kinases and associated modulatory small molecules involved primarily in *tau* phosphorylation events and the development of AD. Discussion is limited to molecules with in cell and/or *in vivo* activity in models of *tau* pathology and to a lesser extent amyloid pathology. Molecules with activity for many of these targets with utility for other central nervous system indications fall outside of the scope of this chapter.

2. INHIBITORS OF DUAL-SPECIFICITY KINASES

2.1. DYRK1A inhibitors

Dual-specificity tyrosine phosphorylation–regulated kinase-1A (DYRK1A) belongs to the family of CMGC (cyclin-dependent kinases, mitogen-activated protein kinases, glycogen synthase kinases and CDK-like kinases) proline/arginine-directed Ser/Thr kinases and has been found to be upregulated in the frontal cortex of AD brains. Mounting evidence strongly links phosphorylation by DYRK1A on multiple substrates in various cell signaling pathways with neuronal degeneration and death.[8–10] DYRK1A phosphorylates *tau* on 11 different Ser/Thr residues, many of which are detected in NFTs, and the phosphorylation action of DYRK1A at Thr212 has been shown to prime *tau* for further phosphorylation by GSK-3β at Ser308.[11] Additionally, DYRK1A phosphorylates the alternative splicing factor as well as a regulator of calcineurin-1, which is also suggested to promote *tau* hyperphosphorylation and NFT formation.[12,13] Furthermore, DYRK1A phosphorylates the amyloid precursor protein (APP) at Thr668 and prenesilin-1

(PS-1), and both of these phosphorylation events are correlated with increased cleavage of APP by β/γ secretases, leading to the formation of neurotoxic β-amyloid peptides ($A\beta$) and accumulation of senile plaques.[14,15] Interestingly, $A\beta$ is suggested to be involved in a positive feedback loop for further promoting the expression of DYRK1A.[16] DYRK1A can also activate signaling cascades in the brain, including PI3K/Akt and ASK/JNK1 pathways, that are suggested to lead to neuronal death under pathological conditions.[17,18] In support, the cognitive deficits of DYRK1A overexpression have been demonstrated in numerous *in vivo* models.[19,20]

Harmine (**1**, Fig. 5.1) is a β-carboline alkaloid known to potently inhibit DYRK1A (reported IC_{50}s of 33 and 80 nM), with additional data suggesting high selectivity against structurally related kinases.[21–24] Promisingly, **1** and its analogs have been shown to block *tau* phosphorylation at sites found in NFTs *in vitro* ($IC_{50} = 700$ nM) and H4 neuroglioma cells.[25,26]

EGCg (**2**, Fig. 5.1) is a natural polyphenolic catechin found in green tea that is reported to be an adenosine triphosphate (ATP) noncompetitive inhibitor of DYRK1A ($IC_{50} = 330$ nM).[27,28] Interestingly, despite anticipated low bioavailability of **2**, a study in which DYRK1A overexpressing mice were treated with green tea, containing **2** as the major component, demonstrated substantial improvement in cognitive functioning and synaptic plasticity compared to untreated mice.[29a] The molecule is undergoing evaluation in patients with multiple sclerosis and early AD (Phase 2, patient recruitment still underway, NCT00951834), although studies suggest lowering of $A\beta$ aggregation is also intimately associated with its mode of action.[29b]

The natural product DYRK1A inhibitors leucettamine B (**3a**, $IC_{50} = 60$ nM) and a synthetic derivative leucettine L41 (**3b**, $IC_{50} = 95$ nM) (Fig. 5.1) are both moderately selective inhibitors. **3b** was also able to inhibit DYRK1A in a dose-dependent manner in HT22 cells in addition to

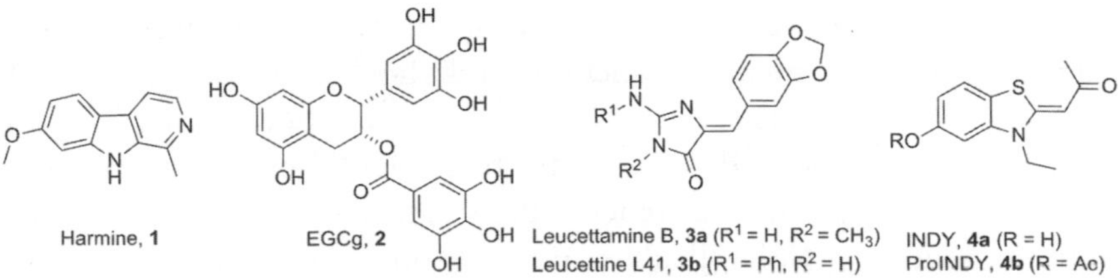

Figure 5.1 Structures of selected DYRK1A inhibitors.

providing protection against neurodegenerative and apoptotic effects induced by glutamate and APP.[23]

A high-throughput screening campaign with a biochemical assay identified the novel benzothiazole DYRK1A inhibitor **4a** ($IC_{50} = 240$ nM) (Fig. 5.1). The functional activity of **4a** was subsequently confirmed in COS-7 cells, and **4a** was able to fully block *tau* phosphorylation on Thr212 at a concentration of 30 µM. Additionally, treatment of DYRK1A overexpressing *Xenopus laevis* tadpole embryos with a prodrug of **4b** (Fig. 5.1, 2.5 µM) visually restored normal development in 86% of the sample population.[30] Interestingly, **1**, **3b**, and **4a** are three of the four small molecule inhibitors of DYRK1A so far cocrystallized in the active site of the enzyme (PDB ID 3ANR, 4AZE, and 3ANQ, respectively).

Recently, several novel series of DYRK1A inhibitors have been published, including meridianins, pyrazines, chromenoindoles, pyrazolo-quinazolines, and thienopyrimidines, with associated *in vitro* activity and significant selectivity over structurally related kinases.[31,32]

3. INHIBITORS OF SERINE–THREONINE KINASES

3.1. GSK-3β inhibitors

Research supports a key role for glycogen synthase kinase-3β (GSK-3β), a proline-directed Ser/Thr kinase, in neurodegeneration and various processes underlying the pathogenesis of AD.[33–35] Upregulation of GSK-3β activity is observed in the frontal cortex and hippocampus of AD patients and is believed to contribute to various phenomena observed in AD brains, including formation of the characteristic pathological lesions (NFTs and Aβ senile plaques), but also neuronal apoptosis, axonal inflammation, insulin, and cholinergic deficits.[36–43] GSK-3β is thought to contribute to Aβ plaque formation by binding PS-1, likely leading to the enhancement of γ-secretase-mediated cleavage of APP.[44] Additionally, GSK-3β directly phosphorylates various *tau* residues, promoting the formation of NFTs.[38] Indeed, GSK-3β transgenic mice display *tau* hyperphosphorylation and a reduction in spatial memory function, and those treated with GSK-3β inhibitors exhibit a reduction in NFTs.[45,46] Furthermore, a link between neurotoxic Aβ plaques and NFTs has been established such that an increase in Aβ plaques in the neurons reduces PI3K signaling, blocking a negative regulatory effect on GSK-3β and resulting in an increase in *tau* phosphorylation and NFT formation.[47,48]

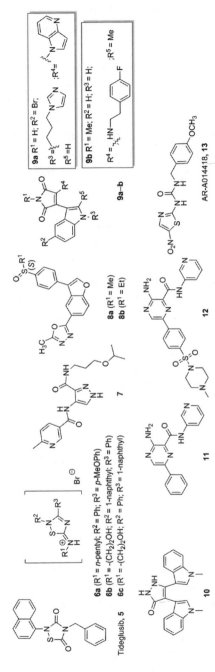

Figure 5.2 Selected structures of GSK-3β inhibitors.

Originally described in 2002 as a GSK-3β inhibitor with low μM potency, thiadiazolidindione Tideglusib (**5**, Fig. 5.2) is reported to function as an ATP noncompetitive, irreversible inhibitor, yet no definitive evidence of covalent adduction by **5** exists in the primary literature.[49,50] Oral administration of **5** to double APP–*tau* transgenic mice strongly demonstrates its ability to prevent associated memory deficits and protect neurons from cell death through reduction of *tau* phosphorylation, amyloid deposition, and plaque-associated astrocytic proliferation.[51] Notably, in 2006 the clinical development of Tideglusib commenced for the treatment of AD and progressive supranuclear palsy, and the molecule is currently in Phase 2.[52]

Structurally related to **5** are the 5-imino-1,2,4-thiadiazoles **6a–c** (Fig. 5.2), reported to be reversible, substrate competitive inhibitors of GSK-3β (IC$_{50}$s ~300 nM).[53] This class of molecule was predicted to significantly passively diffuse across the blood-brain barrier (BBB) after evaluation in the standard surrogate BBB permeability assay PAMPA (parallel artificial membrane permeability assay). Indeed, they exhibited both anti-inflammatory and neuroprotective activity, as witnessed by reduction of nitrite production from primary glial cells, astrocytes, and microglia after pretreatment with lipopolysaccharide. The compounds also exhibited neurogenic activity.[53]

Recently, pyrazolopyridine **7** (Fig. 5.2), emerged as a potent and selective GSK-3β inhibitor (IC$_{50}$ = 2 nM),[54] exhibiting *in vivo* efficacy in JNPL3 mice via reduction of pathological aggregated *tau*. Chronic oral administration of **7** in older JNPL3 mice also led to a reduction of late-stage *tau* pathology.[54]

A series of 1,3,4-oxadiazoles were originally reported as GSK-3β inhibitors.[55] Interestingly, insertion of a (S)-sulfinyl group in the parent compounds afforded analogs **8a** and **8b** with similar biochemical activity (**8a**, IC$_{50}$ = 34 nM and **8b**, IC$_{50}$ = 20 nM), and these analogs inhibited *tau* hyperphosphorylation at Ser396 in a cold water stress model when administered orally.[56,57]

Further study of the pharmacological effects of **8a** revealed inhibition of *tau* phosphorylation in both rat primary neural cell cultures and C57BL/6N mice.[58] Moreover, a study of the effects of the molecule in 3xTg-AD mice also demonstrated concomitant reduction of *tau* phosphorylation. Interestingly in the same model, APP phosphorylation was not affected, yet the molecule elicited improvements of both memory and cognitive function.

4-Azaindolyl-maleimide **9a** (Fig. 5.2) also possesses GSK-3β inhibitory activity (IC$_{50}$ = 140 nM) and leads to a reduction of *tau* phosphorylation at Ser396 in SH-SY5Y human neuroblastoma cells.[59] Structurally related indolylmaleimide **9b** ($IC_{50} = 53$ nM) (Fig. 5.2) exhibits a number of cellular effects, including increasing β-catenin levels in human neural progenitor cells, thus promoting neural differentiation.[60]

Due to the lack of selectivity of the maleimide-based GSK-3β inhibitors **9a** and **9b**, a postulated "redundant" carbonyl group was removed to afford pyrazolone **10** ($IC_{50} = 34$ nM) (Fig. 5.2) with a much improved selectivity profile.[61] **10** also displayed superior neuroprotective activity when compared to SB-216763, an earlier reported indolylmaleimide analog, in a homocysteic acid–induced oxidative stress neuronal cell model.[61a]

Pyrazine **11** (Fig. 5.2) was identified as a GSK-3β inhibitor via high-throughput screening, and subsequent optimization of activity produced the congener **12** ($IC_{50} = 76$ nM), shown to inhibit *tau* phosphorylation in T3T cells. Encouragingly, several of the pyrazine-derived GSK-3β inhibitors seem suitable for *in vivo* studies after evaluation in surrogate assays predictive of passive BBB diffusion.[62]

AR-A014418 **13** (Fig. 5.2), a nitro-thiazole, completes the suite of AD-related GSK-3β inhibitors and is claimed to potently and specifically inhibit the target protein (IC_{50} not reported). **13** was administered in a mouse model of spinal cord trauma (4 mg/kg, *iv*) and induced acceleration of neurologic improvement by preventing mitochondrial apoptosis and reducing inflammation.[63]

Among additional GSK-3β inhibitors derived from natural sources, the alkaloid manzamine A exhibits moderate inhibitory activity ($IC_{50} = 10.2$ μM), and exposure to SH-SY5Y neuroblastoma cells confirmed decreased *tau* phosphorylation levels at Ser396.[64]

The natural product Ginsenoside Rd is also reported to protect Sprague–Dawley rats and cultured cortical neurons against okadaic acid–induced toxicity via decreases in *tau* hyperphosphorylation, and recent literature suggests that this is due to GSK-3β inhibitory activity.[65] Increased activity of phosphatase 2A, a key phosphatase involved in *tau* dephosphorylation, was also observed, thus providing a further rationale for the observed neuroprotective activity.[65] The natural product was also able to inhibit *tau* phosphorylation at multiple sites in Aβ-treated cultured cortical neurons, in both a rat and transgenic mouse models.[66]

3.2. CDK5/p25 inhibitors

Cyclin-dependent kinase 5 (CDK5), a proline-directed Ser/Thr kinase, is activated and tightly regulated upon binding to neuron-specific proteins,

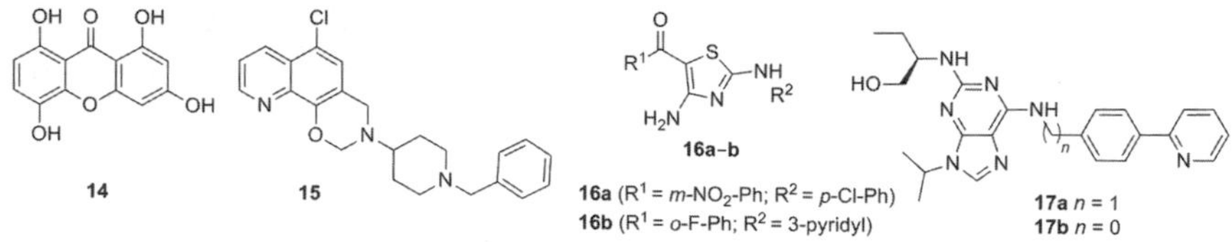

Figure 5.3 Selected structures of CDK5/p25 inhibitors.

primarily p35. In the early 1990s, CDK5 was recognized as one of the major kinases involved in the phosphorylation of residues often found to be phosphorylated in NFTs.[67,68] Upon cellular stress, p35 is cleaved by calpain to give p25, which can bind to and activate CDK5 in a similar but more stable way than p35, resulting in a hyperactive form of CDK5.[69] Indeed, increased expression of CDK5 and p25 has been observed in AD brains, and treatment with Aβ peptide and/or induction of oxidative stress or excitotoxicity promotes the accumulation of p25 coincident with neurodegeneration.[70] It has also been shown that *in vitro* CDK5/p25 is able to phosphorylate *tau* much more efficiently than CDK5/p35, indicating that the truncated activator protein likely alters the substrate specificity of CDK5 to favor those contributing to neurotoxicity.[71]

Ahn and coworkers have reported three CDK5/p25 inhibitors able to reduce *tau* phosphorylation at Ser396 and Ser404, namely Bellidin (**14**, $IC_{50} = 0.2\ \mu M$, ATP competitive), diazaphenanthrene **15** ($IC_{50} = 2.0\ \mu M$, substrate competitive), and 4-aminothiazole **16a** ($IC_{50} = 17\ \mu M$, ATP competitive) (Fig. 5.3). Compounds were also tested for effects in IMR-32 neuroblastoma cells and rat primary neurons, showing nominal effects on inhibition of *tau* phosphorylation (**15**, $EC_{50} < 20\ \mu M$, **16a** $EC_{50} < 50\ \mu M$).[72]

Interestingly, 4-aminothiazole **16b** (Fig. 5.3) was reported to possess CDK5 inhibitory activity ($IC_{50} = 30\ nM$), with additional activity detected for CDK2 ($IC_{50} = 25\ nM$) and GSK3-β ($IC_{50} = 45 nM$). **16b** inhibited *tau* phosphorylation in rat primary neurons ($EC_{50} = 5.5\ \mu M$); however, the mechanistic source of cellular activity or attempts at delineation were not reported.[73]

Roscovitine derivatives **17a/b** (Fig. 5.3) have been shown to inhibit both CDK5 (**17a**, $IC_{50} = 130\ nM$, **17b**, $IC_{50} = 80\ nM$) and CK1 (**17a**, $IC_{50} = 600\ nM$, **17b**, $IC_{50} = 14\ nM$), and both compounds showed inhibitory effects on Aβ40 plaque production in N2A cells.[74] A series of purine-based triazoles have also been reported as CDK5 inhibitors.[75]

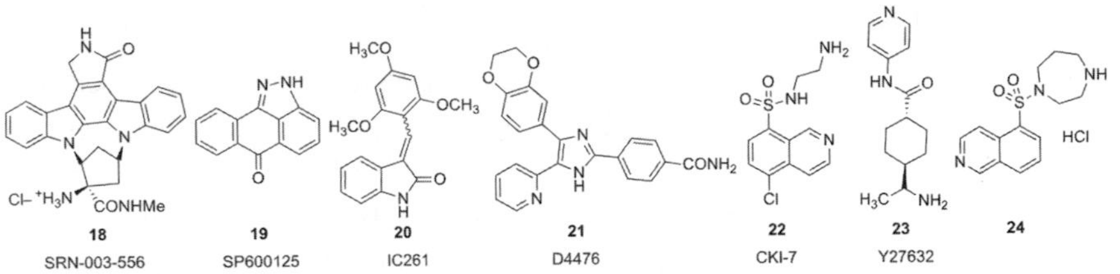

Figure 5.4 Other kinase inhibitors herein discussed.

3.3. ERK2 inhibitors

Various mitogen–activated protein kinases have been identified as key contributors to neurodegenerative processes, and among them extracellular signal-regulated kinase-2 (ERK2) is unique in its ability to stoichiometrically phosphorylate all 17 Ser/Thr-Pro motifs in *tau* protein. Additionally, in an immunocytochemical study of the activation states of mitogen-activated protein kinases in postmortem normal and AD brains, ERK2 was found to be activated in all AD neurons in which the accumulation of hyperphosphorylated *tau* and NFTs was observed.[76–79]

SRN–003-556 (**18**, Fig. 5.4) inhibits ERK2 ($IC_{50} = 0.6\ \mu M$), and in adult rat hippocampal slices, reduced okadaic acid–induced *tau* hyperphosphorylation at AT8 ($IC_{50} = 0.6\ \mu M$) and AP422 ($IC_{50} = 0.6\ \mu M$) epitope sites. Furthermore, **18** significantly delayed the development of motor deficiencies when tested in JNPL3 transgenic mice expressing mutant human P301L 4R0N *tau*.[80]

3.4. JNK3 inhibitors

C-Jun N-terminal kinase-3 (JNK3), the brain-specific isoform of the JNK family of kinases, has been found to be overactive in human AD brains and shown to phosphorylate *tau* at sites found in paired helical filaments (PHFs), as well as at Thr668 on APP in mouse models.[81] JNK3 has also been shown to colocalize with Aβ deposits and NFTs in various animal models.[82,83] Additionally, JNK3-deficient cortical neurons and JNK3-null mice are significantly less susceptible to Aβ-induced neuronal apoptosis.[84–87]

A well-documented JNK inhibitor, SP600125 (**19**, Fig. 5.4), invoked inhibition of okadaic acid–induced phosphorylation in a primary cultured hippocampal neuron model system and was shown to reduce *tau* phosphorylation at Ser422 ($IC_{50} = 7.2\ \mu M$) and other sites.[88]

3.5. Casein kinase-1 inhibitors

Casein kinases-1 (CK1s) are ubiquitously expressed Ser/Thr kinases that mediate various processes in the brain, including possible involvement in the glutamate deficiency associated with AD and other neurodegenerative diseases.[89–91] Of the seven known isoforms, CK1δ and CK1ε seem to be the most pharmacologically relevant in terms of neurodegeneration, although all CK1s are reportedly upregulated in AD brains and all seem to show an increase in activity in response to Aβ treatment.[92–94] Commercially available CK1 inhibitors IC261 (**20**), D4476 (**21**), and CKI-7 (**22**, Fig. 5.4) were shown to reduce endogenous Aβ40/42 peptide production in a dose-dependent manner in N2A cells transiently expressing CK1ε. All three inhibitors were also shown to significantly reduce Aβ40/42 formation (5–68%) after 3 h at 50 μM concentration in N2A cells expressing APP-695, thus further supporting the pivotal role of CK1 in Aβ formation.[95]

3.6. Inhibitors of ROCK kinase

Rho GTPase and its kinase ROCK are responsible for regulating actin–cytoskeleton organization and maintaining neuron morphology.[96,97] Brains of rats affected by AD have exhibited upregulation of the Rho-kinase pathway, with a corresponding increase in collapsing response mediator protein-2 phosphorylation. Y27632 (**23**, Fig. 5.4), a known ROCK inhibitor, reduced collapsing response mediator protein-2 phosphorylation in Aβ-treated SH-SY5Y cells, as detected by immunoprecipitation experiments.[98] Also, Fasudil hydrochloride (**24**, $IC_{50} = 10.7$ μM) (Fig. 5.4) has been shown to improve synaptic structures through the enhancement of synaptophysin expression in the hippocampus of streptozotocin-treated Sprague–Dawley rats.[99]

4. CONCLUSIONS

With 24.3 million people affected in 2005 and an estimated rise to 40 million in 2020, dementia is currently a leading unmet medical need and a costly burden on public health. Targeting kinases, in particular those associated with promotion of *tau*-related pathology, represent a new high value alternate approach to AD treatment, which in recent years has been dominated in the private sector by efforts to modulate APP through inhibition of the proteases β- and γ-secretase. The recent clinical Phase 3 failures of Semagacestat (NCT00762411) and Tarenflurbil, both targeting γ-secretase, further highlight the need to investigate new targets for AD. As such, this chapter provides an account of AD-relevant kinases and their associated inhibitors, focusing on molecules that exhibit clear functional activity. Encouragingly, several of the compounds reviewed herein exhibit promising cellular activity, in addition to significant activity in animal models of AD and cognitive deficits. In summary, promising advances in inhibitor design have been made in recent years that will undoubtedly continue. Indeed, it will be particularly intriguing to monitor the progress of Tideglusib (**5**), currently undergoing Phase 2 clinical evaluation. As most of the molecules herein reviewed, **5** displayed high selectivity toward its target kinase. In particular, selectivity has always represented a major issue in the area of kinase modulation, due to the high similarity in the architecture of the active sites of the 518 kinases so far characterized. Despite this, lack of selectivity is not always a major drawback in the area of AD treatment, as inhibition of two or more kinases involved in the development of *tau* pathology may represent a valuable approach to future clinical treatments.

REFERENCES

1. Mekk, P. D.; McKeithan, K.; Schumock, G. T. *Pharmacotherapy* **1998**, *18*, 68.
2. Thies, W. *Alzheimers Dement.* **2011**, 7, 175.
3. Trinczek, B.; Biernat, J.; Baumann, K.; Mandelkow, E. M.; Mandelkow, E. *Mol. Biol. Cell* **1995**, *6*, 1887.
4. Arriagada, P. V.; Growdon, J. H.; Hedley-Whyte, E. T.; Hyman, B. T. *Neurology* **1992**, *42*, 631.
5. Savage, M. J.; Gingrich, D. E. *Drug Dev. Res.* **2009**, *70*, 125.
6. Lloyd, M.; Mokdsi, G.; Spielthenner, D. *Pharm. Pat. Analyst* **2012**, *1*, 437.
7. Mangialasche, F.; Solomon, A.; Winblad, B.; Mecocci, P.; Kivipelto, M. *Lancet* **2010**, *9*, 702.
8. Ferrer, I.; Barrachina, M.; Puig, B.; Martinez de Lagran, M.; Marti, E.; Avila, J.; Dierssen, M. *Neurobiol. Dis.* **2005**, *20*, 392.
9. Dierssen, M.; Martinez de Lagran, M. *Sci. World J.* **1911**, *2006*, 6.
10. Becker, W.; Joost, H. G. *Prog. Nucleic Acid Res.* **1999**, *62*, 1.
11. Wegiel, J.; Kaczmarski, W.; Barua, M.; Kuchna, I.; Nowicki, A.; Wang, K. C.; Wegiel, J.; Yang, S. M.; Frackowiak, J.; Mazur-Kolecka, B.; Silverman, W. P.; Reisberg, B.; Monteiro, I.; De Leon, M.; Wisniewski, T.; Dalton, A.; Lai, F.; Hwang, Y. W.; Adayev, T.; Liu, F.; Iqbal, K.; Iqbal, I. G.; Gong, C. X. *J. Neuropath. Exp. Neur.* **2011**, *70*, 36.
12. Shi, J.; Zhang, T.; Zhou, C.; Chohan, M. O.; Gu, X.; Wegiel, J.; Zhou, J.; Hwang, Y. W.; Iqbal, K.; Grundke-Iqbal, I.; Gong, C. X.; Liu, F. *J. Biol. Chem.* **2008**, *283*, 28660.
13. Jung, M. S.; Park, J. H.; Ryu, Y. S.; Choi, S. H.; Yoon, S. H.; Kwen, M. Y.; Oh, J. Y.; Song, W. J.; Chung, S. H. *J. Biol. Chem.* **2011**, *286*, 40401.
14. Ryoo, S. R.; Cho, H. J.; Lee, H. W.; Jeong, H. K.; Radnaabazar, C.; Kim, Y. S.; Kim, M. J.; Son, M. J.; Seo, H.; Chung, S. H.; Song, W. J. *J. Neurochem.* **2008**, *104*, 1333.

15. Ryu, Y. S.; Park, S. Y.; Jung, M. S.; Yoon, S. H.; Kwen, M. Y.; Lee, S. Y.; Choi, S. H.; Radnabaazar, C.; Kim, M. K.; Kim, H.; Kim, K.; Song, W. J.; Chung, S. H. *J. Neurochem.* **2010**, *115*, 574.

16. Kimura, R.; Kamino, K.; Yamamoto, M.; Nuripa, A.; Kida, T.; Kazui, H.; Hashimoto, R.; Tanaka, T.; Kudo, T.; Yamagata, H.; Tabara, Y.; Miki, T.; Akatsu, H.; Kosaka, K.; Funakoshi, E.; Nishitomi, K.; Sakaguchi, G.; Kato, A.; Hattori, H.; Uema, T.; Takeda, M. *Hum. Mol. Genet.* **2007**, *16*, 15.

17. Guedi, F.; Pereira, P. L.; Najas, S.; Barallobre, M. J.; Chabert, C.; Souchet, B.; Sebrie, C.; Verney, C.; Herault, Y.; Arbones, M.; Delabar, J. M. *Neurobiol. Dis.* **2012**, *46*, 190.

18. Choi, H. K.; Chung, K. C. *Exp. Neurobiol.* **2011**, *20*, 35.

19. Ahn, K. J.; Jeong, H. K.; Choi, H. S.; Ryoo, S. R.; Kim, Y. J.; Goo, J. S.; Choi, S. Y.; Han, J. S.; Ha, I.; Song, W. J. *Neurobiol. Dis.* **2006**, *22*, 463.

20. Altafaj, X.; Dierssen, M.; Baamonde, C.; Marti, E.; Visa, J.; Guimera, J.; Oset, M.; Gonzalez, J. R.; Florez, J.; Fillat, C.; Estevill, X. *Hum. Mol. Genet.* **2001**, *10*, 1915.

21. Bain, J.; Plater, L.; Elliott, M.; Shpiro, N.; Hastie, J.; McLauchlan, H.; Klevernic, I.; Arthur, J. S. C.; Alessi, D. R.; Cohen, P. *Biochem. J.* **2007**, *408*, 297.

22. Göckler, N.; Guillermo, J.; Papadopoulos, C.; Soppa, U.; Tejador, F. J.; Becker, W. *FEBS J.* **2009**, *276*, 6324.

23. Tahtouh, T.; Elkins, J. M.; Filippakopoulos, P.; Soundararajan, M.; Burgy, G.; Durieu, E.; Cochet, C.; Schmid, R. S.; Lo, D. C.; Delhommel, F.; Oberholzer, A. E.; Pearl, L. H.; Carreaux, F.; Bazureau, J. P.; Knapp, S.; Meijer, L. J. *Med. Chem.* **2012**, *55*, 9312.

24. Grabher, P.; Durieu, E.; Kouloura, E.; Halabalaki, M.; Skaltsounis, L. A.; Meijer, L.; Hamburger, M.; Potterat, O. *Planta Med.* **2012**, *78*, 951.

25. Azorsa, D. O.; Robeson, R. H.; Frost, D.; Meechoovet, B.; Brautigam, G. R.; Dickey, C.; Beaudry, C.; Basu, G. D.; Holz, D. R.; Hernandez, J. A.; Bisanz, K. M.; Gwinn, L.; Grover, A.; Rogers, J.; Reiman, E. M.; Hutton, M.; Stephan, D. A.; Mousses, D.; Dunckley, T. *BMC Genomics* **2010**, *11*, 25.

26. Frost, D.; Meechoovet, B.; Wang, T.; Gately, S.; Giorgetti, M.; Shcherbakova, I.; Dunckley, T. *PLoS One* **2011**, *6*, e19264.

27. Adayev, T.; Chen-Hwang, M. C.; Murakami, N.; Wegiel, J.; Hwang, Y. W. *Biochemistry* **2006**, *45*, 12011.

28. Bain, J.; McLauchlan, H.; Elliott, M.; Cohen, P. *Biochem. J.* **2003**, *371*, 199.

29. (a) Guedj, F.; Sébrié, C.; Rivals, I.; Ledru, A.; Paly, E.; Bizot, J. C.; Smith, D.; Rubin, E.; Gillet, B.; Arbones, M.; Delabar, J. M. *PLoS One* **2009**, *4*, e4606. (b) Mandel, S. A.; Amit, T.; Kalfon, L.; Reznichenko, L.; Weinreb, O.; Youdim, M. B. *J. Alzheimers Dis.* **2008**, *15*, 211.

30. Ogawa, Y.; Nonaka, Y.; Goto, T.; Ohnishi, E.; Hiramatsu, T.; Kii, I.; Yoshida, M.; Ikura, T.; Onogi, H.; Shibuya, H.; Hosoya, T.; Ito, N.; Hagiwara, M. *Nat. Commun.* **2010**, *86*, 1.

31. Smith, B.; Medda, F.; Gokhale, V.; Dunckley, T.; Hulme, C. *ACS Chem. Neurosci.* **2012**, *3*, 857.

32. Becker, W.; Sippl, W. *FEBS J.* **2011**, *278*, 246.

33. Benbow, J. W.; Helal, C. J.; Kung, D. W.; Wager, T. T. *Annu. Rep. Med. Chem.* **2005**, *40*, 135.

34. Hooper, C.; Killick, R.; Lovestone, S. *J. Neurochem.* **2008**, *104*, 1433.

35. Takashima, A. *J. Pharmacol. Sci.* **2009**, *109*, 174.

36. Sanna, P. P.; Cammalleri, M.; Berton, F.; Simpson, C.; Lutjens, R.; Bloom, F. E.; Rancesconni, W. *J. Neurosci.* **2002**, *22*, 3359.

37. Chen, J.; Park, C. S.; Tang, S. J. *J. Biol. Chem.* **2006**, *281*, 11910.

38. Hanger, D. P.; Hughes, K.; Woodgett, J. R.; Brion, J. P.; Anderton, B. H. *Neurosci. Lett.* **1992**, *147*, 58.

39. Takashima, A.; Noguchi, K.; Sato, K.; Hoshino, T.; Imahori, K. *Proc. Natl. Acad. Sci. U.S.A.* **1993**, *90*, 7789.
40. Turenne, G. A.; Price, B. D. *BMC Cell Biol.* **2001**, *2*, 12.
41. Jope, R. S.; Yuskaitis, C. J.; Beurel, E. *Neurochem. Res.* **2007**, *32*, 577.
42. De la Monte, S. M.; Tong, M.; Lester-Coll, N.; Plater, M.; Wands, J. R. J. *J. Alzheimers Dis.* **2006**, *10*, 89.
43. Hoshi, M.; Takashima, A.; Noguchi, K.; Murayama, M.; Sato, M.; Kondo, S.; Saitoh, Y.; Ishiguro, K.; Hoshino, T.; Imahori, K. *Proc. Natl. Acad. Sci. U.S.A.* **1996**, *93*, 2719.
44. Takashima, A.; Murayama, M.; Murayama, O.; Kohno, T.; Honda, T.; Yasutake, K.; Nihonmatsu, N.; Mercken, M.; Yamaguchi, H.; Sugihara, S.; Wolozin, B. *Proc. Natl. Acad. Sci. U.S.A.* **1998**, *95*, 9637.
45. Takashima, A. *J. Alzheimers Dis.* **2006**, *9*, 309.
46. Noble, W.; Planel, E.; Zehr, C.; Olm, V.; Meyerson, J.; Suleman, F. *Proc. Natl. Acad. Sci. U.S.A.* **2005**, *102*, 6990.
47. Takashima, A.; Noguchi, K.; Michel, G.; Mercken, M.; Hoshi, M.; Ishiguro, K.; Imahori, K. *Neurosci. Lett.* **1996**, *203*, 33.
48. Takashima, A.; Honda, T.; Tasutake, K.; Michel, G.; Murayama, O.; Murayama, M.; Ishiguro, K.; Yamaguchi, H. *Neurosci. Res.* **1998**, *31*, 317.
49. Martinez, A.; Alonso, M.; Castro, A.; Pérez, C.; Moreno, F. J. *J. Med. Chem.* **2002**, *45*, 1292.
50. Domínguez, J. M.; Fuertes, A.; Orozco, L.; Del Monte-Millá, M.; Delgado, E.; Medina, M. *J. Biol. Chem.* **2012**, *287*, 893.
51. Serenó, L.; Coma, M.; Rodríguez, M.; Sánchez-Ferrer, P.; Sánchez, M. B.; Gich, I.; Agulló, J. M.; Pérez, M.; Avila, J.; Guardia-Laguarta, C.; Clarimón, J.; Lleó, A.; Gómez-Isla, T. *Neurobiol. Dis.* **2009**, *35*, 359.
52. Del Ser, T. *Alzheimers Dement.* **2010**, *6*, 147.
53. Palomo, V.; Perez, D. I.; Perez, C.; Morales-Garcia, J. A.; Soteras, I.; Alonso-Gil, S.; Encinas, A.; Castro, A.; Campillo, N. E.; Perez-Castillo, A.; Gil, C.; Martinez, A. *J. Med. Chem.* **2012**, *55*, 1645.
54. Uno, Y.; Iwashita, H.; Tsukamoto, T.; Uchiyama, N.; Kawamoto, T.; Kori, M.; Nakanishi, A. *Brain Res.* **2009**, *1296*, 148.
55. Saitoh, M.; Kunitomo, J.; Kimura, E.; Hayase, Y.; Kobayashi, H.; Uchiyama, N.; Kawamoto, T.; Tanaka, T.; Mol, C. D.; Dougan, D. R.; Textor, G. S.; Snell, G. P.; Itoh, F. *Bioorg. Med. Chem.* **2009**, *17*, 2017.
56. Khanfar, M. A.; Hill, R. A.; Kaddoumi, A.; El Sayed, K. A. *J. Med. Chem.* **2010**, *53*, 8543.
57. Saitoh, M.; Kunitomo, J.; Kimura, E.; Iwashita, H.; Uno, Y.; Onishi, T.; Uchiyama, N.; Kawamoto, T.; Tanaka, T.; Mol, C. D.; Dougan, D. R.; Textor, G. P.; Snell, G. P.; Takizawa, M.; Itoh, F.; Kori, M. *J. Med. Chem.* **2009**, *52*, 6270.
58. Onishi, T.; Iwashita, H.; Uno, Y.; Kunitomo, J.; Saitoh, M.; Kimura, E.; Fujita, H.; Uchiyama, N.; Kori, M.; Takizawa, M. *J. Neurochem.* **2011**, *119*, 1330.
59. Ye, Q.; Xu, G.; Lv, D.; Cheng, Z.; Li, J.; Hu, Y. *Bioorg. Med. Chem.* **2009**, *17*, 4302.
60. Schmöle, A. C.; Brennführer, A.; Karapetyan, G.; Jaster, R.; Pews-Davtyan, A.; Hübner, R.; Ortinau, S.; Beller, M.; Rolfs, A.; Frech, M. J. *Bioorg. Med. Chem.* **2010**, *18*, 6785.
61. (a) Chen, W.; Gaisina, I. N.; Gunosewoyo, H.; Malekiani, S. A.; Hanania, T.; Kozikowski, A. P. *ChemMedChem* **2011**, *6*, 1587. (b) Coghlan, M. P.; Culbert, A. A.; Cross, D. A. E.; Corcoran, S. L.; Yates, J. W.; Pearce, N. J.; Rausch, O. L.; Murphy, G. J.; Carter, P. S.; Cox, L. R.; Mills, D.; Brown, M. J.; Haigh, D.; Ward, R. W.; Smith, D. G.; Murray, K. J.; Reith, A. D.; Holder, J. C. *Chem. Biol.* **2000**, *7*, 793.

62. Berg, S.; Bergh, M.; Hellberg, S.; Högdin, K.; Lo-Alfredsson, Y.; Söderman, P.; Von Berg, S.; Weigelt, T.; Ormö, M.; Xue, Y.; Tucker, J.; Neelissen, J.; Jerning, E.; Nilsson, Y.; Bhat, R. *J. Med. Chem.* **2012**, *55*, 9107.

63. Tunçdemir, T.; Yıldırım, A.; Karaoğlan, A.; Akdemir, O.; Öztürk, M. *Neurocirugia* **2013**, *24*, 22.

64. Hamann, M.; Alonso, D.; Martín-Aparicio, E.; Fuertes, A.; Pérez-Puerto, M. J.; Castro, A.; Morales, S.; Navarro, M. L.; del Monte-Millán, M.; Medina, M.; Pennaka, H.; Balaiah, A.; Peng, J.; Cook, J.; Wahyuono, S.; Martínez, A. *J. Nat. Prod.* **2007**, *70*, 1397.

65. Li, L.; Liu, J.; Yan, X.; Qin, K.; Shi, M.; Lin, T.; Zhu, Y.; Kang, T.; Zhao, G. *J. Ethnopharmacol.* **2011**, *48*, 97.

66. Li, L.; Liu, Z.; Liu, J.; Tai, X.; Hu, X.; Liu, X.; Wu, Z.; Zhang, G.; Shi, M.; Zhao, G. *Neurobiol. Dis.* **2013**, *1*, 52. http://dx.doi.org/10.1016/j.ndb.2013.01.002.

67. Baumann, K.; Mandelkow, E. M.; Biernat, J.; Piwnica-Worms, H.; Mandelkow, E. *FEBS Lett.* **1993**, *336*, 417.

68. Paudel, H. K.; Lew, J.; Ali, Z.; Wang, J. H. *J. Biol. Chem.* **1993**, *268*, 23512.

69. Nath, R.; Davis, M.; Probert, A. W.; Kupina, N. C.; Ren, X.; Schielke, G. P.; Wang, K. K. *Biochem. Biophys. Res. Commun.* **2000**, *274*, 16.

70. Patrick, G. N.; Zuckerberg, L.; Nikolic, M.; De la Monte, S.; Dikkes, P.; Tsai, L. H. *Nature* **1999**, *402*, 615.

71. Hashiguchi, M.; Saito, T.; Hisanaga, S.; Hashiguchi, T. *J. Biol. Chem.* **2002**, *277*, 44525.

72. Ahn, J. S.; Radhakrishnan, M. L.; Mapelli, M.; Choi, S.; Tidor, B.; Cuny, G. D.; Musacchio, A.; Yeh, L. A.; Kosik, K. S. *Chem. Biol.* **2005**, *12*, 811.

73. Laha, J. K.; Zhang, X.; Qiao, L.; Liu, M.; Chatterjee, S.; Robinson, S.; Kosik, K. S.; Cuny, G. D. *Bioorg. Med. Chem. Lett.* **2011**, *21*, 2098.

74. Oumata, N.; Bettayeb, K.; Ferandin, Y.; Demange, L.; Lopez-Giral, A.; Goddard, M. L.; Myrianthopoulos, V.; Mikros, E.; Flajolet, M.; Greengard, P.; Meijer, L.; Galons, H. *J. Med. Chem.* **2008**, *51*, 5229.

75. Nair, N.; Kudo, W.; Smith, M. A.; Abrol, R.; Goddard, W. A., 3rd.; Reddy, V. P. *Bioorg. Med. Chem. Lett.* **2011**, *21*, 3957.

76. Drewes, G.; Lichtenberg-Kraag, B.; Döring, F.; Mandelkow, E. M.; Biernat, J.; Goris, J.; Dorée, M.; Mandelkow, E. *EMBO J.* **1992**, *11*, 2131.

77. Veeranna, N.; Amin, N. D.; Ahn, N. G.; Jaffe, H.; Winters, C. A.; Grant, P.; Pant, H. C. *J. Neurosci.* **1998**, *18*, 4008.

78. Perry, G.; Roder, H.; Nunomura, A.; Atsushi, T.; Friedlich, A. L.; Zhu, X.; Raina, A. K.; Holbrook, N.; Siedlak, S. L.; Harris, P. L. R.; Smith, M. A. *Neuroreport* **1999**, *10*, 2411.

79. Pei, J. J.; Braak, H.; An, W. L.; Winblad, B.; Cowburn, R. F.; Iqbal, K.; Iqbal, K. *Mol. Brain Res.* **2002**, *109*, 45.

80. Le Corre, S.; Klafki, H. W.; Plesnila, N.; Hübinger, G.; Obermeier, A.; Sahagún, H.; Monse, B.; Seneci, P.; Lewis, J.; Eriksen, J.; Zehr, C.; Yue, M.; McGowan, E.; Dickson, D. W.; Hutton, M.; Roder, H. M. *Proc. Natl. Acad. Sci. U.S.A.* **2006**, *103*, 9673.

81. Borsello, T.; Forloni, G. *Curr. Pharm. Des.* **2007**, *13*, 1875.

82. Zhu, X.; Riana, A. K.; Rottkamp, C. A.; Aliev, G.; Perry, G.; Boux, H.; Smith, M. A. *J. Neurochem.* **2001**, *76*, 435.

83. Shoji, M.; Iwakami, N.; Takeuchi, S.; Waragi, M.; Suzuki, M.; Kanazawa, I.; Lippa, C. F.; Ono, S.; Okazawa, H. *Brain Res. Mol. Brain Res.* **2000**, *85*, 221.

84. Otth, C.; Mendoza-Naranjo, A.; Mujica, L.; Zambrano, A.; Concha, I. I.; Maccioni, R. B. *Neuroreport* **2003**, *14*, 2403.

85. Hwang, D. Y.; Cho, J. S.; Lee, S. H.; Chae, K. R.; Lim, H. J.; Min, S. H.; Seo, S. J.; Song, Y. S.; Song, C. W.; Paik, S. G.; Sheen, Y. Y.; Kim, Y. K. *Exp. Neurol.* **2004**, *186*, 20.

86. Bozyczko-Coyne, D.; Saporito, M. S.; Hudkins, R. L. *Curr. Drug Targets CNS Neurol. Disord.* **2002**, *1*, 31.

87. Morishima, Y.; Gotoh, Y.; Zieg, J.; Barret, T.; Takano, H.; Flavel, R.; Davis, R. J.; Shirasaki, Y.; Greenberg, M. E. *J. Neurosci.* **2001**, *21*, 7551.

88. Vogel, J.; Anand, V. S.; Ludwig, B.; Nawoschik, S.; Dunlop, J.; Braithwaite, S. P. *Neuropharmacology* **2009**, *57*, 539.

89. Perez, D. I.; Gil, C.; Martinez, A. *Med. Res. Rev.* **2011**, *31*, 924.

90. Gross, S. D.; Anderson, R. A. *Cell. Signal.* **1998**, *10*, 699.

91. Cherugi, K.; Svenningsson, P.; Greengard, P. *J. Neurosci.* **2005**, *25*, 6601.

92. Churcher, I. *Curr. Top. Med. Chem.* **2006**, *6*, 579.

93. Yasojima, K.; Kuret, J.; DeMaggio, A. J.; McGeer, E.; McGeer, P. L. *Brain Res.* **2000**, *865*, 116.

94. Flaherty, D. B.; Soria, J. P.; Tomasiewicz, H. G.; Wood, J. G. *J. Neurosci. Res.* **2000**, *62*, 463.

95. Flajolet, M.; He, G.; Heiman, M.; Lin, A.; Nairn, A. C.; Greengard, P. *Proc. Natl. Acad. Sci. U.S.A.* **2007**, *104*, 4159.

96. Fukata, Y.; Amano, M.; Kaibuchi, K. *Trends Pharmacol. Sci.* **2001**, *22*, 32.

97. Luo, L. *Annu. Rev. Cell Dev. Biol.* **2002**, *18*, 601.

98. Petratos, S.; Li, Q. X.; George, A. J.; Hou, X.; Kerr, M. L.; Unabia, S. E.; Hatzinisiriou, I.; Maksel, D.; Aguilar, M. I.; Small, D. H. *Brain* **2008**, *131*, 90.

99. Hou, Y.; Zhou, L.; Yang, Q. D.; Du, X. P.; Li, M.; Yuan, M.; Zhou, Z. W. *Neuroscience* **2012**, *200*, 120.

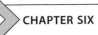

CHAPTER SIX

Orexin Receptor Antagonists in Development for Insomnia and CNS Disorders

Scott D. Kuduk*, **Christopher J. Winrow†**, **Paul J. Coleman***

*Department of Medicinal Chemistry, Merck Research Laboratories, West Point, Pennsylvania, USA
†Neuroscience Department, Merck Research Laboratories, West Point, Pennsylvania, USA

Contents

1. INTRODUCTION

Since the discovery of the orexin neuropeptides and their receptors almost 15 years ago, and their demonstrated role in regulating sleep/wake and vigilance, there has been a significant interest in targeting the orexin receptors. Using a combination of genetic and biochemical approaches, two teams separately reported the characterization of a novel neuropeptide secreted from the hypothalamus, naming the peptide hypocretin[1] and orexin,[2] with the latter terminology found most frequently in the patent literature and adopted as the IUPHAR standard nomenclature.[3] Orexin neuropeptides (OX-A and OX-B) are derived from a common prepropeptide secreted from orexinergic neurons localized in the lateral hypothalamus, and signal through activation of two G-protein coupled receptors (GPCRs), orexin receptor 1 (OX_1R) and orexin receptor 2 (OX_2R). The peptides show

73

differential selectivity with OX-A binding equally to both receptors and OX-B showing ~10-fold greater affinity for OX_2R.[3]

Orexin receptors are distributed widely throughout the brain, with key innervation of regions responsible for governing wake, vigilance, and reward seeking behaviors. In all species examined, orexin signaling is most active during the normal wake period and falls silent during the sleep period.[4–6] Administration of exogenous orexin peptides promotes wakefulness and increases locomotor activity.[7,8] Efforts to identify small molecule agonists have not yet been successful. As a result of foundational work, several companies identified antagonists targeting both OX_1R and OX_2R, or each receptor individually. These antagonists offer promise for areas of unmet medical need as many are progressing through late stage clinical development.[9–11] Several comprehensive historical reviews on the chemistry and pharmacology of the orexin system and antagonists are available.[12–15] The present chapter will cover the most recent developments in the field of orexin pharmacology, highlighting important advances in the discovery of chemical series selectively targeting the orexin receptors.

2. CLINICAL PROGRAMS

Numerous companies have been engaged in the discovery of both selective orexin receptor antagonists (SORAs) and dual orexin receptor antagonists (DORAs). Early efforts focused on glycine sulfonamides[16] **1** (OX_2R/OX_1R $IC_{50} = 4/6$ nM) and dihydropyrazoloquinazolinone ureas[17] **2** (OX_2R/OX_1R $IC_{50} = 16/12$ nM) as DORAs, respectively. Tetrahydroisoquinoline DORAs were investigated from a high throughput screening lead **3** which was further optimized using combinatorial chemistry methods.[18] This resulted in the discovery of almorexant **4**, a DORA (OX_2R/OX_1R $IC_{50} = 8/13$ nM) with potent *in vivo* activity.[19] Almorexant dose–dependently increased both rapid eye movement (REM) and non-REM (NREM) sleep in Wistar rats and reduced wakefulness in dogs when administered in the active period. Cessation of dosing did not produce residual or rebound sleep effects in rats or dogs, and did not elicit cataplexy or generate other tolerability signals of concern.[20,21]

1

2

3

4: Almorexant

Almorexant was advanced into human clinical studies and half-lives ranging from 13.1 to 19.0 h were noted. No cataplexy-related side effects were seen with almorexant nor were other significant tolerability issues noted. In double-blind studies with zolpidem as an active control, almorexant increased sleep efficiency, reduced sleep latency, and increased total sleep time (TST) at doses greater than 200 mg in healthy volunteers.[22] A Phase II proof-of-concept study with almorexant established efficacy for the primary endpoint of increased sleep efficiency. Significant effects on secondary endpoints included reduced latency to persistent sleep (LPS) and reduced wake after sleep onset (WASO) at the 400-mg dose.[23] In 2007, Actelion initiated a Phase III study in adults with primary insomnia (RESTORA1). In this study, almorexant met the primary endpoint of superiority compared with placebo on both objective and subjective measures of WASO. However, an undisclosed human tolerability issue resulted in the termination of Phase III clinical development in January 2009.[24]

Analogs of almorexant have been described and reviewed.[25] A derivative of almorexant was described in which subtle changes could provide a potent analog **5** (OX$_1$R IC$_{50}$ = 27 nM) with >20-fold selectivity over OX$_2$R.[26] Further simplified analogs in which the piperidine nitrogen is directly acylated such as **6** (OX$_2$R IC$_{50}$ = 13 nM, OX$_1$R IC$_{50}$ = 7 nM) have been reported as DORAs.[27] Finally, some completely different non–almorexant type pyrazole amides have been described as highly potent OX_2R antagonists (**7**: OX_2R $IC_{50} = 4$ nM), although OX_1R data were not provided.[28] Lastly, proline **8** is a DORA (OX_2R $IC_{50} = 5$ nM; OX_1R $IC_{50} = 30$ nM)[29] that showed sedative effects, altered effects on morphine-induced locomotor sensitization, and reduced fear-potentiated startle in rodent.

Early efforts in the field focused on heterocyclic ureas such as SB-334867 (**9**, OX_1R $IC_{50} = 40$ nM, OX_2R $IC_{50} = 1995$ nM) that were characterized as OX_1R selective antagonists with good brain exposure in rodents.[30] Intraperitoneal administration of this antagonist provides high plasma and brain exposures for several hours after dosing. Indeed, the pharmacology associated with this OX_1R antagonist was explored in various *in vivo* assays including an orexin-A stimulated food intake assay. In those studies, administration of orexin-A increased food intake in rodents that could be blocked by SB-334867.[31]

Direct intracerebroventricular injection of orexin-A to rats reduces REM sleep duration in the active phase, and these effects can be blocked by direct IP administration of SB-334867.[32] In a follow-up study, IP administration of SB-334867 decreased wakefulness and increased NREM sleep.

Interpretation of these data however is complicated by the fact that SB-334867 is not entirely selective for OX_1R and several potent off-target CNS activities have been reported.[3]

Piperidine carboxamide DORA structures have been reported,[33] and of particular importance is SB-649868 (**10**), a DORA (OX_1R pK_i 9.5, OX_2R pK_i 9.4) that was reported to potently promote sleep in rats and nonhuman primates in preclinical studies.[34] It was announced that SB-649868 had entered Phase II clinical studies in 2007 and was well tolerated in early clinical studies with plasma half-lives of 4–7 h with a rapid T_{max} (1–3 h).

10

In polysomnography studies in healthy volunteers,[35] significant improvements in TST were realized when measured versus placebo. SB-649868 also improved sleep induction and maintenance parameters without producing impairments in cognitive function as measured by digit symbol substitution test. SB-649868 was reported to be a potent inhibitor of CYP3A4 *in vitro* and subsequent clinical studies have shown that SB-649868 increases exposure of coadministered simvastatin in a drug–drug interaction study.[36] SB-649868 was placed on clinical hold in late 2007 from Phase II studies due to the emergence of a preclinical toxicity and is no longer reported as active in development.[37]

More recently, analogs of SB-649868 have emerged in which the benzofuran amide has been replaced with heterocycles. Imidazopyrimidines such as **11** and **12** have been described with a range of potencies at OX_2R and OX_1R, although data provided for **12** showed it is a DORA (OX_2R pK_i 8.3, OX_1R pK_i 8.2). Similarly, related imidazopyridazines[38] such as **13** and imidazopyridines such as **14** bearing non-thiazole amides were described with a range of potencies (pK_i 5.8–9.1) at one or both of the orexin receptors. Also, 2-subsituted piperidines bearing a fused cyclopropane on the piperidine such as **15**[39] and **16** have been described with pK_i of 7.1 and 9.5 on OX_1R and between 6.7 and 8.9 on OX_2R suggesting these scaffolds portend more OX_1R selective.[40] Lastly, piperidine derivative GSK1059865 (**17**) was described as a potent OX_1R preferring ligand (OX_1R pK_i 8.8, OX_2R pK_i 6.9).[41]

11

12

13

14

15

16

17

A series of N,N-disubstituted-1,4-diazepanes were developed and optimized to produce DORAs with improved receptor potency and physicochemical properties.[42] Specifically, 5-methyl substitution on the diazepane core was optimized to enhance potency and improve pharmacokinetics. Further structural modifications produced molecules with reduced metabolic liabilities. From these efforts, suvorexant (MK-4305, **18**) was selected as a potent DORA with excellent potency in cell-based assays (OX_1R $IC_{50} = 50$ nM, OX_2R $IC_{50} = 56$ nM).[43] Suvorexant is orally bioavailable and has good brain penetrance in preclinical species. In rodent sleep studies, suvorexant dose-dependently reduced active wake and increased REM and NREM sleep.

18

In Phase I studies, suvorexant was well tolerated with peak plasma levels achieved at ~2 h and a mean terminal plasma half-life of ~12.2 h.[44] In healthy volunteers, dose-dependent observations of somnolence were evident. Results from the Phase IIb study demonstrated that suvorexant was superior to placebo in improving sleep efficiency on the first night of treatment as well as at the end of 4 weeks in patients with primary insomnia, and improvement in sleep efficiency was noted at all doses.[45] Suvorexant also provided improvements in the secondary endpoints of reduced WASO and reduced LPS at a dose of 80 mg. In 2010, suvorexant entered into Phase III development and it was reported that an NDA had been filed with the FDA in 2012.[46]

The structure and preclinical pharmacology of MK-6096 (**19**) has been reported as a potent DORA (OX_1R $IC_{50} = 15$ nM, OX_2R $IC_{50} = 16$ nM) which is efficacious in promoting sleep in rats and dogs.[47] MK-6096 was recounted to have entered Phase II clinical studies in 2009. Detailed reviews have been published around the structure–activity relationship (SAR) of both the aforementioned diazepane and piperidine DORAs.[12]

19

3. PRECLINICAL PROGRAMS

A novel class of cyclopropane amides appeared in a patent publication with the majority having significant antagonism of OX_2R ($IC_{50} = 4$ and 2 nM for **20** and **21**, respectively).[48] Related antagonists were described[49] with some level of OX_2R selectivity as **22** demonstrated excellent potency (OX_2R $IC_{50} = 3$ nM) and 33-fold selectivity over OX_1R in a fluorometric imaging plate reader (FLIPR) assay, and also exhibited efficacy in a mouse sleep model.

20	**21**	**22**

JNJ-10397049 (**23**) is a potent (OX_2R $K_i = 5$ nM), urea-derived OX_2R SORA with 600-fold selectivity over OX_1R.[50] SORA **23** demonstrated sleep efficacy similar to a DORA in the rat. Coadministration of **23** with an OX_1R SORA (SB-408124) attenuated rat sleep efficacy when compared with the OX_2R SORA alone. These authors indicate that blockade of OX_2R in rat is sufficient to initiate and prolong sleep, and suggest that simultaneous inhibition of OX_1R may attenuate the sleep promoting effects of OX_2R inhibition.[51] However, differentiation of OX_2R SORAs compared to DORAs has yet to be clinically validated. More recently, the same group published patents describing bicyclic fused piperidines/pyrrolidines.[52,53] Bis–octahydropyrrole **24** (OX_2R $K_i = 9$ nM) was highlighted which exhibits 96-fold binding selectivity for OX_2R.[54]

23	**24**

A series of indoles[55] have been described as represented by EP-009-0249 (**25**), which was a potent (OX_2R $IC_{50} = 23$ nM) ORA with 50-fold selectivity over OX_1R, but was highly lipophilic with poor intrinsic clearance.[56]

Indazole EP-009-0403 (**26**) was highlighted as a less lipophilic congener (OX_2R $IC_{50} = 14$ nM) that maintained $\sim 70 \times$ selectivity over OX_1R.

25   **26**

Heterocyclic-fused pyridines such as **27** were claimed as potent DORAs (OXR $IC_{50}s < 10$ nM).[57] Diazaspiro[5,5]undecanes such as **28** were reported as moderately selective SORAs (OX_2R $IC_{50} = 57$ nM, OX_1R $IC_{50} = 856$ nM).[58,59] In a mouse EEG study, **28** was reported to achieve a similar profile to suvorexant with respect to total sleep, active wake, and quiet wake, although it was noted that the length of REM sleep was longer for **28** by 15 min. This difference in REM sleep was presented as an advantage over suvorexant. A related regioisomeric compound **29** was also claimed in the aforementioned case, and found to be a potent antagonist (OX_2R $IC_{50} = 1$ nM), with a more balanced DORA-like profile ($OX_1R = 13$ nM).

27   **28**   **29**

Benzoxazepine orexin antagonists have been discovered and **30** is a potent SORA (OX_2R $IC_{50} = 27$ nM) with ~ 110-fold selectivity over OX_1R.[60] Additionally, **30** was orally bioavailable in rat, exhibited adequate brain exposure, and caused a dose–dependent reduction of OX-A-induced hyperlocomotion in rats. The same group also reported spiropiperidine OX_2R SORAs as exemplified by **31**.[61] Optimized SORA **31** has excellent OX_2R potency ($IC_{50} = 2$ nM) and >400-fold selectivity over OX_1R, but unlike the benzoxazepines, no *in vivo* evaluation of this series was described.

30   31

The identification of pyrazole amide DORAs has been reported, describing the discovery of **32** and **33**.[62] These representative examples were reported with functional potency of 1 nM at both receptors. Piperidines bearing a 3-benzoxazole group as represented by **34** have also been reported as potent DORAs (OX$_2$R/OX$_1$R IC$_{50}$ = 2.1/0.7 nM).[63]

32   33   34

Further examples of 2-subsituted piperidines that bear a cyclopropane, similar to **15** and **16**, but at the 4-position have been reported that are selective for OX$_1$R.[64] Heterocyclic amines **35** (OX$_1$R pK_B = 8) and **36** (OX$_1$R pK_B = 8.8) are both potent for OX$_1$R, but have pK_Bs < 5 at OX$_2$R.

35   36

4. POTENTIAL INDICATIONS BEYOND SLEEP

The majority of clinical efforts have focused on the development of DORAs for the treatment of insomnia, demonstrating clinical proof-of-concept for antagonizing OX$_1$R and OX$_2$R simultaneously.[9–11] There is some debate as to the merits of DORAs or selective OX$_2$R antagonists to promote sleep based on preclinical data,[4,9,65] However, no clinical data have been presented with OX_2R receptor antagonists to date. As noted in this review, a number of antagonists with selectivity for OX_1R over OX_2R, including SB-108424, SB-334867, and GSK-1059865, have been evaluated in preclinical models for a range of indications including anxiety, depression, pain, obesity, and addictive behaviors. In addition, genetic deletion or pharmacological antagonism of OX_1R shows modest effects on sleep in nonclinical studies.[66,67] The most widely studied OX_1R antagonist is SB-334867, which has been evaluated in over 150 publications. Although intriguing, caution should be used in interpreting some studies with SB-334867, as this compound exhibits hydrolytic instability,[68] is only modestly selective (~50×) over OX2R, and has several off-target activities ($K_i < 4$ µM) including adenosine, monoamine and norepinephrine transporters, and serotonin receptors.[14] Given these limitations of SB-334867, a number of studies have combined the use of peptide agonists, genetics, and multiple structurally diverse orexin receptor antagonists to build the case for involvement of orexin signaling across therapeutic areas.

Outside of sleep disorders, the most widely investigated opportunity has been in the area of addiction.[69] Orexinergic neurons innervate brain regions involved in reward processing, and administration of orexin receptor ligands can promote drug seeking behaviors in preclinical models. Conversely, pharmacological studies, primarily with OX_1R antagonists, have demonstrated blockade of addictive behavioral responses to cocaine, amphetamine, alcohol, and opiates, indicating the potential utility of orexin receptor antagonists for treating drug addiction and relapse.

There are several lines of evidence that might indicate a role for orexin receptor activity in anxiety and mood disorders.[70] Clinically, variation in CSF orexin levels has been observed in patients with posttraumatic stress and anxiety disorders.[71] Preclinical studies indicate an anxiogenic role for orexin peptides[72,73] and therefore antagonists might serve to reduce anxiety, although it is not clear whether dual or receptor selective antagonists would be most effective. Orexin signaling has similarly been implicated in preclinical depression models, but both agonism and antagonism have been reported to improve the performance depending on the model used, and confounding effects on vigilance and locomotor activity must be considered.[74,75]

Finally, the potential for disrupted orexin signaling in pain and migraine has been examined. Given the close integration between sleep, pain, and migraine, a role for orexin activity in nociception and pain responses would not be unexpected. Combined approaches using peptide and small molecule ligands and genetic models have been applied to investigate orexin signaling in this area. The results have been mixed, with administration of both activating peptides and orexin receptor antagonists showing efficacy in different nonclinical models of pain and migraine.[76,77] Furthermore, mice with dysregulated orexin signaling have also shown distinct responses to pain.[6,78] Given the limitations of nonclinical pain models, these responses will need to be better characterized in order to understand the potential clinical impact of modulating orexin signaling for pain and migraine.

5. CONCLUDING REMARKS

Since the first identification of the key role of the orexin signaling pathway in regulating sleep/wake and vigilance, a number of drug candidates have entered clinical development targeting the orexin receptors. Among them, suvorexant has been reported to be filed as an NDA with the FDA in 2012 with the potential to be the first approved orexin receptor antagonist for the treatment of insomnia. A number of additional dual antagonists as well as selective antagonists have been identified to further elucidate the roles of each individual subtype. Beyond sleep disorders, targeting orexin receptors may offer the potential for other therapeutic indications such as migraine, neuropathic pain, depression, anxiety, and addiction.

REFERENCES

1. de Lecea, L.; Kilduff, T. S.; Peyron, C.; Gao, X. B.; Foye, P. E.; Danielson, P. E.; Fukuhara, C.; Battenberg, E. L. F.; Gautvik, V. T.; Bartlett, F. S.; Frankel, W. N.; van den Pol, A. N.; Bloom, F. E.; Gautvik, K. M.; Sutcliffe, J. G. *PNAS* **1998**, *95*, 322.
2. Sakurai, T.; Amemiya, A.; Ishii, M.; Matsuzaki, I.; Chemelli, R. M.; Tanaka, H.; Williams, S. C.; Richardson, J. A.; Kozlowski, G. P.; Wilson, S.; Arch, J. R. S.; Buckingham, R. E.; Haynes, A. C.; Carr, S. A.; Annan, R. S.; McNulty, D. E.; Liu, W. S.; Terrett, J. A.; Elshourbagy, N. A.; Bergsma, D. J.; Yanagisawa, M. *Cell* **1998**, *92*, 573.
3. Gotter, A. L.; Webber, A. L.; Coleman, P. J.; Renger, J. J.; Winrow, C. J. *Pharmacol. Rev.* **2012**, *64*, 389.
4. Grady, S. P.; Nishino, S.; Czeisler, C. A.; Hepner, D.; Scammell, T. E. *Sleep* **2006**, *29*, 295.
5. Taheri, S.; Sunter, D.; Dakin, C.; Moyes, S.; Seal, L.; Gardiner, J.; Rossi, M.; Ghatei, M.; Bloom, S. *Neurosci. Lett.* **2000**, *279*, 109.
6. Xie, X. M.; Wisor, J. P.; Hara, J.; Crowder, T. L.; LeWinter, R.; Khroyan, T. V.; Yamanaka, A.; Diano, S.; Horvath, T. L.; Sakurai, T.; Toll, L.; Kilduff, T. S. *J. Clin. Invest.* **2008**, *118*, 2471.
7. Deadwyler, S. A.; Porrino, L.; Siegel, J. M.; Hampson, R. E. *J. Neurosci.* **2007**, *27*, 14239.
8. Piper, D. C.; Upton, N.; Smith, M. I.; Hunter, A. J. *Eur. J. Neurosci.* **2000**, *12*, 726.

9. Bettica, P.; Squassante, L.; Zamuner, S.; Nucci, G.; Danker-Hopfe, H.; Ratti, E. *Sleep* **2012**, *35*, 1097.
10. Hoever, P.; de Haas, S. L.; Dorffner, G.; Chiossi, E.; van Gerven, J. M.; Dingemanse, J. *J. Psychopharmacol.* **2012**, *26*, 1071.
11. Herring, W. J.; Snyder, E.; Budd, K.; Hutzelmann, J.; Snavely, D.; Liu, K.; Lines, C.; Roth, T.; Michelson, D. *Neurology* **2012**, *79*, 2265.
12. Coleman, P. J.; Cox, C. D.; Roecker, A. J. *Curr. Top. Med. Chem.* **2011**, *11*, 696.
13. Rufoff, C.; Cao, M.; Guilleminault, C. *Curr. Pharm. Des.* **2011**, *17*, 1476.
14. Gotter, A. L.; Roecker, A. J.; Hargreaves, R.; Coleman, P. J.; Winrow, C. J.; Renger, J. J. *Prog. Brain Res.* **2012**, *198*, 163.
15. Mieda, M.; Sakurai, T. *CNS Drugs* **2013**, *27*, 83.
16. Aissaoui, H.; Clozel, M.; Weller, T.; Koberstein, R.; Sifferlen, T.; Fischli, W. WO 2004/033418, 2004.
17. Aissaoui, H.; Clozel, M.; Fischli, W.; Koberstein, R.; Sifferlen, T. Patent Application WO2004/004733, 2004.
18. Koberstein, R.; Aissaoui, H.; Bur, D.; Clozel, M.; Fischli, W.; Jenck, F.; Mueller, C.; Nayler, O.; Sifferlen, T.; Treiber, A.; Weller, T. *Chimia* **2003**, *57*, 270.
19. Boss, C.; Brisbare-Roch, C.; Jenck, F. *J. Med. Chem.* **2009**, *52*, 891.
20. Brisbare-Roch, C.; Dingemanse, J.; Koberstein, R.; Hoever, P.; Aissaoui, H.; Flores, S.; Mueller, C.; Nayler, O.; van Gerven, J.; deHaas, S.; Hess, P.; Qiu, C.; Buchmann, S.; Scherz, M.; Weller, T.; Fischli, W.; Clozel, M.; Jenck, F. *Nat. Med.* **2007**, *13*, 150.
21. Jenck, F.; Fischer, C.; Qiu, C.; Hess, P.; Koberstein, R.; Brisbare-Roch, C. World Sleep Congress Poster Presentation, Cairns, Australia, September 3–6, 2007.
22. Dingemanse, J.; Dorffner, G.; Hajak, G.; Benes, H.; Danker-Hopfe, H.; Polo, O.; Saletu, B.; Barbanoj, M. J.; Pillar, G.; Penzel, T.; Chiossi, E.; Hoever, P. World Sleep Congress Poster Presentation, Cairns, Australia, September 3–6, 2007.
23. Hoever, P.; Dorffner, G.; Benes, H.; Penzel, T.; Danker-Hopfe, H.; Barbanoj, M. J.; Pillar, G.; Saletu, B.; Polo, O.; Kunz, D.; Zeitlhofer, J.; Berg, S.; Partinen, M.; Bassetti, C. L.; Hoegl, B.; Ebrahim, I. O.; Holsboer-Trachsler, E.; Bengtsson, H.; Peker, Y.; Hemmeter, U.-M.; Chiossi, E.; Hajak, G.; Dingemanse, J. *Clin. Pharmacol. Ther.* **2012**, *91*, 975.
24. Actelion. Actelion and GSK discontinue clinical development of almorexant (http://www1.actelion.com/en/our-company/news-and-events/index.page?newsId=1483135). Accessed 4 Dec 2012.
25. Gatfield, J.; Brisbare-Roch, C.; Jenck, F.; Boss, C. *Chem. Med. Chem.* **2010**, *5*, 1197.
26. Boss, C.; Brisbare-Roch, C.; Jenck, F.; Steiner, M. EP 2012/2402322, 2012.
27. Bolli, M.; Boss, C.; Brotschi, C.; Heidmann, B.; Sifferlen, T.; Trachsel, D.; Williams, J. T. WO 2012/114252, 2012.
28. Bolli, M., Boss, C., Brotschi, C., Gude, M., Heidmann, B., Sifferlen, T., Williams, J. T. WO 2012/110986, 2012.
29. Boss, C.; Brotschi, C.; Gatfield, J.; Gude, M.; Heidmann, B.; Sifferlen, T.; Wiliams, J. WO 2012/025877, 2012.
30. Porter, R. A.; Chan, W. N.; Coulton, S.; Johns, A.; Hadley, M. S.; Widdowson, K.; Jerman, J. C.; Brough, S. J.; Coldwell, M.; Smart, D.; Jewitt, F.; Jeffrey, P.; Austin, N. *Bioorg. Med. Chem. Lett.* **2001**, *11*, 1907.
31. Haynes, A. C.; Jackson, B.; Chapman, H.; Tadayyon, M.; Johns, A.; Porter, R. A.; Arch, J. R. *Regul. Pept.* **2000**, *96*, 45.
32. Jones, D. N. C.; Gartlon, J.; Parker, F.; Taylor, S. G.; Routledge, C.; Hemmati, P.; Munton, R. P.; Ashmeade, T. E.; Hatcher, J. P.; Johns, A.; Porter, R. A. *Psychopharmacology,* **2001**, *153*, 210.
33. Di Fabio, R.; Pellacani, A.; Faedo, S.; Roth, A.; Piccoli, L.; Gerrard, P.; Porter, R. A.; Johnson, C. N.; Thewlis, K.; Donati, D.; Stasi, L.; Spada, S.; Stemp, G.; Nash, D.;

Branch, C.; Kindon, L.; Massagrande, M.; Poffe, A.; Braggio, S.; Chiarparin, E.; Marchioro, C.; Ratti, E.; Corsi, M. *Bioorg. Med. Chem. Lett.* **2011**, *21*, 5562.

34. Gerrard, P. A.; Porter, R. A.; Holland, V.; Masagrande, M.; Poffe, A.; Piccoli, L.; Bettica, P.; Corsi, M.; Hagan, F.; Ratti, E. *Sleep* **2009**, *32*, A42 (Abstract Supplement).

35. Bettica, P.; Nucci, G.; Pyke, C.; Squassante, L.; Zamuner, S.; Ratti, E.; Gomeni, R.; Alexander, R. *J. Psychopharmacol.* **2012**, *26*, 1058.

36. Bettica, P.; Squassante, L.; Groeger, J. A.; Gennery, B.; Winsky-Sommerer, R.; Dijk, D. J. *Neuropsychopharmacology* **2012**, *37*, 1224.

37. http://www.gsk.com/content/dam/gsk/globals/documents/pdf/GSK%202013% 20Pipeline.pdf.

38. Alvaro, G.; Amantini, D. Patent Application WO 2010/072722, 2010.

39. Di Fabio, R. Patent Application WO2012/089606, 2012.

40. Di Fabio, R. Patent Application WO2012/089607, 2012.

41. Gozzi, A.; Turrini, G.; Piccoli, L.; Massagrande, M.; Amantini, D.; Antolini, M.; Martinelli, P.; Cesari, N.; Montanari, D.; Tessari, M.; Corsi, M.; Bifone, A. *PLoS One* **2011**, *1*, e16406.

42. Cox, C. D.; Breslin, M. J.; Whitman, D. B.; Brashear, K. M.; Roecker, A. J.; Bogusky, M.; Bednar, R. A.; Lemaire, W.; Bruno, J. G.; Hartman, G. D.; McGaughey, G.; Reiss, D. R.; Harrell, C. M.; Doran, S. M.; Garson, S. L.; Kraus, R. L.; Li, Y.; Prueksaritanont, T.; Li, C.; Winrow, C. J.; Koblan, K. S.; Renger, J. J.; Coleman, P. J. *ChemMedChem* **2009**, *4*, 1069.

43. Cox, C. D.; Breslin, M. J.; Whitman, D. B.; Schreier, J.; McGauhey, G. B.; Bogusky, M. J.; Roecker, A. J.; Mercers, S. P.; Bednar, R. A.; Lemaire, W.; Bruno, J. G.; Reiss, D. R.; Harell, C. M.; Murphy, K. L.; Garson, S. L.; Doran, S. M.; Prueksaritanont, T.; Anderson, W. B.; Tang, C.; Roller, S.; Cabalu, T. D.; Cui, D.; Hartman, G. D.; Young, S. D.; Koblan, K. S.; Winrow, C. J.; Renger, J. J.; Coleman, P. J. J. *Med. Chem.* **2010**, *53*, 5320.

44. Sun, H.; Kennedy, W. P.; Wilbraham, D.; Lewis, N.; Calder, N.; Li, X.; Ma, J.; Yee, K. L.; Ermlich, S.; Mangin, E.; Lines, C.; Rosen, L.; Chodakewitz, J.; Murphy, G. M. *Sleep* **2013**, *36*, 259.

45. Herring, W. J.; Budd, K. S.; Hutzelmann, E.; Snyder, D.; Snavely, D.; Liu, K.; Lines, C.; Michelson, D.; Roth, T. Association of Professional Sleep Societies Annual Meeting, San Antonio, TX, 2010; p A0591.

46. http://finance.yahoo.com/news/merck-announces-fda-acceptance-drug-133000354.html.

47. Coleman, P. J.; Schreier, J. D.; Cox, C. D.; Breslin, M. J.; Whitman, D. B.; Bogusky, M. J.; McGaughey, G. B.; Bednar, R. A.; Lemaire, W.; Doran, S. M.; Fox, S. V.; Garson, S. L.; Gotter, A. L.; Harrell, C. M.; Reiss, D. R.; Cabalu, T. D.; Cui, D.; Prueksaritanont, T.; Stevens, J.; Tannenbaum, P. L.; Ball, R. G.; Stellabott, J.; Young, S. D.; Hartman, G. D.; Winrow, C. J.; Renger, J. J. *ChemMedChem* **2012**, 7, 415.

48. Terauchi, T.; Takemura, A.; Doko, T.; Yoshida, Y.; Tanaka, T.; Sorimachi, K. Naoe, Y.; Beuchmann, C.; Kazuta, Y. Patent Application WO 2012/039371, 2012.

49. Terauchi, T.; Takemura, A.; Doko, T.; Yoshida, Y.; Tanaka, T.; Sorimachi, K. Naoe, Y.; Beuchmann, C.; Kazuta, Y. Patent Application WO 2012/165339, 2012.

50. McAtee, L. C.; Sutton, S. W.; Rudolph, D. A.; Li, X.; Aluisio, L. E.; Phuong, V. K.; Dvorak, C. A.; Lovenberg, T. W.; Carruthers, N. I.; Jones, T. K. *Bioorg. Med. Chem. Lett.* **2004**, *14*, 4225.

51. Dugovic, C.; Shelton, J. E.; Aluisio, L. E.; Fraser, I. C.; Jiang, X.; Sutton, S. W.; Bonaventure, P.; Yun, S.; Li, X.; Lord, B.; Dvorak, C. A.; Carruthers, N. I.; Lovenburg, T. W. *JPET* **2009**, *330*, 142.

52. Branstetter, B. J.; Letavic, M. A.; Ly, K. S.; Rudolph, D. A.; Savall, B. M.; Shah, C. R.; Shireman, B. T. Patent Application WO2011/050200, 2011.

53. Branstetter, B. J.; Letavic, M. A.; Ly, K. S.; Rudolph, D. A.; Savall, B. M.; Shah, C. R.; Shireman, B. T. Patent Application WO2011/050202, 2011.
54. Chai, W.; Letavic, M. A.; Ly, K. S.; Pippel, D. J.; Rudolph, D. A.; Sappey, K. C.; Savall, B. M.; Shat, C. R.; Shireman, B. T.; Soyode-Johnson, A.; Stocking, E. M.; Swanson, D. M. Patent Application WO2011/050198, 2011.
55. Bentley, J. M.; Davenport, T.; Hallett, D. J. Patent Application WO 2011/0138266, 2011.
56. Bentley, J.; Heifetz, A. 4th RSC/SCI Symposium on GPCR's in Medicinal Chemistry, Erl Wood, Windlesham, Surrey, UK.
57. Badiger, S.; Behnke, D.; Betschart, C.; Chaudhari, V.; Chebrolu, M.; Contesta, S.; Hintermann, S.; Pandit, C. Patent Application WO2011/076744, 2011.
58. Badiger, S.; Behnke, D.; Betschart, C.; Chaudhari, V.; Contesta, S.; Hinrichs, J. H.; Ofner, S.; Pandit, C. Patent Application WO 2011/076747, 2011.
59. Hintermann, S. 244th ACS National Meeting, Philadelphia, PA, 2012; Abstract MEDI-221.
60. Fujimoto, T.; Tomata, Y.; Kunitomo, J.; Hirozane, M.; Marui, S. *Bioorg. Med. Chem. Lett.* **2011**, *21*, 6414.
61. Fujimoto, T.; Tomata, Y.; Kunitomo, J.; Hirozane, M.; Marui, S. *Bioorg. Med. Chem. Lett.* **2011**, *21*, 6409.
62. Abe, M.; Futamura, A.; Suzuki, R.; Nozawa, D.; Ohta, H.; Arkaki, Y.; Asamura, K. Patent Application WO 2012/081692, 2012.
63. Abe, M.; Futamura, A.; Suzuki, R.; Nozawa, D.; Ohta, H.; Arkaki, Y.; Asamura, K. Patent Application WO 2013/005755, 2013.
64. Stasi, L. P.; Rovati, L. WO 2011/006960, 2011.
65. Morairty, S. R.; Revel, F. G.; Malherbe, P.; Moreau, J. L.; Valladao, D.; Wettstein, J. G.; Kilduff, T. S.; Borroni, E. *PLoS One* **2012**, 7, 39131.
66. Dugovic, C.; Shelton, J.; Sutton, S.; Yun, S.; Li, X.; Dvorak, C.; Carruthers, N.; Atack, J.; Lovenberg, T. *Sleep* **2008**, *31*, A33.
67. Willie, J. T.; Chemelli, R. M.; Sinton, C. M.; Yanagisawa, M. *Annu. Rev. Neurosci.* **2001**, *24*, 429–458.
68. McElhinny, C. J.; Lewin, A. H.; Mascarella, S. W.; Runyon, S.; Brieaddy, L.; Carroll, F. I. *Bioorg. Med. Chem. Lett.* **2012**, *22*, 6661.
69. Aston-Jones, G.; Smith, R. J.; Sartor, G. C.; Moorman, D. E.; Massi, L.; Tahsili-Fahadan, P.; Richardson, K. A. *Brain Res.* **2010**, *16*, 74.
70. Johnson, P. L.; Molosh, A.; Fitz, S. D.; Truitt, W. A.; Shekhar, A. *Prog. Brain Res.* **2012**, *198*, 133.
71. Johnson, P. L.; Truitt, W.; Fitz, S. D.; Minick, P. E.; Dietrich, A.; Sanghani, S.; Traskman-Bendz, L.; Goddard, A. W.; Brundin, L.; Shekhar, A. *Nat. Med.* **2010**, *16*, 111–115.
72. Suzuki, M.; Beuckmann, C. T.; Shikata, K.; Ogura, H.; Sawai, T. *Brain Res.* **2005**, *1044*, 116.
73. Li, Y. H.; Li, S.; Wei, C. G.; Wang, H. Y.; Sui, N.; Kirouac, G. J. *Psychopharmacology* **2010**, *212*, 251.
74. Ito, N.; Yabe, T.; Gamo, Y.; Nagai, T.; Oikawa, T.; Yamada, H.; Hanawa, T. *Neuroscience* **2008**, *157*, 720.
75. Lutter, M.; Krishnan, V.; Russo, S. J.; Jung, S.; McClung, C. A.; Nestler, E. J. J. *Neuroscience* **2008**, *28*, 3071.
76. Chiou, L. C.; Lee, H. J.; Ho, Y. C.; Chen, S. P.; Liao, Y. Y.; Ma, C. H.; Fan, P. C.; Fuh, J. L.; Wang, S. J. *Curr. Pharm. Des.* **2010**, *16*, 3089.
77. Holland, P.; Goadsby, P. J. *Headache* **2007**, *47*, 951.
78. Watanabe, S.; Kuwaki, T.; Yanagisawa, M.; Fukuda, Y.; Shimoyama, M. *Neuroreport* **2005**, *16*, 5.

SECTION 2

Cardiovascular and Metabolic Diseases

Section Editor: Robert L. Dow
Pfizer R&D, Cambridge, Massachusetts

Discovery and Development of Prolylcarboxypeptidase Inhibitors for Cardiometabolic Disorders

Sarah Chajkowski Scarry*, John M. Rimoldi[†]
*Department of Chemistry, Boston University, Boston, Massachusetts, USA
[†]Department of Medicinal Chemistry, University of Mississippi, University, Mississippi, USA

Contents

1. INTRODUCTION

The serine protease prolylcarboxypeptidase (PRCP; EC 3.4.16.2), initially identified in kidney extract 40 years ago, is a member of the serine hydrolase superfamily of enzymes characterized by their capacity to catalyze the hydrolysis of amide or ester bonds by virtue of a base-activated serine nucleophile.[1] Of the approximately 240 serine hydrolases identified, the S28 exopeptidase family represents an exclusive class with only two members, human dipeptidyl peptidase-7 (DPP7) and PRCP. These two enzymes (termed an "enzymatic odd couple")[2] share only 40% sequence identity and 55% sequence similarity to each other and low sequence identity to all other proteins.[3] PRCP's uniqueness may bode well in the context of drug discovery efforts, where challenges associated with protease

Annual Reports in Medicinal Chemistry, Volume 48
ISSN 0065-7743
http://dx.doi.org/10.1016/B978-0-12-417150-3.00007-7
91

selectivity have hampered the development of new small-molecule inhibitors.[4] Aberrant PRCP levels have been linked to hypertension, inflammation, and metabolic disorders such as obesity and diabetes. PRCP is a validated drug target that has a regulatory function in the renin–angiotensin system (RAS), the kallikrein–kinin system (KKS), and in hypothalamic pro-opiomelanocortin (POMC) downstream processing of alpha-melanocyte-stimulating hormone (α-MSH$_{1-13}$). It is in this latter arena where the development of PRCP inhibitors is being aggressively pursued in the quest for targeted antiobesity agents.[5] A number of excellent reviews have been published recently on the function and regulation of PRCP in the cardiovascular[6,7] and central nervous systems.[8-11]

2. PRCP FUNCTION AND STRUCTURE

PRCP catalyzes the hydrolysis of the vasoactive peptides angiotensin II (AngII), angiotensin III (AngIII), and protected dipeptides (CBz–Pro–X) with preferential cleavage of carboxyl-terminal amino acids (hydrophobic, X = Ala, Val, Leu, Phe) linked to a penultimate proline.[12] Cloning and sequencing experiments established that human PRCP (hPRCP) is comprised of 496 amino acids containing a 30-amino acid N-terminal signal peptide and a 15-amino acid propeptide.[13] PRCP is widely expressed in tissues including lung, liver, kidney, and brain and is preferentially expressed in the endothelium and in cultured endothelial cells.[14] hPRCP has been cloned and expressed using *Drosophila melanogaster* S2 cells,[15] baclovirus Sf21 cells, and mammalian CHO cells, the latter expression system giving rise to crystallization-grade protein with high purity.[16]

The X-ray structure of hPRCP has been solved and represents a tremendous advance in deciphering active site requirements for catalysis and substrate-binding features responsible for peptide recognition and inhibitor design.[3] Globally, the structure of PRCP consists of an α/β hydrolase domain constructed from two interrupted segments (residues 46–204 and 405–491) in addition to a novel helical bundle motif (SKS domain) covering the active site. The active site catalytic triad (Ser179–Asp430–His455) is structurally similar to other serine α/β hydrolases, and the reputed oxyanion hole is formed from the backbone amide NH of Tyr180 and Gly181. An unanticipated feature of the active site is the presence of a rare charge-relay system that includes an additional histidine (His456) and arginine (Arg460) which may modulate the pK_a of His455 via hydrogen bonding interactions (Fig. 7.1) and help to explain the catalytic dependency on pH.[12] Structural

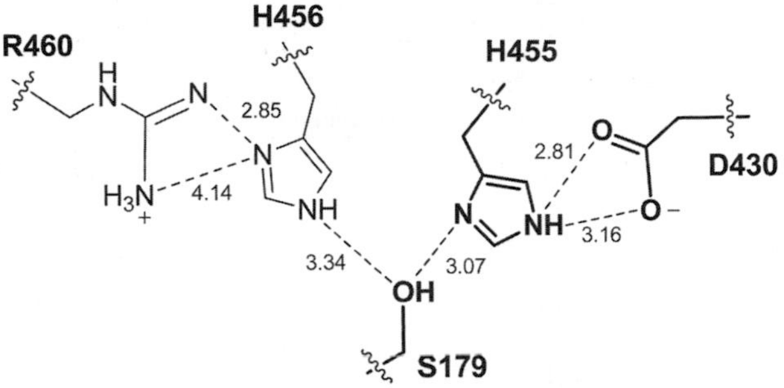

Figure 7.1 Active site depiction of hPRCP from X-ray crystal structure.

alignments with dipeptidyl peptidase-4 (DPP4) and prolylendopeptidase suggest that the S1 substrate recognition pocket contains key hydrophobic residues Trp432, Trp459, Met183, and Met369. Induced-fit docking analysis of bradykinin 1–8 (BK$_{1-8}$) with PRCP (homology model and X-ray structure) has also been accomplished, providing for the first time putative docking modes for oligopeptide substrate binding.[17]

3. PRCP: PEPTIDE PROCESSING AND REGULATION

The processing of peptides/proteins by PRCP under normal or altered cellular and physiological conditions has been demonstrated in (1) the RAS in the inactivation of AngII and AngIII, (2) the KKS in kallikrein activation and the hydrolysis of BK$_{1-8}$, (3) the POMC system in the inactivation of the anorexigenic peptide α-MSH$_{1-13}$. A new peptide substrate for PRCP has also been discovered. Using MS-profiling, the peptide YPRPIHPA, derived from the extracellular portion of human endothelin B receptor-like protein 2, was identified in human cerebrospinal fluid and represents a new candidate target engagement marker for PRCP activity in the CNS.[18]

3.1. The renin–angiotensin system

Targeting specific enzyme and receptors in the RAS has led to the successful and widespread clinical use of angiotensin-converting enzyme (ACE) inhibitors, angiotensin receptor type 1 (AT1) blockers, and renin inhibitors for the treatment of cardiovascular diseases. The understanding of RAS has expanded beyond circulation,[19] particularly with the discovery and evolution of intracellular RAS, tissue RAS, new functional roles of AngII-derived peptides, the pro(renin) receptor, and other regulatory RAS enzymes like

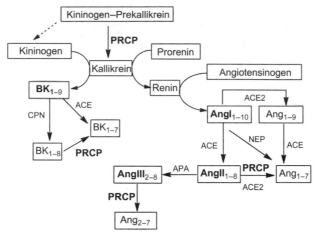

Figure 7.2 Role of PRCP in RAS and KKS.

PRCP.[7] Figure 7.2 illustrates the salient elements of the RAS cascade in the production of vasoactive angiotensin peptides resulting from iterative hydrolytic cleavages. The action of the aspartyl protease renin on angiotensinogen liberates the decapeptide AngI which is further processed by the enzymes ACE, ACE2, and neutral endopeptidase to AngII, Ang_{1-9}, and Ang_{1-7}, respectively. The formation of the key effector peptides AngII and Ang_{1-7} is of particular significance with AngII mediating vasoconstriction, inflammation, and fibrosis principally via AT_1 receptors, and Ang_{1-7} exerting opposing effects through Mas receptor activation pathways.[20] AngII is converted by aminopeptidase A (APA) to AngIII, a peptide product with physiological effector profiles similar to AngII.

PRCP provides a regulatory role in the RAS by deactivating the peptides AngII and AngIII to their corresponding phenylalanine cleavage products Ang_{1-7} and Ang_{2-7}, respectively.[12,21] The conversion of AngII to Ang_{1-7} by PRCP represents a redundancy in the RAS; the same transformation is catalyzed more efficiently by ACE2. Like ACE2, PRCP may attenuate the RAS from AngII-AT1 to Ang_{1-7} Mas receptor activation, promoting vasoprotection and antithrombotic effects. The PRCP-catalyzed inactivation of AngIII to Ang_{2-7} represents an additional governing pathway for the deactivation of AngIII, initially produced from the action of APA on AngII. Using an LC–MS-based assay, it was shown that AngI is a weak competitive inhibitor of PRCP *in vitro*; the significance of this is unknown.[21] PRCP expression is linked to blood pressure regulation. PRCP hypomorph

($PRCP^{gt/gt}$) mice have elevated blood pressure and vascular dysfunction, reversed by treatment with mitochondrial oxidants suggesting reactive oxygen species generation in this phenotype.[22] The hPRCP gene is also considered a candidate gene for essential hypertension[23]; polymorphisms (E112D allele) have been linked to preeclampsia[24] and to variable response in short-term benazepril use in Chinese hypertensive patients.[25] These data suggest that therapeutic modulation of peripheral PRCP activity with inhibitors may result in a hypertensive state.

3.2. The kallikrein–kinin system

The KKS is a complex system also involved in the regulation of cardiovascular function, coronary vascular tone in humans, and hemostasis[26] that largely counterbalances the RAS. Mechanistically, when the complex of high molecular weight kininogen and prekallikrein (PK) bind to endothelium membranes, PRCP catalyzes the conversion of PK to kallikrein (Fig. 7.2). The liberated kallikrein cleaves kininogen with the generation of bradykinin (BK), resulting ultimately in induction of vasodilation through BK action on B2 receptors and subsequent NO and prostaglandin release.[27] PRCP also catalyzes the inactivation of the kinin BK_{1-8} yielding BK_{1-7}.[17] It was shown that the upregulation of PRCP in lipopolysaccharide stimulated endothelial cells increased activation of kallikrein from PK and promotes an inflammatory response. Inhibitors of peripheral PRCP may find utility as anti-inflammatory agents.[28]

3.3. α-Melanocyte stimulating hormone

Melanocortin neuropeptides, derived from prohormone POMC, have been shown to be major regulators in energy homeostasis, metabolic disorders, and inflammation.[11,29] POMC is synthesized in the pituitary gland and in neurons of the hypothalamus, and is processed sequentially by enzymes ultimately leading to the production of $\alpha\text{-}MSH_{1-13}$. In hypothalamic neurons, cleavage of POMC by proconvertase 1 (PC1) results in the release of $ACTH_{1-39}$ (adrenocorticotropic hormone), which is further processed by proconvertase 2 (PC2) yielding $ACTH_{1-17}$.[8] Production of mature, functional $\alpha\text{-}MSH_{1-13}$ from $ACTH_{1-17}$ is catalyzed by the sequential actions of carboxypeptidase E, peptidyl α-amidating monooxygenase, and N-acetyltransferase, as illustrated in Fig. 7.3.[11]

The anorexic hormone leptin plays a crucial role in energy homeostasis, including appetite and metabolism, and promotes the synthesis of

Figure 7.3 POMC-derived peptides.

α-MSH$_{1-13}$ which stimulates melanocortin-4 receptors in the hypothalamus suppressing appetite and increasing energy expenditure.[8] The regulation of the short-lived anorexigenic peptide α-MSH$_{1-13}$ was not established until recently, where it was demonstrated that PRCP expressed in the hypothalamus catalyzes its inactivation to α-MSH$_{1-12}$.[30] This study further demonstrated that peripheral or central administration of PRCP inhibitors (N–Boc–Pro-Prolinal or CBz–Pro-Prolinal) decreases food intake in both wild-type (WT) and obese mice. Elevated levels of α-MSH$_{1-13}$ in the hypothalamus were observed in PRCP-null mice, which were leaner than the WT control mice and were resistant to high-fat diet (HFD)–induced obesity. Additional studies were conducted to characterize the metabolic responses to standard and HFD in PRCP-null mice versus WT controls. PRCP-null mice exhibited a reduction of body weight gain and fat mass accumulation, a reduction in HFD-induced hepatic steatosis, and an improvement in glucose metabolism when exposed to HFD feeding, regardless of adiposity.[31] Ghrelin is an orexigenic hormone involved in the melanocortin-signaling pathway that is produced by the stomach and released into circulation when fasting. Ghrelin stimulates the liberation of neuropeptide Y and agouti-related protein, inhibits the anorexigenic POMC neurons, thus resulting in increased food intake.[32] Recently, it was discovered that ghrelin administration to mice centrally or peripherally upregulates hypothalamic PRCP expression representing an additional mechanism to decrease melanocortin signaling.[33]

Antibodies raised against PRCP purified from human neutrophils resulted in the establishment of an ELISA assay, used to measure plasma PRCP levels in three cohorts, with or without diabetes mellitus or obesity. PRCP levels were found to be significantly higher in diabetic or obese patients versus the control cohorts, providing a potential marker for assessing metabolic conditions.[34]

4. PRCP INHIBITORS AS ANTIOBESITY AGENTS

Drug discovery and development efforts have resulted in the evolution of several classes of small-molecule inhibitors of PRCP as antiobesity agents, with an emphasis placed on brain-penetrating agents. Small-molecule PRCP inhibitors may find therapeutic value in the treatment of cardiometabolic disorders including obesity and diabetes. The first class of PRCP inhibitors developed, represented by compound **4**, was derived from *N*-acylpyrrolidine hits previously designed as general mechanism-based serine protease inhibitors (Fig. 7.4). Thus, initial screening using a fluorometric PRCP assay led to the discovery of compound **1** (hPRCP $IC_{50} = 0.3$ μM) with moderate potency against PRCP. The potential for inherent poor selectivity for PRCP over other serine proteases conferred by the reactive carbonyl toward the catalytic serine, coupled with the challenge of rapidly optimizing this scaffold, resulted in the strategy to move away from such warheads toward structures consistent with noncovalent inhibition. While compounds **2** (hPRCP 16% inhib. at 5 μM) and **3** (hPRCP $IC_{50} = 5$ μM) were not particularly potent but were highly amenable to structure–activity relationship (SAR) and were considered a reasonable starting point for optimization. The SAR was rapidly explored through the use of parallel synthesis/high-throughput purification, taking advantage of the modularity of the SAR where combining optimal groups from the left (biphenyl) and right (benzimidazole) sides resulted in at least additive potency gains. While not completely understood, significant improvement in pharmacokinetic properties was demonstrated in molecules with an alpha-amino butyrate

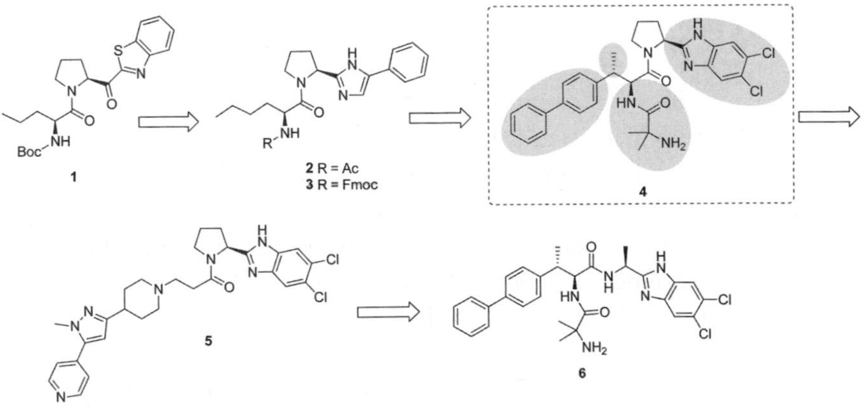

Figure 7.4 Evolution of PRCP inhibitors from lead prolinyl compound **4**.

group. In addition, the beta–methyl group sufficiently disrupted specific binding to mouse serum albumin to reduce serum shift by $>10 \times$ relative to the non–methyl congener. Due to its good pharmacokinetic profile, *in vitro* potency, and selectivity for PRCP, compound **4** (hPRCP $IC_{50} = 1$ nM; mAngIII $IC_{50} = 26$ nM) was selected for *in vivo* studies. Specifically, oral dosing of **4** (100 mg/kg) for 5 days to the established diet-induced obesity (eDIO) PRCP WT mice showed a 5% body weight loss compared to vehicle group. The observed weight loss was linked to peripherally PRCP inhibition, since compound **4** lacked brain exposure in mice.[35]

The benzimidazole pyrrolidinyl amide **5** was constructed using a simplified scaffold, with replacement of the biphenyl group with a piperidine group tethered to polar heteroaryls (i.e., pyridine or pyrazole). Compound **5** (hPRCP $IC_{50} = 1.4$ nM; mAngIII $IC_{50} = 6.7$ nM) demonstrated excellent PRCP inhibition. However, *in vivo* studies revealed that **5** showed similar efficacy in the PRCP KO eDIO mice indicating that weight loss in WT eDIO mice was due to an off–target effect.[36] In an effort to develop PRCP inhibitors with improved pharmacokinetic profiles, amino acid derivatives derived from sample collections were screened, with alanine-based compounds demonstrating the best PRCP inhibitory activity. Structure editing of compound **4** (deletion of proline ring) and subsequent replacement with an alanine-scaffold resulted in compound **6** (hPRCP $IC_{50} = 0.37$ nM, mAngIII $IC_{50} = 22$ nM), which exhibited excellent hPRCP inhibitory activity, and slightly improved pharmacokinetic parameters over its proline-based counterpart **4** when administered orally to mice.[37]

Replacement of the dichlorobenzimidazole-substituted pyrrolidine group of **5** with (*S*)-1-(4-chlorophenyl)-2-methylpropan-1-amide resulted in analogs represented by compound (*S*)-**7** (hPRCP $IC_{50} = 1.5$ nM; mAngIII $IC_{50} = 12.6$ nM). Using the eDIO mouse model, continuous SC infusion dosing of (*S*)-**7** at 10 mg/kg failed to demonstrate significant weight loss compared to vehicle. In addition, significant weight loss was not achieved in a subchronic weight loss study in WT and PRCP KO mice. The results obtained in this study, which achieved good exposure (terminal plasma concentrations $>20 \times$ mAngIII IC_{50}, brain:plasma ratios >2.5) corresponding to $>85\%$ *in vivo* inhibition of PRCP activity, are in contrast to those seen for nonbrain-penetrant analog **4** that demonstrated weight loss efficacy at somewhat lower drug coverage, but likely significantly better plasma PRCP inhibition (*vide supra*). Unfortunately, due to the poor solubility of (*S*)-**7**, skin-related adverse effects were prominent at the injection site at doses >10 mg/kg.[38]

The pyrimidine-4,6-diamine scaffold was selected as a viable template to create PRCP inhibitors with enhanced CNS penetration.[39] Optimization of functionality at both the 4- and 6-amino positions of the pyrimidine scaffold led to the synthesis of 40 analogs; three exhibited excellent inhibitory activity against hPRCP. Additionally, these were active in the whole plasma mAngIII assay, while displaying reduced Pgp efflux potential. Compound **8** (hPRPC $IC_{50} = 43$ nM; mAngIII $IC_{50} = 350$ nM) was selected and further evaluated in CF1 and WT mice (SC administration); a brain/plasma ratio of 1.4 was achieved, representing a strong effort in creating CNS-active PRCP inhibitors.[39]

Benzodihydroisofuran scaffolds were elaborated to produce CNS-penetrating PRCP inhibitors represented by *ent*-**9** (hPRCP $IC_{50} = 0.44$ nM; mAngIII $IC_{50} = 23$ nM). Although *ent*-**9** had good pharmacokinetic properties and inhibited PRCP in both plasma (88%) and brain (61%) of eDIO mice, it failed to induce significant weight loss in eDIO mice in a 5-day study. This study combined with the benzimidazole and non-benzimidazole pyrrolidinyl amide classes of PRCP inhibitors suggests that complete inhibition of PRCP in the plasma is required to display weight loss in the eDIO mouse model.[36,38,40]

High-throughput screening efforts led to the discovery of a new class of potent hPRCP inhibitors, represented by compound **10** (hPRCP $IC_{50} = 0.9$ nM), containing a highly functionalized pyrrolidine core scaffold.[41] SAR analysis revealed that hPRCP inhibitory activity was optimal with 3,4-disbutituted aromatic substituents attached to a triazole or methyl-pyrazole ring, with preservation of a 3-pyrrolidine substituted 2,4-difluorobenzene group. Deletion of the *N-tert*-butyl substituent also resulted in analogs that displayed excellent hPRCP inhibition. Incubation of **10** with mouse liver microsomes led to oxidative metabolism of the *N-tert*-butyl group and pyrrolidine ring. Further evaluation of compound **10** and select analogs revealed minimal plasma serum shifts, potent hPRCP inhibition, and complete *ex vivo* plasma target engagement in mice following oral dosing. In an effort to obviate the metabolic liabilities and poor pharmacokinetic properties associated with **10**, further structure editing was warranted. The pyrrolidine group was transformed into an amino-cyclopentane, and the triazole replaced with a pyrazole ring, represented by compounds **11** (hPRCP IC$_{50}$ = 7 nM) and **12** (hPRCP IC$_{50}$ = 2 nM) which showed complete *ex vivo* plasma target engagement and low CNS penetration following oral dosing in mice.[42]

13 R = H
14 R = NH–C$_{12}$H$_{25}$

15, UM8190

A class of reversible and selective proline-based dipeptide PRCP inhibitors was recently synthesized and evaluated using a rPRCP assay and a human pulmonary artery endothelial cell (HPAEC) assay. Simple CBz–Pro–Pro–NH–amide derivatives of CBz–Pro–Prolinal, **13** a known weak inhibitor of PRCP,[43] were examined. From this initial screening, analogs containing *N*-benzyl substituents on the proline B-ring proved to be modest inhibitors of PRCP while the homologous alkyl series, represented by dodecyl substituted compound **14** (hrPRCP K_i = 62 µM), were shown to have improved inhibitory activity and served as the lead molecule for SAR analysis. Interestingly, compound **15** (UM8190, hrPRCP K_i = 43 µM) lacking the proline A-ring carbamate and containing the dodecyl substituted proline B-ring proved to have the highest affinity and best selectivity for PRCP against a panel of proteases *in vitro* and was further evaluated *in vivo*. In a cell-based assay, **15** also demonstrated inhibition of PRCP-dependent activation on HPAEC cells (hrPRCP K_i = 34 µM). Furthermore, **15** blocked BK generation, BK-induced permeability, and showed an anorexigenic effect when administered intraperitoneally (10 or 100 mg/kg) to fasting mice, with a dose- and time-dependent reduction in food intake.[44]

The recent patent literature describes the synthesis and *in vitro* evaluation of heteroarylpiperdine-based hPRCP inhibitors for the treatment of obesity and related metabolic disorders. Some of the most potent compounds in this series were **16** ($IC_{50} = 2$ nM), *ent*-**17** ($IC_{50} = 0.035$ nM), and **18** ($IC_{50} = 0.65$ nM). The potency of these compounds against hPRCP was determined by a fluorescence intensity kinetic assay utilizing an internally quenched fluorescent substrate.[45–47]

5. CONCLUSIONS

PRCP has multiple regulatory roles in RAS, KKS, and POMC systems with aberrant expression and/or activity associated with hypertension, inflammation, diabetes, and obesity. The development of PRCP-selective small-molecule inhibitors is still in its infancy stage and will remain a challenge since PRCP is widely expressed in many tissues and cell types in both the CNS and periphery. With the structure of hPRCP solved drug discovery efforts should be accelerated when coupled to high-throughput screening programs. Over the last few years small-molecule inhibitors targeting PRCP have shown potential as antiobesity agents due to their ability to block the inactivation of α-MSH$_{1-13}$ and regulate food intake. Recent developments of additional and novel functions of PRCP are also surfacing including its regulation in cancer cells. For example, PRCP was found to be a regulator of 4-hydroxytamoxifen (4-OHTAM) resistance in MCF7 estrogen receptor-positive breast cancer cells.[48] Overexpression of PRCP enhanced cell proliferation and autophagy while protecting MCF7 cells against 4-OHTAM-induced cytotoxicity. PRCP KO or inhibition with an inhibitor reversed these effects further supporting the notion that inhibition of cytoprotective autophagy in conjunction with standard chemotherapy may provide a therapeutic advantage by enhancing chemosensitivity and antitumor efficacy.[49] Although PRCP was identified over 40 years ago, its function in several complex systems is just now becoming fully elucidated.

REFERENCES

1. Bachovchin, D. A.; Cravatt, B. F. *Nat. Rev. Drug Discov.* **2012**, *11*, 52.
2. Kozarich, J. *BMC Biol.* **2010**, *8*, 87.
3. Soisson, S.; Patel, S.; Abeywickrema, P.; Byrne, N.; Diehl, R.; Hall, D.; Ford, R.; Reid, J.; Rickert, K.; Shipman, J.; Sharma, S.; Lumb, K. *BMC Struct. Biol.* **2010**, *10*, 16.
4. Drag, M.; Salvesen, G. S. *Nat. Rev. Drug Discov.* **2010**, *9*, 690.
5. Dietrich, M.; Horvath, T. *Nat. Rev. Drug Discov.* **2012**, *11*, 675.
6. Mallela, J.; Yang, J.; Shariat-Madar, Z. *Int. J. Biochem. Cell Biol.* **2009**, *41*, 477.
7. Shariat-Madar, B.; Taherian, M.; Shariat-Madar, Z. On the Mechanism of Action of Prolylcarboxypeptidase. In *Recent Advances in Cardiovascular Risk Factors*; Atiq, M., Ed.; InTech North America: New York, 2012; http://dx.doi.org/10.5772/30890, 978-953-51-0321-9.
8. Palmiter, R. D. *J. Clin. Invest.* **2009**, *119*, 2130.
9. Shariat-Madar, B.; Kolte, D.; Verlangieri, A.; Shariat-Madar, Z. *Diabetes Metab. Syndr. Obes.* **2010**, *3*, 67.
10. Jeong, J. K.; Diano, S. *Trends Endocrinol. Metab.* **2013**, *24*, 61.
11. Sabrina, D. *Front. Neuroendocrinol.* **2011**, *32*, 70.
12. Odya, C. E.; Marinkovic, D. V.; Hammon, K. J.; Stewart, T. A.; Erdos, E. G. *J. Biol. Chem.* **1978**, *253*, 5927.
13. Tan, F.; Morris, P. W.; Skidgel, R. A.; Erdos, E. G. *J. Biol. Chem.* **1993**, *268*, 16631.
14. Shariat-Madar, Z.; Mahdi, F.; Schmaier, A. H. *J. Biol. Chem.* **2002**, *277*, 17962.
15. Shariat-Madar, Z.; Mahdi, F.; Schmaier, A. H. *Blood* **2004**, *103*, 4554.
16. Abeywickrema, P. D.; Patel, S. B.; Byrne, N. J.; Diehl, R. E.; Hall, D. L.; Ford, R. E.; Rickert, K. W.; Reid, J. C.; Shipman, J. M.; Geissler, W. M.; Pryor, K. D.; SinhaRoy, R.; Soisson, S. M.; Lumb, K. J.; Sharmaa, S. *Acta Crystallogr. Sect. F Struct. Biol. Cryst. Commun.* **2010**, *66*, 702.
17. Chajkowski, S. M.; Mallela, J.; Watson, D. E.; Wang, J.; McCurdy, C. R.; Rimoldi, J. M.; Shariat-Madar, Z. *Biochem. Biophys. Res. Commun.* **2011**, *405*, 338.
18. Zhao, X.; Southwick, K.; Cardasis, H. L.; Du, Y.; Lassman, M. E.; Xie, D.; El-Sherbeini, M.; Geissler, W. M.; Pryor, K. D.; Verras, A.; Garcia-Calvo, M.; Shen, D.-M.; Yates, N. A.; Pinto, S.; Hendrickon, R. C. *Proteomics* **2010**, *10*, 2882.
19. Nguyen Dinh Cat, A.; Touyz, R. M. *Peptides* **2011**, *32*, 2141.
20. Iwai, M.; Horiuchi, M. *Hypertens. Res.* **2009**, *32*, 533.
21. Mallela, J.; Perkins, R.; Yang, J.; Pedigo, S.; Rimoldi, J. M.; Shariat-Madar, Z. *Biochem. Biophys. Res. Commun.* **2008**, *374*, 635.
22. Adams, G. N.; Stavrou, G. A. L. E.; Zhou, Y.; Nieman, M. T.; Jacobs, G. H.; Cui, Y.; Lu, Y.; Jain, M. K.; Mahdi, F.; Shariat-Madar, Z.; Okada, Y.; D'Alecy, L. G.; Schmaier, A. H. *Blood* **2011**, *117*, 3929.
23. Watson, B., Jr.; Nowak, N. J.; Myracle, A. D.; Shows, T. B.; Warnock, D. G. *Genomics* **1997**, *44*, 365.
24. Wang, L.; Feng, Y.; Zhang, Y.; Zhou, H.; Jiang, S.; Niu, T.; Wei, L.-J.; Xu, X.; Xu, X.; Wang, X. *Am. J. Obstet. Gynecol.* **2006**, *195*, 162.
25. Zhang, Y.; Hong, X.; Xing, H.; Li, J.; Huo, Y.; Xu, X. *Chin. Med. J. (Peking)* **2009**, *122*, 2461.
26. Su, J. B. *J. of Renin-Angiotensin-Aldosterone Syst.*, 1470320312474854, first published on February 5, **2013**.
27. Dielis, A. W. J. H.; Smid, M.; Spronk, H. M. H.; Hamulyak, K.; Kroon, A. A.; ten Cate, H.; de Leeuw, P. W. *Hypertension* **2005**, *46*, 1236.
28. Ngo, M.-L.; Mahdi, F.; Kolte, D.; Shariat-Madar, Z. *J. Inflammation (London, U.K.)* **2009**, *6*, 3.
29. Shariat-Madar, B.; Kolte, D.; Verlangieri, A.; Shariat-Madar, Z. *Diabetes Metab. Syndr. Obes.* **2012**, *3*, 67.

30. Wallingford, N.; Perroud, B.; Gao, Q.; Coppola, A.; Gyengesi, E.; Liu, Z.-W.; Gao, X.-B.; Diament, A.; Haus, K. A.; Shariat-Madar, Z.; Mahdi, F.; Wardlaw, S. L.; Schmaier, A. H.; Warden, C. H.; Diano, S. *J. Clin. Invest.* **2009**, *119*, 2291.

31. Jeong, J. K.; Szabo, G.; Raso, G. M.; Meli, R.; Diano, S. *Am. J. Physiol. Endocrinol. Metab.* **2012**, *302*, E1502.

32. Ferrini, F.; Salio, C.; Lossi, L.; Merighia, A. *Curr. Neuropharmacol.* **2009**, 7, 37.

33. Kwon Jeong, J.; Dae Kim, J.; Diano, S. *Mol. Metab.* **2013**, 2, 23.

34. Xu, S.; Lind, L.; Zhao, L.; Lindahl, B.; Venge, P. *Clin. Chem.* **2012**, *58*, 1110.

35. Zhou, C.; Garcia-Calvo, M.; Pinto, S.; Lombardo, M.; Feng, Z.; Bender, K.; Pryor, K. D.; Bhatt, U. R.; Chabin, R. M.; Geissler, W. M.; Shen, Z.; Tong, X.; Zhang, Z.; Wong, K. K.; Roy, R. S.; Chapman, K. T.; Yang, L.; Xiong, Y. J. *J. Med. Chem.* **2010**, *53*, 7251.

36. Shen, H. C.; Ding, F.-X.; Zhou, C.; Xiong, Y.; Verras, A.; Chabin, R. M.; Xu, S.; Tong, X.; Xie, D.; Lassman, M. E.; Bhatt, U. R.; Garcia-Calvo, M. M.; Geissler, W.; Shen, Z.; Chen, D.; SinhaRoy, R.; Hale, J. J.; Tata, J. R.; Pinto, S.; Shen, D.-M.; Colletti, S. L. *Bioorg. Med. Chem. Lett.* **2011**, *21*, 1299.

37. Wu, Z.; Yang, C.; Xiong, Y.; Feng, Z.; Lombardo, M.; Verras, A.; Chabin, R. M.; Xu, S.; Tong, X.; Xie, D.; Lassman, M. E.; Bhatt, U. R.; Garcia-Calvo, M. M.; Geissler, W.; Shen, Z.; Chen, Q.; Sinharoy, R.; Hale, J. J.; Tata, J. R.; Pinto, S.; Shen, D.-M.; Colletti, S. L. *Bioorg. Med. Chem. Lett.* **2012**, *22*, 1774.

38. Graham, T. H.; Shen, H. C.; Liu, W.; Xiong, Y.; Verras, A.; Bleasby, K.; Bhatt, U. R.; Chabin, R. M.; Chen, D.; Chen, Q.; Garcia-Calvo, M.; Geissler, W. M.; He, H.; Lassman, M. E.; Shen, Z.; Tong, X.; Tung, E. C.; Xie, D.; Xu, S.; Colletti, S. L.; Tata, J. R.; Hale, J. J.; Pinto, S.; Shen, D.-M. *Bioorg. Med. Chem. Lett.* **2012**, *22*, 658.

39. Wu, Z.; Yang, C.; Graham, T. H.; Verras, A.; Chabin, R. M.; Xu, S.; Tong, X.; Xie, D.; Lassman, M. E.; Bhatt, U. R.; Garcia-Calvo, M. M.; Shen, Z.; Chen, Q.; Bleasby, K.; Sinharoy, R.; Hale, J. J.; Tata, J. R.; Pinto, S.; Colletti, S. L.; Shen, D.-M. *Bioorg. Med. Chem. Lett.* **2012**, *22*, 1727.

40. Shen, H. C.; Ding, F.-X.; Jiang, J.; Verras, A.; Chabin, R. M.; Xu, S.; Tong, X.; Chen, Q.; Xie, D.; Lassman, M. E.; Bhatt, U. R.; Garcia-Calvo, M. M.; Geissler, W.; Shen, Z.; Murphy, B. A.; Gorski, J. N.; Wiltsie, J.; SinhaRoy, R.; Hale, J. J.; Pinto, S.; Shen, D.-M. *Bioorg. Med. Chem. Lett.* **2012**, *22*, 1550.

41. Graham, T. H.; Liu, W.; Verras, A.; Sebhat, I. K.; Xiong, Y.; Bleasby, K.; Bhatt, U. R.; Chen, Q.; Garcia-Calvo, M.; Geissler, W. M.; Gorski, J. N.; He, H.; Lassman, M. E.; Lisnock, J.; Li, X.; Shen, Z.; Tong, X.; Tung, E. C.; Wiltsie, J.; Xiao, J.; Xie, D.; Xu, S.; Hale, J. J.; Pinto, S.; Shen, D.-M. *Bioorg. Med. Chem. Lett.* **2012**, *22*, 2811.

42. Graham, T. H.; Liu, W.; Verras, A.; Reibarkh, M.; Bleasby, K.; Bhatt, U. R.; Chen, Q.; Garcia-Calvo, M.; Geissler, W. M.; Gorski, J. N.; He, H.; Lassman, M. E.; Lisnock, J.; Li, X.; Shen, Z.; Tong, X.; Tung, E. C.; Wiltsie, J.; Xie, D.; Xu, S.; Xiao, J.; Hale, J. J.; Pinto, S.; Shen, D.-M. *Bioorg. Med. Chem. Lett.* **2012**, *22*, 2818.

43. Moreira, C. R.; Schmaier, A. H.; Mahdi, F.; da Motta, G.; Nader, H. B.; Shariat-Madar, Z. *FEBS Lett.* **2002**, *523*, 167.

44. Rabey, F. M.; Gadepalli, R. S. V. S.; Diano, S.; Cheng, Q.; Tabrizian, T.; Gailani, D.; Rimoldi, J. M.; Shariat-Madar, Z. *Curr. Med. Chem.* **2012**, *19*, 4194.

45. Ding, F. -X.; Jiang, J.; Shen, D. -M.; Shi, Z. -C.; Shu, M.; Yang, C. U.S. 2013/0059830, 2013.

46. Graham, T. H.; Shen, D. -M.; Shu, M. U.S. 2013/0030019, 2013.

47. Hale, J. J.; Jiang, J.; Shen, D. -M.; Shi, Z. -C.; Shu, M.; Wu, Z.; Yang, C. U.S. 2013/0040929, 2013.

48. Duan, L.; Motchoulski, N.; Danzer, B.; Davidovich, I.; Shariat-Madar, Z.; Levenson, V. V. *J. Biol. Chem.* **2011**, *286*, 2864.

49. Yang, Z. J.; Chee, C. E.; Huang, S.; Sinicrope, F. A. *Cancer Biol. Ther.* **2011**, *11*, 169.

CHAPTER EIGHT

Molecular Targeting of Imaging and Drug Delivery Probes in Atherosclerosis

Cesare Casagrande[*], Daniela Arosio[†], Claudia Kusmic[‡],
Leonardo Manzoni[†], Luca Menichetti[‡], Antonio L'Abbate[§]
[*]Department of Chemistry, University of Milan, Milan, Italy
[†]CNR-Institute of Molecular Science and Technologies (ISTM), Milan, Italy
[‡]CNR-Institute of Clinical Physiology (IFC), Pisa, Italy
[§]Scuola Superiore Sant'Anna, Pisa, Italy

Contents

1. INTRODUCTION

Selective targeting of biomolecules specifically expressed in pathological tissues offers substantial advantages in the development of novel diagnostic and therapeutic agents. In the cardiovascular system, molecular targeting of imaging probes represents a significant advancement with respect to the established diagnostic techniques of morphological, functional, or metabolic imaging.[1,2] However, research on molecular imaging/targeting in cardiovascular diseases is at a very early stage, almost exclusively preclinical.

Most cardiovascular diseases are of ischemic nature and thought to stem from atherosclerosis. Atherosclerosis is characterized by segmental arterial wall thickening to form an atherosclerotic plaque, a process in which cholesterol deposition, inflammation, extracellular–matrix (ECM) formation, calcium deposition, and thrombosis play important roles. Organ damage, and symptoms when present, is caused by progressive or sudden occlusion of an artery resulting in a shortage of blood supply and consequent ischemic

Annual Reports in Medicinal Chemistry, Volume 48
ISSN 0065-7743
http://dx.doi.org/10.1016/B978-0-12-417150-3.00008-9

injury of the downstream, under-perfused tissue. Ischemia can affect any organ, with the heart and the brain being the most commonly damaged. Today, both invasive and non-invasive imaging techniques are employed in the diagnosis and prognostic stratification of patients with cardiovascular diseases.[3–5] Non-invasive techniques, which are preferred for broad clinical use, provide information on cardiac and vascular anatomy, as well as functional parameters such as vessel and intra-cardiac flows, tissue perfusion, microcirculation, and mechanical performance of the heart.[6–10]

In atherosclerosis, imaging of morphological alterations of large artery walls is the only direct diagnostic evidence provided by available techniques, and only indirect information can be derived for the possible effects of the atherosclerotic vessel on the anatomy and/or function of the fed organ. It is not currently possible to observe the fine cellular and molecular processes that are responsible for the fate of the plaque, either remaining stable for the patient lifespan or suddenly becoming unstable and life-threatening. Post-mortem studies have shown that atherosclerosis is an endemic pathological condition in developed countries, affecting a great majority of people from a young age; however, only a minority progress to acute or chronic ischemic disease involving heart, brain, kidney, or peripheral circulation. Thus, atherosclerosis cannot be equated to ischemic disease either in diagnostic or prognostic terms, and diagnostic imaging should focus on the factors involved in transformation of the relatively benign condition into malign atherosclerosis, and on the recognition of 'active' (or 'unstable') plaques.

Because altered vessel morphology is ambiguous, the ability to evaluate molecular and cellular processes, such as endothelial activation, local inflammation, ECM alterations, cellular damage, and activation of pro-thrombotic mechanisms becomes crucial in terms of prognosis and preventive treatment. New molecular imaging approaches, progressing in parallel with advancements in molecular pathology and with refinement of imaging techniques, will provide valuable contributions to fundamental knowledge of these processes, and to development of diagnostic tools. Targetable biomolecules, which are specific of the initiation and progression of atherosclerosis include cellular adhesion molecules (CAM), such as integrins, selectins, cadherins, and IgSF CAM, and markers of cellular dysfunction and apoptosis, such as proteolytic enzymes (cathepsins, matrix metalloproteases (MMP)), oxidized low density lipoproteins (LDL), and phosphatidyl serine, as well as novel pathologically relevant bio-markers that are being discovered and validated by proteomic studies.[11–14]

▷ 2. CHEMICAL AND IMAGING APPROACHES

Targeting of biomolecular markers of atherosclerosis initially employed specific antibodies, linked to imaging probes by streptavidin–biotin coupling.[15] This approach afforded essential knowledge on a number of markers and on relevant mechanisms, constituting a sound basis for further progress. The present review focuses on innovative medicinal chemistry approaches relying on synthetic ligands with high affinity for biomolecular targets, and to their chemical conjugation with imaging probes and drug delivery systems, as well as with systems combining both imaging and therapeutic entities ("theranostics").[16] Targeting moieties of small molecular size can overcome the intrinsic problems of immunogenicity and tolerability of antibodies and other proteins, thus easing the development of cost-effective and safe products for broad clinical application. Efficient methods of conjugation allow the appropriate combination of properties such as high affinity for the targeted molecules with diagnostic sensitivity, or with effectiveness of drug delivery. For an optimal balance of properties, conjugations can be designed to set up constructs with different numerical combinations ("multivalency")[17] of the targeting and diagnostic/therapeutic moieties. Furthermore, these constructs may range in size from small- and medium-sized molecules up to nano- and micro-particles with increased capacities for drug loading. We have recently reviewed studies on nanoparticles targeted to adhesion molecules and their potential application in cancer and cardiovascular diseases,[18] and we shall focus herein on selected examples suggestive of future developments in cardiovascular disease.

A short analysis of the techniques employed in constructs targeting biomolecular markers of atherosclerosis, along with the results in early studies in animal models (Section 3), can be of help in the evaluation of the approaches deserving advanced research with a view to future clinical translation. The first of these are single photon emission computed tomography (SPECT) and positron emission tomography (PET) which have picomolar sensitivities, but limited spatial resolution of 0.3–3 mm and 1–4 mm, respectively. Typical SPECT radionuclides, emitting γ photons, include ^{99m}Tc ($t_{1/2}=6$ h), ^{111}In ($t_{1/2}=2.8$ days), ^{123}I ($t_{1/2}=13.2$ h), and ^{125}I ($t_{1/2}=59.5$ days). PET differs in that it relies on positron emitters (β^{+}) with shorter half-lives, such as ^{18}F ($t_{1/2}=109.7$ min), ^{68}Ga(III) ($t_{1/2}=67.7$ min), ^{64}Cu(II) ($t_{1/2}=12.7$ h), ^{124}I ($t_{1/2}=4.18$ days), and ^{11}C ($t_{1/2}=20$ min). ^{18}F labeling is generally achieved via nucleophilic substitution, using $^{18}F^-$, directly obtainable from a cyclotron, in synthons suitable for short synthetic routes toward ^{18}F-probes. For instance, a novel analog of ^{18}FDG which is in early clinical investigation in cancer, ^{18}F-galacto-RGD, is molecularly targeted to the integrin $\alpha_V\beta_3$.[19] A promising alternative to ^{18}F is $^{68}Ga(III)$, a positron emitter that does not require an on-site accelerator, being produced in a transferable generator containing the parent radionuclide ^{68}Ge, and allows simple labeling by last step complexation in pre-synthesized constructs carrying suitable metal chelating moieties.[2,20]

Magnetic resonance imaging (MRI), characterized by superior spatial resolution (50–250 µm), uses ^1H nuclei as the imaging signal with intensity dependent upon water content in tissues. Image contrast depends on two tissue-specific parameters: the longitudinal relaxation time, T_1 (spin–lattice relaxation time), and the transverse relaxation time, T_2 (spin–spin relaxation time). Two strategies are employed in contrast agents that magnetically modify the proton spin environment: positive enhancement (T_1-targeted) by gadolinium chelates, or negative enhancement (T_2- and T_2^*-targeted probes) by paramagnetic iron oxide nanoparticles. The latter are taken up in various cell types, typically through phagocytosis by macrophages infiltrating the atherosclerotic plaque, resulting in strong relaxation times detectable by T_2/T_2^*-weighted MRI, or by off-resonance positive contrast imaging. This approach provides a unique method for characterizing plaque morphology and tissue composition. However, there is substantial room for improvement of molecular targeting of macrophage and vascular cell markers by MRI alone or in combination with other techniques. These enhancements could lead to a better understanding of the inflammatory processes and differential characterization of unstable plaques, thus improving prognosis and targeted treatments. Examples of invasive imaging by combined MRI and near infrared (NIR) optical imaging have also been reported.[21,22] However, expectations for future clinical applications need to be directed towards non-invasive techniques. Bimodal PET–MRI imaging, taking advantage of newly developed hybrid scanners, is currently in a proof-of-principle phase, but is attracting great interest for its unique ability to simultaneously collect and correlate molecular and morphological information.[23]

Contrast enhanced ultrasound (CEUS) extends echographic ultrasound techniques to the exploitation of gas-filled microspheres ("microbubbles", MB) as an ultrasound contrast medium. MB consist of a lipidic or polymeric shell filled with a low-boiling perfluorinated alkane, and are commercially available for clinical use in cardiovascular imaging, being confined by their size (1–4 μm) to the intravascular space. Their proven clinical tolerability, along with the advantages of real-time imaging, high spatial resolution (0.05–0.5 mm), and, not least, the relatively low cost of equipment renders molecular targeting of MB an attractive option for future development from its current preclinical stage.[24–29]

Optical imaging using fluorescence in the visible or NIR regions provides high sensitivity *in vitro*. NIR imaging can also be performed *in vivo*, but only in small animals because of the limited tissue penetration of infrared waves. In vascular applications, NIR optical-coherence tomography (OCT) is now proposed for the cross-sectional imaging of the vessel walls with very high resolution (4–16 μm) and 2–3 mm penetration. Interesting examples of NIR imaging with molecularly-targeted probes,[30,31] including bimodal NIR–MRI experiments,[21,32] have been reported. However, future clinical development remains dependent upon the use of vascular catheters carrying the NIR probes.[33]

Molecularly-targeted constructs have to be validated by experiments defining their distribution upon intravenous administration, stability in plasma/blood, and ability to reach and bind vascular targets in high concentration with respect to other tissues. Thus, rapid systemic clearance accompanied by relatively longer on-target persistence is desirable, and fast urinary excretion of the unmodified molecule represents an optimal distribution profile. In the case of radionuclide- and Gd-based probes, these properties, along with the choice of chelating moieties to afford very stable complexes, can minimize the risk of fixation of radioactivity or toxic metals. Also the use of constructs based on microparticles confined to the intravascular space can lower the risks of organ toxicity.

3. TARGETING OF VASCULAR ADHESION MOLECULES AND MARKERS OF INFLAMMATION

Molecular targeting of atherosclerosis will require suitable imaging techniques in the early and/or late stages of the disease. Markers of endothelial activation and leukocyte adhesion are of relevance in the early stages, whereas macrophage density, oxidized lipoproteins, phosphatidyl serine, activity of proteases, and signals of apoptosis or necrosis become increasingly important in advanced stages for the identification of unstable plaques.[14,34]

The targeting of adhesion molecules as signals of inflammation mainly exploits constructs based on ligands of integrins or of their counter-receptors.[18,35] Integrin $\alpha_V\beta_3$, expressed in endothelial cells, is the most investigated marker of neoangiogenesis in tumors and ischemic organs, whereas in atherogenesis its expression is related to inflammatory reactions and growth of new vessels which feed the developing plaques and promote intraplaque hemorragies.[36–39] On the other hand, VCAM-1 and ICAM-1, as counter-receptors of the leukocyte integrins $\alpha_4\beta_1$ (VLA-4) and $\alpha_L\beta_2$ (LFA-1), are markers of endothelial inflammation and monocyte recruitment and play a role in the initiation of atherosclerotic plaques.

Preclinical *in vivo* studies have been performed with probes carrying specific antibodies for VCAM-1 on paramagnetic particles for MRI imaging[40,41] (including dual labeling with P-selectin and VCAM-1 antibodies) and on MB for CEUS imaging.[29,42] Linear and cyclic peptide ligands of VCAM or ICAM-1 were recently synthesized and conjugated with imaging probes, for example, the ICAM-1 binding disulphide peptide cyclo-(1,12)-PenITDGEATDSGC (cLABL),[43] derived from α_L domain of LFA-1 integrin, and the VCAM-1 binding heptapeptide VHSPNKK, a homolog of an epitope of the α-chain of $\alpha_4\beta_1$. The VHSPNKK sequence was incorporated into a synthetic bis-cisteinyl, FITC-marked peptide, which was conjugated on a cross-linked iron oxide (CLIO) nanoparticle, also carrying the NIR probe Cy5.5. *In vivo* MRI of the construct **1** (Fig. 8.1) in mice showed accumulation in atherosclerotic lesions, which was confirmed by *ex vivo* MRI and fluorescence observation of the excised aortas.[21] A related sequence, VHPKQHR, was multimerically loaded on CLIO nanoparticles along with the Cy5.5 NIR probe.[32] This construct was taken up in vessels of atherosclerotic mice *in vivo*. The uptake was decreased, in parallel with reduced VCAM-1, in animals treated with atorvastatin.

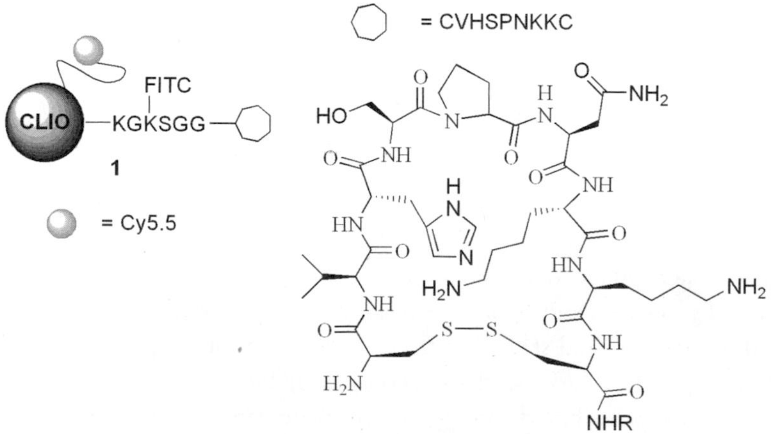

Figure 8.1 Bimodal MRI/NIR nanoparticles targeting VCAM-1.[21]

More recently the same peptide was incorporated into a ^{18}F-PET imaging tool, [18]F-4V, **2** (Fig. 8.2) which provided similar results.[44] Together, these studies suggest that probes containing small-molecule ligands could open the way towards clinically applicable, non-invasive MRI or PET imaging of VCAM-1, whereas an invasive clinical analysis[14] of advanced plaques could be allowed by novel NIR detectors positioned on intravascular catheters,[33] already investigated in rabbit atherosclerosis models.[45]

A number of peptidic and non–peptidic ligands of $\alpha_V\beta_3$, based on recognition of the RGD sequence in ECM proteins and tumor cells, have been synthesized and widely investigated as anti-angiogenic agents in cancer,[46] and for anti-atherosclerotic action in some animal studies.[47–49] A non-peptide quinolone ligand of integrin $\alpha_V\beta_3$ has been linked to PEG-distearoylethanolamine and incorporated in the lipid shell of a novel small particle, enclosing a core of liquid perfluoroctyl bromide (PFC particle).[50,51] These particles are smaller (~250 nm) than gas-filled microspheres and less sensitive to acoustic detection. Because of good retention in the intravascular space they are suitable for vascular imaging and drug delivery. Incorporation in the shell of a large amount of lipid linked Gd–chelates is also possible and affords high sensitivity in MRI detection (Fig. 8.3).[51] The

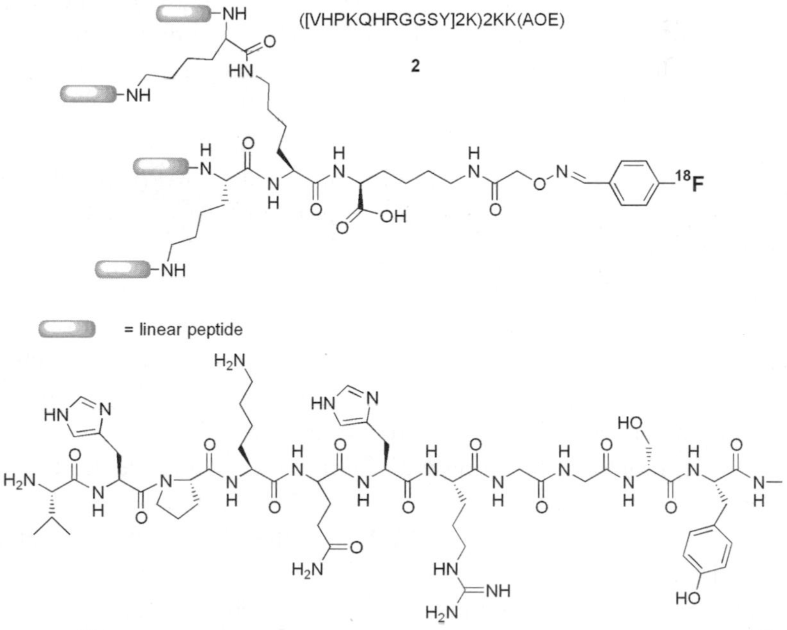

Figure 8.2 ^{18}F-PET imaging tool carrying a tetrameric VCAM-1 peptidic ligand.[44]

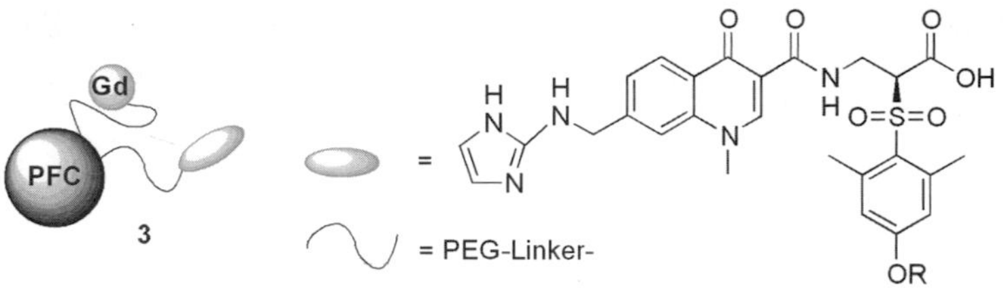

Figure 8.3 Integrin $\alpha_V\beta_3$ targeted and Gd loaded PFC particles.[51]

integrin targeted PFC particles were also exploited as a delivery system for fumagillin (as an experimental tool for anti-angiogenetic activity), or with rapamycin, taking advantage of the incorporation of these lipophilic drugs in the lipid shell. Particles loaded with fumagillin or rapamicin were respectively shown to slow the progression of either atherosclerosis or restenosis in rabbits.[50,51]

A novel type of paramagnetic particles, that is; MPIO (micro-sized particles of iron oxide), differ from ultrasmall (20–50 nm) and small (60–250 nm) magnetic nanoparticle (NP) based on their larger size (0.9–4.5 μm) and higher load of iron per particle. MPIO appear convenient for intravascular targeting because they are confined in vessels and cleared with 50–100 s half-lives (*vs* 24–48 h for NP).[40,41] Covalent binding was successfully achieved by direct reaction of a VCAM antibody to a MPIO carrying tosyl groups. The construct was suitable to detect cardiac and cerebral vascular inflammation in mouse models.[41] MPIO dually labeled with P-selectin and VCAM-1 antibodies were more effective than single-label particles in the imaging of plaques in mice,[40] whereas MPIO labeled with an antibody of integrin $\alpha_{IIb}\beta_3$ detected platelet deposition at sites of vascular injury.[52,53]

Both MPIO and PFC particles appear to share a favorable vascular distribution profile with MB and are deserving of further investigation; however, targeted MB seem to represent the best candidate for clinical translation, since non-targeted MB are validated for clinical use.[25,26] A number of preclinical studies with targeted MB have been reported. For example, molecular targeting with antibodies against vascular adhesion molecules, including ICAM-1, VCAM-1, and P-selectin, and also by dual-targeting with antibodies to ICAM-1 and the oligosaccharide selectin ligand[24] sialyl Lewis[X], have been successfully investigated in models of cardiac transplant rejection, hindlimb ischemia, and vascular (specifically cerebrovascular) inflammation.[24–29] The targeting of VCAM under high shear stress conditions has been enhanced by extracorporeal magnetic driving of iron nanoparticles loaded in MB.[54] In addition to antibodies, the conjugation of small targeting molecules, including sialyl Lewis[X], echistatin (a polypeptidic non-selective integrin ligand from snake venoms[55]), or phosphatidyl serine[26] have been reported.

Thus, further progress in the design and conjugation of synthetic ligands could open the way to clinical translation for targeted MB, which is also appealing from the drug delivery perspective. In fact the membranes of drug-loaded MB can be disrupted to release a therapeutic agent by ultrasound burst with energy higher than that needed for imaging.[56] A notable example of this capability is provided by the administration of thrombus-targeted MB in pigs with coronary occlusion, followed by local disruption, which improved epicardial recanalization and microvascular recovery[57] when compared with non-targeted MB. A commentary on this study suggested that "sonothrombolysis" could represent a future best technique in acute treatment of myocardial infarction and stroke.[58] An example of delivery by local MB disruption in the ischemic myocardium has been reported in rats treated with non-targeted MB loaded with angiogenic plasmids, resulting in improved neovascularization, myocardial perfusion, and ventricular function.[59]

Ferromagnetic NPs have been successfully employed in MRI imaging of atherosclerotic plaques, where they are taken up by macrophages *via* non-specific mechanisms. Up to now, attempts at specific targeting did not afford a clear advantage over non-targeted NP.[41] Recently, polymeric micelles carrying Cy5.5 and multiple Gd-chelates[22] for bimodal imaging, and targeted to phosphatidylserine *via* annexin V (Fig. 8.4), revealed selective uptake in apoptotic macrophages and cells in mice by *in vivo* MRI and *ex vivo* NIR. A SPECT probe with ^{99m}Tc-labeled annexin V has been investigated in animals,[60] and in early clinical trials,[61] allowing the detection of apoptotic macrophages in atherosclerotic plaques.

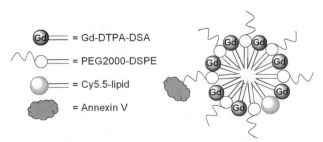

Figure 8.4 Polymeric micelles carrying annexin V, Cy5.5, and multiple Gd-chelates.[22]

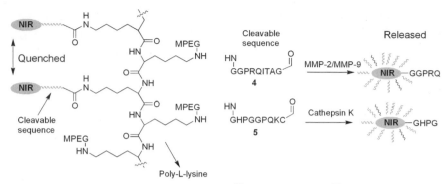

Figure 8.5 NIR probes carrying MMP-2/-9[30] or cathepsin K[31] cleavable peptide sequences.

NIR probes have been targeted to plaque proteases in mouse experiments using fluorogenic peptides cleavable by matrix metalloprotease MMP-2 and -9 (**4**, Fig. 8.5),[30] or by cysteine protease cathepsin K (**5**, Fig. 8.5).[31] The constructs consisted of a number of repeats (on the order of 20) of the cleavable peptide bound to the NIR fluorophore, Cy5.5, and grafted on a polylysine scaffold. Enzymatic cleavage and release of soluble moieties markedly enhanced the fluorescence signal. Using this technique, MMP activity was assessed non-invasively *in vivo* and also *ex vivo*,[30] whereas cathepsin activity was assessed invasively *in vivo* and *ex vivo*,[31] as well as in specimens of human atherosclerotic aortas.[31] *In vivo* assessment of cathepsin activity in atherosclerotic rabbits was achieved with an intravascular probe.[45]

The 18-kDa translocator protein (TSPO), also named "peripheral benzodiazepine receptor", is markedly expressed in macrophages and represents a valuable marker of inflammation in atherosclerosis.[62] The ligand [11C]-(R)-PK11195, **6** (Fig. 8.6) was recently tested in a preliminary PET study in patients with inflammatory vasculitis[63] and was suggested as a non-invasive tool for detection of intraplaque inflammation. This is an elegant example of a small molecule with both targeting and imaging properties; however, [11C]-compounds, derived from cyclotron generation, have short half-lives (20.4 min) and the synthetic scope is restricted to methylations using [11C]-methyl iodide or triflate.[2]

Perspectives of clinical applications appear more favorable for PET probes carrying ^{18}F or [68]Ga.[2,20] The first, *in vivo* cardiovascular study was recently reported, using combined microPET-CT in a rat model of

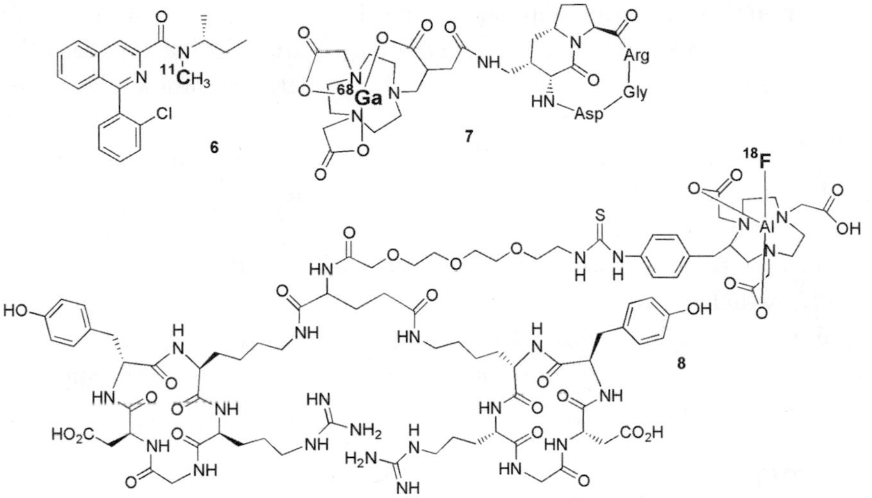

Figure 8.6 PET probes carrying ^{11}C, ^{68}Ga, and ^{18}F radionuclides.

myocardial infarction with a novel RGD–mimetic probe **7** (Fig. 8.6) targeting $\alpha_V\beta_3$.[64] At four weeks from coronary ligation, marked neoangiogenesis was observed in the peri-infarct area. The RGD–coupled ^{18}F-probe **8** (Fig. 8.6) was recently prepared by direct, last-step chelation to a DOTA-cage from an extemporaneous mixture of ^{18}F$^-$ and Al^{3+} solutions, thus avoiding multiple steps with ^{18}F-synthons. *In vivo* imaging showed delayed accumulation in the infarct area of rats at 1–3 weeks after coronary ligation and reperfusion.[65]

4. CONCLUSIONS

The challenge of improving the diagnostic armamentarium for atherosclerosis in order to improve risk stratification and personalized therapy poses both scientific and technical problems: fundamental biology and pathology on one side, and physical and chemical methodologies on the other. As for the contribution of medicinal chemistry, a number of thought-provoking examples of molecular targeting of markers of atherosclerosis are reported in the present survey. Although important in proving new principles, these contributions are still in an exploratory stage. Future studies should be performed in animal species and/or models endowed with increased power of predicting human efficacy and safety, thus warranting clinical development in a demanding regulatory environment.

Among the explored approaches, the non-invasive ones are the most promising for development, with special consideration for the bimodal technique of PET–MRI which benefits from synthetically simple constructs, such as combining magnetic iron particles with easily accessible radionuclides. However, among invasive techniques, intravascular imaging by NIR probes on catheters deserves attention as they are potentially suitable for detection of unstable plaques and also for focal drug delivery.

Finally, particles with micrometer size, such as MB, PFC particles, and MPIO which are retained in the vascular compartment, are of interest for cardiovascular applications. MB are already in clinical use for CEUS, and targeted MB could in the near future be promoted to clinical investigation for diagnosis and/or drug release by on site disruption.

REFERENCES

1. Choudhury, R. P.; Fuster, V.; Fayad, Z. A. *Nat. Rev. Drug Discov.* **2004**, *3*, 913.
2. Anderson, C. J.; Bulte, J. W. M.; Kai Chen, K.; Chen, X.; Khaw, B. A.; Shokeen, M.; Wooley, K. L.; VanBrocklin, H. F. *J. Nucl. Med.* **2010**, *51*, 3S.
3. van Werkhoven, J. M.; Bax, J. J.; Nucifora, G.; Jukema, J. W.; Kroft, L. J.; de Roos, A.; Schuijf, J. D. *J. Nucl. Cardiol.* **2009**, *16*, 970.
4. Camici, P. G.; Rimoldi, O. E.; Gaemperli, O.; Libby, P. *Eur. Heart J.* **2012**, *33*(11), 1309.
5. Joshi, F. R.; Lindsay, A. C.; Obaid, D. R.; Falk, E.; Rudd, J. H. *Eur. Heart J. Cardiovasc. Imaging* **2012**, *13*(3), 205.
6. Galiuto, L.; Natale, L.; Leccisotti, L.; Locorotondo, G.; Giordano, A.; Bonomo, L.; Crea, F. *J. Nucl. Cardiol.* **2009**, *16*, 811.
7. Klein, R.; Beanlands, R. S.; deKemp, R. A. *J. Nucl. Cardiol.* **2010**, *17*, 555.
8. Yoshinaga, K.; Manabe, O.; Tamaki, N. *J. Nucl. Cardiol.* **2011**, *18*, 486.
9. Arrighi, J. A.; Dilsizian, V. *Curr. Cardiol. Rep.* **2012**, *14*, 234.
10. von Knobelsdorff-Brenkenhoff, F.; Schulz-Menger, J. *J. J. Magn. Reson. Imaging* **2012**, *36*, 20.
11. Sanz, J.; Fayad, Z. A. *Nature* **2008**, *451*, 953.
12. Matter, C. M.; Stuber, M.; Nahrendorf, M. *Eur. Heart J.* **2009**, *30*, 2566.
13. Libby, P.; DiCarli, M.; Weissleder, R. *J. Nucl. Med.* **2010**, *51*, 33S.
14. ten Kate, G. L.; Sijbrands, E. J.; Valkema, R.; ten Cate, F. J.; Feinstein, S. B.; van der Stehen, A. F.; Daemen, M. J.; Schinkel, A. F. *J. Nucl. Cardiol.* **2010**, *17*, 897.
15. Haugland, R. P.; Bhalgat, M. K. *Methods Mol. Biol.* **2008**, *418*, 1.
16. Xie, J.; Lee, S.; Chen, X. *Adv. Drug Deliv. Rev.* **2010**, *62*, 1164.
17. Mammen, M.; Choi, S.-K.; Whitesides, G. M. *Angew. Chem. Int. Ed.* **1998**, *37*, 2754.
18. Arosio, D.; Casagrande, C.; Manzoni, L. *Curr. Med. Chem.* **2012**, *19*, 3128.
19. Beer, A. J.; Kessler, H.; Wester, H. J.; Schwaiger, M. *Theranostics* **2011**, *1*, 48.
20. Morrison, A. R.; Sinusas, A. J. *J. Nucl. Cardiol.* **2010**, *17*, 116.
21. Kelly, K. A.; Allport, J. R.; Tsourkas, A.; Shinde-Patil, V. R.; Josephson, L.; Weissleder, R. *Circ. Res.* **2005**, *96*, 327.
22. van Tilborg, G. A.; Vucic, E.; Strijkers, G. J.; Cormode, D. P.; Mani, V.; Skajaa, T.; Reutelingsperger, C. P.; Fayad, Z. A.; Mulder, W. J.; Nicolay, K. *Bioconjug. Chem.* **2010**, *21*(10), 1794.
23. Rischpler, C.; Nekolla, S. G.; Dregely, I.; Schwaiger, M. *J. Nucl. Med.* **2013**, *54*(3), 402.

24. Weller, G. E.; Villanueva, F. S.; Tom, E. M.; Wagner, W. R. *Biotechnol. Bioeng.* **2005**, *92*, 780.
25. Villanueva, F. S.; Jankowski, R. J.; Klibanov, S.; Pina, M. L.; Alber, S. M.; Watkins, S. C. *Circulation* **1998**, *98*, 1.
26. Lindner, J. R. *JACC Cardiovasc. Imaging* **2010**, *3*, 204.
27. Yan, Y.; Liao, Y.; Yang, L.; Wu, J.; Du, J.; Xuan, W.; Ji, L.; Huang, Q.; Liu, Y.; Bin, J. *Cardiovasc. Res.* **2011**, *89*(1), 175.
28. Kaufmann, B. A.; Sanders, J. M.; Davis, C.; Xie, A.; Aldred, P.; Sarembock, I. J.; Lindner, J. R. *Circulation* **2007**, *116*, 276.
29. Kaufmann, B. A.; Lewis, C.; Xie, A.; Mirza-Mohd, A.; Lindner, J. R. *Eur. Heart J.* **2007**, *28*, 2011.
30. Deguchi, J. O.; Aikawa, M.; Tung, C. H.; Aikawa, E.; Kim, D. E.; Ntziachristos, V.; Weissleder, R.; Libby, P. *Circulation* **2006**, *114*(1), 55.
31. Jaffer, F. A.; Kim, D. E.; Quinti, L.; Tung, C. H.; Aikawa, E.; Pande, A. N.; Kohler, R. H.; Shi, G. P.; Libby, P.; Weissleder, R. *Circulation* **2007**, *115*, 2292.
32. Nahrendorf, M.; Jaffer, F. A.; Kelly, K. A.; Sosnovik, D. E.; Aikawa, E.; Libby, P.; Weissleder, R. *Circulation* **2006**, *114*, 1504.
33. Razansky, R. N.; Rosenthal, A.; Mallas, G.; Razansky, D.; Jaffer, F. A.; Ntziachristos, V. *Opt. Express* **2010**, *18*, 11372.
34. Gallino, A.; Stuber, M.; Crea, F.; Falk, E.; Corti, R.; Lekakis, J.; Schwitter, J.; Camici, P.; Gaemperli, O.; Di Valentino, M.; Prior, J.; Garcia-Garcia, H. M.; Vlachopoulos, C.; Cosentino, F.; Windecker, S.; Pedrazzini, G.; Conti, R.; Mach, F.; De Caterina, R.; Libby, P. *Atherosclerosis* **2012**, *224*, 25.
35. Cox, D.; Brennan, M.; Moran, N. *Nat. Rev. Drug Discov.* **2010**, *9*(19), 804.
36. Wickline, S. A.; Neubauer, A. M.; Winter, P. M.; Caruthers, S. D.; Lanza, G. M. *J. Magn. Reson. Imaging* **2007**, *25*, 667.
37. McCarthy, J. R. *Adv. Drug Deliv. Rev.* **2010**, *62*, 1023.
38. McCarthy, J. R. *Curr. Cardiovasc. Imaging Rep.* **2010**, *3*, 42.
39. Godin, B.; Sakamoto, J. H.; Serda, R. E.; Grattoni, A.; Bouamrani, A.; Ferrari, M. *Trends Pharmacol. Sci.* **2010**, *31*, 199.
40. McAteer, M. A.; Akhtar, A. M.; von zur Muhlen, C.; Choudhury, R. P. *Atherosclerosis* **2010**, *209*, 18.
41. McAteer, M. A.; Choudhury, R. P. *Vascul. Pharmacol.* **2013**, *58*, 31.
42. Behm, C. Z.; Kaufmann, B. A.; Carr, C.; Lankford, M.; Sanders, J. M.; Rose, C. E.; Kaul, S.; Lindner, J. R. *Circulation* **2008**, *117*, 2902.
43. Dunehoo, A. L.; Anderson, M.; Majumdar, S.; Kobayashi, N.; Berkland, C.; Siahaan, T. J. *J. Pharm. Sci.* **2006**, *95*(9), 1856.
44. Nahrendorf, M.; Keliher, E.; Panizzi, P.; Zhang, H.; Hembrador, S.; Figueiredo, J. L.; Aikawa, E.; Kelly, K.; Libby, P.; Weissleder, R. *JACC Cardiovasc. Imaging* **2009**, *2*, 1213.
45. Jaffer, F. A.; Calfon, M. A.; Rosenthal, A.; Mallas, G.; Razansky, R. N.; Mauskapf, A.; Weissleder, R.; Libby, P.; Ntziachristos, V. *J. Am. Coll. Cardiol.* **2011**, *57*, 2516.
46. Desgrosellier, J. S.; Cheresh, D. A. *Nat. Rev. Cancer* **2010**, *10*(1), 9.
47. Elitok, S.; Brodsky, S. V.; Patschan, D.; Orlova, T.; Lerea, K. M.; Chander, P.; Goligorsky, M. S. *Am. J. Physiol. Renal Physiol.* **2006**, *290*, F159.
48. Bishop, G. G.; McPherson, J. A.; Sanders, J. M.; Hesselbacher, S. E.; Feldman, M. J.; McNamara, C. A.; Gimple, L. W.; Powers, E. R.; Mousa, S. A.; Sarembock, I. J. *Circulation* **2001**, *103*, 1906.
49. Maile, L. A.; Busby, W. H.; Nichols, T. C.; Bellinger, D. A.; Merricks, E. P.; Rowland, M.; Veluvolu, U.; Clemmons, D. R. *Sci. Transl. Med.* **2010**, *2*, 18ra11.
50. Caruthers, S. D.; Cyrus, T.; Winter, P. M.; Wickline, S. A.; Lanza, G. M. *Wiley Interdiscip. Rev. Nanomed. Nanobiotechnol.* **2009**, *1*, 311.

51. Lanza, G. M.; Winter, P. M.; Caruthers, S. D.; Hughes, M. S.; Hu, G.; Schmieder, A. H.; Wickline, S. A. *Angiogenesis* **2010**, *13*, 189.
52. von zur Muhlen, C.; von Elverfeldt, D.; Moeller, J. A.; Choudhury, R. P.; Paul, D.; Hagemeyer, C. E.; Olschewski, M.; Becker, A.; Neudorfer, I.; Bassler, N.; Schwarz, M.; Bode, C.; Peter, K. *Circulation* **2008**, *118*, 258.
53. von zur Muhlen, C.; Peter, K.; Ali, Z. A.; Schneider, J. E.; McAteer, M. A.; Neubauer, S.; Channon, K. M.; Bode, C.; Choudhury, R. P. *J. Vasc. Res.* **2009**, *46*(1), 6.
54. Wu, J.; Leong-Poi, H.; Bin, J.; Yang, L.; Liao, Y.; Liu, Y.; Cai, J.; Xie, J.; Liu, Y. *Radiology* **2011**, *260*(2), 463.
55. Leong-Poi, H.; Christiansen, J.; Heppner, P.; Lewis, C. W.; Klibanov, A. L.; Kaul, S.; Lindner, J. R. *Circulation* **2005**, *111*, 3248.
56. Sirsi, S.; Borden, M. *Bubble Sci. Eng. Technol.* **2009**, *1*, 3.
57. Xie, F.; Lof, J.; Matsunaga, T.; Zutshi, R.; Porter, T. R. *Circulation* **2009**, *119*(10), 1378.
58. Kaul, S. *Circulation* **2009**, *119*(10), 1358.
59. Fujii, H.; Li, S. H.; Wu, J.; Miyagi, Y.; Yau, T. M.; Rakowski, H.; Egashira, K.; Guo, J.; Weisel, R. D.; Li, R. K. *Eur. Heart J.* **2011**, *32*, 2075.
60. Johnson, L. L.; Schofield, L.; Donahay, T.; Narula, N.; Narula, J. *J. Nucl. Med.* **2005**, *46*(7), 1186.
61. Kietselaer, B. L.; Reutelingsperger, C. P.; Heidendal, G. A.; Daemen, M. J.; Mess, W. H.; Hofstra, L.; Narula, J. *N. Engl. J. Med.* **2004**, *350*(14), 1472.
62. Fujimura, Y.; Hwang, P. M.; Trout, I. H.; Kozloff, L.; Imaizumi, M.; Innis, R. B.; Fujita, M. *Atherosclerosis* **2008**, *201*, 108.
63. Pugliese, F.; Gaemperli, O.; Kinderlerer, A. R.; Lamare, F.; Shalhoub, J.; Davies, A. H.; Rimoldi, O. E.; Mason, J. C.; Camici, P. G. *J. Am. Coll. Cardiol.* **2010**, *56*, 653.
64. Menichetti, L.; Kusmic, C.; Panetta, D.; Arosio, D.; Petroni, D.; Matteucci, M.; Salvadori, P. A.; Casagrande, C.; L'Abbate, A.; Manzoni, L. *Eur. J. Nucl. Med. Mol. Imaging* **2013**, *40*, 1265.
65. Gao, H.; Lang, L.; Guo, N.; Cao, F.; Quan, Q.; Hu, S.; Kiesewetter, D. A.; Niu, G.; Chen, X. *Eur. J. Nucl. Med. Mol. Imaging* **2012**, *39*, 683.

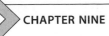

CHAPTER NINE

Oral GLP-1 Modulators for the Treatment of Diabetes

David J. Edmonds, David A. Price
Pfizer Worldwide Research and Development, Cambridge, MA, USA

Contents

1. INTRODUCTION

Incretin hormones were postulated following the observation that oral administration of glucose leads to higher and more sustained insulin levels than an equivalent glucose infusion. Two such hormones were subsequently discovered, named glucagon-like peptide 1 (GLP-1) and gastric inhibitory peptide (also sometimes referred to as glucose-dependent insulinotropic peptide), and found to stimulate insulin release from pancreatic beta cells. The main circulating form of GLP-1 is a 30-amino acid product of the proglucagon gene, GLP-1[7−36]NH_2 (**1**). It is produced in L-cells in the ileum and colon in response to nutrients in the gut and acts on the GLP-1 receptor (GLP-1R) on pancreatic beta cells to stimulate insulin release, suppress glucagon production, and promote beta cell proliferation. Crucially, the effect is glucose dependent and insulin release ceases as plasma glucose falls, thus preventing hypoglycemia. GLP-1 is rapidly cleared ($t_{1/2}=2$ min) from the bloodstream by cleavage of the first two residues by the protease dipeptidyl peptidase IV (DPP-IV). The GLP-1R is also expressed in the gut, brain, and other tissues, although its role in these tissues is less well understood. The receptor is a class B GPCR, a family which combines a typical seven-transmembrane helical bundle with a globular

Annual Reports in Medicinal Chemistry, Volume 48
ISSN 0065-7743
http://dx.doi.org/10.1016/B978-0-12-417150-3.00009-0

extracellular domain, which is involved in binding to the large peptidic ligand.[1] The large surface area and complex mode of ligand binding has made this family of receptors particularly challenging with respect to small molecule modulation.

1. HAEGTFTSDVSSYLEGQAAKEFIAWLVKGR-NH$_2$
2. HGEGTFTSDLSKQMEEEAVRLFIEWLKNGGPSSGAPPPS-NH$_2$

The enormous potential of the GLP-1 pathway for the treatment of type 2 diabetes mellitus (T2DM) has been realized in two distinct approaches. Inhibitors of DPP-IV serve to prolong the action of GLP-1, and several successful small molecule inhibitors are currently marketed for T2DM. A second approach has been to deliver supraphysiological levels of a GLP-1R agonist via injection. This approach has demonstrated significant weight loss in addition to HbA1c lowering, which provides added benefit to diabetes patients. Exenatide (synthetic exendin-4, **2**) is a 39-residue peptide which shares 53% sequence identity with GLP-1. The glycine at position 2 renders exenatide resistant to DPP-IV, giving it a more useful half-life in plasma. Exenatide was approved for the treatment of T2DM as a *bid* or *qw* subcutaneous (sc) injection, the latter using an extended release formulation. An alternative approach has been to extend the half-life of GLP-1 through increased binding to serum albumin. Liraglutide (**3**) achieves this profile through addition of a lipid side chain, and is marketed as a *qd* injection.[2] Unsurprisingly, several other agents are also undergoing clinical evaluation. In this chapter, we review the next evolution of the GLP-1 field, which is the invention of GLP-1R modulators that have the convenience of oral administration and the efficacy profile of the injectable peptides.

3: R^1=H; R^2=X
4: R^1=Me; R^2=Y

2. FORMULATION OF PEPTIDIC AGONISTS

The challenge of formulating peptides such as liraglutide or exenatide to deliver reasonable oral bioavailability with low interpatient variability cannot be overestimated. Proteolytic degradation, in the stomach and at the intestinal brush border, and an almost complete lack of membrane permeability combine to prevent absorption of such peptides from the gut. That portion of the peptide that reaches the blood is then subject to further proteolytic metabolism and also renal clearance. The development of agonists such as exenatide and liraglutide has gone some way to address the problem of the short half-life of GLP-1, leaving gut permeability as the key problem.

Absorption via the paracellular route is an attractive option for peptides and similar drugs, as the aqueous environment of the cellular junctions lacks proteolytic enzymes and avoids the need for the hydrophilic compounds to pass through cell membranes. However, peptides such as GLP-1 are too large to be effectively absorbed by the paracellular route. Tight junction modifiers, which are often surfactants, have been the subject of considerable research, despite safety concerns associated with disrupting the integrity of the intestinal membrane. This area has been comprehensively reviewed and will not be covered in detail here.[3,4] An alternative approach to enhancing permeability is to use "carriers" or "delivery agents". These compounds, typically salts of medium chain fatty acid derivatives, are thought to bind noncovalently to the peptide drug to enable passage through biological membranes. These excipients used are typically food additives, which are listed as generally regarded as safe and have an established safety record in rats, dogs, and humans.[5] The sodium salt of N-[8-(2-hydroxybenzoyl) amino]caprylate (SNAC, **5**) has been reported as useful for the delivery of several biopharmaceutical agents, including insulin, and has been marketed as part of the Eligen® delivery system.[6]

5

A 2010 report described the pharmacodynamic evaluation of ORMD-0901, a formulation of exenatide with varying quantities of undisclosed carrier and adjuvant components, in porcine and canine models.[7] Pharmacokinetics were not reported; however, a pharmacodynamic response was observed in the form of reduced glucose AUC following an oral glucose tolerance test (OGTT) in both species. For example, an oral dose of a formulation including 100 μg of exenatide demonstrated similar reduction in glucose AUC to that observed with a 2.5 μg exenatide sc injection in dogs. A poster disclosure, also in 2010, described an FIH study with a similar formulation of ORMD-0901, in which six healthy volunteers were dosed first with placebo and subsequently with ORMD-0901. The treatment was reported to be well tolerated; however, efficacy was limited. Insulin levels following an OGTT were reported for four patients, with a very modest increase in insulin AUC observed in the treated group, together with notable interpatient variability.[8] The formulation is reportedly undergoing further clinical evaluation, scheduled to be completed in early 2013.

A recent patent application described the combination of lipidated GLP-1 analogs, particularly semaglutide (4), in combination with SNAC-type permeation enhancers for the oral treatment of diabetes and related diseases. Semaglutide is a lipidated analog of GLP-1, similar to liraglutide, with an additional mutation of Ala8 to Aib (2-aminoisobutyric acid) to confer resistance to DPP-IV. The patent disclosed several versions of the formulation together with oral PK data in dogs. The optimal formulation reported contained 20 mg of semaglutide and 300 mg of SNAC along with several other excipients, and resulted in a mean oral bioavailability of 1.4% ($n = 8$). As might be expected with such low bioavailability, there was considerable variability in exposure, with a standard deviation in the %F of 0.95 in the best case reported. Other PK parameters reported included T_{max} (1.3 h) and C_{max} (94 nM).[9] Half-life was not disclosed in these studies, although semaglutide is currently being investigated as a once-weekly injectable, and has a reported[10] human $t_{1/2}$ of 160 h. An oral formulation of semaglutide is currently undergoing multidose clinical trials in Europe, albeit at doses (up to 60 mg) considerably higher than used via the subcutaneous route.[11]

3. ACTIVE TRANSPORT OF PEPTIDIC AGONISTS

The challenge of achieving passive permeability with a GLP-1R agonist has led several groups to explore active uptake processes by conjugating the peptide pharmacophore to transporter substrates. Biotin is an essential nutrient, which is absorbed from the diet by active uptake via the sodium-dependent multivitamin transporter (SMVT), and GLP-1–biotin conjugates have been designed as potential oral agents.[12] Reaction of

GLP-1[7−36]NH$_2$ with biotin N-hydroxysuccinimide ester (biotin–NHS) led to the K^{26},K^{34}-bisacylated conjugate **6** (DB-GLP-1). Alternatively, GLP-1[7−36]NH$_2$ could be selectively acylated with a PEGylated biotin at Lys34, allowing for the preparation of a K^{26}-biotin-K^{34}-PEGbiotin-GLP-1 derivative (DBP-GLP-1). DB-GLP-1 and DBP-GLP-1 both stimulated insulin release from pancreatic islets, and showed 2.4- and 9.9-fold improvements, respectively, in stability toward DPP-IV. Pharmacokinetic studies revealed that DBP-GLP-1, but not DB-GLP-1, could be detected in plasma following oral administration in rats, although absolute bioavailability was not reported. DBP-GLP-1 also showed improved glycemic control following glucose tolerance tests in rodents. The improved activity of the PEGylated conjugate as compared to the biotin only compound was attributed to increased absorption and plasma stability.[12,13] Similar results have also been reported for biotinylated exendin-4 analogs.[14] It should be noted that the mechanism of uptake was not clearly demonstrated to be due to active uptake by the SMVT.

HAEGTFTSDVSSYLEGQAA–K–EFIAWLV–K–GR-NH$_2$

6

Vitamin B$_{12}$ is a large organometallic complex which cannot be absorbed passively from the gut. Rather, it is scavenged from dietary sources by an intricate system of three chaperone proteins. Beginning in the mouth, haptocorrin is released into saliva and forms a complex with the vitamin. This strong interaction shields the vitamin molecule and, potentially, any conjugated peptide, from the harsh conditions of the stomach. In the small intestine, the B$_{12}$–haptocorrin complex breaks down as pH rises and the vitamin is bound by intrinsic factor (IF). The IF–B$_{12}$ complex is taken up into enterocytes via endocytosis mediated by cubulin (the IF–B$_{12}$ receptor). Within the enterocyte, the IF–B$_{12}$ complex breaks down and B$_{12}$ is secreted into the bloodstream bound to a third chaperone protein, transcobalamin II. The potential to exploit this pathway for the active uptake of complex drug substances has been recognized, with several groups attempting to deliver peptides through this pathway.[15,16] GLP-1 is an attractive target to pursue via B_{12}-mediated active uptake. There is good precedent in the form of liraglutide and others that the pharmacophore will tolerate quite large conjugates, and the target is extracellular. Furthermore, the GLP-1 peptide is highly potent, which offsets the limited (nmoles/dose) capacity of the vitamin B_{12} uptake pathway. Conjugation of GLP-1 to vitamin B_{12} has recently been demonstrated. Following the precedent of liraglutide, K34R GLP-1 [7–36]NH_2, was selectively derivatized at Lys26 to provide conjugate **7**. In an *in vitro* activity assay, the carefully purified conjugate was shown to activate the GLP-1R (transfected into HEK cells) with an EC_{50} equivalent to that of K34R GLP-1[7–36]NH_2 in the same assay, indicating that the conjugation was indeed tolerated in the pharmacophore of GLP-1. Further assessment of this conjugate for oral absorption and stimulation of insulin release *in vivo* will be of significant interest in the field of oral GLP-1 agonist development.[17]

HAEGTFTSDVSSYLEGQAA–K–EFIAWLVRGR-NH$_2$

7

The work described so far in this report has focused on the delivery of full–length (i.e., 30 or more amino acid) peptidic GLP-1 analogs. An attractive alternative, which may facilitate oral delivery while reducing the cost of goods, is to develop a truncated peptide agonist, and, to that end, a series of 11–mer GLP-1 analogs have been reported. The N-terminal nine residues of the GLP-1 sequence, which are thought to be essential for activity, were initially held constant. The remaining helical portion of GLP-1 was replaced with two biphenylalanine residues, to afford peptide **8** (Table 9.1),[18] which retained modest activity in a cAMP assay. The activity was improved first by optimizing substituents on the biphenyl groups to give **9**, followed by addition of quaternary amino acids at positions 2 and 6, and finally, fluorination of the phenylalanine derivative. Notably, this series of relatively small additions afforded cumulative potency gains of over 5000-fold between **8** and the optimized peptide **10**. The α–Me groups served to induce helical character in the truncated peptide, which was confirmed by 2D NMR studies. Peptide **10** displayed *in vitro* potency comparable to GLP-1 itself despite a molecular weight reduction of more than 50%. Upon subcutaneous dosing, peptide **10** showed a similar ability to stimulate insulin release and lower glucose levels to exendin-4 in ob/ob mice following an intraperitoneal glucose tolerance test (IPGTT), albeit at rather higher doses. Subsequent reports disclosed further optimization of positions 2 and 11, with particular flexibility at the 11 position with respect to potency.[19,20] The pharmacokinetic profile of **10** was promising, suggesting the compound would be suitable for once daily dosing in humans. Drug levels in plasma were detectable for at least 24 h after subcutaneous dosing in mice (460 μg/kg) and 8 h in dogs (67 μg/kg). A closely related peptide was reported to have no oral bioavailability;[21] nonetheless, the development of these substantially truncated agonists represents a milestone achievement in the development of the next generation of GLP-1 agonists.

Table 9.1 11-mer GLP-1R agonists

Cpd	R^1	R^2	R^3	R^4	R^5	R^6	hGLP-1R cAMP EC$_{50}$ (nM)
8	H	Me	H	Bn	BIP	BIP	545
9	H	Me	H	Bn	BIP(2′-Et,4′-OMe)	BIP(2′-Me)	7.0
10	Me	Me	Me	2-F-Bn	BIP(2′-Et,4′-OMe)	BIP(2′-Me)	0.087

A recent patent describes similar peptides, in which the optimized 11mer sequence is combined with an N-terminal valine, designed to improve bioavailability through active uptake by the PEPT1 transporter.[22] It is interesting to note that the agonist sequence is tolerant of the N-terminal addition, with many of the reported 12mers showing good potency in a cAMP assay. However, these GLP-1R agonists are considerably larger than the usual di- and tripeptide substrate scope of the PEPT1 transporter,[23] and no data are reported in the patent to demonstrate active uptake. The clinical candidate ZYOG1 (**11**) demonstrated good oral efficacy in several preclinical models of diabetes, including insulin release in ob/ob mice at 10 mg/kg similar to that achieved with subcutaneous exenatide.[24] In clinical studies, the compound showed dose-proportional exposure with a half-life of approximately 2–3 h. Absolute bioavailability was not disclosed, although it is expected to be rather low given the reported C_{max} (~2.5 nM following a 50 mg dose) and half-life.

11

4. SMALL MOLECULE AGONISTS OF THE GLP-1R

The cyclobutane derivative known as Boc5 (**12**) was reported in 2007 as a nonpeptidic GLP-1R agonist,[25,26] which was discovered through screening of >48,000 compounds against the rat GLP-1R using a luciferase reporter system. Boc5 was reported to be a full agonist in that assay format (vs. GLP-1 set at 100%), albeit with significantly reduced potency ($EC_{50} = 1$ μM) compared to GLP-1 ($EC_{50} = 68$ pM). Parenteral administration of the compound led to antidiabetic effects similar to those observed with GLP-1 in db/db mice, including improved response to an IPGTT, suppression of food intake, and reduction of HbA1c on chronic dosing. Suppression of food intake was also observed following a 10 mg oral dose, and was completely antagonized by pretreatment with peptidic GLP-1R antagonist exendin[9–39]. Oral bioavailability data were not presented.[25,27] The intriguing C_2-symmetrical structure of Boc5 reportedly arose from precursor **13** via dimerization in the DMSO solution used for screening. A recent publication described further chemical characterization of Boc5, together with optimization of the cyclobutane substituents for agonist activity. The optimized compound (**14**) showed approximately fourfold improvement in potency against the rat GLP-1R.[28]

12: R = Boc
14: R = C(O)*i*-Pr

13

Two recent patent applications describe a series of phenylalanine derivatives, exemplified by **15**, as oral agonists of the GLP-1R.[29,30] These compounds are reminiscent of half of the dimeric structure of Boc5. Although pharmacokinetic data are not presented, the molecular weight and polarity of the compounds place them outside the chemical space of typical oral drugs. Competition binding against ^{125}I-GLP-1[7–36]NH$_2$ and cAMP responses in GLP-1R-transfected CHO cells were used to characterize the reported GLP-1R agonists, and while it is difficult to assess the best compounds from the data presented, compound **15** is prepared on large scale and both enantiomers are exemplified.

Another recently published patent application describes compounds such as **16**. As with those described above, these compounds can be loosely classified as phenylalanine derivatives. In this case, the compounds are claimed as GLP-1R modulators and stabilizers, and a use as stabilizers for structural biology studies is specified. The biological activity reported is for the allosteric modulation of the receptor toward GLP-1.[31]

rac-15   **16**

A further class of small molecules reported in the patent literature is exemplified by TTP-054 (**17**). This class of GLP-1 agonists was first disclosed in a patent application in 2009. GLP-1R agonism was assessed using a cAMP assay in transfected CHO and HEK cells, with compound **17** showing a 65 nM EC_{50} and a maximal response of 82% relative to GLP-1[7–36] NH_2.[32] A subsequent patent application[33] described a variety of oral formulations specifically for **17**. Given the expected properties of this large (MW = 880) and lipophilic compound, it is perhaps unsurprising that considerable formulation effort would be needed to secure acceptable oral exposure. A related series has also been published in the patent literature, in which the left-hand morpholinone ring of the tricyclic core is replaced with a dihydropyran.[34] The most potent compounds have single digit nanomolar EC_{50} values; however, maximal responses are not disclosed, rendering a meaningful comparison of the series impossible.

17

A very recent publication discussed the downstream signaling of **12** and **17**, along with truncated peptide agonist **10**.[35] That study found that **12** and **17** differed from GLP-1[7–36]NH_2, in that each showed reduced signaling through cAMP as compared to their pERK pathway response. β–Arrestin recruitment was found to be impaired, with no response recorded for peptide **10** or **12**. Interestingly, while the potency of the cAMP response of the nonpeptidic compounds was broadly similar to that reported previously,[18,28,32] these authors found truncated peptide **10** to be somewhat right shifted as compared to GLP-1. In addition, the cAMP responses of **10**, **17**, and, to a lesser extent, **12** were found to be enhanced by known positive allosteric modulators of the GLP-1R.[35] This study serves to highlight the ongoing challenge of identifying an orally bioavailable compound capable of fully reproducing the effects of GLP-1 itself.

5. CONCLUSION

The impact of drugs that modulate the GLP-1 pathway on the quality of patients' lives is high and this area of research is far from complete. In this review, we have focused on the current state of orally bioavailable GLP-1R agonists, which is only a small portion of the field. Further innovation and creativity will surely deliver agents that are safe and possess the efficacy of the subcutaneously administered agents currently available with the convenience of oral administration. There is still a high unmet medical need from the patient population that will spur the drug discovery community forward to deliver these agents.

REFERENCES

1. Drucker, D. J. *Cell Metab.* **2006**, *3*, 153.
2. Drucker, D. J.; Nauck, M. A. *Lancet* **2006**, *368*, 1696.
3. Maher, S.; Brayden, D. J.; Feighery, L.; McClean, S. *Crit. Rev. Ther. Drug Carrier Syst.* **2008**, *25*, 117.
4. Salamat-Miller, N.; Johnston, T. P. *Int. J. Pharm.* **2005**, *294*, 201.
5. Maher, S.; Leonard, T. W.; Jacobsen, J.; Brayden, D. J. *Adv. Drug Delivery Rev.* **2009**, *61*, 1427.
6. Arbit, E.; Majuru, S.; Gomez-Orellana, I. In: *Protein Formulation and Delivery*; McNally, E., Hastedt, J. E., Eds.; Information Healthcare USA Inc: New York, NY, 2008; pp 285–303.
7. Eldor, R.; Kidron, M.; Greenberg-Shushlav, Y.; Arbit, E. *J. Diabetes Sci. Technol.* **2010**, *4*, 1516.
8. Eldor, R.; Kidron, M.; Arbit, E. 70th Scientific Sessions of the American Diabetes Association, Orlando, FL, 2010, Abstract 6-LB.
9. Drustrup, J.; Huus, K.; Balschmidt, P. Patent Application. WO 2012/098187, 2012.
10. Nauck, M. A.; Petrie, J. R.; Sesti, G.; Mannucci, E.; Courrèges, J.-P.; Atkin, S.; Düring, M.; Jensen, C. B.; Heller, S. 48th EASD Annual Meeting of the European Association for the Study of Diabetes, Berlin, Germany, 2012, Abstract OP-01-2.
11. http://www.clinicaltrials.gov/ct2/show/NCT01686945?term=nn9924&rank=5.
12. Chae, S. Y.; Jin, C.-H.; Shin, H. J.; Youn, Y. S.; Lee, S.; Lee, K. C. *Bioconjugate Chem.* **2008**, *19*, 334.
13. Youn, Y. S.; Chae, S. Y.; Lee, S.; Kwon, M. J.; Shin, H. J.; Lee, K. C. *Eur. J. Pharm. Biopharm.* **2008**, *68*, 667.
14. Jin, C.-H.; Chae, S. Y.; Son, S.; Kim, T. H.; Um, K. A.; Youn, Y. S.; Lee, S.; Lee, K. C. *J. Controlled Release* **2009**, *133*, 172.
15. Petrus, A. K.; Fairchild, T. J.; Doyle, R. P. *Angew. Chem. Int. Ed.* **2009**, *48*, 1022.
16. Clardy, S. M.; Allis, D. G.; Fairchild, T. J.; Doyle, R. P. *Expert Opin. Drug Deliv.* **2011**, *8*, 127.
17. Clardy-James, S.; Chepurny, O. G.; Leech, C. A.; Holz, G. G.; Doyle, R. P. *ChemMedChem* **2013**, *8*, 582.
18. Mapelli, C.; Natarajan, S. I.; Meyer, J.-P.; Bastos, M. M.; Bernatowicz, M. S.; Lee, V. G.; Pluscec, J.; Riexinger, D. J.; Sieber-McMaster, E. S.; Constantine, K. L.; Smith-Monroy, C. A.; Golla, R.; Ma, Z.; Longhi, D. A.; Shi, D.; Xin, L.;

Taylor, J. R.; Koplowitz, B.; Chi, C. L.; Khanna, A.; Robinson, G. W.; Seethala, R.; Antal-Zimanyi, I. A.; Stoffel, R. H.; Han, S.; Whaley, J. M.; Huang, C. S.; Krupinski, J.; Ewing, W. R. *J. Med. Chem.* **2009**, *52*, 7788.

19. Haque, T. S.; Lee, V. G.; Riexinger, D.; Lei, M.; Malmstrom, S.; Xin, L.; Han, S.; Mapelli, C.; Cooper, C. B.; Zhang, G.; Ewing, W. R.; Krupinski, J. *Peptides* **2010**, *31*, 950.

20. Haque, T. S.; Martinez, R. L.; Lee, V. G.; Riexinger, D. G.; Lei, M.; Feng, M.; Koplowitz, B.; Mapelli, C.; Cooper, C. B.; Zhang, G.; Huang, C.; Ewing, W. R.; Krupinski, J. *Peptides* **2010**, *31*, 1353.

21. Qian, F.; Mathias, N.; Moench, P.; Chi, C.; Desikan, S.; Hussain, M.; Smith, R. L. *Int. J. Pharm.* **2009**, *366*, 218.

22. Bahekar, R.; Jain, M. R.; Patel, P. R. Patent Application. WO 2011/048614, 2011.

23. Bailey, P. D.; Boyd, C. A. R.; Bronk, J. R.; Collier, I. D.; Meredith, D.; Morgan, K. M.; Temple, C. S. *Angew. Chem. Int. Ed.* **2000**, *39*, 506.

24. Jain, M. R.; Bahekar, R.; Joharapurkar, A.; Bandhyopadhyay, D.; Sundar, R.; Patel, H.; Pawar, V. 72nd Scientific Sessions of the American Diabetes Association, Philadelphia, PA, 2012, Abstract 57-OR.

25. Chen, D.; Liao, J.; Li, N.; Zhou, C.; Liu, Q.; Wang, G.; Zhang, R.; Zhang, S.; Lin, L.; Chen, K.; Xie, X.; Nan, F.; Young, A. A.; Wang, M.-W. *Proc. Natl. Acad. Sci. U.S.A.* **2007**, *104*, 943.

26. He, M.; Guan, N.; Gao, W.; Liu, Q.; Wu, X.; Ma, D.; Zhong, D.; Ge, G.; Li, C.; Chen, X.; Yang, L.; Liao, J.; Wang, M. *Acta Pharmacol. Sin.* **2012**, *33*, 148.

27. Su, H.; He, M.; Li, H.; Liu, Q.; Wang, J.; Wang, Y.; Gao, W.; Zhou, L.; Liao, J.; Young, A. A.; Wang, M.-W. *PLoS One* **2008**, *3*, e2892.

28. Liu, Q.; Li, N.; Yuan, Y.; Lu, H.; Wu, X.; Zhou, C.; He, M.; Su, H.; Zhang, M.; Wang, J.; Wang, B.; Wang, Y.; Ma, D.; Ye, Y.; Weiss, H.-C.; Gesing, E. R. F.; Liao, J.; Wang, M.-W. *J. Med. Chem.* **2012**, *55*, 250.

29. Liao, J.; Hong, Y.; Wang, Y.; Von Geldern, T. W.; Zhang, K. E. Patent Application. WO 2011/094890, 2011.

30. Liao, J.; Hong, Y.; Wang, Y; Von Geldern, T. W.; Zhang, K. E. Patent Application. WO 2011/097300, 2011.

31. Boehm, M. F.; Martinborough, E.; Moorjani, M.; Huang, L.; Tamiya, J.; Griffith, M. T.; Fowler, T.; Novak, A.; Knaggs, M.; Meghani, P. Patent Application. WO 2011/156655, 2011.

32. Mjalli, A. M. M.; Polisetti, D.R.; Yokum, T. S.; Kalpathy, S.; Guzel, M.; Behme, C.; Davis, S. T.; Patent Application. WO 2009/111700, 2009.

33. Polisetti, D.R.; Benjamin, E.; Quada, J. Patent Application. WO 2011/031620, 2011.

34. Mjalli, A. M. M.; Behme, C.; Christen, D. P.; Polisetti, D. R.; Quada, J.; Santosh, K.; Bondlela, M.; Guzel, M.; Yarragunta, R. R.; Gohimukkula, D. R.; Andrews, R. C.; Davis, S. T.; Yokum, T. S.; Freeman, J. L. R. Patent Application. WO 2010/114824, 2010.

35. Wootten, D.; Savage, E. E.; Willard, F. S.; Bueno, A. B.; Sloop, K. W.; Christopoulos, A.; Sexton, P. M. *Mol. Pharmacol.* **2013**, *83*, 822.

SECTION 3

Inflammation Pulmonary GI Diseases

Section Editor: David S. Weinstein
Bristol-Myers Squibb R&D, Princeton, New Jersey

CHAPTER TEN

Recent Advances in the Discovery and Development of CCR1 Antagonists

Penglie Zhang, Daniel J. Dairaghi, Juan C. Jaen, Jay P. Powers

ChemoCentryx Inc., Mountain View, California, USA

Contents

1. INTRODUCTION

1.1. CCR1 and its chemokine ligands

Chemokines are a group of small proteins that are best known for their ability to mediate leukocyte trafficking from blood into tissues during homeostasis and in response to injury and other pro-inflammatory stimuli. The more than 40 chemokines identified to date are divided into four subfamilies of CC, CXC, CX_3C, and C chemokines based on the number and position of the invariant cysteines in their amino acid sequences.

CCR1, the first CC chemokine receptor to be cloned,[1] is expressed on a range of cell types, including monocytes, macrophages, T lymphocytes, immature dendritic cells, neutrophils, basophils, eosinophils, natural killer (NK) cells, mast cells, osteoclast (OC) precursors, and mature OCs. The known CC chemokines that bind to CCR1 include CCL3 (MIP-1α), CCL5 (RANTES), CCL7 (MCP-3), CCL15 (leukotactin 1), and CCL23

Annual Reports in Medicinal Chemistry, Volume 48
ISSN 0065-7743
http://dx.doi.org/10.1016/B978-0-12-417150-3.00010-7

(CKβ8). In addition to leukocyte trafficking, activation of CCR1 can lead to upregulation of the integrin CD11b and subsequent increased adhesion of leukocytes to the endothelium, enhanced T cell activation, regulation of Th-1/Th-2 polarization, and stimulation of macrophage function and protease secretion.[2]

1.2. Clinical relevance and potential therapeutic indications

Proper functioning of the chemokine system plays an important role in host defense against pathogenic organisms. However, inappropriate activation of the chemokine network is associated with multiple human diseases.[3] Chemokines and leukocytes expressing the corresponding receptors are typically elevated in diseased organs and nearby environments. For example, high levels of CCR1-expressing monocytes/macrophages and CCR1 chemokines such as CCL3 and CCL5 have been identified in the synovial fluid and tissues of patients with rheumatoid arthritis (RA).[4–6] Genetic deletion and/or pharmacological intervention using antibodies or small molecule antagonists that target chemokines or chemokine receptors have demonstrated benefits in many animal models of human disease.[7] Consistent with this, treatment with anti-CCL3 antibodies demonstrated a delayed onset of arthritis and a reduction of disease severity in the mouse collagen-induced arthritis (CIA) model.[8] Studies of human multiple sclerosis (MS) brain lesions suggest that CCR1$^+$ monocytes can migrate into the CNS, where cytokines such as tumor necrosis factor α (TNF-α) and interleukin 1β (IL-1β) are secreted and promote MS pathology. Also, the presence of CCR1 chemokines leads to retention of CCR1-expressing monocytes in the CNS.[9] In an experimental autoimmune encephalomyelitis (EAE) model of MS, CCR1$^{-/-}$ mice had significantly reduced incidence of disease compared to wild type mice with minimal spinal cord inflammation.[10] Other human diseases with an inflammatory component reported to involve CCR1 and its chemokine ligands include psoriasis,[11] chronic obstructive pulmonary disease (COPD),[12] endometriosis,[13] allergic asthma,[14] sepsis,[15] progressive kidney disease,[16] atherosclerosis,[17] Alzheimer's disease,[18] and organ transplant rejection.[19]

Recent results from *in vitro* and *in vivo* experiments also continue to support targeting CCR1 or its ligands in the treatment of multiple myeloma (MM) and associated osteolytic bone disease (OBD). OC and MM cells support and nourish each other *in vitro* and *in vivo*, and CCR1 is expressed on the surface of OC precursors. Some MM cells modestly express CCR1, but all constitutively secrete high levels of CCL3, which stimulates OC activity and inhibits osteoblast formation. Serum levels of CCL3 correlate with the extent of lytic bone lesions and MM patient survival.[20] *In vitro* experiments have demonstrated that the small molecule CCR1 antagonist MLN3897 (**9**) blocked OC development and function by inhibiting differentiation of OC precursors.[21] Moreover, **9** abrogated MM cell migration and adhesion to OCs. Finally, **9** overcame the protective effect of OCs on MM cell survival and proliferation, thus inhibiting the interactive loop between OCs and MM cells. CCX721, another mouse-active CCR1 antagonist, produced a profound decrease in tumor burden and osteolytic damage in a murine model of MM–OBD at a dose selected to provide maximal inhibition of CCR1.[22]

2. SMALL MOLECULE CCR1 ANTAGONISTS
2.1. Early compounds

The topic of small molecule CCR1 antagonists was last reviewed in this publication in 2004.[23] A number of compounds covered there continue to receive considerable attention, and reports of their preclinical and clinical profiles have recently emerged. Among them is the quinoxaline amide derivative CP-481,715 (**1**). Structure-activity relationship (SAR) studies leading to the discovery of **1** have been described,[24] and process development work for its large scale (15 kg) synthesis has been reported.[25] The *in vitro* CCR1 antagonist pharmacology of **1** has also been well characterized.[26] In human CCR1 knock-in mice, **1** inhibited CCL3-induced neutrophil infiltration into skin or into an air pouch with an ED_{50} of 0.2 mg/kg.[27] In a delayed-type hypersensitivity model, compound **1** significantly inhibited footpad swelling and decreased inflammatory cytokines (IFN-γ and IL-2) produced by isolated spleen cells from sensitized animals. Compound **1** was evaluated clinically in healthy volunteers and in two completed trials in RA patients (*vide infra*).

1

2

3

4

A recent publication identified **2** as AZD4818,[28] a CCR1 antagonist that advanced into clinical trials for the treatment of moderate to severe COPD (*vide infra*). AZD4818 was reported to potently inhibit CCL3 binding to human, rat, mouse, and dog CCR1 receptors, as well as the chemotaxis of human monocytes to CCL3.[29] Studies in rats showed a significant reduction of neutrophil influx and TNF-α levels in bronchoalveolar lavage (BAL) following intratracheal administration of AZD4818 (1 mg/kg) prior to lipopolysaccharide (LPS) challenge. In a chronic LPS animal model, inhaled nebulized AZD4818 (0.01 mg/mL and 1 mg/mL) once daily for 5 days per week for a period of 4 or 8 weeks significantly reduced infiltrating BAL macrophage cell numbers.[29]

The *in vitro* CCR1 antagonist properties of BX-471 (**3**) have been described.[30] A greater than 10,000-fold selectivity for CCR1 over a panel of 28 G protein-coupled receptors (GPCRs), including related chemokine receptors, was observed. BX-471 was reported to have a half-life of ~3 h and an oral bioavailability of 60% in fasted dogs following oral gavage administration. The PK profile of BX-471 in other preclinical species has not been disclosed, although the compound displayed efficacy in rodent models of various human diseases when dosed subcutaneously.[31] Interpretation of the efficacy results in animal models is challenging due to the marked drop in potency against rodent CCR1 receptors (215-fold drop for mouse CCR1 and 120-fold drop for rat CCR1).[30,31] BX-471 was evaluated in clinical trials for the treatment of MS, endometriosis, and psoriasis (*vide infra*).

CP-865,569 (**4**) maintains most of the structural features of BX-471 with an additional methyl substituent on the piperazine and a benzylic sulfonic acid in place of the aryl urea. A large scale synthesis of **4** (96.2 kg) has been reported,[32] and a patent application describes the characterization of associated salt and crystal forms.[33] However, the preclinical *in vitro* and *in vivo* profile of **4** has not been disclosed and its status in clinical development is not clear at this time.

2.2. Recently published/disclosed compounds

Recent scientific reports and patent applications indicate a continued intense interest in the discovery and development of small molecule CCR1 antagonists. New compound series that are structurally related to **3** continue to emerge. Bioisosteric substitutions of select moieties in **3** have been pursued to improve the PK profile, while retaining high anti-CCR1 activity.

Compound **5** employs a diaminocyclobutenedione moiety as a urea mimetic.[34] Reported IC_{50} values of compound **5** are 4 nM (calcium mobilization), 7 nM (chemotaxis), 10 nM (human CCR1 binding), 260 nM (mouse CCR1 binding), and 20 nM (rat CCR1 binding). The half-life of **5** in male BALB/c mice and Sprague–Dawley rats is 2.3 h and 2.6 h, respectively (dosed 50 mg/kg p.o.). At this dose, treatment with **5** resulted in a statistically significant reduction of disease score in a mouse CIA model.

Compound **6** and its analogues incorporate a cinnamide functional group.[35] In binding assays using radiolabeled CCL3 ligand and CHO-K1 cells transfected with CCR1 receptors, **6** displayed equally high potency across species ($IC_{50} = 30$ nM for human and rat and 58 nM for mouse). The PK profile of **6** in rat is significantly improved over **1** (half-life: 16.5 h *vs.* 1.9 h) when dosed orally. Compound **6** displayed efficacy in a mouse CIA model at an oral dose of 60 mg/kg. At 100 mg/kg in the same model, **6** significantly reduced histological scores of inflammation, pannus formation, cartilage degradation, and bone resorption.

CCR1 antagonists based on the cinnamide chemotype (e.g., **7**) were the subject of an additional patent application.[36] Compound **7** inhibited [^{125}I]-CCL3 binding to CCR1-transfected chinese hamster ovary (CHO) cells with an IC_{50} of 14 nM. In compound **8** (binding $IC_{50} = 12$ nM), the chloro cinnamide moiety is replaced with a 5-bromo indole.[37] Compound **8** has a reported plasma half-life of 7.0 h in mice and is well-absorbed, with an oral bioavailability of 53%.

The generic structure of 4-hydroxy-4-aryl-piperidine derivatives that contain a tricyclic moiety as potent CCR1 antagonists has also been previously reviewed in this publication.[23] Related to this, compound **9** is the focus of a more recent application that described its synthesis and salt form selection.[38] It is suggested that **9** is MLN3897,[28] an orally bioavailable CCR1 antagonist that advanced into a Phase 2 clinical trial to evaluate its effect in subjects with RA on a stable dose of methotrexate (MTX). Few details about the preclinical profile of MLN3897 are available. It was shown to be cleared mainly by CYP3A in *in vitro* phenotyping assays, necessitating a clinical drug–drug interaction (DDI) study.[39] The closely related compound **10** is the focus of another patent application,[40] in which its racemate was described as a CCR1 antagonist with a K_i of 3 nM (THP-1 cells). The compound produced 99% inhibition of cell infiltration at 2.5 mg/kg (dosed p.o.) following a challenge of CCL3 in a guinea pig model of neutrophil recruitment. It is stated that compound **10** showed greater selectivity compared to an earlier compound when assayed on other GPCRs and ion channels.

High-throughput screening followed by SAR studies led to the discovery of **11** and **12** as selective and orally bioavailable CCR1 antagonists.[41,42] CCX354 (structure not disclosed) is a compound from the same research group that has demonstrated clinical efficacy in a recently completed Phase 2 trial in RA (*vide infra*). The preclinical characterization of CCX354 has been described.[43] The compound potently inhibits CCL15-mediated THP-1 cell chemotaxis ($IC_{50} = 1.4$ nM). No inhibition of other chemotactic receptors was observed at concentrations up to 10 μM. Saturation binding experiments with [^{125}I]-CCL15 on human monocytes identified CCX354 as a competitive CCR1 antagonist ($K_i = 1.5$ nM). Chemotaxis assays with human monocytes in 100% human serum displayed a typical bell-shaped dose–response curve with respect to the CCR1 chemokines CCL3, CCL5, CCL15, and CCL23. A marked right shift (~10-fold) of the dose–response curves was observed in the presence of 250 nM of CCX354 in human serum, with this shift (A_{10}) representing ~90% CCR1 blockade. THP-1 chemotaxis toward synovial fluid from 35 subjects with RA was significantly blocked by CCX354, but not by the CCR2 antagonist MK0812. CCX354 was evaluated in two animal models of inflammatory leukocyte trafficking: thioglycollate-induced peritonitis in rats and LPS-induced synovitis in rabbits. Rats receiving 100 mg/kg CCX354 had significantly reduced inflammatory cell counts as compared to vehicle-treated animals. These animals experienced >90% CCR1 coverage on blood leukocytes at all times during the study. At 10 mg/kg and 1 mg/kg, 90% blockade of blood leukocytes was not attained in a sustained manner and the animals displayed no reduction in the recruitment of inflammatory cells. Likewise, in a rabbit model of joint inflammation, total cell counts were performed on knee synovial lavage samples collected 16 h after the administration of LPS injection. Rabbits receiving 100 mg/kg or 20 mg/kg CCX354 showed significant reductions of inflammatory cell counts as compared to vehicle-treated animals, where those receiving 4 mg/kg CCX354 showed no difference from vehicle-treated animals. As in the rat model, efficient blockade of inflammatory cell recruitment required >90% CCR1 blockade on blood leukocytes at all times.[43]

Screening followed by iterative parallel synthesis of small libraries of potential CCR1 inhibitors led to the identification of compound **13** with a CCR1 binding IC_{50} of 28 nM (THP-1 cells, [^{125}I]-CCL3).[28] Further optimization of properties including metabolic stability and hERG inhibition led to BMS-817399 (structure not disclosed), a compound currently in a Phase 2 trial for the treatment of RA (*vide infra*).[44] The structure of BMS-817399 may be related to **14** based on recent patent activities around **14** regarding its *in vitro* profile and prodrugs.[45,46] Compound **14** was described to have a K_i of 0.7 nM (THP-1 cells and [^{125}I]-CCL3), an $EC_{50} > 30$ μM in PXR induction, and a 29% inhibition of the hERG channel at 30 μM.[45]

13

14

15

16

In a recent report, the benzylpiperidineamine moiety in compound **2** was replaced with a spirocycle in a novel series exemplified by **15** and **16**.[47] Compound **15** ($IC_{50}=0.36$ nM) is nearly 10-fold more potent than **2** in CCR1 binding affinity, is metabolically stable in human hepatocytes, and exhibits good cell permeability. These properties were retained in compound **16**, in which the secondary hydroxyl was replaced with an amino group (binding $IC_{50}=0.79$ nM). The *in vivo* profiles of **15** and **16** have not been described.

Compound **17** has been described as a CCR1 antagonist lead discovered by optimization of a screening hit.[48] Heterocyclic replacements for the amide moiety in compound **17** led to compound **18** with a reported IC_{50} of 2.9 nM for inhibition of THP-1 cell chemotaxis to CCL3.[49] In rat, compound **18** has a plasma half-life of 3.8 h, a low clearance of 2.8 mL/min/kg, and an oral bioavailability of 26%. PS031291 (structure not disclosed), a compound originating from the same research group, has been described as a potential development candidate in a research update.[50] PS031291 has a binding IC_{50} of 7 nM (THP-1 and [^{125}I]-CCL3), potently inhibits chemotaxis ($IC_{50}=0.6$ nM, cell type and ligand not described) and blocks CCR1 internalization ($IC_{50}=4$ nM in human whole blood).[50] Further development of PS031291 may be on hold, as no status updates have been reported over the past few years.

17   **18**

Compounds **19–22** exemplify a series of novel CCR1 antagonists that has been disclosed in a string of recent patent applications.[51–57] Compound **19** inhibits CCR1-mediated calcium release with an IC_{50} of 0.3 nM. The pyrazole ring in **19** is incorporated into a bicyclic moiety in **20** and **21**, retaining high potency ($IC_{50}=0.4$ nM for **20**, and 2 nM for **21**). Compound **22** features a partially saturated bicyclic spacer and a reported IC_{50} of 0.2 nM. The development status of this series of compounds has not been disclosed.

19

20

21

22

3. CLINICAL UPDATE

There have been at least nine CCR1 antagonists that have advanced into human clinical trials. Among them, C-6448 and C-4462 (structures not disclosed) were described to be in Phase 2 trials for MS and RA, respectively.[58] Little has been disclosed about the compound class and the development status since 2004. MLN3701 (structure not disclosed) was tested in Phase 1 with MS described as the potential clinical indication,[59] however, further updates have not come to light.

BX-471 has a relatively short half-life (2.3 h) in humans, and an extended release tablet formulation was developed for clinical dosing.[60] The 105 patients enrolled in the Phase 2 study of BX-471 in relapse-remitting MS received either 600 mg BX-471 three times daily (for a total of 1800 mg/day) or placebo for 16 weeks.[60] This study failed to show reductions in the number of new inflammatory CNS lesions detected by magnetic resonance imaging with BX-471 treatment despite plasma concentrations that were reported to remain >100 ng/mL throughout the dosing period. BX-471 was also ineffective in the treatment of psoriasis and endometriosis in Phase 2 clinical trials.[61]

AZD4818 (structure not formally disclosed) was evaluated in a Phase 2 trial for tolerability and efficacy in moderate to severe COPD patients.[29] A total of 65 patients received either 300 µg AZD4818 ($n=33$) twice daily as a dry powder for inhalation or matching placebo ($n=32$) for 4 weeks. No statistically significant differences were found between treatment groups with respect to any of the pre-specified efficacy assessments. It was also disclosed that in another study, AZD4818 at 800 µg twice daily for 7 days in healthy males did not reduce the numbers of neutrophils, monocytes, and other white blood cells or the levels of inflammatory markers in induced sputum and blood after LPS challenge, despite a promising prior observation following acute AZD4818 exposure in a rat LPS challenge model.[29,62]

CP-481,715 (**1**) provided a key piece of clinical evidence in support of CCR1 blockade in the treatment of RA in a small Phase 1b trial.[63,64] The compound displayed linear pharmacokinetics up to 300 mg once daily, with a reported half-life of 2–3 h.[63] Subjects with active RA (total $n = 16$) were randomized 3:1 to active:placebo treatment for 14 days.[64] The CP-481,715 treated group (300 mg three times daily for a total of 900 mg/day) showed statistically significant clinical improvement in the number of tender joints ($p = 0.021$), swollen joints ($p = 0.001$), quality of life as measured by a health assessment questionnaire (HAQ; $p = 0.037$), and in disease activity score (DAS; $p = 0.012$). One third of the 12 drug-treated RA patients fulfilled the American College of Rheumatology 20% (ACR20) criteria for improvement after 2 weeks and demonstrated marked decreases in the number of macrophages and other CCR1$^+$ cells in synovial biopsies. In a larger Phase 2 study, 95 patients were randomized 47:48 to receive either CP-481,715 (400 mg three times daily for a total of 1200 mg/day) plus MTX or MTX only for 6 weeks.[2] Reported plasma levels were in line with the previous study in RA patients. At 6 weeks, there was no statistical difference found between groups and the ACR20 response rate was higher for the placebo group (47.9%) as compared with the CP-481,715 treatment group (34.0%), seemingly inconsistent with the Phase 1b trial results. There were also no statistically significant differences between treatment groups at week 2 and week 4 in the ACR response rate or any of the ACR components. Synovial tissue biopsies were not obtained in this study.

MLN3897 (structure not formally disclosed) was evaluated in a multinational Phase 2 trial in RA patients at 27 study centers.[65] Subjects were randomized to receive either 10 mg MLN3897 once daily orally or matching placebo for 12 weeks while continuing their existing once-weekly MTX regime. In the intention-to-treat (ITT) population, ACR20 response rates on day 84 in the MLN3897 and placebo-treated groups were 35% and 33%, respectively ($p = 0.72$). In both groups, four patients achieved an ACR50 response and no patient achieved an ACR70 response in either group. There was no significant change in markers of inflammation, including serum levels of IL-6, TNF-α, sIL-2R, MMP-3, CCL2, and CCL3, as a result of MLN3897 treatment.

CCX354 (structure not disclosed) was studied in healthy volunteers in single- and multiple-dose studies (1–300 mg/day oral dose).[43] The compound displayed a linear dose-exposure profile, with half-life values of 5.2 h and 6.5 h at the 100 mg and 300 mg doses, respectively. A whole-blood *ex vivo* assay using specific Alexa647–CCL3 binding as a measure of free CCR1 indicated that after a single 100 mg dose of CCX354, monocyte CCR1 receptors remained markedly blocked after 12 h in all subjects. The average extent of CCR1 blockade (94%) at 12 h after dosing was consistent with expectations based on the average plasma CCX354 concentrations (440 ± 68 nM, $n = 5$) in those samples and the *in vitro* potency of the drug against Alexa647–CCL3 binding ($K_i = 15$ nM in serum). In a Phase 2 trial in RA subjects, CCX354 was dosed at 100 mg b.i.d. ($n = 53$) or 200 mg q.d. ($n = 53$) for 12 weeks together with a placebo group ($n = 54$).[66] All patients continued to receive their regular MTX dose. In the pre-specified population of patients satisfying C-reactive protein (CRP) (>5 mg/L) and tender/swollen joint count (SJC) (>8) eligibility criteria at the screening and day 1 (predose) visits, ACR20 response at week 12 was 30% in the placebo group, 44% in the 100 mg b.i. d. group ($p = 0.17$), and 56% in the 200 mg q.d. group ($p = 0.01$). In day 1 eligible subjects that were naïve to biological treatment, statistically significant ACR20 response was observed in the 200 mg q.d. group (62% *vs.* 26% for placebo, $p = 0.002$) while only trend-toward-efficacy was observed in the 100 mg b.i.d. group (41% *vs.* 26%, $p = 0.18$). The CCX354 200 mg q.d. group also showed significantly decreased CRP compared with placebo at week 12. Other RA efficacy measures including tender joint count (TJC), erythrocyte sedimentation rate, DAS28, SJC, subject's assessment of RA and pain, and physician's assessment of RA all showed numerically greater improvements with the 200 mg q.d. compared to placebo. Additionally, bone turnover markers C-terminal telopeptide (CTx), osteocalcin, and amino-terminal propetide of type I collagen showed statistically significant treatment effects, evident as early as 1 week into the study, supporting the bone antiresorptive efficacy of CCX354 in RA. This finding is consistent with the role ascribed to CCR1 and its chemokine ligands on OC precursors and mature OCs (*vide supra*), and indicates that chronic treatment with CCR1 antagonists may help protect against bone erosion in patients with RA.

BMS-817399 (structure not disclosed) completed a Phase 1 study assessing the ascending single and multiple oral doses (20–2000 mg q.d. or 40–1000 mg b.i.d.) as oral suspension or capsules.[67] PK/PD details from this study have not been reported. An ongoing proof-of-concept Phase 2 trial that started in September 2011 is testing the effect of BMS-817399

(200 mg b.i.d. and 400 mg b.i.d.) in combination with MTX in subjects with moderate and severe RA.[44] The trial is expected to complete in May 2013.

4. OUTLOOK

To date, drugs that inhibit chemokine receptors have only been approved for the treatment of HIV infection (Maraviroc® for CCR5) and stem cell mobilization (Mozobil® for CXCR4). The promise of chemokine receptor antagonists for the treatment of chronic inflammatory disorders and autoimmune diseases has not yet been fully realized.[61] Clinical trials with CCR1 antagonists in the treatment of MS, endometriosis, psoriasis, and COPD have thus far been unsuccessful, and it is possible that CCR1 plays a minimal role in some of those diseases despite strong evidence of effectiveness in animal models. However, the compounds that have previously advanced into clinical trials, while highly potent in *in vitro* assays, likely have limitations especially in the area of pharmacokinetics as indicated by dosing levels, dosing frequency, and routes of administration. In the area of RA, the experiences have been mixed. CP-481,715 and MLN3897 failed to show therapeutic benefit in Phase 2 trials despite signs of clinical efficacy displayed by CP-481,715 in an early study. On the other hand, CCX354 has demonstrated proof-of-concept in a recently completed Phase 2 RA trial. While it is tempting and convenient to suggest that redundancy within the complex chemokine system or disease heterogeneity makes it difficult to demonstrate clinical efficacy with an antagonist selective to a specific chemokine receptor, this notion has been challenged based on emerging preclinical and clinical observations.[68] For example, monocyte migration induced by RA synovial fluid can be inhibited with either a CCR1-blocking antibody or a small molecule CCR1 antagonist, but not by CCR2- or CCR5-blocking antibodies.[69] Comparative analysis of three CCR1 antagonists tested in RA indicates that high levels of receptor coverage need to be maintained (90% or higher at all times) to achieve efficacy.[43] As small molecule CCR1 antagonists with more favorable preclinical profiles in potency, PK, and safety continue to emerge, sufficient receptor coverage under physiological conditions might be conveniently achieved without safety concerns so that various clinical indications can be tested unambiguously. The outcome of the ongoing RA trial with BMS-817399, if positive, will provide additional support for targeting CCR1 as an effective therapy in inflammatory diseases.

REFERENCES

1. Neote, K.; DiGregorio, D.; Mak, J. Y.; Horuk, R.; Schall, T. *Cell* **1993**, *72*, 415–425.
2. Gladue, R. P.; Brown, M. F.; Zwillich, S. H. *Curr. Top. Med. Chem.* **2010**, *10*, 1268–1277, and references therein.
3. Gerard, C.; Rollins, B. J. *Nat. Immunol.* **2001**, *2*, 108–115.
4. Jenkins, J. K.; Hardy, K. J.; McMurray, R. W. *Am. J. Med. Sci.* **2002**, *323*, 171–180.
5. Koch, A. E.; Kunkel, S. L.; Harlow, L. A.; Mazarakis, D. D.; Haines, G. K.; Burdick, M. D.; Pope, R. M.; Strieter, R. M. *J. Clin. Invest.* **1994**, *93*, 921–928.
6. Volin, M. V.; Shah, M. R.; Tokuhira, M.; Haines, G. K.; Woods, J. M.; Koch, A. E. *Clin. Immunol. Immunopathol.* **1998**, *89*, 44–53.
7. Johnson, Z.; Schwarz, M.; Power, C. A.; Wells, T. N. C.; Proudfoot, A. E. I. *Trends Immunol.* **2005**, *26*, 268–274.
8. Kasama, T.; Strieter, R. M.; Lukacs, N. W.; Lincoln, P. M.; Burdick, M. D.; Kunkel, S. L. *J. Clin. Invest.* **1995**, *95*, 2868–2876.
9. Trebst, C.; Sorensen, T. L.; Kivisakk, P.; Cathcart, M. K.; Hesselgesser, J.; Horuk, R.; Sellebjerg, F.; Lassmann, H.; Ransohoff, R. M. *Am. J. Pathol.* **2001**, *159*, 1701–1710.
10. Rottman, J. B.; Slavin, A. J.; Silva, R.; Weiner, H. L.; Gerard, C.; Hancock, W. W. *Eur. J. Immunol.* **2000**, *30*, 2372–2377.
11. Raychaudhuri, S. P.; Jiang, W. Y.; Farber, E. M.; Schall, T. J.; Ruff, M. R.; Pert, C. B. *Acta Derm. Venereol.* **1999**, *79*, 9–11.
12. Hartl, D.; Krauss-Etschmann, S.; Koller, B.; Hordijk, P. L.; Kuijpers, T. W.; Hoffmann, F.; Hector, A.; Eber, E.; Marcos, V.; Bittmann, I.; Eickelberg, O.; Griese, M.; Roos, D. *J. Immunol.* **2008**, *181*, 8053–8067.
13. Hornung, D.; Ryan, I. P.; Chao, V. A.; Vigne, J. L.; Schriock, E. D.; Taylor, R. N. *J. Clin. Endocrinol. Metab.* **1997**, *82*, 1621–1628.
14. Miyazaki, D.; Nakamura, T.; Toda, M.; Cheung-Chau, K. W.; Richardson, R. M.; Ono, S. J. *J. Clin. Invest.* **2005**, *115*, 434–442.
15. He, M.; Horuk, R.; Moochhala, S. M.; Bhatia, M. *Am. J. Physiol. Gastrointest. Liver Physiol.* **2007**, *292*, G1173–G1180.
16. Ninichuk, V.; Anders, H. J. *Am. J. Nephrol.* **2005**, *25*, 365–375.
17. Jordan, N. J.; Watson, M. L.; Williams, R. J.; Roach, A. G.; Yoshimura, T.; Westwick, J. *Br. J. Pharmacol.* **1997**, *122*, 749–757.
18. Halks-Miller, M.; Schroeder, M. L.; Haroutunian, V.; Moenning, U.; Rossi, M.; Achim, C.; Purohit, D.; Mahmoudi, M.; Horuk, R. *Ann. Neurol.* **2003**, *54*, 638–646.
19. Gao, W.; Topham, P. S.; King, J. A.; Smiley, S. T.; Csizmadia, V.; Lu, B.; Gerard, C. J.; Hancock, W. W. *J. Clin. Invest.* **2000**, *105*, 35–44.
20. Terpos, E.; Politou, M.; Szydlo, R.; Goldman, J. M.; Apperley, J. F.; Rahemtulla, A. *Br. J. Haematol.* **2003**, *123*, 106–109.
21. Vallet, S.; Raje, N.; Ishitsuka, K.; Hideshima, T.; Podar, K.; Chhetri, S.; Pozzi, S.; Breitkreutz, I.; Kiziltepe, T.; Yasui, H.; Ocio, E. M.; Shiraishi, N.; Jin, J.; Okawa, Y.; Ikeda, H.; Mukherjee, S.; Vaghela, N.; Cirstea, D.; Ladetto, M.; Boccadoro, M.; Anderson, K. C. *Blood* **2007**, *110*, 3744–3752.
22. Dairaghi, D. J.; Oyajobi, B. O.; Gupta, A.; McCluskey, B.; Miao, S.; Powers, J. P.; Seitz, L. C.; Wang, Y.; Zeng, Y.; Zhang, P.; Schall, T. J.; Jaen, J. C. *Blood* **2012**, *120*, 1449.
23. Carson, K. G.; Jaffee, B. D.; Harriman, G. C. B. *Ann. Rep. Med. Chem.* **2004**, *39*, 149–158.
24. Brown, M. F.; Avery, M.; Kath, J. C.; Brissette, W. H.; Chang, J. H.; Colizza, K.; Conklyn, M.; DiRico, A. P.; Gladue, R. P.; Kath, J. C.; Krueger, S. S.; Lira, P. D.; Lillie, B. M.; Lundquist, G. D.; Mairs, E. N.; McElroy, E. B.; McGlynn, M. A.; Paradis, T. J.; Poss, C. S.; Rossulek, M. I.; Shepard, R. M.; Sims, J.;

Strelevitz, T. J.; Truesdell, S.; Tylaska, L. A.; Yoon, K.; Zheng, D. *Bioorg. Med. Chem. Lett.* **2004**, *14*, 2175–2179.

25. Li, B.; Andresen, B.; Brown, M. F.; Buzon, R. A.; Chiu, C. K. F.; Couturier, M.; Dias, E.; Urban, F. J.; Jasys, V. J.; Kath, J. C.; Kissel, W.; Le, T.; Li, Z. J.; Negri, J.; Poss, C. S.; Tucker, J.; Whiitenour, D.; Zandi, K. *Org. Process Res. Dev.* **2005**, *9*, 466–471.

26. Gladue, R. P.; Tylaska, L. A.; Brissette, W. H.; Lira, P. D.; Kath, J. C.; Poss, C. S.; Brown, M. F.; Paradis, T. J.; Conklyn, M. J.; Ogborne, K. T.; McGlynn, M. A.; Lillie, B. M.; DiRico, A. P.; Mairs, E. N.; McElroy, E. B.; Martin, W. H.; Stock, I. A.; Shepard, R. M.; Showell, H. J.; Neote, K. *J. Biol. Chem.* **2003**, *278*, 40473–40480.

27. Gladue, R. P.; Cole, S. H.; Roach, M. L.; Tylaska, L. A.; Nelson, R. T.; Shepard, R. M.; McNeish, J. D.; Ogborne, K. T.; Neote, K. S. *J. Immunol.* **2006**, *176*, 3141–3148.

28. Cavallaro, C. L.; Briceno, S.; Chen, J.; Cvijic, M. E.; Davies, P.; Hynes, J., Jr.; Liu, R. Q.; Mandlekar, S.; Rose, A. V.; Tebben, A. J.; Kirk, K. V.; Watson, A.; Wu, H.; Yang, G.; Carter, P. H. *J. Med. Chem.* **2012**, *55*, 9643–9653.

29. Kerstjens, H. A.; Bjermer, L.; Eriksson, L.; Dahlstrom, K.; Vestbo, J. *Respir. Med.* **2010**, *104*, 1297–1303.

30. Liang, M.; Mallari, C.; Rosser, M.; Ng, H. P.; May, K.; Monahan, S.; Bauman, J. G.; Islam, I.; Ghannam, A.; Buckman, B.; Shaw, K.; Wei, G. P.; Xu, W.; Zhao, Z.; Ho, E.; Shen, J.; Oanh, H.; Subramanyam, B.; Vergona, R.; Taub, D.; Dunning, L.; Harvey, S.; Snider, R. M.; Hesselgesser, J.; Morrissey, M. M.; Perez, H. D.; Horuk, R. *J. Biol. Chem.* **2000**, *275*, 19000–19008.

31. Horuk, R. *Mini Rev. Med. Chem.* **2005**, *5*, 791–804.

32. Belecki, K.; Berliner, M.; Bibart, R. T.; Meltz, C.; Ng, K.; Phillips, J.; Ripin, D. H. B.; Vetelino, M. *Org. Process Res. Dev.* **2007**, *11*, 754–761.

33. Berliner, M. A.; Hayward, M. M.; Li, Z. J.; Meltz, C. N.; Ng, K. K. Patent Application WO 2004/113,311, 2004.

34. Xie, Y. F.; Lake, K.; Ligsay, K.; Komandla, M.; Sircar, I.; Nagarajan, G.; Li, J.; Xu, K.; Parise, J.; Schneider, L.; Huang, D.; Liu, J.; Dines, K.; Sakurai, N.; Barbosa, M.; Jack, R. *Bioorg. Med. Chem. Lett.* **2007**, *17*, 3367–3372.

35. Revesz, L.; Bollbuck, B.; Buhl, T.; Eder, J.; Esser, R.; Feifel, R.; Heng, R.; Hiestand, P.; Jachez-Demange, B.; Loetscher, P.; Sparrer, H.; Schlapbach, A.; Waelchli, R. *Bioorg. Med. Chem. Lett.* **2005**, *15*, 5160–5164.

36. Wellner, E.; Sandin, H. Patent Application WO 2005/080,362, 2005.

37. Wellner, E.; Sandin, H. Patent Application WO 2005/080,336, 2005.

38. Neves, C.; Chevalier, A.; Billot, P. Patent Application WO 2006/066,200, 2006.

39. Lu, C.; Balani, S. K.; Qian, M. G.; Prakash, S. R.; Ducray, P. S.; von Moltke, L. L. *J. Pharmacol. Exp. Ther.* **2010**, *332*, 562–568.

40. Carson, K. G.; Harriman, G. C. B. Patent Application WO 2004/043,965, 2004.

41. Pennell, A. M. K.; Aggen, J. B.; Sen, S.; Chen, W.; Xu, Y.; Sullivan, E.; Li, L.; Greenman, K.; Charvat, T.; Hansen, D.; Dairaghi, D. J.; Wright, J. J. K.; Zhang, P. *Bioorg. Med. Chem. Lett.* **2013**, *23*, 1228–1231.

42. Zhang, P.; Pennell, A. M. K.; Aggen, J. B.; Chen, W.; Li, L.; Xu, Y.; Li, Y.; Leleti, M.; Hansen, D.; Dairaghi, D. J.; Wright, J. J. K. In 33rd National Medicinal Chemistry Symposium, Tucson, AZ, 2012, Abstract 55.

43. Dairaghi, D. J.; Zhang, P.; Wang, Y.; Seitz, L. C.; Johnson, D. A.; Miao, S.; Ertl, L. S.; Zeng, Y.; Powers, J. P.; Pennell, A. M.; Bekker, P.; Schall, T. J.; Jaen, J. C. *Clin. Pharmacol. Ther.* **2011**, *89*, 726–734.

44. http://www.clinicaltrials.gov/ct2/show/NCT01404585.

45. Santella, J. B. Patent Application WO 2009/158,452, 2009.

46. Hynes, J.; Carter, P. H.; Cornelius, L. A. M.; Dhar, T. G. M.; Duncia, J. V.; Nair, S.; Santella, J. B.; Warrier, J.; Wu, H. Patent Application WO 2011/044,309, 2011.

47. Hossain, N.; Ivanova, S.; Bergare, J.; Eriksson, T. *Bioorg. Med. Chem. Lett.* **2013**, *23*, 1883–1886.

48. Merritt, J. R.; Liu, J.; Quadros, E.; Morris, M. L.; Liu, R.; Zhang, R.; Jacob, B.; Postelnek, J.; Hicks, C. M.; Chen, W.; Kimble, E. F.; Rogers, W. L.; O'Brien, L.; White, N.; Desai, H.; Bansal, S.; King, G.; Ohlmeyer, M. J.; Appell, K. C.; Webb, M. L. *J. Med. Chem.* **2009**, *52*, 1295.

49. Merritt, J. R.; James, R.; Paradkar, V. M.; Zhang, C.; Liu, R.; Liu, J.; Jacob, B.; Chiriac, C.; Ohlmeyer, M. J.; Quadros, E.; Wines, P.; Postelnek, J.; Hicks, C. M.; Chen, W.; Kimble, E. F.; O'Brien, L.; White, N.; Desai, H.; Appell, K. C.; Webb, M. L. *Bioorg. Med. Chem. Lett.* **2010**, *20*, 5477–5479.

50. http://www.secinfo.com/d11MXs.u2gp4.b.html.

51. Cook, B. N.; Harcken, C.; Lee, T. W.; Liu, P.; Mao, C.; Lord, J.; Mao, W.; Raudenbush, B. C.; Razavi, H.; Sarko, C.; Swinamer, A. D. Patent Application WO 2009/137338, 2009.

52. Disalvo, D.; Kuzmich, D.; Mao, C.; Razavi, H.; Sarko, C.; Swinamer, A. D.; Thomson, D.; Zhang, Q. Patent Application WO 2009/134,666, 2009.

53. Cook, B. N.; Disalvo, D.; Fandrick, D. R.; Harcken, C.; Kuzmich, D.; Lee, T. W.; Liu, P.; Lord, J.; Mao, C.; Neu, J.; Raudenbush, B. C.; Razavi, H.; Reeves, J. T.; Song, J.; Swinamer, A. D.; Tan, Z. Patent Application WO 2010/036,632, 2010.

54. Cook, B. N.; Kuzmich, D.; Mao, C.; Razavi, H. Patent Application WO 2011/049,917, 2011.

55. Cook, B.N.; Kuzmich, D. Patent Application WO 2011/056,440, 2011.

56. Huber, J. D. Patent Application WO 2011/137,109, 2011.

57. Betageri, R.; Cook, B. N.; Disalvo, D.; Harcken, C.; Kuzmich, D.; Liu, P.; Lord, J.; Mao, C.; Razavi, H. Patent Application WO 2012/087,782, 2012.

58. http://www.merck.com/finance/annualreport/ar2004/research_pipline.

59. http://library.corporate-ir.net/library/80/801/80159/items/297170/MLNM_AR2007.pdf.

60. Zipp, F.; Hartung, H. P.; Hillert, J.; Schimrigk, S.; Trebst, C.; Stangel, M.; Infante-Duarte, C.; Jakobs, P.; Wolf, C.; Sandbrink, R.; Pohl, C.; Filippi, M. *Neurology* **2006**, *67*, 1880–1883.

61. Pease, J.; Horuk, R. *J. Med. Chem.* **2012**, *55*, 9363–9392.

62. Mo, J. A.; Jansson, A. H.; Morin, E.; Smailagic, A.; Nelinder, A.; Walles, K.; Andersson, A. M. D.; Kvist-Reimer, M.; Stenvall, K.; Strandberg, P.; Vissing, H.; Eriksson, T.; Hansson, J. In 19th European Respiratory Society Annual Congress, Vienna, Austria, 2009, Abstract 3243.

63. Clucas, A.; Shah, A.; Zhang, Y. D.; Chow, V. F.; Gladue, R. P. *Clin. Pharmacokinet.* **2007**, *46*, 757–766.

64. Haringman, J. J.; Kraan, M. C.; Smeets, T. J. M.; Zwinderman, K. H.; Tak, P. P. *Ann. Rheum. Dis.* **2003**, *62*, 715–721.

65. Vergunst, C. E.; Gerlag, D. M.; von Moltke, L.; Karol, M.; Wyant, T.; Chi, X.; Matzkin, E.; Leach, T.; Tak, P. P. *Arthritis Rheum.* **2009**, *60*, 3572–3581.

66. Tak, P. P.; Balanescu, A.; Tseluyko, V.; Bojin, S.; Drescher, E.; Dairaghi, D.; Miao, S.; Marchesin, V.; Jaen, J.; Schall, T. J.; Bekker, P. *Ann. Rheum. Dis.* **2013**, *72*, 337–344.

67. https://www.anzctr.org.au/Trial/Registration/TrialReview.aspx?ACTRN=12609000788279.

68. Schall, T. J.; Proudfoot, A. E. I. *Nat. Rev. Immunol.* **2011**, *11*, 355–363.

69. Lebre, M. C.; Vergunst, C. E.; Choi, I. Y. K.; Aarrass, S.; Oliveira, A. S. F.; Wyant, T.; Horuk, R.; Reedquist, K. A.; Tak, P. P. *PLoS One* **2011**, *6*, 1–7.

CHAPTER ELEVEN

Emerging Targets for the Treatment of Idiopathic Pulmonary Fibrosis

Matthew C. Lucas*, David C. Budd†
*Cubist Pharmaceuticals, Lexington, Massachusetts, USA
†GlaxoSmithKline, Medicines Research Centre, Stevenage, Herts, UK

Contents

1. INTRODUCTION

1.1. Idiopathic pulmonary fibrosis (IPF)

Idiopathic pulmonary fibrosis (IPF) is a chronic, progressive lung disease characterized by scarring of the lung interstitium that likely occurs because of repetitive wounding of the alveolar epithelium followed by incomplete

Annual Reports in Medicinal Chemistry, Volume 48
ISSN 0065-7743
http://dx.doi.org/10.1016/B978-0-12-417150-3.00011-9
149

healing of the injury, although the exact etiology of the disease is unknown. IPF represents the severest form of fibrosing interstitial lung diseases with progressive loss of lung function and a survival rate of only approximately 50% five years after initial diagnosis.[1] The annual incidence of IPF in the United States has been estimated to be 6.8–8.8 per 100,000 population using narrow case definitions and 16.3–17.4 per 100,000 population based on broader case definitions.[2] Alarmingly, some IPF patients present with an exceptionally aggressive disease and exhibit extremely rapid disease progression resulting in death within 12 months of diagnosis.[3] Acute exacerbations of IPF, characterized by rapid episodic declines in lung function, are also a feature of this disease and predict extremely poor outcomes.[4] The factors that drive these acute episodes are currently unknown.[5]

The mechanisms driving interstitial lung fibrosis in IPF are the subject of intensive research. The currently accepted dogma is that repetitive injury to the lung alveolar epithelial cell surface and ineffectual restitution of this cellular population sets in train a sequence of events leading to accumulation and activation of interstitial lung fibroblasts. The outcome of this is excessive production of extracellular matrix (ECM) that may be aberrantly remodeled, leading to tissue scarring, reduction in tissue compliance, respiratory failure, and death.[6] Controversy exists as to the mechanisms leading to accumulation of interstitial lung fibroblasts and the role of inflammation in the disease. Recent studies have suggested that bone marrow-derived cells such as fibrocytes may be attracted to the injured lung via the bloodstream and contribute directly or indirectly to the pool of interstitial lung fibroblasts in the IPF lung.[7] In addition, the minimal efficacy of corticosteroids has lead to the belief that the disease may proceed in the absence of significant inflammation,[8] although cells of the innate immune system such as alveolar macrophages may act to orchestrate the fibrotic process within the IPF lung.[9,10]

1.2. Current treatment options

IPF is a disease of high, unmet medical need due to an exceptionally poor prognosis combined with a lack of effective treatment options capable of altering its progression. Recommended therapeutic regimens for the treatment of IPF continue to evolve, with evidence-based treatments varying from patient to patient. The currently preferred treatment options for acute exacerbations in IPF include limited pharmacological-based therapies (corticosteroids) and non-pharmacological-based approaches (long-term oxygen therapy and lung transplantation).[11] The strategy for developing pharmacological therapies that are specifically aimed at attenuating the fibrotic response gained acceptance, while the development of anti-inflammatory approaches has fallen out of favor in the last decade. Pirfenidone (5-methyl-1-phenyl-2-[1*H*]-pyridone), a small molecule with reported antifibrotic properties but with minimal understanding of the mechanism of action (although at least some of its antifibrotic properties are reported to be due to P38 inhibition),[12] is currently approved for the treatment of IPF in the European Union and Japan.[13] Pirfenidone has not yet been approved in the United States. There were inconclusive findings on the mean change from baseline in percentage-predicted forced vital capacity in two separate Phase III studies conducted in North America.[14,15] Thus, there is clearly an urgent need for the development of novel pharmacological agents capable of providing significant efficacy while modifying the natural course of the disease. In this review, we will describe emerging targets for the treatment of IPF and progress in the development of small-molecule modulators against them. Targets for which there is limited pharmacologic validation or against which only biologic modalities have been reported have been largely excluded.

2. G PROTEIN-COUPLED RECEPTORS (GPCRs)

2.1. CXCR4

The chemokine CXCL12/stromal cell–derived factor-1 agonizes two closely related 7-transmembrane receptors, CXCR4 and CXCR7, which are coexpressed in certain cell types.[16] The biology of the CXCR4/CXCR7/CXCL12 pathway is complex with potential for heterodimerization between the receptor pairs or reciprocal regulation of receptor activity.[15,16] CXCR4 has been widely recognized to have an important role in the retention of hematopoietic stem cells (HSCs) in the bone marrow.[17]

A role for this pathway in driving IPF and other interstitial lung fibrosis indications is supported by evidence pointing toward dysregulated expression of CXCL12. The levels of CXCL12 are elevated in bronchoalveolar lavage fluid and in the peripheral circulation of bleomycin-treated mice.[18,19] Similarly, the expression of CXCL12 has been demonstrated to be elevated in lung tissue from patients with usual interstitial pneumonia and fibrotic nonspecific interstitial pneumonia.[18] The balance of experimental evidence also suggests that CXCL12, acting through the CXCR4 receptor, may be involved in driving the pathophysiology of lung fibrosis in IPF and experimental models of pulmonary fibrosis. For instance, the isolation and characterization of human fibrocytes (CD45(+), collagen I(+)) have demonstrated cell surface expression of CXCR4 and induction of migration *in vitro* following application of CXCL12.[19] Even more relevant to IPF, injection of human fibrocytes into the tail vein of SCID mice following intratracheal bleomycin challenge resulted in the migration and incorporation of these cells into the fibrotic lung of these animals, and this response could be attenuated with neutralizing anti-CXCL12 antibodies.[19] Studies with small-molecule CXCR antagonists such as plerixafor (AMD3100), **1**, are consistent with a role for CXCR4 in driving interstitial lung fibrosis and support its potential as a therapeutic target. Compound **1** attenuates lung injury following bleomycin challenge in rodents, possibly via suppression of fibrocyte recruitment from bone marrow to the injured lung.[20] This latter study is at odds with the well-documented function of CXCR4 in the retention of progenitor cells within the bone marrow since systemic CXCR4 antagonists might be expected to stimulate bone marrow egress of progenitor cells, including fibrocytes, into the peripheral circulation.[17] Thus, **1** likely attenuates the extravasation of fibrocytes from the peripheral blood into injured lung but would be expected to stimulate the release of these cells from the bone marrow.

Compound **1** was first identified as an inhibitor of HIV replication.[21] Further work exploring the antiviral mechanism of **1** uncovered an essential role for CXCR4 in permitting viral entry into cells and identified this molecule as a potent CXCR4 receptor antagonist.[22] Subsequently, **1** was developed to stimulate mobilization of HSCs from the bone marrow to the bloodstream following short-term drug administration.[17] Compound **1** has been approved by the FDA as the first small-molecule CXCR4 antagonist for use in combination with granulocyte-colony stimulating factor to mobilize HSCs to the bloodstream for collection and subsequent autologous transplantation in patients with non-Hodgkin's lymphoma and multiple myeloma. Interestingly, **1** has also been shown to agonize the CXCR7 receptor.[23] It has been demonstrated that activation of the CXCR7 receptor can attenuate the expression and activity of CXCR4.[24] The identification of compounds with differential CXCR4/CXCR7 selectivity profiles may help elucidate whether CXCR4 antagonism alone is sufficient to provide efficacy in experimental models of pulmonary fibrosis and IPF. More importantly, chronic administration of CXCR4 antagonists for IPF is likely to be complicated by dose-limiting cardiotoxicities which have been observed in clinical studies with small-molecule antagonists containing the bicyclam core. Efforts to identify noncyclam containing CXCR4 antagonists have led to the development of quinolines, guanidines, indoles, cyclic pentapeptides, p-xylyl-enediamines, and pyrimidine-based CXCR4 antagonists.[25]

A structure-based approach using the phenylenebis(methylene) bridged bicyclam 1 and TN14003, a peptide-based CXCR4 antagonist, combined with iterative pharmacological testing, led to the identification of series of dipyrimidine amine-based CXCR4 inhibitors. The affinity for binding to the CXCR4 receptor is dependent on the central aromatic ring and the distance between it and the nitrogens of the acyclic linker. These studies led to the identification of MSX–122, 2.[26]

Using published crystallographic structures of CXCR4 bound to the small-molecule inhibitor IT1t,[27] an in silico docking approach demonstrated that 2 was capable of interacting with the CXCR4 receptor at regions distinct from the CXCL12 binding site, thus possibly antagonizing the receptor without competing against the natural ligand.[28] In support of the in silico docking studies, a low concentration of MSX-122 (10 nM) was capable of antagonizing inhibition of cAMP production induced by high concentrations of CXCL12 (150 ng/mL) whereas 1, a CXCR4 antagonist that competes with CXCL12 for binding to the receptor, was considerably less potent.[28] In line with allosteric antagonism of CXCR4 by 2, the compound was not capable of blocking all CXCR4-mediated activities including the inhibition of T-tropic HIV infection and attenuation of calcium flux induction following CXCL12 activation, which is distinct to competitive antagonists like 1. Still, 2 was found to be highly efficacious in the bleomycin model of lung fibrosis. The authors have postulated that its novel mode of antagonism provided efficacy by inhibiting the extravasation of cells from peripheral blood to the lung while not mobilizing pathological cell populations such as fibrocytes from the bone marrow.[28]

2.2. RXFP1

Relaxin (H2 Relaxin, Relaxin-2), a 6-kDa peptidic hormone, exerts its biological effects through activation of the structurally related G-protein-coupled receptors RXFP1 and RXFP2.[29,30] Cell-based studies using recombinant relaxin have demonstrated that relaxin exerts antifibrotic effects directly on lung fibroblasts by inhibiting cell contractility through inhibition of the RhoA/Rho kinase pathway. It has been demonstrated *in vivo* that exogenously administered relaxin inhibits the enhanced contractility observed in the lungs of bleomycin-treated animals by a similar mechanism.[31] Clinical studies have also been conducted to examine the effect of subcutaneous infusion of recombinant relaxin.[32] Small-molecule agonists **3** and **4**, each containing a 2-acetamido–*N*-phenylbenzamide core, were identified during a high throughput screen utilizing a homogeneous time-resolved fluorescence assay of cAMP production by HEK293T cells stably transfected with the human RXFP1 gene.[33] Both **3** and **4** exhibited micromolar potency and reasonable efficacy (80%) in driving cAMP generation.[33]

Selectivity for the RXFP1 receptor is supported by the observation that both compounds failed to induce cAMP in native (i.e., non–RXFP1 expressing) HEK293T cells or HEK293T cells expressing the RXFP2 or vasopressin 1b receptors.[33] The identification of small molecules possessing RXFP1 receptor agonist activity is remarkable given the large size of the natural peptide ligand (6 kDa) and the identification of multiple ligand–receptor interaction sites necessary for full activation of RXFP1 receptor activity.[34,35] It has yet to be demonstrated that this class of small molecules possesses antifibrotic activity in assays using cells of pathophysiological relevance to IPF nor that they are efficacious in animal models of disease, as has been shown with synthetic peptide agonists.[36]

3. EXTRACELLULAR CROSSLINKING ENZYMES

For many decades, the ECM was considered a passive, structural component of tissues. There is now, however, accumulating evidence that the constituents and structural features of the ECM dictate the phenotype of cells and tissues to a remarkable extent. In fact, many of the abnormal features of IPF lung fibroblasts observed in comparative studies against lung fibroblasts from normal individuals, such as differences in proliferation rates and contractility, can be attenuated through alterations in the rigidity of the

ECM.[37] In the case of the lung, where the work of breathing is severely impaired by reduced tissue pliability, changes in ECM rigidity should also exhibit a significant influence on organ function.[38] Consistent with a growing appreciation for the role of the ECM in connective tissue diseases, a number of enzymes involved in crosslinking of ECM components which increase the rigidity of the matrix are being pursued as therapeutic targets in IPF and other connective tissue diseases.

3.1. Transglutaminase 2 (TG2)

There has been progress in the development of antibodies to ECM modifying enzymes, attractive targets due to their localization in the interstitial space.[39] The development of small-molecule inhibitors of a matrix modifying enzyme, transglutaminase 2 (TG2), has received much attention.[40] TG2 is an intracellular and extracellular localized transamidase of which there are eight family members and which catalyzes the post-translational modification of proteins by the formation of isopeptide bonds.[40] This occurs either through protein crosslinking via epsilon-(gamma-glutamyl)lysine bonds or through incorporation of primary amines at selected peptide-bound glutamine residues, ECM crosslinking activities which have led TG2 to be dubbed "nature's biological glue."[41] Recent preclinical studies have highlighted a role for TG2 in mediating the migration and contraction of human lung fibroblasts and the induction of ECM components.[42] TG2 levels are elevated in the lungs of animals following bleomycin challenge and TG2 appears to be upregulated in normal lung fibroblasts following transforming growth factor-β1 (TGFβ1) stimulation.[42,43] TG2 levels and the isopeptide bonds which it generates are elevated in the lungs of IPF patients.[42] Since intracellular TG2 has also been associated with diverse biological processes including controlling the susceptibility of cells to undergo cell death,[44] attention has shifted toward the development of small molecules which lack appreciable cell permeability, thereby allowing for selective inhibition of extracellular TG2 activity.

To achieve non-cell permeable TG2 inhibitors, efforts have been directed at increasing the water solubility of a series of sulfonium peptidylmethylketones **5**.[45]

5
R: Cbz > Fmoc > Boc
A: Pro = Gly = Phe > Ala = Asp > Lys > Ile

Modification of the amino acid core (A) in addition to changes in the carbamate moiety at the R position altered the potency of inhibition of TG2. Analogous to previous reports of irreversible inhibition of proteases by sulfonium peptidyl methylketones, these compounds irreversibly inhibited TG2.[45,46] Consistent with a lack of cell permeability, a member of this series (A = phenylalanine, R = Cbz) was not cytotoxic.[46] More interestingly, this molecule has also been assessed in an animal model of renal scarring and has demonstrated marked reduction in organ scarring and significant protection of kidney function.[47] This underscores the potential of this approach in treating chronic wound healing conditions after systemic administration. This molecule was not assessed in animal models of pulmonary fibrosis but might reasonably be expected to exhibit efficacy given the aforementioned study.

4. KINASE INHIBITORS

Several kinases, including focal adhesion kinase (FAK), activin receptor-like kinase 5 (ALK5), homeodomain-interacting protein kinase 2, MEK1/2, Rho-associated protein kinase, and phosphoinositol 3 kinase (PI3K), have been implicated in fibrotic diseases such as IPF. Herein, we highlight FAK and PI3K, because although these kinases have been the subject of medicinal chemistry efforts at disease intervention for several years, it is only recently that fibrosis has emerged as a potential therapeutic indication.

4.1. Focal adhesion kinase (FAK) inhibitors

FAK is a nonreceptor tyrosine kinase that was identified in the early 1990s.[48] FAK is ubiquitously expressed and could be considered a signaling node due to its integrating signals from numerous signaling pathways, making it a potentially important target for pharmacological intervention. FAK is implicated in cancer cell invasion, metastasis, and survival underscoring the pathophysiological importance of this kinase. As such, the majority of the literature surrounding FAK is related to its potential role in cancer.[49] Recent evidence supporting a significant role for FAK in fibrotic diseases has emerged.[50] As described above, IPF may be triggered by repetitive microinjury to the alveolar epithelial surface of the lung, defective repair, and ultimately the formation of fibrotic lesions. Myofibroblasts are considered key pathogenic mediators of lung fibrosis because they generate the excessive production of ECM within the lung interstitial space. FAK contributes to the synthesis of the ECM by inducing fibroblast to myofibroblast transition through transduction of signals from profibrotic growth factors such as TGFβ. Both pharmacological and genetic inactivation of FAK have been reported to attenuate lung fibrosis in the bleomycin–induced mouse model of fibrosis. Lagares and colleagues also found that FAK expression and activity were upregulated in tissue obtained from lung fibrosis patients.[51]

At least three small-molecule ATP-competitive FAK inhibitors have entered clinical trials for a variety of oncology indications and have been reviewed.[52] All are closely related in structure and based on an amino-pyrimidine hinge-binding core (e.g., PF-573,228, **6**).[53] The claimed genuses of many recent patent applications closely resemble these scaffolds. Scaffolds with greater diversity that offer alternative binding modes have also emerged. For example, compound **7** belongs to a pyrrolo[2,3-*b*]-pyridine class of inhibitors which was identified through a process the authors describe as a "knowledge based fragment growing approach." Kinetic characterization of the inhibitors determined that slow off-rates were required in order to achieve good cellular activity. However, the authors reported difficulty in obtaining good kinase selectivity which may limit the utility of these molecules as drugs.[54] Allosteric inhibitors of FAK based on a 1,5-dihydropyrazolo[4,3-*c*][2,1]benzothiazine scaffold (e.g., **8**) were also recently described.[55] Compound **8** shows slow dissociation from the unphosphorylated (inactive) form of the enzyme and was found to not be ATP competitive, binding to a previously unobserved site. The potency for allosteric inhibition was reduced by phosphorylation of FAK, and allosteric binding with accompanying selectivity over other kinases was only realized when the fused pyrazole was *N*-alkylated. As neither of these new series appear to have been assessed in animal models of fibrosis, they offer potential new starting points for further investigation of the role of FAK in pulmonary fibrosis.

4.2. Phosphatidylinositol 3-kinase (PI3K) inhibitors

PI3Ks play important roles in a variety of inflammatory signaling pathways. They are lipid kinases that convert the second-messenger phospho-inositol(4,5)-bisphosphate (PIP2) to phosphoinositol(3,4,5)-trisphosphate

(PIP3). This in turn leads to activation of Akt and amplification of downstream signaling events that lead to increased cell proliferation, growth, and migration. There are three classes of PI3K, defined based on their primary structure, mode of regulation, and *in vitro* substrate specificity. Class I PI3Ks have been extensively studied and consist of four subtypes of high homology (PI3Kα, PI3Kβ, PI3Kγ, PI3Kδ).[56] Several PI3K inhibitors have entered clinical trials in oncology.[57] PI3Kα and PI3Kβ are ubiquitously expressed, while PI3Kγ and PI3Kδ were initially thought to be primarily expressed in leukocytes. The limited expression of PI3Kγ and PI3Kδ made inhibitors of these subtypes widely sought after as safer treatments for a number of indications, including inflammatory diseases. While anti-inflammatory therapies have fallen short in their ability to treat IPF patients effectively (*vide supra*), there is emerging evidence to suggest there might be an additional role for PI3Ks. During the progression of IPF, fibrocytes are trafficked to the lung where they may contribute to the pool of lung resident myofibroblasts responsible for the deposition of ECM. Recently, PI3Kγ and PI3Kδ were found to be expressed in human lung fibroblasts as well as in leukocytes, suggesting that inhibitors of these isoforms might have utility in retarding progression of IPF.[58] Injury by bleomycin induces fibrosis in human cells, and the activation of the PI3K/Akt pathway is implicated in the fibroproliferative and collagen-inducing effects of bleomycin. The authors disclosed that the pan-PI3K inhibitors LY294002 or wortmannin were able to completely inhibit the activation of fibroblasts.[59] In addition, Lamouille and colleagues showed that the PI3K/Akt pathway plays a role in the TGFβ-induced epithelial–mesenchymal transition (EMT).[60] Use of selective inhibitors of each of the PI3K isoforms as well as a pan-PI3K inhibitor revealed that each individual isoform retarded the proliferative effect of TGFβ. Although only the pan-PI3K inhibitor LY294002 completely suppressed proliferation, the PI3Kγ and PI3Kα isoforms appeared to play a highly significant role. This implies that PI3K isoforms may act in a complementary fashion, which will influence the strategic approach taken when seeking therapeutic PI3K inhibitors to treat IPF. In another study, Russo *et al.* reported an important role for the PI3Kγ isoform in experimental pulmonary fibrosis, with genetic knockout mice showing greater survival, reduced weight loss, and less fibrosis than wildtype counterparts after bleomycin challenge.[61] Thus, specific PI3K isoforms have emerged as pharmacological targets for the treatment of IPF, although whether selective PI3K isoform inhibition will be sufficient to demonstrate efficacy in human IPF patients requires further research.

New PI3K inhibitor scaffolds have also recently emerged in the literature, including pan–PI3K inhibitors such as **9**.[62] As alluded to above, pan inhibitors might be required to reach full efficacy, but these inhibitors might not show sufficient therapeutic windows for the chronic treatment of IPF. The availability of PI3K–isoform–specific inhibitors with acceptable ADMET properties will help determine the utility of specific isoform inhibitors to treat IPF. Noteworthy, recent additions to the portfolio of tools available to face these important challenges include compounds such as the PI3Kα selective, targeted covalent inhibitor **10**.[63] The PI3Kα enzyme was potently and specifically inhibited by targeting an isoform-specific cysteine residue. Interestingly, this irreversible α-selective molecule is closely related to a scaffold that has most recently delivered PI3Kδ selective, reversible inhibitors (e.g., compound **11**).[64] Murray and colleagues reported that this δ isoform selective scaffold derives its selectivity primarily through strong interactions with a specific tryptophan residue in the PI3Kδ binding pocket. As well as single isoform–specific PI3K inhibitors, specific dual inhibitors have also recently emerged. These include PI3Kβ/δ inhibitors such as compound **12**, which showed efficacy in animal models of inflammation,[65] and PI3Kγ/δ inhibitors, which showed acute activity in an asthma model.[66] While none of the compounds described here were investigated in models of pulmonary fibrosis, they may enable some of these studies in the near future.

9

10

11

12

5. ENZYMES

TGFβ1 is a cytokine that plays a critical role in IPF.[67] TGFβ is involved in the transition of fibroblasts into myofibroblasts. Inhibiting the TGFβ pathway might therefore lead to improved outcomes in IPF patients. However, within this approach lie inherent risks, because TGFβ is expressed in so many cell types and tightly controls cell proliferation, differentiation, and immunity. Indeed, recent studies have demonstrated that blocking this pathway through ALK5 kinase inhibitors leads to efficacy in animal models of pulmonary fibrosis but also results in significant dose–limiting cardiotoxicities.[68,69] Two enzyme targets have emerged recently, the inhibition of which might effect a reduction in TGFβ-triggered fibrosis while maintaining a sufficient therapeutic window.

5.1. NADPH oxidase 4 (NOX4) inhibitors

NADPH oxidase (NOX) enzymes are membrane-bound proteins that catalyze the conversion of oxygen to superoxide. There are a number of isoforms, NOX1-5 and dual oxidases 1 and 2 (DUOX1-2), which have diverse functions. The discovery that NOX isoforms are expressed in pulmonary cells including fibroblasts has led to the investigation of their role in the pathology of pulmonary fibrosis. The NOX4 isoform, in particular, seems to be the most relevant to IPF.[70] In lung fibroblasts, TGFβ1 increases reactive oxygen species by upregulating NOX4 expression, leading to the differentiation of this cell type to synthetic myofibroblasts. NOX4 has been reported to be highly expressed in pulmonary fibroblasts obtained from patients with IPF. siRNA-induced gene knockdown of NOX4 in these IPF patient fibroblasts as well as in control cells inhibited myofibroblast differentiation.[71] In addition, NOX4 is highly expressed in alveolar epithelial cells obtained from IPF patients, and NOX4 knockout mice are refractory to bleomycin-induced lung injury.[72]

The small-molecule GKT-137831, **13**, a dual inhibitor of NOX1 and NOX4, is the first NOX inhibitor to have reached the clinic. Compound **13** has been evaluated as a treatment of diabetic nephropathy, but may have application in a broad range of indications, including IPF. The discovery of **13** and structure–activity relationships for the scaffold have been described in detail.[73] Several new, albeit structurally related, pyrazolopyrido-diazapine, -pyrazine, and -oxazine dione derivatives were recently reported by the same research group. The authors determined that the pyrazolopyrido-diazapines showed the best selectivity for NOX4 and NOX1. Notably, compound **14** was highly efficacious in the mouse model of bleomycin-induced pulmonary fibrosis and showed significant superiority over the benchmark compound, pirfenidone.[74]

13   14

5.2. Galectin-3 inhibitors

Galectins are β-galactoside-binding proteins and members of a subfamily of the lectin family of carbohydrate-binding proteins. There are 15 known mammalian galectins that participate in a range of biological processes. This has led to significant interest in their relevance to various diseases, including fibrosis. Galectin-3 is highly expressed in various fibrotic tissues, including fibrotic lung tissue. The link between galectin-3 and IPF was first demonstrated when bronchoalveolar lavage fluid obtained from IPF patients was shown to exhibit increased levels of galectin-3 compared to that obtained from control subjects.[75] More recently, high levels of galectin-3 in lung biopsy specimens and in serum from IPF patients have been reported.[76] Galectin-3$^{-/-}$ fibroblasts were also reportedly protected from collagen synthesis with reduced differentiation to myofibroblasts. Similarly, galectin-3$^{-/-}$ airway epithelial cells showed reduced EMT. Less lung fibrosis developed in galectin-3 knockout mice challenged with adenoviral TGFβ1 delivery to the lung. The authors also reported that TDI139, **15**, a high affinity, selective inhibitor of the galectin-1 and galectin-3 carbohydrate-binding domains ($K_d = 10$ and 14 nM, respectively) slowed the progression of lung fibrosis in mice after bleomycin treatment. Compound **15** comes from a scaffold for which SAR was recently disclosed.[77] No new small-molecule inhibitors of galectin-3 other than **15** have emerged. This is partly explained by the X-ray structure of galectin-3 bound to a carbohydrate ligand. The crystal structure shows a long, polar, and shallow binding site, and it will be challenging to identify inhibitors that are not derived from carbohydrate-like scaffolds.[78] GCS-100, a modified citrus pectin carbohydrate and selective galectin-3 inhibitor, is reported to have advanced to through phase I clinical trials for cancer and renal transplantation.[79] Most recently, recruiting began for the treatment of chronic kidney disease.[80]

15

6. LYSOPHOSPHOLIPID MEDIATORS

6.1. Lysophosphatidic acid receptor-1 antagonists

There are six isoforms of lysophosphatidic acid (LPA) receptors. Tager and colleagues showed that LPA levels increase in bronchoalveolar lavage fluid following lung injury in the bleomycin model of pulmonary fibrosis. The authors also showed that mice lacking the LPA1 receptor were protected from fibrosis and mortality in the same model, and that LPA levels were also elevated in the BALF of human IPF patients.[81] The role of LPA in chronic inflammatory disorders has been reviewed.[82] AM095, **16**, and AM966, **17**, were the first nonlipid, orally active, LPA1 receptor antagonists to be reported in the peer-reviewed literature.[83,84] Triazole **18** was subsequently reported to improve on the selectivity profiles of isoxazoles **16** and **17**.[85] **18** inhibited the proliferation and contraction of normal human lung fibroblasts following LPA stimulation and further highlights the potential for utility of LPA1 antagonists in IPF.[85]

16 (R = H), **17** (R = Cl)   **18**

7. CONCLUSIONS

IPF is a debilitating and rapidly progressing fibrotic disease with a high associated mortality. After initial diagnosis, the patient's quality of life rapidly deteriorates, the ability to participate in normal physical activities declines, and ultimately patients require supplemental oxygen. In 2008, it was estimated that as many as 5 million people were affected by the disease worldwide.[86] Currently, the options available to physicians for the treatment of patients with IPF are extremely limited. A small molecule, pirfenidone, is available in Japan and Europe and it is not clear whether it will be approved in the United States. There is significant unmet medical need, and while the task is particularly challenging due to the incomplete understanding of the etiology of the disease, it is clear that a number of interesting new targets have emerged in recent years which might serve as basis to meet this need.

REFERENCES

1. Su, R.; Bennett, M.; Jacobs, S.; Hunter, T.; Bailey, C.; Krishnan, E.; Rosen, G.; Chung, L. *J. Rheumatol.* **2011**, *38*, 693.
2. Nalysnyk, L.; Cid-Ruzafa, J.; Rotella, P.; Esser, D. *Eur. Respir. Rev.* **2012**, *21*, 355.
3. Boon, K.; Bailey, N. W.; Yang, J.; Steel, M. P.; Groshong, S.; Kervitsky, D.; Brown, K. K.; Schwarz, M. I.; Schwartz, D. A. *PLoS One* **2009**, *4*, e5134.
4. Tachikawa, R.; Tomii, K.; Ueda, H.; Nagata, K.; Nanjo, S.; Sakurai, A.; Otsuka, K.; Kaji, R.; Hayashi, M.; Katakami, N.; Imai, Y. *Respiration* **2012**, *83*, 20.
5. Kim, D. S. *Clin. Chest Med.* **2012**, *33*, 59.
6. King, T. E.; Pardo, A.; Selman, M. *Lancet* **2011**, *378*, 1949.
7. Maharaj, S. S.; Baroke, E.; Gauldie, J.; Kolb, M. R. *Pulm. Pharmacol. Ther.* **2012**, *25*, 263.
8. Fiorucci, E.; Lucantoni, G.; Paone, G.; Zotti, M.; Li, B. E.; Serpilli, M.; Regimenti, P.; Cammarella, I.; Puglisi, G.; Schmid, G. *Eur. Rev. Med. Pharmacol. Sci.* **2008**, *12*, 105.
9. Gibbons, M. A.; MacKinnon, A. C.; Ramachandran, P.; Dhaliwal, K.; Duffin, R.; Phythian-Adams, A. T.; van Rooijen, N.; Haslett, C.; Howie, S. E.; Simpson, A. J.; Hirani, N.; Gauldie, J.; Iredale, J. P.; Sethi, T.; Forbes, S. J. *Am. J. Respir. Crit. Care Med.* **2011**, *184*, 569.
10. Morimoto, K.; Janssen, W. J.; Terada, M. *Respir. Med.* **2012**, *106*, 1800.
11. Cerri, S.; Spagnolo, P.; Luppi, F.; Richeldi, L. *Clin. Chest Med.* **2012**, *33*, 85.
12. Potts, J.; Yogaratnam, D. *Ann. Pharmacother.* **2013**, *47*, 361.
13. Hilberg, O.; Simonsen, U.; du Bois, R.; Bendstrup, E. *Clin. Respir. J.* **2012**, *6*, 131.
14. Noble, P. W.; Albera, C.; Bradford, W. Z.; Costabel, U.; Glassberg, M. K.; Kardatzke, D.; King, T. E., Jr.; Lancaster, L.; Sahn, S. A.; Szwarcberg, J.; Valeyre, D.; du Bois, R. M. *Lancet* **2011**, *377*, 1760.
15. Richeldi, L.; du Bois, R.M. *Expert. Rev. Respir. Med.* **2011**, *5*, 473–481.
16. Heinrich, E. L.; Lee, W.; Lu, J.; Lowy, A. M.; Kim, J. *J. Transl. Med.* **2012**, *10*, 68.
17. Broxmeyer, H. E.; Orschell, C. M.; Clapp, D. W.; Hangoc, G.; Cooper, S.; Plett, P. A.; Liles, W. C.; Li, X.; Graham-Evans, B.; Campbell, T. B.; Calandra, G.; Bridger, G.; Dale, D. C.; Srour, E. F. *J. Exp. Med.* **2005**, *201*, 1307.

18. Shimizu, Y.; Dobashi, K.; Endou, K.; Ono, A.; Yanagitani, N.; Utsugi, M.; Sano, T.; Ishizuka, T.; Shimizu, K.; Tanaka, S.; Mori, M. *Int. J. Immunopathol. Pharmacol.* **2010**, *23*, 449.
19. Phillips, R. J.; Burdick, M. D.; Hong, K.; Lutz, M. A.; Murray, L. A.; Xue, Y. Y.; Belperio, J. A.; Keane, M. P.; Strieter, R. M. *J. Clin. Invest.* **2004**, *114*, 438.
20. Song, J. S.; Kang, C. M.; Kang, H. H.; Yoon, H. K.; Kim, Y. K.; Kim, K. H.; Moon, H. S.; Park, S. H. *Exp. Mol. Med.* **2010**, *42*, 465.
21. De Clercq, E.; Yamamoto, N.; Pauwels, R.; Balzarini, J.; Witvrouw, M.; De Vreese, K.; Debyser, Z.; Rosenwirth, B.; Peichl, P.; Datema, R. *Antimicrob. Agents Chemother.* **1994**, *38*, 668.
22. Donzella, G. A.; Schols, D.; Lin, S. W.; Este, J. A.; Nagashima, K. A.; Maddon, P. J.; Allaway, G. P.; Sakmar, T. P.; Henson, G.; De Clercq, E.; Moore, J. P. *Nat. Med.* **1998**, *4*, 72.
23. Kalatskaya, I.; Berchiche, Y. A.; Gravel, S.; Limberg, B. J.; Rosenbaum, J. S.; Heveker, N. *Mol. Pharmacol.* **2009**, *75*, 1240.
24. Uto-Konomi, A.; Wirtz, J.; Sato, Y.; Takano, A.; Nanki, T.; Suzuki, S. *Biochem. Biophys. Res. Commun.* **2013**, *431*, 772.
25. Debnath, B.; Xu, S.; Grande, F.; Garofalo, A.; Neamati, N. *Theranostics* **2013**, *3*, 47.
26. Zhu, A.; Zhan, W.; Liang, Z.; Yoon, Y.; Yang, H.; Grossniklaus, H. E.; Xu, J.; Rojas, M.; Lockwood, M.; Snyder, J. P.; Liotta, D. C.; Shim, H. *J. Med. Chem.* **2010**, *53*, 8556.
27. Wu, B.; Chien, E. Y.; Mol, C. D.; Fenalti, G.; Liu, W.; Katritch, V.; Abagyan, R.; Brooun, A.; Wells, P.; Bi, F. C.; Hamel, D. J.; Kuhn, P.; Handel, T. M.; Cherezov, V.; Stevens, R. C. *Science* **2010**, *330*, 1066.
28. Liang, Z.; Zhan, W.; Zhu, A.; Yoon, Y.; Lin, S.; Sasaki, M.; Klapproth, J. M.; Yang, H.; Grossniklaus, H. E.; Xu, J.; Rojas, M.; Voll, R. J.; Goodman, M. M.; Arrendale, R. F.; Liu, J.; Yun, C. C.; Snyder, J. P.; Liotta, D. C.; Shim, H. *PLoS One* **2012**, 7, e34038.
29. Cernaro, V.; Lacquaniti, A.; Lupica, R.; Buemi, A.; Trimboli, D.; Giorgianni, G.; Bolignano, D.; Buemi, M. *Med. Res. Rev.* **2013**, http://dx.doi.org/10.1002/med.21277. [Epub ahead of print].
30. Hsu, S. Y.; Nakabayashi, K.; Nishi, S.; Kumagai, J.; Kudo, M.; Sherwood, O. D.; Hsueh, A. J. *Science* **2002**, *295*, 671.
31. Huang, X.; Gai, Y.; Yang, N.; Lu, B.; Samuel, C. S.; Thannickal, V. J.; Zhou, Y. *Am. J. Pathol.* **2011**, *179*, 2751.
32. Khanna, D.; Clements, P. J.; Furst, D. E.; Korn, J. H.; Ellman, M.; Rothfield, N.; Wigley, F. M.; Moreland, L. W.; Silver, R.; Kim, Y. H.; Steen, V. D.; Firestein, G. S.; Kavanaugh, A. F.; Weisman, M.; Mayes, M. D.; Collier, D.; Csuka, M. E.; Simms, R.; Merkel, P. A.; Medsger, T. A.; Sanders, M. E.; Maranian, P.; Seibold, J. R. *Arthritis Rheum.* **2009**, *60*, 1102.
33. Chen, C. Z.; Southall, N.; Xiao, J.; Marugan, J. J.; Ferrer, M.; Hu, X.; Jones, R. E.; Feng, S.; Agoulnik, I. U.; Zheng, W.; Agoulnik, A. I. *J. Biomol. Screen.* **2013**, *18*, 670.
34. Sudo, S.; Kumagai, J.; Nishi, S.; Layfield, S.; Ferraro, T.; Bathgate, R. A.; Hsueh, A. J. *J. Biol. Chem.* **2003**, *278*, 7855.
35. Halls, M. L.; Bathgate, R. A.; Sudo, S.; Kumagai, J.; Bond, C. P.; Summers, R. J. *Ann. N. Y. Acad. Sci.* **2005**, *1041*, 17.
36. Pini, A.; Shemesh, R.; Samuel, C. S.; Bathgate, R. A.; Zauberman, A.; Hermesh, C.; Wool, A.; Bani, D.; Rotman, G. *J. Pharmacol. Exp. Ther.* **2010**, *335*, 589.
37. Marinkovic, A.; Liu, F.; Tschumperlin, D. J. *Am. J. Respir. Cell Mol. Biol.* **2013**, *48*, 422.
38. Kolb, M. R. J.; Gauldie, J. *Am. J. Respir. Crit. Care Med.* **2011**, *184*, 627.
39. Barry-Hamilton, V.; Spangler, R.; Marshall, D.; McCauley, S.; Rodriguez, H. M.; Oyasu, M.; Mikels, A.; Vaysberg, M.; Ghermazien, H.; Wai, C.; Garcia, C. A.; Velayo, A. C.; Jorgensen, B.; Biermann, D.; Tsai, D.; Green, J.; Zaffryar-Eilot, S.;

Holzer, A.; Ogg, S.; Thai, D.; Neufeld, G.; Van Vlasselaer, P.; Smith, V. *Nat. Med.* **2010**, *16*, 1009.

40. Siegel, M.; Khosla, C. *Pharmacol. Ther.* **2007**, *115*, 232.
41. Griffin, M.; Casadio, R.; Bergamini, C. M. *Biochem. J.* **2002**, *368*, 377.
42. Olsen, K. C.; Sapinoro, R. E.; Kottmann, R. M.; Kulkarni, A. A.; Iismaa, S. E.; Johnson, G. V.; Thatcher, T. H.; Phipps, R. P.; Sime, P. J. *Am. J. Respir. Crit. Care Med.* **2011**, *184*, 699.
43. Oh, K.; Park, H. B.; Byoun, O. J.; Shin, D. M.; Jeong, E. M.; Kim, Y. W.; Kim, Y. S.; Melino, G.; Kim, I. G.; Lee, D. S. *J. Exp. Med.* **2011**, *208*, 1707.
44. Gundemir, S.; Johnson, G. V. *PLoS One* **2009**, *4*, e6123.
45. Shaw, E. *J. Biol. Chem.* **1988**, *263*, 2768.
46. Baumgartner, W.; Golenhofen, N.; Weth, A.; Hiiragi, T.; Saint, R.; Griffin, M.; Drenckhahn, D. *Histochem. Cell Biol.* **2004**, *122*, 17.
47. Johnson, T. S.; Fisher, M.; Haylor, J. L.; Hau, Z.; Skill, N. J.; Jones, R.; Saint, R.; Coutts, I.; Vickers, M. E.; El Nahas, A. M.; Griffin, M. *J. Am. Soc. Nephrol.* **2007**, *18*, 3078.
48. Hanks, S. K.; Calalb, M. B.; Harper, M. C.; Patel, S. K. *Proc. Natl. Acad. Sci. U.S.A.* **1992**, *89*, 8487.
49. Griffin, M.; Mongeot, A.; Collighan, R.; Saint, R. E.; Jones, R. A.; Coutts, I. G.; Rathbone, D. L. *Bioorg. Med. Chem. Lett.* **2008**, *18*, 5559.
50. Parsons, J. T. *J. Cell Sci.* **2003**, *116*, 1409.
51. Lagares, D.; Kapoor, M. *BioDrugs* **2013**, *27*, 15.
52. Lagares, D.; Busnadiego, O.; García-Fernández, R. A.; Kapoor, M.; Liu, S.; Carter, D. E.; Abraham, D.; Shi-Wen, X.; Carreira, P.; Fontaine, B. A.; Shea, B. S.; Tager, A. M.; Leask, A.; Lamas, S.; Rodríguez-Pascual, F. *Arthritis Rheum.* **2012**, *64*, 1653.
53. Schultze, A.; Fiedler, W. *Anticancer Agents Med. Chem.* **2011**, *11*, 593.
54. Heinrich, T.; Seenisamy, J.; Emmanuvel, L.; Kulkarni, S. S.; Bomke, J.; Rohdich, F.; Greiner, H.; Esdar, C.; Krier, M.; Grädler, U.; Musil, D. *J. Med. Chem.* **2013**, *56*, 1160.
55. Iwatani, M.; Iwata, H.; Okabe, A.; Skene, R. J.; Tomita, N.; Hayashi, Y.; Aramaki, Y.; Hosfield, D. J.; Hori, A.; Baba, A.; Miki, H. *Eur. J. Med. Chem.* **2013**, *61*, 49.
56. Ihle, N. T.; Powis, G. *Curr. Opin. Drug Discov. Dev.* **2010**, *13*, 41.
57. Rodon, J.; Dienstmann, R.; Serra, V.; Tabernero, J. *Nat. Rev. Clin. Oncol.* **2013**, *10*, 143.
58. Conte, E.; Fruciano, M.; Fagone, E.; Gili, E.; Caraci, F.; Iemmolo, M.; Crimi, N.; Vancheri, C. *PLoS One* **2011**, *6*, e24663.
59. Lu, Y.; Azad, N.; Wang, L.; Iyer, A. K.; Castranova, V.; Jiang, B. H.; Rojanasakul, Y. *Am. J. Respir. Cell Mol. Biol.* **2010**, *42*, 432.
60. Lamouille, S.; Derynck, R. *Cells Tissues Organs* **2011**, *193*, 8.
61. Russo, R. C.; Garcia, C. C.; Barcelos, L. S.; Rachid, M. A.; Guabiraba, R.; Roffê, E.; Souza, A. L.; Sousa, L. P.; Mirolo, M.; Doni, A.; Cassali, G. D.; Pinho, V.; Locati, M.; Teixeira, M. M. *J. Leukoc. Biol.* **2011**, *89*, 269.
62. Norman, M. H.; Andrews, K. L.; Bo, Y. Y.; Booker, S. K.; Caenepeel, S.; Cee, V. J.; D'Angelo, N. D.; Freeman, D. J.; Herberich, B. J.; Hong, F. T.; Jackson, C. L.; Jiang, J.; Lanman, B. A.; Liu, L.; McCarter, J. D.; Mullady, E. L.; Nishimura, N.; Pettus, L. H.; Reed, A. B.; Miguel, T. S.; Smith, A. L.; Stec, M. M.; Tadesse, S.; Tasker, A.; Aidasani, D.; Zhu, X.; Subramanian, R.; Tamayo, N. A.; Wang, L.; Whittington, D. A.; Wu, B.; Wu, T.; Wurz, R. P.; Yang, K.; Zalameda, L.; Zhang, N.; Hughes, P. E. *J. Med. Chem.* **2012**, *55*, 7796.
63. Nacht, M.; Qiao, L.; Sheets, M. P.; St Martin, T.; Labenski, M.; Mazdiyasni, H.; Karp, R.; Zhu, Z.; Chaturvedi, P.; Bhavsar, D.; Niu, D.; Westlin, W.; Petter, R. C.; Medikonda, A. P.; Singh, J. *J. Med. Chem.* **2013**, *56*, 712.

64. Murray, J. M.; Sweeney, Z. K.; Chan, B. K.; Balazs, M.; Bradley, E.; Castanedo, G.; Chabot, C.; Chantry, D.; Flagella, M.; Goldstein, D. M.; Kondru, R.; Lesnick, J.; Li, J.; Lucas, M. C.; Nonomiya, J.; Pang, J.; Price, S.; Salphati, L.; Safina, B.; Savy, P. P.; Seward, E. M.; Ultsch, M.; Sutherlin, D. P. *J. Med. Chem.* **2012**, *55*, 7686.
65. Gonzalez-Lopez de Turiso, F.; Shin, Y.; Brown, M.; Cardozo, M.; Chen, Y.; Fong, D.; Hao, X.; He, X.; Henne, K.; Hu, Y. L.; Johnson, M. G.; Kohn, T.; Lohman, J.; McBride, H. J.; McGee, L. R.; Medina, J. C.; Metz, D.; Miner, K.; Mohn, D.; Pattaropong, V.; Seganish, J.; Simard, J. L.; Wannberg, S.; Whittington, D. A.; Yu, G.; Cushing, T. D. *J. Med. Chem.* **2012**, *55*, 7667.
66. Ellard, K.; Sunose, M.; Bell, K.; Ramsden, N.; Bergamini, G.; Neubauer, G. *Bioorg. Med. Chem. Lett.* **2012**, *22*, 4546.
67. Son, J. Y.; Kim, S. Y.; Cho, S. H.; Shim, H. S.; Jung, J. Y.; Kim, E. Y.; Lim, J. E.; Park, B. H.; Kang, Y. A.; Kim, Y. S.; Kim, S. K.; Chang, J.; Park, M. S. *Lung* **2013**, *191*, 199.
68. Higashiyama, H.; Yoshimoto, D.; Kaise, T.; Matsubara, S.; Fujiwara, M.; Kikkawa, H.; Asano, S.; Kinoshita, M. *Exp. Mol. Pathol.* **2007**, *83*, 39.
69. Anderton, M. J.; Mellor, H. R.; Bell, A.; Sadler, C.; Pass, M.; Powell, S.; Steele, S. J.; Roberts, R. R.; Heier, A. *Toxicol. Pathol.* **2011**, *39*, 916.
70. Crestani, B.; Besnard, V.; Boczkowski, J. *Int. J. Biochem. Cell Biol.* **2011**, *43*, 1086.
71. Amara, N.; Goven, D.; Prost, F.; Muloway, R.; Crestani, B.; Boczkowski, J. *Thorax* **2010**, *65*, 733.
72. Carnesecchi, S.; Deffert, C.; Donati, Y.; Basset, O.; Hinz, B.; Preynat-Seauve, O.; Guichard, C.; Arbiser, J. L.; Banfi, B.; Pache, J. C.; Barazzone-Argiroffo, C.; Krause, K. H. *Antioxid. Redox Sign.* **2011**, *15*, 607.
73. Laleu, B.; Gaggini, F.; Orchard, M.; Fioraso-Cartier, L.; Cagnon, L.; Houngninou-Molango, S.; Gradia, A.; Duboux, G.; Merlot, C.; Heitz, F.; Szyndralewiez, C.; Page, P. *J. Med. Chem.* **2010**, *53*, 7715.
74. Gaggini, F.; Laleu, B.; Orchard, M.; Fioraso-Cartier, L.; Cagnon, L.; Houngninou-Molango, S.; Gradia, A.; Duboux, G.; Merlot, C.; Heitz, F.; Szyndralewiez, C.; Page, P. *Bioorg. Med. Chem.* **2011**, *19*, 6989.
75. Nishi, Y.; Sano, H.; Kawashima, T.; Okada, T.; Kuroda, T.; Kikkawa, K.; Kawashima, S.; Tanabe, M.; Goto, T.; Matsuzawa, Y.; Matsumura, R.; Tomioka, H.; Liu, F. T.; Shirai, K. *Allergol. Int.* **2007**, *56*, 57.
76. Mackinnon, A. C.; Gibbons, M. A.; Farnworth, S. L.; Leffler, H.; Nilsson, U. J.; Delaine, T.; Simpson, A. J.; Forbes, S. J.; Hirani, N.; Gauldie, J.; Sethi, T. *Am. J. Respir. Crit. Care Med.* **2012**, *185*, 537.
77. Salameh, B. A.; Cumpstey, I.; Sundin, A.; Leffler, H.; Nilsson, U. J. *Bioorg. Med. Chem.* **2010**, *18*, 5367.
78. Seetharaman, J.; Kanigsberg, A.; Slaaby, R.; Leffler, H.; Barondes, S. H.; Rini, J. M. *J. Biol. Chem.* **1998**, *273*, 13047.
79. http://www.ljpc.com/pipeline-main.html; http://www.ljpc.com/news/2013/012813. html.
80. ClinicalTrials.gov Identifier: NCT01717248.
81. Tager, A. M.; LaCamera, P.; Shea, B. S.; Campanella, G. S.; Selman, M.; Zhao, Z.; Polosukhin, V.; Wain, J.; Karimi-Shah, B. A.; Kim, N. D.; Hart, W. K.; Pardo, A.; Blackwell, T. S.; Xu, Y.; Chun, J.; Luster, A. D. *Nat. Med.* **2008**, *14*, 45.
82. Sevastou, I.; Kaffe, E.; Mouratis, M. A.; Aidinis, V. *Biochim. Biophys. Acta* **2013**, *1831*, 42.
83. Swaney, J. S.; Chapman, C.; Correa, L. D.; Stebbins, K. J.; Bundey, R. A.; Prodanovich, P. C.; Fagan, P.; Baccei, C. S.; Santini, A. M.; Hutchinson, J. H.; Seiders, T. J.; Parr, T. A.; Prasit, P.; Evans, J. F.; Lorrain, D. S. *Br. J. Pharmacol.* **2010**, *160*, 1699.

84. Swaney, J. S.; Chapman, C.; Correa, L. D.; Stebbins, K. J.; Broadhead, A. R.; Bain, G.; Santini, A. M.; Darlington, J.; King, C. D.; Baccei, C. S.; Lee, C.; Parr, T. A.; Roppe, J. R.; Seiders, T. J.; Ziff, J.; Prasit, P.; Hutchinson, J. H.; Evans, J. F.; Lorrain, D. S. *J. Pharmacol. Exp. Ther.* **2011**, *336*, 693.
85. Qian, Y.; Hamilton, M.; Sidduri, A.; Gabriel, S.; Ren, Y.; Peng, R.; Kondru, R.; Narayanan, A.; Truitt, T.; Hamid, R.; Chen, Y.; Zhang, L.; Fretland, A. J.; Sanchez, R. A.; Chang, K.-C.; Lucas, M.; Schoenfeld, R. C.; Laine, D.; Fuentes, M. E.; Stevenson, C. S.; Budd, D. C. *J. Med. Chem.* **2012**, *55*, 7920.
86. Eric, B.; Meltzer, E. B.; Noble, P. W. *Orphanet J. Rare Dis.* **2008**, *3*, 8. http://dx.doi.org/10.1186/1750-1172-3-8.

CHAPTER TWELVE

Targeting the Nuclear Hormone Receptor RORγt for the Treatment of Autoimmune and Inflammatory Disorders

T. G. Murali Dhar*, Qihong Zhao*, David W. Markby[†]
*Bristol-Myers Squibb, Research and Development, Princeton, New Jersey, USA
[†]Exelixis Inc., South San Francisco, California, USA

Contents

1. INTRODUCTION

1.1. ROR family of nuclear hormone receptors

RORγt (Retinoic acid–related Orphan Receptor γt), a lymphoid–specific member of the nuclear hormone receptor (NHR) family, regulates the expression of proinflammatory cytokines that have been implicated in multiple inflammatory and autoimmune diseases, including psoriasis, multiple sclerosis (MS), rheumatoid arthritis (RA), and inflammatory bowel disease (IBD).[1,2] NHR family members are modular transcription factors composed of separate DNA and ligand binding domains (LBDs) that regulate gene

Annual Reports in Medicinal Chemistry, Volume 48
ISSN 0065-7743
http://dx.doi.org/10.1016/B978-0-12-417150-3.00012-0

169

expression in response to diverse ligands such as steroids, retinoids, and vitamin D. Ligands modulate target gene transcription by promoting association with upstream DNA response elements and changes in the interaction with associated coactivator and corepressor proteins. As ligand-regulated transcription factors, NHRs are attractive drug targets, and NHR-targeted drugs are effective therapeutics for inflammatory and autoimmune disorders, metabolic diseases, and cancer.[3,4]

The retinoic acid–related orphan receptor family includes three members (RORα, RORβ, and RORγ) that play important roles in organ development, immunity, metabolic regulation, neural function, and circadian rhythms. Although RORα and RORγ bind naturally occurring oxysterols and RORβ binds retinoids, the RORs remain "orphan receptors" with no verified physiological ligands.[1,5] RORγ exists as two splice variants known as RORγ and RORγt that are identical apart from an N-terminal extension of ~20 amino acids present only in RORγ (*vide infra*). RORγ is expressed in several tissues including the kidney, liver, muscle, and adipose, whereas expression of RORγt is restricted to lymphoid cells. RORγt is essential for the development of lymph nodes and for the normal differentiation and/or function of specialized lymphocytes including IL-17 producing T helper (Th17) cells, innate lymphoid cells (ILCs), and γδ cells.[6–8] In non-immune cells, RORγ regulates the expression of genes that control metabolism in skeletal muscle and fat[9] and that regulate circadian rhythms.[10] Studies using RORα- and RORγ-deficient mice show that function of RORα partially overlaps with that of RORγ; however, RORγ plays the dominant role in lymphoid cells.[11] As a therapeutic for inflammatory and autoimmune diseases, selectivity versus other RORs is desirable to reduce potential effects on neural function (RORα and RORβ) and lipid metabolism (RORα).

1.2. RORγt and Th17 cells

Interest in RORγt as a drug target stems from the importance of IL-17 and related cytokines in inflammation and autoimmunity and the role of RORγt as a master regulator of Th17 cells and other IL-17-producing lymphocytes. CD4⁺ T helper cell subsets orchestrate different aspects of adaptive immunity. For example, Th1 cells protect against intracellular bacterial and viral infections by producing IFNγ and inducing cytotoxic T cell responses, and Th2 cells respond to parasites and allergens by secreting cytokines such as IL-4 and inducing immunoglobulin production by B cells.

Until 10 years ago, many inflammatory and autoimmune conditions were attributed to hyperactivation of Th1 cells; however, more recent data point to the importance of a distinct class of CD4$^+$ effector cells, Th17 cells.[12,13]

Th17 cells form part of the adaptive response to specific fungi and bacteria and differentiate from uncommitted precursors upon engagement with antigen presenting cells in the presence of cytokines such as IL-6, TGFβ, IL-1β, and IL-21. Differentiated Th17 cells express high levels of RORγt, the IL-23 receptor (IL-23R), and chemokine receptors such as CCR6.[13] RORγt is essential for normal Th17 cell development and IL-17 expression,[7] and IL-23 promotes Th17 expansion as well as IL-17 expression.[14] Th17 cells reside primarily at mucosal surfaces such as the intestinal lamina propria. When activated, they produce antimicrobial peptides and proinflammatory cytokines including IL-17 (also known as IL-17A), IL-17F, and IL-22 which induce recruitment of neutrophils and regulate epithelial barrier function.[13] Th17 cells exhibit plasticity and can differentiate to express other cytokines such as IFNγ especially during chronic inflammation.[15]

Overproduction of Th17 cytokines is associated with several human autoimmune and inflammatory diseases, including MS, RA, psoriasis, IBD, and asthma. In murine models of these diseases, inhibition of IL-17 function by neutralizing antibodies or genetic disruption of IL-17 or the IL-17RA receptor can ameliorate the disease course or clinical symptoms.[16] The IL-23/IL-17 axis has also been validated in these disease models by functional inhibition of IL-23, a key cytokine for the expansion and maintenance of Th17 cells and for the expression of IL-17 by RORγ$^+$ lymphocytes.[17]

1.3. RORγt$^+$ innate lymphoid cells and γδ T cells

In addition to orchestrating the production of cytokines (especially IL-17 and IL-22) by Th17 cells, RORγt also mediates expression of such cytokines by ILCs and γδ T cells.[8,18]

ILCs are derived from common lymphoid precursors but do not express an antigen receptor. ILCs include Natural Killer (NK) cells and the more recently identified lymphoid tissue inducer (LTi) cells and NK22 cells. ILCs are similar to CD4$^+$ T helper cells because they produce such cytokines as IFNγ, IL-17, IL-22, or IL-13; yet they are different because they are not antigen specific and can expand and respond quickly to specific alert molecules, thus serving as early defense mechanisms for the host. LTi cells express high levels of IL-17, and LTi and NK22 cells express high levels of IL-22. These two cell populations require RORγt for their development and function. They are abundant in the intestinal lamina propria where they defend against intestinal pathogens and maintain epithelial integrity.[19]

RORγt also mediates IL-17 production by γδ T cells in response to cytokines, especially IL-23, IL-1β, and IL-18.[8] RORγt⁺ γδ T cells are a critical player not only in the mucosal defense against bacterial infection but also in the pathogenesis of psoriasis in mice and presumably in humans.[20,21]

2. TARGET VALIDATION OF RORγ/γt IN DISEASE MODELS

2.1. Genetic validation

As summarized in Table 12.1, numerous studies with RORγ/RORγt-deficient mice have shown that RORγ/γt is a critical component in the pathogenesis of multiple diseases. These indications cover autoimmunity (MS, IBDs, psoriasis, and graft vs. host disease), allergy (asthma), metabolic disorders (obesity and type 2 diabetes), and cancer. In autoimmune disease models, the reduction in disease severity is accompanied by a decrease in Th17-related cytokines; for allergy, the reduction in disease severity is associated with a decrease in eosinophilia and Th2 cytokines. In summary, genetic data from these murine models provide compelling evidence that targeting RORγ/γt might have broad clinical utility.

2.2. IL-17 targeted therapies

Although RORγt antagonists have not yet entered clinical trials, other therapeutics targeting the IL-23/IL-17 axis have shown promise for treating inflammatory and autoimmune diseases. The most advanced clinical agents are monoclonal antibodies to the common p40 subunit of IL-12/IL-23; therefore, they inhibit both IL-12-driven Th1 responses and IL-23-driven Th17 responses. Ustekinumab is approved to treat moderate to severe plaque psoriasis and is in clinical trials for additional inflammatory indications as are other IL-23-targeted antibodies.[30] Therapies that specifically inhibit IL-17 signaling include the monoclonal antibodies ixekizumab and secukinumab which target IL-17 and brodalumab which targets the IL-17RA receptor. All three antibodies were efficacious in Phase 2 studies of moderate to severe plaque psoriasis and are undergoing further evaluation.[31,32] In other trials reported to date, secukinumab did not reach the

Table 12.1 Genetic validation of RORγ/γt in murine disease models

Disease	Mouse model	Effect of deleting RORγ and/or RORγt
MS	Experimental autoimmune encephalomyelitis (EAE)[7,11]	Reduced disease severity and reduced infiltrating Th17 cells and associated cytokines in spinal cords
Psoriasis	Imiquimod induced skin lesions[20]	Reduced skin thickness and reduced CD45$^+$ cells expressing IL-17F and/or IL-22
IBD	1. CD4$^+$CD25− T cell colitis in RAG KO mice 2. Anti-CD40 induced colitis in RAG KO mice[22,23]	1. Reduced disease severity and colonic infiltration of dendritic cells and neutrophils Decreased splenic production of IL-17A, IL-6, GM-CSF, and IFNγ 2. Reduced disease severity and colonic expression of IL-22, IFNγ, TNFα, and IL-23R
Nephritis	Anti-glomerular basement membrane induced nephritis[24]	Reduced disease severity and renal expression of IL-17A and IL-17F
Graft versus host disease (GVHD)	Acute GVHD[25]	Attenuated acute disease and reduced pathology of colon, lung, and liver
Asthma	Ovalbumin-induced lung inflammation[26]	Attenuated allergic response and eosinophilia. Reduced CD4+ T cells and Th2 cytokines in the lung
Obesity/type 2 diabetes	Diet-induced obesity/T2DM[27,28]	1. Protection from hyperglycemia and insulin resistance in response to diet-induced obesity 2. Reduced adipocyte size, MMP3 expression, and reduced body weight gain
Cancer	Subcutaneous injection of melanoma line B16F10[29]	Increased survival and reduced tumor volume, increased IL-9 expression

primary endpoint in a Phase 2 study for RA[33] and showed no activity in a Phase 2 study for Crohn's disease with a higher rate of adverse events than placebo.[34] Compared to antibodies that specifically target IL-17 signaling, an RORγ antagonist will block the production of additional proinflammatory cytokines and offer a distinct clinical profile.

3. SAFETY CONCERNS TARGETING RORγt

Based on studies of RORγ-deficient mice, an antagonist of RORγt may present several safety concerns. These potential risks include lymphoma formation, dysregulation of intestinal homeostasis, and metabolic liabilities. First, RORγ-deficient mice exhibit an increased incidence of thymic lymphoma[35] that is most likely due to dysregulated differentiation and proliferation of thymocytes.[6] Second, RORγt helps maintain Th17 and ILC functions during homeostasis.[7,19] This protective function was demonstrated using RORγt-deficient mice which enhanced intestinal damage following acute injury with dextran sodium sulfate.[36,37] However, other studies have shown that RORγt is an important player in the pathogenesis of colitis induced in T/B cell-deficient mice by $CD4^+$ T cell transfer and by anti-CD40 agonistic antibody, demonstrating that both RORγt-expressing T cells and ILCs contribute to disease progression.[22,23] Finally, concern over metabolic liabilities is based on the finding that RORγ controls the expression of genes that regulate muscle and fat mass and lipid homeostasis.[9] However, recent studies have shown that RORγ-deficient mice are protected from hyperglycemia, insulin resistance, and weight gain following diet-induced obesity.[27,28] In humans, RORγ expression is higher in visceral adipose tissue from subjects with morbid obesity and high insulin resistance, and there is a positive correlation between RORγ expression and many components of metabolic syndrome.[38] Whether pharmacological inhibition of RORγt will increase lymphoma formation, impair intestinal homeostasis, or impact obesity and insulin resistance remains to be determined.

4. RORγt STRUCTURE

As discussed earlier, RORγt is an isoform of the RORγ protein that is 497 amino acids in length.[5] Similar to most NHRs, RORγt contains four functional domains—a variable N-terminal A/B domain which consists of the constitutively active ligand-independent activation function-1 (AF-1), a highly conserved DNA-binding domain, a hinge domain that connects the DNA-binding domain to the LBD, and the moderately conserved carboxy terminal LBD that consists of the ligand-dependent activation function-2 (AF-2). Similar to those of other NHRs, the LBD of RORγt has 12 α-helices and two additional helices H2′ and H11′.

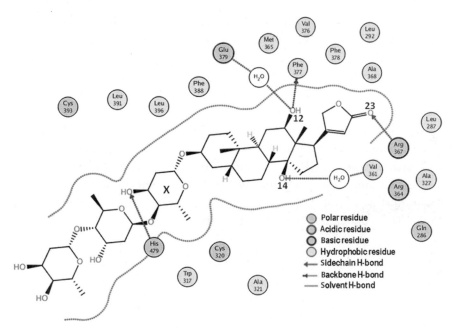

Figure 12.1 Crystal structure of the RORγt ligand binding pocket in complex with digoxin, shown as a two-dimensional ligand interaction diagram.[41] (See color plate.)

The X-ray cocrystal structure of RORγ in the agonist mode has been determined with the coactivator peptide steroid receptor coactivator-2 (SRC2) and three different hydroxycholesterol (HC) ligands—20α-HC, 22R-HC, and 25-HC.[39] The X-ray cocrystal structure of RORγt in the antagonist mode in complex with the cardiac glycoside digoxin at 2.2 Å resolution has been solved recently.[40] Key interactions include H-bond interactions of (a) the lactone carbonyl (O23) with Arg-367 in H5; (b) O12 with the backbone carbonyl of Phe-377 in β-strand 1, and with Glu-379 in loop s1–s2 via a water molecule; (c) the hydroxyl on ring X of the digitoxose moiety with His-479 on H11; and (d) O14 with the backbone carbonyl of Val-361 via a water molecule (Fig. 12.1). Because of the interaction of His-479 on H11 with the hydroxy group on ring X of the sugar moiety of digoxin, His-479 can no longer engage Tyr-502 of H11′ and Phe-506 resulting in an inactive conformation of helix 12.

5. SMALL-MOLECULE ANTAGONISTS OF RORγt

Compelling evidence from both preclinical and clinical studies on the role of IL-17 in inflammatory processes and the importance of RORγt for the induction of IL-17 transcription has led to a significant effort in academia and the pharmaceutical industry to identify small-molecule antagonists of RORγt.[1,2] An overview of small-molecule antagonists that have been reported in the open and patent literature is provided below.

T0901317 (**1**) was the first synthetic RORγt antagonist reported in the literature with a K_i of ~51 nM.[42] It was identified through screening of 65 compounds with known NHR activity at the Scripps Research Molecular Screening Center. However, T0901317 is a promiscuous NHR ligand with activity against LXR, PXR, and FXR. It has a binding K_i of ~132 nM for RORα and is essentially inactive at RORβ. The authors also demonstrated that in a RORα/γ-dependent transactivation assay in HepG2 cells, T0901317 reduced the recruitment of SRC2.

In an attempt to improve the NHR selectivity profile of T0901317, a number of closely related analogs were synthesized which led to the identification of SR1001 (**2**) which has significantly improved NHR selectivity profile while maintaining RORγt potency.[43] In a competitive radioligand binding assay, SR1001 dose-dependently displaced [³H]25–HC binding to RORα and RORγ with K_i values of 172 nM and 111 nM, respectively. In addition, the compound was selective versus 48 other human nuclear receptors in cell-based cotransfection assays. Employing differential hydrogen–deuterium exchange mass spectrometry analysis, the authors provide evidence that SR1001 prevents the recruitment of the SRC2 peptide by the receptor interaction domain. Furthermore, in a mouse model of experimental autoimmune encephalomyelitis (EAE), a preclinical model for MS, SR1001 dosed at 25 mg/kg (BID, i.p.) delayed the onset and clinical severity of EAE. In order to improve the selectivity for this class of compounds versus RORα, modifications to the SR1001 scaffold were undertaken that led to the identification of SR2211 (**3**) with a binding K_i of 105 nM for RORγ in a scintillation proximity assay.[44] The functional transcriptional activity of SR2211 was assessed using a chimeric receptor GAL4-RORγ LBD and a full-length RORγ protein construct. SR2211 was found to be selective for RORγ (IC_{50} ~ 350 nM in the GAL4 assay) with minimal or no activity against RORα, LXRα, FXR, and VP-16. The authors also indicate that lipophilic groups were introduced into SR2211 in order to improve the CNS penetration of the compound compared to SR1001. However, preclinical pharmacokinetic studies documenting exposure levels in the CNS are not disclosed in the paper. A slightly less potent compound SR1555 (**4**) (GAL4 IC$_{50}$ = 1.5 μM) that belongs to the same structural class as SR2211 was recently disclosed,[45] was shown to inhibit Th17 cell development and function and increase the frequency of T regulatory cells (Treg) in an *in vitro* Th17 cell differentiation assay. In contrast, SR2211 (**3**) did not alter Treg cell proliferation.

A screen of 4812 compounds employing the GAL4-RORγ LBD functional assay and a displacement assay using 25-hydroxy cholesterol led to the identification of the cardiac glycoside digoxin (Fig. 12.1) as a RORγt selective ligand with IC$_{50}$ values of ~1.98 μM and 4.1 μM in the functional and ligand binding assays, respectively,[46] while maintaining selectivity against other NHRs. Circular dichroism studies suggested that digoxin directly interacts with the LBD of RORγ. Closely related compounds, digitoxin (compound devoid of the C-12 hydroxyl, see Fig. 12.1) and β-acetyldigoxin (acetyl group on the C-4 hydroxyl of the terminal digitoxose ring), have similar IC$_{50}$ values to digoxin in the GAL4-RORγ LBD assay. In a mouse EAE model, digoxin dosed at 40 μg (i.p., starting from day 2 after disease induction) delayed the onset and severity of disease. Digoxin was also shown to ameliorate inflammation in a rat adjuvant-induced arthritis model when administered (2 μg/g body weight every day) 4 days after disease induction.[47] However, digoxin had no effect in immune cell infiltration within joints.

In a HTS screen conducted at the NIH Chemical Genomics Center employing more than 300,000 compounds, a series of diphenylpropanamide compounds were identified as selective RORγt inhibitors.[48] SR-9805 (**5**) is an example of a compound from this class with an IC$_{50}$ of 76 nM in a RORγt reporter assay. However, because of the narrow SAR, further improvements in potency and pharmacokinetic parameters in this series could not be optimized thus limiting *in vivo* evaluation. A patent application covering this class of compounds has also published.[49] An additional patent application that covers hydantoins (**6**) which were identified as RORγt antagonists from a HTS screen also published recently.[50]

5
SR-9805

6

7
Ursolic acid

A HTS screen of more than 2000 compounds using an *in vitro* human Th17 differentiation assay led to the identification of ursolic acid (**7**) as a Th17 inhibitor (IC$_{50}$=0.56 µM).[51] Ursolic acid was also shown to prevent the recruitment of the SRC2 peptide to RORγt-LBD with an IC$_{50}$ of 0.68 µM. Binding of SRC2 peptide to the RORα-LBD was not inhibited, suggesting that ursolic acid is a RORγ-specific antagonist. In a mouse EAE model, ursolic acid dosed at ∼150 mg of body weight (i.p., every other day) starting from day 2 after disease induction delayed the onset and clinical severity of EAE.

Patent applications[52a–d] claiming thiophene and thiazole amides (e.g., **8** and **9**) as RORγt antagonists have published recently. Compound **8** has a pIC$_{50}$ of 8 in the FRET and reporter cell-based assays and is claimed to reduce disease severity when dosed at 100 mg/kg (PO, BID) in mouse EAE and CIA models.[52c] However, systemic exposures and the extent to which disease severity was reduced are not described in this patent application. Reduction in disease severity was seen with compound **9** when dosed at 100 mg/kg (PO, BID) in mouse EAE and CIA models.[52a]

8: X = CH
9: X = N

Other heterocyclic amide and sulfonamides as antagonists of RORγt have also been claimed in patent applications.[53a–e] FRET and RORγ reporter assays were used to test the ability of the compounds to inhibit RORγ activity. Representative examples from these structural classes are shown below (**10–13**). Compound **11** has a pIC$_{50}$ of 7.97 in a TR-FRET assay.[53d] Compound **12** was tested in a modified mouse EAE model and was found to be effective in reducing IL-17 levels when dosed at 10 mg/kg (BID, i.p.).[53a]

10   11   12   13

Pyrazole–isoxazole compounds that inhibit the production of both IL-17 and IFNγ in *in vitro* assays have been reported recently.[54] Two compounds, SC-92366 and SC-89732, were shown to inhibit RORγ in a FRET-based assay with IC_{50}'s of 166 nM and 115 nM, respectively. Only partial structures of these two compounds, represented by **14**, have been disclosed. Inhibition of IFNγ is unexpected for compounds targeting RORγt and is potentially an off-target effect for this class of compounds.

Certain azole-based fungicides, for example, triflumizole (**15**) and hexaconazole (**16**), have been reported to inhibit IL-17 production in *in vitro* assays via RORα/γ at micromolar concentrations in ROR-reporter-based assays,[55] suggesting that environmental chemicals can act as modulators of IL-17 expression in immune cells. However, binding data for these compounds against RORα/γ were not reported in the paper. Retinoid-based structures represented by LE450 (**17**) have been claimed as RORγt modulators.[56] LE 450 displaced tritiated 25-HC in a radioligand binding assay and inhibited the production of IL-17 in human PBMCs in a dose-dependent manner.

6. CONCLUSIONS

A large body of emerging preclinical literature provides evidence that RORγ/γt is an important player in driving the pathogenesis of multiple diseases and in maintaining homeostasis, especially at mucosal surfaces. In addition, potent and selective small-molecule RORγ/γt antagonists with a diversity of structures have been disclosed. Undoubtedly, availability of potent and selective antagonists of RORγ/γt with good pharmacokinetic properties will be critical to provide pharmacologic validation of the target in chronic models of diseases and to help understand the safety concerns associated with its targeting. Although clinical proof of concept for an antagonist of the RORγ/γt is yet to be obtained, developing such an antagonist for its broad clinical utility holds promise.

REFERENCES

1. Solt, L. A.; Burris, T. P. *Trends Endocrinol. Metab.* **2012**, *23*, 619.
2. Huh, J. R.; Littman, D. R. *Eur. J. Immunol.* **2012**, *42*, 2232.
3. Chen, T. *Curr. Opin. Chem. Biol.* **2008**, *12*, 418.
4. Huang, P.; Chandra, V.; Rastinejad, F. *Annu. Rev. Physiol.* **2010**, *72*, 247.
5. Jetten, A. M. *Nucl. Recept. Signal.* **2009**, 7, e003.
6. Sun, Z.; Unutmaz, D.; Zou, Y. R.; Sunshine, M. J.; Pierani, A.; Brenner-Morton, S.; Mebius, R. E.; Littman, D. R. *Science* **2000**, *288*, 2369.
7. Ivanov, I. I.; McKenzie, B. S.; Zhou, L.; Tadokoro, C. E.; Lepelley, A.; Lafaille, J. J.; Cua, D. J.; Littman, D. R. *Cell* **2006**, *126*, 1121.
8. Sutton, C. E.; Mielke, L. A.; Mills, K. H. *Eur. J. Immunol.* **2012**, *42*, 2221.
9. Raichur, S.; Lau, P.; Staels, B.; Muscat, G. E. *J. Mol. Endocrinol.* **2007**, *39*, 29.
10. Takeda, Y.; Jothi, R.; Birault, V.; Jetten, A. M. *Nucleic Acids Res.* **2012**, *40*, 8519.
11. Yang, X. O.; Pappu, B. P.; Nurieva, R.; Akimzhanov, A.; Kang, H. S.; Chung, Y.; Ma, L.; Shah, B.; Panopoulos, A. D.; Schluns, K. S.; Watowich, S. S.; Tian, Q.; Jetten, A. M.; Dong, C. *Immunity* **2008**, *28*, 29.
12. (a) Zhu, J.; Yamane, H.; Paul, W. E. *Annu. Rev. Immunol.* **2010**, *28*, 445. (b) Ghosh, S.; Lobera, M.; Sundrud, M. S. *Ann. Reports Med. Chem.* **2011**, *46*, 155.
13. Maddur, M. S.; Miossec, P.; Kaveri, S. V.; Bayry, J. *Am. J. Pathol.* **2012**, *181*, 8.
14. McGeachy, M. J.; Cua, D. J. *Semin. Immunol.* **2007**, *19*, 372.
15. Nakayamada, S.; Takahashi, H.; Kanno, Y.; O'Shea, J. J. *Curr. Opin. Immunol.* **2012**, *24*, 297.
16. Hu, Y.; Shen, F.; Crellin, N. K.; Ouyang, W. *Ann. N. Y. Acad. Sci.* **2011**, *1217*, 60.
17. Tang, C.; Chen, S.; Qian, H.; Huang, W. *Immunology* **2012**, *135*, 112.
18. Cherrier, M.; Ohnmacht, C.; Cording, S.; Eberl, G. *Curr. Opin. Immunol.* **2012**, *24*, 277.
19. Sawa, S.; Lochner, M.; Satoh-Takayama, N.; Dulauroy, S.; Bérard, M.; Kleinschek, M.; Cua, D.; Di Santo, J. P.; Eberl, G. *Nat. Immunol.* **2011**, *12*, 320.
20. Pantelyushin, S.; Haak, S.; Ingold, B.; Kulig, P.; Heppner, F. L.; Navarini, A. A.; Becher, B. *J. Clin. Invest.* **2012**, *122*, 2252.
21. Cai, Y.; Shen, X.; Ding, C.; Qi, C.; Li, K.; Li, X.; Jala, V. R.; Zhang, H. G.; Wang, T.; Zheng, J.; Yan, J. *Immunity* **2011**, *35*, 596.
22. Leppkes, M.; Becker, C.; Ivanov, I. I.; Hirth, S.; Wirtz, S.; Neufert, C.; Pouly, S.; Murphy, A. J.; Valenzuela, D. M.; Yancopoulos, G. D.; Becher, B.; Littman, D. R.; Neurath, M. F. *Gastroenterology* **2009**, *136*, 257.
23. Buonocore, S.; Ahern, P. P.; Uhlig, H. H.; Ivanov, I. I.; Littman, D. R.; Maloy, K. J.; Powrie, F. *Nature* **2010**, *464*, 1371.
24. Steinmetz, O. M.; Summers, S. A.; Gan, P. Y.; Semple, T.; Holdsworth, S. R.; Kitching, A. R. *J. Am. Soc. Nephrol.* **2011**, *22*, 472.
25. Fulton, L. M.; Carlson, M. J.; Coghill, J. M.; Ott, L. E.; West, M. L.; Panoskaltsis-Mortari, A.; Littman, D. R.; Blazar, B. R.; Serody, J. S. *J. Immunol.* **2012**, *189*, 1765.
26. Tilley, S. L.; Jaradat, M.; Stapleton, C.; Dixon, D.; Hua, X.; Erikson, C. J.; McCaskill, J. G.; Chason, K. D.; Liao, G.; Jania, L.; Koller, B. H.; Jetten, A. M. *J. Immunol.* **2007**, *178*, 3208.
27. Meissburger, B.; Ukropec, J.; Roeder, E.; Beaton, N.; Geiger, M.; Teupser, D.; Civan, B.; Langhans, W.; Nawroth, P. P.; Gasperikova, D.; Rudofsky, G.; Wolfrum, C. *EMBO Mol. Med.* **2011**, *3*, 637.
28. Upadhyay, V.; Poroyko, V.; Kim, T. J.; Devkota, S.; Fu, S.; Liu, D.; Tumanov, A. V.; Koroleva, E. P.; Deng, L.; Nagler, C.; Chang, E. B.; Tang, H.; Fu, Y. X. *Nat. Immunol.* **2012**, *13*, 947.
29. Purwar, R.; Schlapbach, C.; Xiao, S.; Kang, H. S.; Elyaman, W.; Jiang, X.; Jetten, A. M.; Khoury, S. J.; Fuhlbrigge, R. C.; Kuchroo, V. K.; Clark, R. A.; Kupper, T. S. *Nat. Med.* **2012**, *18*, 1248.

30. Yeilding, N.; Szapary, P.; Brodmerkel, C.; Benson, J.; Plotnick, M.; Zhou, H.; Goyal, K.; Schenkel, B.; Giles-Komar, J.; Mascelli, M. A.; Guzzo, C. *Ann. N. Y. Acad. Sci.* **2012**, *1263*, 1.
31. Spuls, P. I.; Hooft, L. *Br. J. Dermatol.* **2012**, *167*, 710.
32. Papp, K. A.; Langley, R. G.; Sigurgeirsson, B.; Abe, M.; Baker, D. R.; Konno, P.; Haemmerle, S.; Thurston, H. J.; Papavassilis, C.; Richards, H. B. *Br. J. Dermatol.* **2013**, *168*, 412.
33. Genovese, M. C.; Durez, P.; Richards, H. B.; Supronik, J.; Dokoupilova, E.; Mazurov, V.; Aelion, J. A.; Lee, S. H.; Codding, C. E.; Kellner, H.; Ikawa, T.; Hugot, S.; Mpofu, S. *Ann. Rheum. Dis.* **2012**, *72*, 863.
34. Hueber, W.; Sands, B. E.; Lewitzky, S.; Vandemeulebroecke, M.; Reinisch, W.; Higgins, P. D.; Wehkamp, J.; Feagan, B. G.; Yao, M. D.; Karczewski, M.; Karczewski, J.; Pezous, N.; Bek, S.; Bruin, G.; Mellgard, B.; Berger, C.; Londei, M.; Bertolino, A. P.; Tougas, G.; Travis, S. P. *Gut* **2012**, *61*, 1693.
35. Ueda, E.; Kurebayashi, S.; Sakaue, M.; Backlund, M.; Koller, B.; Jetten, A. M. *Cancer Res.* **2002**, *62*, 901.
36. Lochner, M.; Ohnmacht, C.; Presley, L.; Bruhns, P.; Si-Tahar, M.; Sawa, S.; Eberl, G. *J. Exp. Med.* **2011**, *208*, 125.
37. Kimura, K.; Kanai, T.; Hayashi, A.; Mikami, Y.; Sujino, T.; Mizuno, S.; Handa, T.; Matsuoka, K.; Hisamatsu, T.; Sato, T.; Hibi, T. *Biochem. Biophys. Res. Commun.* **2012**, *427*, 694.
38. Tinahones, F. J.; Moreno-Santos, I.; Vendrell, J.; Chacon, M. R.; Garrido-Sanchez, L.; García-Fuentes, E.; Macias-González, M. *Obesity* **2012**, *20*, 488.
39. Jin, L.; Martynowski, D.; Zheng, S.; Wada, T.; Xie, W.; Li, Y. *Mol. Endocrinol.* **2010**, *24*, 923.
40. Fujita-Sato, S.; Ito, S.; Isobe, T.; Ohyama, T.; Wakabayashi, K.; Morishita, K.; Ando, O.; Isono, F. *J. Biol. Chem.* **2011**, *286*, 31409.
41. Figure generated using *Molecular Operating Environment (MOE)* software, 2011.10; Chemical Computing Group Inc., 1010 Sherbooke St. West, Suite #910, Montreal, QC, Canada, H3A 2R7, **2011**.
42. Kumar, N.; Solt, L. A.; Conkright, J. J.; Wang, Y.; Istrate, M. A.; Busby, S. A.; Garcia-Ordonez, R. D.; Burris, T. P.; Griffin, P. R. *Mol. Pharmacol.* **2010**, *77*, 228.
43. Solt, L. A.; Kumar, N.; Nuhant, P.; Wang, Y.; Lauer, J. L.; Liu, J.; Istrate, M. A.; Kamenecka, T. M.; Rousch, W. R.; Vidovic, D.; Schurer, S. C.; Xu, J.; Wagoner, G.; Drew, P. D.; Griffin, P. R.; Burris, T. P. *Nature* **2011**, *472*, 491.
44. Kumar, N.; Lyda, B.; Chang, M. R.; Lauer, J. L.; Solt, L. A.; Burris, T. P.; Kamenecka, T. M.; Griffin, P. R. *ACS Chem. Biol.* **2012**, *7*, 672.
45. Solt, L. A.; Kumar, N.; He, Y.; Kamenecka, T. M.; Griffin, P. R.; Burris, T. P. *ACS Chem. Biol.* **2012**, *7*, 1515.
46. Huh, J. R.; Leung, M. W. L.; Huang, P.; Ryan, D. A.; Krout, M. R.; Malapaka, R. R. V.; Chow, J.; Manel, N.; Ciofani, M.; Kim, S. V.; Cuesta, A.; Santori, F. R.; Lafaille, J. J.; Xu, H. E.; Gin, D. Y.; Rastinejad, F.; Littman, D. R. *Nature* **2011**, *472*, 486.
47. Cascao, R.; Vidal, B.; Raquel, H.; Neves-Costa, A.; Figueiredo, N.; Gupta, V.; Fonseca, J. E.; Moita, L. F. *Autoimmun. Rev.* **2012**, *11*, 856.
48. (a) Khan, P. M.; El-Gendy, B. E. M.; Kumar, N.; Garcia-Ordonez, R.; Lin, L.; Ruiz, C. H.; Cameron, M. D.; Griffin, P. R.; Kamenecka, T. M. *Bioorg. Med. Chem. Lett.* **2013**, *23*, 532. (b) Huh, J. R.; Englund, E. E.; Wang, H.; Huang, R.; Huang, P.; Rastinejad, F.; Inglese, J.; Austin, C. P.; Johnson, R. L.; Huang, W.; Littman, D. R. *ACS Med. Chem. Lett.* **2013**, *4*, 79.
49. Littman, D.; Huh, J. R.; Huang, R.; Huang, W.; Englund, E. E. Patent Application WO 2011/112263, 2011.

50. Littman, D.; Huh, J. R.; Huang, R.; Huang, W. Patent Application WO 2011/112264, 2011.
51. Xu, T.; Wang, X.; Zhong, B.; Nurieva, R. I.; Ding, A.; Dong, C. *J. Biol. Chem.* **2011**, *286*, 22707.
52. (a) Wang, Y.; Cai, W.; Liu, Q.; Xiang, J. Patent Application WO 2012/027965, 2012. (b) Wang, Y.; Cai, W.; Liu, Q.; Xiang, J. Patent Application WO 2012/028100, 2012. (c) Wang, Y.; Yang, T. Patent Application WO 2012/100732, 2012.(d) Wang, Y.; Yang. T. Patent Application WO 2012/100734, 2012.
53. (a) Kinzel, O.; Steeneck, C.; Kleymann, G.; Albers, M.; Hoffmann, T.; Kremoser, C.; Perovicottstadt, S.; Schluter, T. Patent Application WO 2011/107248, 2011. (b) Steeneck, C.; Kinzel, O.; Gege, C.; Kleymann, G.; Hoffmann, T. Patent Application WO 2012/139775, 2012. (c) Glick, G. D.; Toogood, P. L.; Romero, A. G.; Vanhuis, C. A.; Aicher, T. D.; Kaub, C.; Mattson, M. N.; Thomas, W. D.; Stein, K. A.; Krogh-Jespersen, E.; Wang, Z. Patent Application WO 2012/064744, 2012. (d) Karstens, W. F. J.; van der Stelt, M.; Cals, J.; Azevedo, R. C. R. G.; Barr, K. J.; Zhang, H.; Beresis, R. T.; Zhang, D.; Duan, X. Patent Application WO 2012/106995, 2012. (e) Maeba, T.; Maeda, K.; Kotoku, M.; Hirata, K.; Yamanaka, H.; Sakal, T.; Hirashima, S.; Obika, S.; Shiozaki, M.; Yokota, M. Patent Application WO 2012/147916, 2012.
54. Tasler, S. 244th ACS National Meeting, Philadelphia, PA, 2012; Abstract MEDI-10.
55. Kojima, H.; Muromoto, R.; Takahashi, M.; Takeuchi, S.; Takeda, Y.; Jetten, A. M.; Matsuda, T. *Toxicol. Appl. Pharmacol.* **2012**, *259*, 338.
56. Deuschle, U.; Abel, U.; Kremoser, C.; Schlueter, T.; Hoffman, T.; Perovicottstadt, S. Patent Application WO 2010/049144, 2010.

Oncology

Section Editor: Shelli R. McAlpine
School of Chemistry, University of
New South Wales, Sydney, Australia

Recent Advances in Small-Molecule Modulation of Epigenetic Targets: Discovery and Development of Histone Methyltransferase and Bromodomain Inhibitors

Ramzi F. Sweis, Michael R. Michaelides
AbbVie Inc., North Chicago, Illinois, USA

Contents

1. INTRODUCTION

Epigenetics has served as a rich source of druggable targets for discovery efforts over the course of the past decade. Derived from the Greek "epi" meaning over or above, epigenetics encompasses the realm of gene expression based on mechanisms of gene activation/deactivation that do not involve alterations in the underlying DNA sequence. Maturation of this field has, in certain cases, helped to explain the lack of linearity between the genetic code and the diversity of phenotypes. Furthermore, dysregulated epigenetic pathways have had clear implications on disease states, and there is an accumulating body of knowledge linking aberrant epigenetic marks with cancers, neurological disorders, and autoimmune diseases.[1] Several small-molecule drug discovery efforts toward epigenetic targets have been disclosed over the past decade and have been coupled with success, particularly in the area of cancer treatment. Inhibitors of DNA methyltransferases (DNMTs) were the first drugs approved: 5-azacytidine and 5-aza-2-deoxycytidine for myelodysplastic syndrome (approved in 2004 and 2006, respectively).[2,3] Beyond direct DNA methylation, research has been active in the context of histone modification.[4] Inhibitors of histone deacetylases (HDACs) were brought to market soon after, with vorinostat (2006) and romidepsin (2009) both approved for cutaneous T-cell lymphoma. Several recent reviews have covered ongoing discovery efforts in DNMTs and HDACs, as well as investigations into the potential therapeutic applications of inhibitors from these classes.[5,6] In the last 4 years, however, some promising breakthroughs have been achieved targeting other enzymes and proteins that interact with histones. These recent developments point toward the next wave of potential clinical candidates.

2. HISTONE MODIFICATIONS

The basic subunits of chromatin are nucleosomes, which facilitate the dense packing of DNA into cell nuclei. Nucleosomes are comprised of DNA wrapped around octamers of core histones (two copies of H3, H4, H2A, and H2B) separated by segments of unbound DNA.[7] Lysine residues are well-known sites of numerous types of covalent histone modifications including methylation, acetylation, SUMOylation, and ubiquitination.[1] As highlighted in Fig. 13.1, methylation and acetylation of histone lysines are prominent mechanisms by which transcriptional activation and repression can occur. Acetylation by histone acetyltransferases effectively attenuates the positive charge on lysine and hence its association with the negative backbone of DNA. This results in a less compact form of chromatin, euchromatin, from which active transcription occurs. The reverse process of deacetylation is catalyzed by HDACs, leading to gene silencing associated with heterochromatin, the more compact form of chromatin.[8,9] Unlike acetylation, the dependence of gene expression and silencing on lysine methylation is less uniform. Certain methyl marks are associated with gene

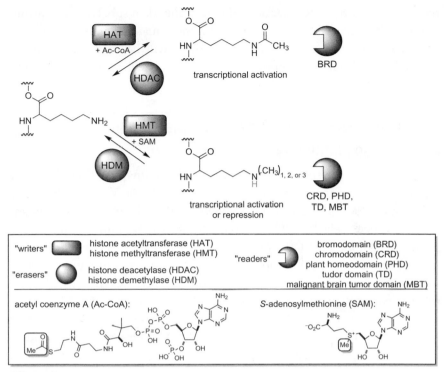

Figure 13.1 Histone lysine modifications and their recognition proteins.

silencing (H3K9Me2/3, H3K27Me3),[a] whereas others are linked to activation of gene expression (H3K4Me3, H3K36Me3, and H3K79Me3).[4,10] Moreover, lysine can be mono-, di-, or trimethylated and the degree of methylation has varying ramifications on observed phenotypes (e.g., H4K20Me: gene expression, H4K20Me3: gene silencing).[11]

The approval of small-molecule HDAC inhibitors as drugs highlights the effort already put forward in targeting these classes of enzymes and proteins for oncology indications. Indeed, there is accumulating evidence for the role of epigenetic dysregulation in tumorigenesis.[9,12] This evidence constitutes an actively growing body of data that has resulted in the identification of different histone "readers," "writers," and "erasers" as candidates for continued research in this context.[13–15] The enzymes responsible for histone lysine modifications, as well as the protein domains that bind to and recognize

---

[a] The designation of an epigenetic methyl mark used herein defines the location of methylation (histone core unit/amino acid residue) and the degree of methylation (mono-, di-, or tri-). That is, H3K9Me3 refers to trimethylation of lysine 9 of histone H3.

these marks, are numerous. Among the most abundant subclasses are histone lysine methyltransferases (>50) and bromodomain-containing proteins (>40). Discovery programs in both of these will be exemplified in this review with an emphasis on the most advanced small-molecule inhibitors to have progressed over the last few years.

3. HISTONE METHYLTRANSFERASE INHIBITORS

3.1. G9a inhibitors

Euchromatic histone methyltransferase 2 (also known as G9a) has been reported to have a role in cancer development. G9a is overexpressed in different cancer types, and its knockdown inhibits the growth of lung and prostate cancer cells.[16,17] This role in disease has brought about interest in the development of G9a inhibitors. As such, a few recent reports have highlighted the G9a inhibitory properties of various compounds derived from natural products and mimetics of the histone methyltransferase (HMT) cofactor/methyl-donor: S-adenosyl-methionine (SAM).[18–20]

The first reported potent and selective small-molecule inhibitor of a histone lysine methyltransferase was BIX01294, **1**.[21] This molecule was identified from a high-throughput screen (HTS) measuring the methyltransferase activity of G9a. Quinazoline **1** exhibited good potency (IC$_{50}$ = 1.7 μM) and selectivity, with no evidence of inhibition of several other lysine and arginine methyltransferases. Reduction of the H3K9Me2 methyl mark was observed in mouse embryonic stem cells, mouse embryonic fibroblast cells, and HeLa cells after a 2-day incubation. It is noteworthy that **1** was reported to be uncompetitive with SAM. Elaborating the template provided by **1**, two analogs were recently reported with improved potency toward G9a. Utilizing information obtained from a crystal structure of **1** bound to a related HMT, G9a-like protein (GLP),[22] a 3-dimethylaminopropoxy-subunit, was installed in place of the 7-methoxy moiety to occupy the lysine-binding channel.[23,24] This key modification led to compound **2** (UNC0224), which exhibited improved potency relative to **1** as measured in multiple assays (i.e., fivefold higher affinity via isothermal titration calorimetry (ITC)). Despite this higher potency *in vitro*, quinazoline **2** was less potent in cellular assays due to lower cell permeability.[25] Further optimization led to compound **3** (UNC0638), which possessed not only similar *in vitro* potency to its predecessor, but also robust cellular potency. In MDA-MB-231 cells, **3** reduced H3K9Me2 levels with an IC$_{50}$ of 81 nM (sixfold more potent than **1**). Moreover, the separation of functional potency (methyl mark reduction) and cell toxicity (as measured via an MTT (3-(4,5-dimethylthiazol-2-yl)-2,5-diphenyl tetrazolium bromide) assay) was significantly improved over **1** (ratio increased by >23-fold). The improved potency and selectivity of **3** over its predecessors render it a useful probe for evaluating the function of G9a more effectively.

1

2

3

3.2. EZH2 inhibitors

Enhancer of zeste homolog 2 (EZH2) is a core component of the polycomb repressive complex 2, which catalyzes the methylation of H3K27. EZH2 has been reported to play a role in tumor progression and is overexpressed in a large number of cancers.[26] In addition, somatic mutations of this HMT have been found in diffuse large B-cell lymphomas (DLBCLs) and follicular lymphomas. The best-characterized point mutation of EZH2 in lymphomas occurs at Tyr641. This results in differential substrate selectivity, with enhanced catalytic activity toward di- and trimethylation of H3K27 relative to monomethylation. In contrast, the wild-type EZH2 is most efficient as a monomethyltransferase. Thus, the combined activity of wild-type and mutant EZH2 results in hypertrimethylation of H3K27, which correlates with aberrant proliferation.[27] The association of EZH2 overexpression and its point mutations with cancer has led to several recent drug discovery efforts targeting the inhibition of this HMT.

3.2.1 4,6-Disubstituted 3-amidomethyl pyridones

The first reports of potent and selective EZH2 inhibitors were a series of 3-aminomethyl-dialkyl-pyridones described in two patent applications (e.g., compounds **4** and **6**).[28,29] Subsequent patent applications covered variations on the central indole and indazole to include other indole- and pyrazolopyridine amides (**5**, **7**).[30,31] These compounds were identified through an HTS of ~2 million compounds.[32] Two different assay formats were used: measuring the methylation of either the H3K27 peptide or the nucleosome. Compound **8** was identified as a submicromolar hit in both screens. Preliminary structure–activity relationships (SAR) suggested that the pyridone moiety, its 4,6-disubstitution pattern, the linking amide, and a branched alkyl at the 1-position of the pyrazolopyridine are all important features of the pharmacophore.

4 X = CH
5 X = N

6

7

8

9

10

One area of fruitful investigation was identified at the R^1 position of the core, as substitution of the cyclopropyl unit in compound **8** with an amine-containing group via an aryl linker provided a large boost in potency. Further optimization of the pyridone 4-position and replacement of the pyrazolopyridine core with an indazole led to compound **9** (GSK343).[33] Indazole **9** is a SAM-competitive inhibitor of EZH2. Modeling of **9** in the SAM site suggests that the pyridone NH and carbonyl group make key hydrogen bonds with the backbone amide of His689, analogous to those seen with the adenine moiety of SAM. Pyridone **9** is a potent EZH2 inhibitor in both enzymatic and cellular assays (enzyme $K_i^{app} = 1.2$ nM/cellular H3K27Me3 $IC_{50} = 174$ nM)[b] but displays high clearance in rat pharmacokinetic studies. It is also very selective (>1000-fold) against 17 other methyltransferases.

[b] The cellular assay was run in HCC1806 breast cancer cells.

Surprisingly, it is also selective over the highly homologous EZH1 (albeit only 60-fold). As such, it has demonstrated utility as an *in vitro* tool.

Indole **10** (GSK126) was derived from the same HTS, and is a potent inhibitor of wild-type and mutant EZH2 methyltransferase activity ($K_i^{app} = 0.5$–3.0 nM).[34] Compound **10** is also selective (>1000-fold) against 20 other methyltransferases and EZH1 (by 150-fold). In a panel of B-cell lymphoma cell lines, DLBCL cell lines were the most sensitive to EZH2 inhibition by **10** ($IC_{50} = 28$–861 nM). Among these, six of the seven lines harbored EZH2 mutations, highlighting the dependence of these cells on EZH2 activity for cell growth. *In vivo* mouse xenograft studies were conducted with two of the most sensitive DLBCL cell lines: KARPAS-422 and Pfeiffer. In both models, 50 mg/kg once daily IP dosing led to decreased levels of H3K27Me3 after 10 days and complete tumor growth inhibition over ~35 days of dosing.

Structurally related compounds were independently discovered via screening of a 175,000 compound chemical diversity library.[35] Similar to the screen that identified compound **8**, many of the hits in this screen proved to be false positives or nonviable due to a variety of reasons (e.g., promiscuity, poor solubility, redox active, or aggregate forming). Additional screening through elaboration of initial hits led to **11**, which after further optimization provided compound **12** (EPZ-005687). As with pyridone **9**, **12** inhibits EZH2 in a SAM-competitive manner with a K_i of 24 nM. Compound **12** has similar affinity for the wild-type and Tyr641 mutant, but it has significantly higher affinity for the A677G-mutant enzyme. The scope of SAR was expanded through the investigation of substituted benzenes in place of the pyrazolopyridine core (as in compound **13**). A recent patent application includes enzyme and cell proliferation data on 194 examples of such compounds.[36] Pyridone **13** (EPZ-6438), which can be viewed as ring-opened analog of **12**, was shown to be active *in vitro* (EZH2 wt $IC_{50} = 13$ nM; Y641F mutant $IC_{50} = 11$ nM; proliferation WSU-DLCL2 cells, 6 days, $IC_{50} = 369$ nM). In addition, *in vivo* activity was demonstrated in two separate tumor models, using a variety of dosing schedules. In a WSU-DLCL2 lymphoma model, **13** showed statistically significant tumor growth inhibition after 28 days at 150 mg/kg BID (58%). Compound **13** was also tested as the trihydrochloride salt in a KARPAS-422 DLBCL model, resulting in regression at 161 and 322 mg/kg BID oral dosing for the same time period.[37]

A third group discovered the same series of inhibitors via HTS, highlighting the specificity of EZH2 toward similar subsets of compounds present in library collections across different sources.[38] In this case, elaboration led to compound **14** which has a similar biochemical profile to **9**, **10**, and **12**.[39]

11 → **12**

13   **14**

3.2.2 Tetramethylpiperidine-4-benzamides

A different chemotype for the inhibition of EZH2 was claimed in a recently published patent application.[40] Enzymatic potency ranges were provided for the EZH2 wild-type and Y641N mutant. The majority of the nearly 600 examples are 2,6-tetramethyl-substituted piperidines with a 4-benzamide substitution (e.g., **15, 16**: $IC_{50} < 5$ μM). Two prophetic examples (**17, 18**) from this series appear to be designed as hybrid molecules with the aforementioned pyridone series. The tetramethyl piperidine moiety in **17** and **18** likely serves as a replacement of the pyridone of compounds **4–14**, where the pyridone is believed to mimic the adenine in the SAM site.

15   **16**

17 X: CH R: Me
18 X: N   R: H

3.3. DOT1L inhibitors

DOT1-like, histone H3 methyltransferase (DOT1L) is believed to play a crucial role in development of mixed-lineage leukemia (MLL). MLL is a genetically distinct form of acute leukemia characterized by a chromosomal translocation of the *MLL* gene, which results in recruitment of DOT1L to the aberrant gene location and subsequent methylation of H3K79.[41] A potent and selective DOT1L inhibitor, **19** (EPZ-5676), is the first HMT inhibitor to enter clinical trials. A Phase I was initiated in September 2012, with the compound being administered as a 21-day continuous intravenous infusion in patients with advanced hematological malignancies.[42] Only acute leukemia patients with rearrangement of the MLL gene will be eligible for the expanded cohort. The structure and properties of **19** were disclosed at the American Society of Hematology meeting in December of 2012.[43] The compound displayed potent inhibition of DOT1L ($K_i = 80$ pM), a long residence time (>24 h), and was highly selective ($>37,000$-fold against representative lysine and arginine HMTs). It demonstrated selective *in vitro* killing of MLL–rearranged cells and caused regression *in vivo* in a rat xenograft model of MLL-fusion leukemia (at 70 mg/kg per day via continuous 21-day iv infusion). Additionally, the compound was shown to inhibit H3K79 methylation in tumor, bone marrow, and PBMCs in the same model. The effective half-life in rats and dogs after iv administration is 0.25 and 1.5 h, respectively.

Compound **19** is structurally related to **20** (EPZ-004777), a potent and selective DOT1L inhibitor that has been extensively characterized.[44,45] The series of compounds represented by **19** and **20** was originally designed as mechanism-guided SAM mimetics.[c] In the course of SAR studies, intermediate **21** was tested and found to have unexpectedly potent activity against DOT1L ($K_i = 20$ μM). Prompted by this finding, additional investigation led to replacement of the Fmoc group with a *tert*-butyl phenyl urea, as well as extension of the alkyl linker. This resulted in compound **22**, a potent DOT1L inhibitor ($K_i = 13$ nM).[46] Replacement of the N-Me group with isopropyl (believed to occupy the lysine-binding pocket) and replacement of a ring nitrogen with carbon led to **20**. This compound showed an additional gain in potency ($K_i = 0.3$ nM). Extensive kinetic studies indicated that this potency enhancement was driven by reduction in the dissociation rate, resulting in large increases in residence time from <1 min to 1 h. The crystal structure of **22** bound to DOT1L shed light on the origin of the unexpected

---

[c] By a "mechanism-guided" approach, inhibitor design is inspired by knowledge of the natural substrate and of the chemical reaction catalyzed by the target.

potency boost. The *tert*-butyl phenyl urea occupies a previously unknown pocket (not seen in the structures with *S*-adenosyl-L-homocysteine (SAH)) immediately adjacent to the amino acid–binding subsite. This hydrophobic pocket is created by significant conformational changes in the enzyme.

The crystal structure of **20** bound to DOT1L was independently solved and published in 2012, with the authors arriving at similar conclusions.[47] It is noteworthy that the DOT1L structure was solved in a complex with four other closely related analogs. All four analogs recapitulated the same binding mode. However, each structure captured the activation and substrate-binding loops in a different conformation, providing further evidence of the high degree of conformational adaptability of this enzyme. Additionally, compound **23**, which incorporated a bromine at the 7-position of the adenine, was postulated to take advantage of a hydrophobic cleft. This resulted in a boost in potency relative to **20** ($K_D = 0.06$ nM vs. 0.25 nM; cellular H3K79Me2 $IC_{50} = 8.8$ nM vs. 84 nM).

A series of related compounds have been reported wherein the N^6 position of the adenine was elaborated with hydrophobic groups.[48] Groups as large as benzyl are tolerated, with compound **24** inhibiting DOT1L ($K_i = 22$ nM) and maintaining selectivity over PRMT1, CARM1, and SUV39H1. These studies were carried out prior to the publication of the crystal structures. As such, the compounds were modeled as bisubstrate inhibitors, binding in the SAH site and extending through the lysine-binding pocket to the substrate site. However, ITC studies indicated that the inhibitors competed with SAM/SAH and not with the substrate nucleosome.

4. BROMODOMAIN INHIBITORS

The previous examples account for the most advanced recent efforts in targeting histone-modifying enzymes or "writers" (HMTs). Concurrently, significant progress has been reported in targeting bromodomains (i.e., "readers"). As highlighted in Fig. 13.1, bromodomains are the protein domains that recognize acetylated lysine residues. Accordingly, inhibition of this recognition event may potentially alter the consequence of a particular histone acetylation. The BET (bromodomain and extra terminal) family of proteins represents the most well-studied bromodomain targets for small-molecule modulation, particularly in the context of oncology, cardiovascular disease, and immunology. This subset consists of four members: BRD2, BRD3, BRD4, and BRDT. Of these, BRD4 has received the most attention. It has been implicated in cancer for several years, most notably as part of a chromosomal translocation with NUT (nuclear protein in testis). This translocation is the genetic hallmark of a particular form of squamous carcinoma known as BRD4–NUT–midline carcinoma (NMC).[49] BRD4 is also implicated in acute myeloid leukemia (AML) as highlighted by an RNAi screen targeting 243 known chromatin regulators.[50]

The first in a series of several recent disclosures of small-molecule inhibitors of BET proteins was a patent application highlighting the use of thienotriazolodiazepines as antitumor agents.[51] Compounds **25** and **26** exhibited potent binding ($IC_{50} < 200$ nM) to BRD4 in a TRFRET assay, as well as robust efficacy in cell growth inhibition across a panel of 11 cell lines representing a spectrum of tumor types (GI_{50} from 0.06 to 1.03 μM). This finding enabled a separate group in the subsequent development and profiling of **27**, known as (+)-JQ1.[52] The binding of compound **27** to BRD4 was determined by differential scanning fluorimetry and isothermal titration calorimetry. Potency was comparable to compounds **25** and **26**, but the *tert*-butyl ester was designed to enhance selectivity over other receptors known to bind benzodiazepines (e.g., gamma-aminobutyric acid receptors). The functional relevance of binding to BRD4 in cells was demonstrated using a fluorescence recovery after photobleaching assay, where **27** effectively displaced BRD4 from chromatin at submicromolar concentrations. In an NMC cell line, **27** induced a differentiation phenotype accompanied by inhibition of proliferation and G1 cell–cycle arrest. In mouse xenograft models of NMC, tumor growth inhibition was observed after 4 days of treatment (50 mg/kg per day via IP dosing).

A closely related subclass of triazolodiazepines represented by **28** was published at a similar time.[53–55] This diazepine derivative, known as I-BET or GSK525762A, was discovered through an HTS of activators of ApoA-1 expression in HepG2 cells.[56] It displayed the highest affinity toward the BET proteins and was effective in suppressing inflammatory gene expression (i.e., LPS-inducible cytokines and chemokines). This compound was advanced to clinical development for treatment of NMC (currently in Phase I/II).[57]

More recently, a series of patent applications were disclosed that expanded the scope of thieno- and benzodiazepines as inhibitors of BRD4. Examples highlighting this are represented by **29**, which contains an isoxazole replacement for the triazole,[58] and compound **30**, which possesses additional saturation on the central ring.[59] The two series represented by these compounds also define new variations on the appendages to the central diazepene core. Both **29** and **30** are potent in an AlphaLISA assay (<0.5 μM) measuring binding to BRD4.

The structural features of both **27** and **28** complexed to BRD4 have been reported, and not surprisingly, both bind in a similar fashion by occupying the acetylated lysine pocket. The triazole mimics acetylated lysine, partaking in the same hydrogen bond interaction with asparagine 140 or 429 (for bromodomain 1 and 2 of BRD4, respectively). Noteworthy in both molecules is the importance of the lone stereocenter, with the more potent enantiomer illustrated.

Since the discovery of **27**, it has served as a useful tool in identifying therapeutic opportunities. It has been particularly valuable in studying bromodomains, since they are often subunits in multifunction proteins or are part of protein complexes. Hence, a selective small-molecule inhibitor would be expected to provide a clearer understanding (compared to a genetic knockdown) of specific phenotypes related to bromodomain inhibition. Illustrating this, two recent reports highlighted the role of BET inhibition in down-regulating MYC transcription in several myeloma, leukemia, and lymphoma cell lines.[60,61] Treatment of a panel of several multiple myeloma (MM) cell lines with **27** resulted in inhibition of cell proliferation. Efficacy of **27** was also demonstrated in mouse models of MM and AML. Compounds **29** and **30** are also effective in reducing mRNA levels of c-Myc in an AML cell line (MV4;11).

In addition to diazepine-containing compounds, dimethylisoxazoles represent another common motif reported in BET inhibitors. Isoxazoloquinoline **31** (I-BET151) is one such compound which was also derived from a screen of ApoA-1 activators and subsequent SAR optimization.[62,63] The potency and selectivity of **31** toward the BET proteins is comparable to **28** (submicromolar IC_{50}s and selectivity against 23 other bromodomain proteins).[64] Compound **31** was used as a tool to probe BET inhibition in the context of MLL-fusion-driven leukemia. It was generally potent (IC_{50}: <200 nM) in a panel of cell lines containing different MLL fusions. In two mouse models of MLL leukemia, **31** effectively slowed disease progression and provided a survival benefit in both.

The dimethylisoxazole subunit has been independently reported as a privileged structural motif for bromodomains arising from evaluation of methyl-substituted heterocycles[65] as well as substructure-based screening efforts.[66,67] Similar to the triazole in compounds **25–28**, the isoxazole serves as an acetylated lysine mimic. It binds via the oxygen accepting a hydrogen bond from the key asparagine residue, the nitrogen being part of a water-bridged interaction with a nearby tyrosine, and the two methyl groups occupying small-lipophilic pockets.

The applicability of such screening approaches to this class of proteins was further demonstrated by the development of quinazolinone **32**.[68] 3-Methyl-3,4-dihydroquinazolin-2(1*H*)-one was chosen as a preferred starting point for optimization based on its potency in an Alphascreen format toward BRD4 (sub–30 μM). In addition, a small degree of selectivity was observed based on lack of activity toward a few other bromodomains. The optimized compound **32** displayed submicromolar potency toward BRD4, which represented a >100-fold boost in potency achieved in only two design cycles and involving the synthesis of less than 250 compounds. Tetrahydroquinoline **33** also was recently reported in a patent application for BET inhibitors (IC$_{50}$ < 400 nM in a fluorescence anisotropy binding assay),[69] and the methyl-tetrahydroquinoline core was originally disclosed as a privileged substructure as well.[52] Additionally, a distinct class of methyl-triazolopyridazines, as depicted by compound **34**, has been reported as potent inhibitors of BRD4 (IC$_{50}$ < 500 nM in an AlphaLISA assay).[70] Finally, quinazolinone **35** (RVX-208) is noteworthy as the most advanced BET inhibitor in the clinic.[71] This molecule is an HDL cholesterol increasing agent which acts by enhancing ApoA-1 expression. It is currently being evaluated in a second Phase IIb clinical trial (ASSURE) for its ability to regress atherosclerotic disease. To date, these examples represent the outcomes of many different strategies employed for inhibitor identification in this class of proteins.

5. CONCLUSIONS AND OUTLOOK

The past few years have seen tremendous growth in the development of small-molecule inhibitors of HMTs and bromodomains, and the efforts have yielded some early promise. Progress has been facilitated by groups such as the Structural Genomics Consortium, from which several compounds highlighted here have been made available as probes to the research community.[72] Notable advances include three BET inhibitors in the clinic. Two have been listed in Phase I clinical trials: GSK525762 (**28**) for NMC and OTX015 for various hematological malignancies. One is currently in Phase IIb for treating atherosclerotic disease: RVX–208 (**35**). From the HMT class, there is a Phase I study for DOT1L inhibitor EPZ-5676 (**19**) for MLL-rearranged AML. The potential for modulating additional epigenetic targets remains to be seen. The "druggability" (i.e., the potential for the function of a protein or enzyme to be modulated by a small-drug-like molecule) of various bromodomains is by no means predicted to be equivalent.[73] This holds true for HMTs as well.[74] In addition, complexities of inhibitor discovery include selectivity within a subclass and substrate/cofactor site inhibition (for HMTs). The abundance of targets in both of these classes with reported implications in cancer and other therapeutic areas suggests that this will remain an active field of investigation for the foreseeable future.[75]

REFERENCES

1. Portela, A.; Esteller, M. *Nat. Biotechnol.* **2010**, *28*, 1057.
2. O'Dwyer, K.; Maslak, P. *Expert Opin. Pharmacother.* **2008**, *9*, 1981.
3. Lyko, F.; Brown, R. *J. Natl. Cancer Inst.* **2005**, *97*, 1498.
4. Jones, P. *MedChemComm* **2012**, *3*, 135.
5. Kelly, T. K.; De Carvalho, D. D.; Jones, P. A. *Nat. Biotechnol.* **2010**, *28*, 1069.
6. Arrowsmith, C. H.; Bountra, C.; Fish, P. V.; Lee, K.; Schapira, M. *Nat. Rev. Drug Discov.* **2012**, *11*, 384.
7. Luger, K.; Mäder, A. W.; Richmond, R. K.; Sargent, D. F.; Richmond, T. J. *Nature* **1997**, *389*, 251.
8. Villar-Garea, A.; Imhof, A. *Biochim. Biophys. Acta* **2006**, *1764*, 1932.
9. Li, B.; Carey, M.; Workman, J. L. *Cell* **2007**, *128*, 707.
10. Chi, P.; Allis, C. D.; Wang, G. G. *Nat. Rev. Cancer* **2010**, *10*, 457.
11. Martin, C.; Zhang, Y. *Nat. Rev. Mol. Cell Biol.* **2005**, *6*, 838.
12. Jones, P. A.; Baylin, S. B. *Cell* **2007**, *128*, 683.
13. Bissinger, E.-M.; Heinke, R.; Sippl, W.; Jung, M. *MedChemComm* **2010**, *1*, 114.
14. Furdas, S. D.; Carlino, L.; Sippl, W.; Jung, M. *MedChemComm* **2012**, *3*, 123.
15. Copeland, R. A.; Moyer, M. P.; Richon, V. M. *Oncogene* **2013**, *32*, 939.
16. Watanabe, H.; Soejima, K.; Yasuda, H.; Kawada, I.; Nakachi, I.; Yoda, S.; Naoki, K.; Ishizaka, A. *Cancer Cell Int.* **2008**, *8*, 15.
17. Kondo, Y.; Shen, L.; Ahmed, S.; Boumber, Y.; Sekido, Y.; Haddad, B. R.; Issa, J. P. *PLoS One* **2008**, *3*, e2037.
18. Greiner, D.; Bonaldi, T.; Eskeland, R.; Roemer, E.; Imhof, A. *Nat. Chem. Biol.* **2005**, *1*, 143.
19. Fujishiro, S.; Dodo, K.; Iwasa, E.; Teng, Y.; Sohtome, Y.; Hamashima, Y.; Ito, A.; Yoshida, M.; Sodeoka, M. *Bioorg. Med. Chem. Lett.* **2013**, *23*, 733.
20. Yuan, Y.; Wang, Q.; Paulk, J.; Kubicek, S.; Kemp, M. M.; Adams, D. J.; Shamji, A. F.; Wanger, B. K.; Schreiber, S. L. *ACS Chem. Biol.* **2012**, *7*, 1152.

21. Kubicek, S.; O'Sullivan, R. J.; August, E. M.; Hickey, E. R.; Zhang, Q.; Teodoro, M. L.; Rea, S.; Mechtler, K.; Kowalski, J. A.; Homon, C. A.; Kelly, T. A.; Jenuwein, T. *Mol. Cell* **2007**, *25*, 473.

22. Chang, Y.; Zhang, X.; Horton, J. R.; Upadhyay, A. K.; Spannhoff, A.; Liu, J.; Snyder, J. P.; Bedford, M. T.; Cheng, X. *Nat. Struct. Mol. Biol.* **2009**, *16*, 312.

23. Liu, F.; Chen, X.; Allali-Hassani, A.; Quinn, A. M.; Wasney, G. A.; Dong, A.; Barsyte, D.; Kozieradzki, I.; Senisterra, G.; Chau, I.; Siarheyeva, A.; Kireev, D. B.; Jadhav, A.; Herold, J. M.; Frye, S. V.; Arrowsmith, C. H.; Brown, P. J.; Simeonov, A.; Vedadi, M.; Jin, J. *J. Med. Chem.* **2009**, *52*, 7950.

24. Liu, F.; Chen, X.; Allali-Hassani, A.; Quinn, A. M.; Wigle, T. J.; Wasney, G. A.; Dong, A.; Senisterra, G.; Chau, I.; Siarheyeva, A.; Norris, J. L.; Kireev, D. B.; Jadhav, A.; Herold, J. M.; Janzen, W. P.; Arrowsmith, C. H.; Frye, S. V.; Brown, P. J.; Simeonov, A.; Vedadi, M.; Jin, J. *J. Med. Chem.* **2010**, *53*, 5844.

25. Vedadi, M.; Barsyte-Lovejoy, D.; Liu, F.; Rival-Gervier, S.; Allali-Hassani, A.; Labrie, V.; Wigle, T. J.; DiMaggio, P. A.; Wasney, G. A.; Siarheyeva, A.; Dong, A.; Tempel, W.; Wang, S.-C.; Chen, X.; Chau, I.; Mangano, T. J.; Huang, X.; Simpson, C. D.; Pattenden, S. G.; Norris, J. L.; Kireev, D. B.; Tripathy, A.; Edwards, A.; Roth, B. L.; Janzen, W. P.; Garcia, B. A.; Petronis, A.; Ellis, J.; Brown, P. J.; Frye, S. V.; Arrowsmith, C. H.; Jin, J. *Nat. Chem. Biol.* **2011**, *7*, 566.

26. Chang, C. J.; Hung, M. C. *Br. J. Cancer* **2012**, *106*, 243.

27. Yap, D. B.; Chu, J.; Berg, T.; Schapira, M.; Cheng, S. W.; Moradian, A.; Morin, R. D.; Mungall, A.; Meissner, B.; Boyle, M.; Marquez, V. E.; Marra, M. A.; Gascoyne, R. D.; Humphries, R. K.; Arrowsmith, C. H.; Morin, G. B.; Aparicio, S. A. *Blood* **2011**, *117*, 2451.

28. Duquenne, C.; Johnson, N.; Knight, S. D.; LaFrance, L.; Miller, W. H.; Newlander, K.; Romeril, S.; Rouse, M. B.; Tian, X.; Verma, S. K. Patent Application WO 2011/140325, 2011.

29. Brackley, J.; Burgess, J. L.; Grant, S.; Johnson, N.; Knight, S. D.; LaFrance, L.; Miller, W. H.; Newlander, K.; Romeril, S.; Rouse, M. B.; Tian, X.; Verma, S. K. Patent Application WO 2011/140324, 2011.

30. Burgess, J. L.; Johnson, N.; Knight, S.; LaFrance, L.; Miller, W. H.; Newlander, K.; Romeril, S.; Rouse, M. B.; Tian, X.; Verma, S. K.; Suarez, D. Patent Application WO 2012/005805, 2012.

31. Burgess, J. L.; Johnson, N.; Knight, S.; LaFrance, L.; Miller, W. H.; Newlander, K.; Romeril, S.; Rouse, M. B.; Tian, X.; Verma, S. K.; Suarez, D. Patent Application WO 2012/075080, 2012.

32. Diaz, E.; Machutta, C. A.; Chen, S.; Jiang, Y.; Nixon, C.; Hofmann, G.; Key, D.; Sweitzer, S.; Patel, M.; Wu, Z.; Creasy, C. L.; Kruger, R. G.; LaFrance, L.; Verma, S. K.; Pappalardi, M. B.; Le, B.; Van Aller, G. S.; McCabe, M. T.; Tummino, P. J.; Pope, A. J.; Thrall, S. H.; Schwartz, B.; Brandt, M. *J. Biomol. Screen.* **2012**, *17*, 1279.

33. Verma, S. K.; Tian, X.; LaFrance, L. V.; Duquenne, C.; Suarez, D. P.; Newlander, K. A.; Romeril, S. P.; Burgess, J. L.; Grant, S. W.; Brackley, J. A.; Graves, A. P.; Scherzer, D. A.; Shu, A.; Thompson, C.; Ott, H. M.; Van Aller, G. S.; Machutta, C. A.; Diaz, D.; Jiang, Y.; Johnson, N. W.; Knight, S. D.; Kruger, R. G.; McCabe, M. T.; Dhanak, D.; Tummino, P. J.; Creasy, C. L.; Miller, W. H. *ACS Med. Chem. Lett.* **2012**, *3*, 1091.

34. McCabe, M. T.; Ott, H. M.; Ganji, G.; Korenchuk, S.; Thompson, C.; Van Aller, G. S.; Liu, Y.; Graves, A. P.; Della Pietra, A., III.; Diaz, E.; LaFrance, L. V.; Mellinger, M.; Duquenne, C.; Tian, X.; Kruger, R. G.; McHugh, C. F.; Brandt, M.; Miller, W. H.; Dhanak, D.; Verma, S. K.; Tummino, P. J.; Creasy, C. L. *Nature* **2012**, *492*, 108.

35. Knutson, S. K.; Wigle, T. J.; Warholic, N. M.; Sneeringer, C. J.; Allain, C. J.; Klaus, C. R.; Sacks, J. D.; Raimondi, A.; Majer, C. R.; Song, J.; Scott, M. P.; Jin, L.; Smith, J. J.; Olhava, E. J.; Chesworth, R.; Moyer, M. P.; Richon, V. M.; Copeland, R. A.; Keilhack, H.; Pollock, R. M.; Kuntz, K. W. *Nat. Chem. Biol.* **2012**, *8*, 890.

36. Kuntz, K. W.; Chesworth, R.; Duncan, K. W.; Keilhack, H.; Warholic, N.; Klaus, C.; Zheng, W.; Seki, M.; Shirotori, S.; Kawano, S. Patent Application WO 2012/142504, 2012.

37. Knutson, S. K.; Warholic, N. M.; Wigle, T. J.; Klaus, C. R.; Allain, C. J.; Raimondi, A.; Scott, M. P.; Chesworth, R.; Moyer, M. P.; Copeland, R. A.; Richon, V. M.; Pollock, R. M.; Kuntz, K. W.; Keilhack, H. *Proc. Natl. Acad. Sci. U.S.A.* **2013**, *110*, 7922.

38. Qi, W.; Chan, H.; Teng, L.; Li, L.; Chuai, S.; Zhang, R.; Zeng, J.; Li, M.; Fan, H.; Lin, Y.; Gu, J.; Ardayfio, O.; Zhang, J. H.; Yan, X.; Fang, J.; Mi, Y.; Zhang, M.; Zhou, T.; Feng, G.; Chen, Z.; Li, G.; Yang, T.; Zhao, K.; Liu, X.; Yu, Z.; Lu, C. X.; Atadja, P.; Li, E. *Proc. Natl. Acad. Sci. U.S.A.* **2012**, *109*, 21360.

39. A more recent example of a bioavailable inhibitor that is chemically related to these: Konze, K. D.; Ma, A.; Li, F.; Barsyte-Lovejoy, D.; Parton, T.; MacNevin, C. J.; Liu, F.; Gao, C.; Huang, X.-P.; Kuznetsova, E.; Rougie, M.; Jiang, A.; Pattenden, S. G.; Norris, J. L.; James, L. I.; Roth, B. L.; Brown, P. J.; Frye, S. V.; Arrowsmith, C. H.; Hahn, K. M.; Wang, G. G.; Vedadi, M.; Jin, J. *ACS Chem. Biol.* **2013**, *8*, 1324.

40. Albrecht, B. K.; Audia, J. A.; Gagnon, A.; Harmange, J. -C.; Nasveschuk, C. G. Patent Application WO 2012/068589, 2012.

41. Kritchov, A. V.; Armstrong, S. A. *Nat. Rev. Cancer* **2007**, *7*, 823.

42. A phase 1, open-label, dose-escalation and expanded cohort, continuous IV infusion, multi-center study of the safety, tolerability, PK and PD of EPZ-5676 in treatment relapsed/refractory patients with leukemias involving translocation of the MLL gene at 11q23 or advanced hematologic malignancies (NCT01684150) ClinicalTrials.gov Web Site September 06, 2012.

43. Pollock, R. M.; Daigle, S. R.; Therkelsen, C. A.; Basavapathruni, A.; Jin, L.; Allain, C. J.; Klaus, C. R.; Raimondi, A.; Scott, M. P.; Chesworth, R.; Moyer, M. P; Copeland, R. A.; Richon, V. M.; Olhava, E. J. American Society of Hematology Annual Meeting and Exposition, 2012, 54th: December 09 (Abs. 2379).

44. Daigle, S. R.; Olhava, E. J.; Therkelsen, C. A.; Majer, C. R.; Sneeringer, C. J.; Song, J.; Johnston, L. D.; Scott, M. P.; Smith, J. J.; Xiao, Y.; Jin, L.; Kuntz, K. W.; Chesworth, R.; Moyer, M. P.; Bernt, K. M.; Tseng, J. C.; Kung, A. L.; Armstrong, S. A.; Copeland, R.; Richon, V.; Pollock, R. M. *Cancer Cell* **2011**, *20*, 53.

45. Travers, J.; Blagg, J.; Workman, P. *Nat. Chem. Biol.* **2011**, *7*, 663.

46. Basavapathruni, A.; Jin, L.; Daigle, S. R.; Majer, C. R.; Therkelsen, C. A.; Wigle, T. J.; Kuntz, K. W.; Chesworth, R.; Pollock, R. M.; Scott, M. P.; Moyer, M. P.; Richon, V. M.; Copeland, R. A.; Olhava, E. J. *Chem. Biol. Drug Des.* **2012**, *80*, 971.

47. Yu, W.; Chory, E. J.; Wernimont, A. K.; Tempel, W.; Scopton, A.; Federation, A.; Marineau, J. J.; Qi, J.; Barsyte-Lovejoy, D.; Yi, J.; Marcellus, R.; Iacob, R. E.; Engen, J. R.; Griffin, C.; Aman, A.; Wienholds, E.; Li, F.; Pineda, J.; Estiu, G.; Shatseva, T.; Hajian, T.; Al-Awar, R.; Dick, J. E.; Vedadi, M.; Brown, P. J.; Arrowsmith, C. H.; Bradner, J. E.; Schapira, M. *Nat. Commun.* **2012**, *3*, 1288.

48. Anglin, J. L.; Deng, L.; Yao, Y.; Cai, G.; Liu, Z.; Jiang, H.; Cheng, G.; Chen, P.; Dong, S.; Song, Y. *J. Med. Chem.* **2012**, *55*, 8066.

49. French, C. A.; Miyoshi, I.; Kubonishi, I.; Grier, H. E.; Perez-Atayde, A. R.; Fletcher, J. A. *Cancer Res.* **2003**, *63*, 304.

50. Zuber, J.; Shi, J.; Wang, E.; Rappaport, A. R.; Herrmann, H.; Sison, E. A.; Magoon, D.; Qi, J.; Blatt, K.; Wunderlich, M.; Taylor, M. J.; Johns, C.; Chicas, A.; Mulloy, J. C.; Kogan, S. C.; Brown, P.; Valent, P.; Bradner, J. E.; Lowe, S. W.; Vakoc, C. R. *Nature* **2011**, *478*, 524.

51. Miyoshi, S.; Ooike, S.; Iwata, K.; Hikawa, H.; Sugahara, K. Patent Application WO 2009/084693, 2009.

52. Filippakopoulos, P.; Qi, J.; Picaud, S.; Shen, Y.; Smith, W. B.; Fedorov, O.; Morse, E. M.; Keates, T.; Hickman, T. T.; Felletar, I.; Philpott, M.; Munro, S.; McKeown, M. R.; Wang, Y.; Christie, A. L.; West, N.; Cameron, M. J.; Schwartz, B.; Heightman, T. D.; La Thangue, N.; French, C. A.; Wiest, O.; Kung, A. L.; Knapp, S.; Bradner, J. E. *Nature* **2010**, *468*, 1067.

53. Nicodeme, E.; Jeffrey, K. L.; Schaefer, U.; Beinke, S.; Dewell, S.; Chung, C.-W.; Chandwani, R.; Marazzi, I.; Wilson, P.; Coste, H.; White, J.; Kirilovsky, J.; Rice, C. M.; Lora, J. M.; Prinjha, R. K.; Lee, K.; Tarakhovsky, A. *Nature* **2010**, *468*, 1119–1123.

54. Chung, C.-W.; Coste, H.; White, J. H.; Mirguet, O.; Wilde, J.; Gosmini, R. L.; Delves, C.; Magny, S. M.; Woodward, R.; Hughes, S. A.; Boursier, E. V.; Flynn, H.; Bouillot, A. M.; Bamborough, P.; Brusq, J.-M. G.; Gellibert, F. J.; Jones, E. J.; Riou, A. M.; Homes, P.; Martin, S. L.; Uings, I. J.; Toum, J.; Clément, C. A.; Boullay, A.-B.; Grimley, R. L.; Blandel, F. M.; Prinjha, R. K.; Lee, K.; Kirilovsky, J.; Nicodeme, E. *J. Med. Chem.* **2011**, *54*, 3827.

55. Bailey, J.; Gosmini, R. L. M.; Mirguet, O.; Witherington, J. Patent Application WO 2011/054845, 2011.

56. Gosmini, R. L. M.; Mirguet, O. Patent Application WO 2011/054553, 2011.

57. A phase I/II open-label, dose-escalation study to investigate the safety, pharmacokinetics, pharmacodynamics, and clinical activity of GSK-525762 in subjects with NUT midline carcinoma (NMC) (NCT01587703) ClinicalTrials.gov Web Site April 03, 2012.

58. Albrecht, B. K.; Audia, J. E.; Côté, A.; Gehling, V. S.; Harmange, J. -c.; Hewitt, M. C.; Leblanc, Y. Patent Application WO 2012/075383, 2012.

59. Albrecht, B. K.; Gehling, V. S.; Hewitt, M. C.; Taylor, A. M.; Harmange, J. -C. Patent Application WO 2012/151512, 2012.

60. Delmore, J. E.; Issa, G. C.; Lemieux, M. E.; Rahl, P. B.; Shi, J.; Jacobs, H. M.; Kastritis, E.; Gilpatrick, T.; Paranal, R. M.; Qi, J.; Chesi, M.; Schinzel, A. C.; McKeown, M. R.; Heffernan, T. P.; Vakoc, C. R.; Bergsagel, P. L.; Ghobrial, I. M.; Richardson, P. G.; Young, R. A.; Hahn, W. C.; Anderson, K. C.; Kung, A. L.; Bradner, J. E.; Mitsiades, C. S. *Cell* **2011**, *146*, 904.

61. Mertz, J. A.; Conery, A. R.; Bryant, B. M.; Sandy, P.; Balasubramanian, S.; Mele, D. A.; Bergeron, L.; Sims, R. J. *Proc. Natl. Acad. Sci.* **2011**, *108*, 16669.

62. Mirguet, O.; Lamotte, Y.; Donche, F.; Toum, J.; Gellibert, F.; Bouillot, A.; Gosmini, R.; Nguyen, V.-L.; Delannée, D.; Seal, J.; Blandel, F.; Boullay, A.-B.; Boursier, E.; Martin, S.; Brusq, J.-M.; Krysa, G.; Riou, A.; Tellier, R.; Costaz, A.; Huet, P.; Dudit, Y.; Trottet, L.; Kirilovsky, J.; Nicodeme, E. *Bioorg. Med. Chem. Lett.* **2012**, *22*, 2963.

63. Seal, J.; Lamotte, Y.; Donche, F.; Bouillot, A.; Mirguet, O.; Gellibert, F.; Nicodeme, E.; Krysa, G.; Kirilovsky, J.; Beinke, S.; McCleary, S.; Rioja, I.; Bamborough, P.; Chung, C.-W.; Gordon, L.; Lewis, T.; Walker, A. L.; Cutler, L.; Lugo, D.; Wilson, D. M.; Witherington, J.; Lee, K.; Prinjha, R. K. *Bioorg. Med. Chem. Lett.* **2012**, *22*, 2968.

64. Dawson, M. A.; Prinjha, R. K.; Dittmann, A.; Giotopoulos, G.; Bantscheff, M.; Chan, W.-I.; Robson, S. C.; Chung, C.-W.; Hopf, C.; Savitski, M. M.; Huthmacher, C.; Gudgin, E.; Lugo, D.; Beinke, S.; Chapman, T. D.; Roberts, E. J.;

Soden, P. E.; Auger, K. R.; Mirguet, O.; Doehner, K.; Delwel, R.; Burnett, A. K.; Jeffrey, P.; Drewes, G.; Lee, K.; Huntly, B. J. P.; Kouzarides, T. *Nature* **2011**, *478*, 529.

65. Hewings, D. S.; Wang, M.; Philpott, M.; Fedorov, O.; Uttarkar, S.; Filippakopoulos, P.; Picaud, S.; Vuppusetty, C.; Marsden, B.; Knapp, S.; Conway, S. J.; Heightman, T. D. *J. Med. Chem.* **2011**, *54*, 6761.

66. Chung, C.-W.; Dean, A. W.; Woolven, J. M.; Bamborough, P. *J. Med. Chem.* **2012**, *55*, 576.

67. Bamborough, P.; Diallo, H.; Goodacre, J. D.; Gordon, L.; Lewis, A.; Seal, J. T.; Wilson, D. M.; Woodrow, M. D.; Chung, C.-W. *J. Med. Chem.* **2012**, *55*, 587.

68. Fish, P. V.; Filippakopoulos, P.; Bish, G.; Brennan, P. E.; Bunnage, M. E.; Cook, A. S.; Federov, O.; Gerstenberger, B. S.; Jones, H.; Knapp, S.; Marsden, B.; Nocka, K.; Owen, D. R.; Philpott, M.; Picaud, S.; Primiano, M. J.; Ralph, M. J.; Sciammetta, N.; Trzupek, J. D. *J. Med. Chem.* **2012**, *55*, 9831.

69. Demont, E. R., Gosmini, R. L. M. Patent Application WO 2011/054848, 2011.

70. Albrecht, B. K.; Harmange, J. -c.; Côté, A.; Taylor, A. WO 2012/174487, 2012.

71. Phase IIb multi-center, double-blind, randomized, parallel group, placebo-controlled clinical trial for the assessment of coronary plaque changes with RVX000222 as determined by intravascular ultrasound (NCT01067820) ClinicalTrials.gov Web Site February 10, 2010.

72. www.theSGC.org.

73. Vidler, L. R.; Brown, N.; Knapp, S.; Hoelder, S. *J. Med. Chem.* **2012**, *55*, 7346.

74. Campagna-Slater, V.; Mok, M. W.; Nguyen, K. T.; Feher, M.; Najmanovich, R.; Schapira, M. *J. Chem. Inf. Model.* **2011**, *51*, 612.

75. Prior to the publication of this chapter, a phase I clinical trail for EZH2 inhibitor EPZ-6438 (compound **13**) was also announced: An Open-Label, Multicenter, Phase 1/2 Study of E7438 (EZH2 Histone Methyl Transferase [HMT] Inhibitor) as a Single Agent in Subjects With Advanced Solid Tumors or With B Cell Lymphomas (NCT01897571) ClinicalTrials.gov Web Site June 21, 2013.

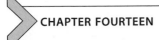

CHAPTER FOURTEEN

Inhibition of Ubiquitin Proteasome System Enzymes for Anticancer Therapy

David Wustrow, Han-Jie Zhou, Mark Rolfe
Cleave Biosciences, Burlingame, California, USA

Contents

1. INTRODUCTION

All proteins within a cell have a life and death cycle of synthesis and degradation. Half-lives of proteins can vary from minutes to days and it is this variation that controls many important aspects of cellular phenotype. Additionally, owing to environmental insults and the inherent error prone nature of the protein synthetic machinery, the cell has to cope with a burden of many damaged, mutated, and misfolded proteins.

There are two major degradative pathways that control the destruction of regulatory and mutated or misfolded intracellular proteins; the ubiquitin proteasome system (UPS) and the autophagy lysosome system.[1,2] Here, we will discuss drug discovery opportunities within the UPS. The UPS tags

Annual Reports in Medicinal Chemistry, Volume 48
ISSN 0065-7743
http://dx.doi.org/10.1016/B978-0-12-417150-3.00014-4

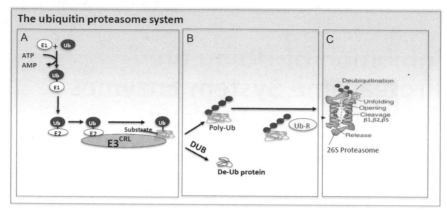

Scheme 14.1 The ubiquitin proteasome system (adapted with permission[3]). (A) Unwanted proteins are tagged for degradation through attachment of ubiquitin (Ub) by a cascade of enzymes (E1, E2, and E3s). (B) The ubiquitin label is removed by a deubiquitinating enzyme (DUB) or the protein tagged with a poly-Ub chain is chaperoned to the proteasome. (C) Before the tagged protein can enter the core of the proteasome, the poly-Ub chain must be removed by a DUB enzyme present in the proteasome. The protein is then unfolded and cleaved by beta subunits in the core of the proteasome. (See color plate.)

proteins that are destined for degradation with the small protein ubiquitin (Ub) via an enzymatic cascade (Scheme 14.1A) that includes two ubiquitin–activating enzymes (E1s), approximately forty ubiquitin conjugating enzymes (E2s), and greater than five hundred ubiquitin ligases (E3s). In the initial step of this cascade, the C-terminal glycine of Ub is activated by attachment to an E1 enzyme. This activated Ub is transferred to an E2 enzyme, which in concert with an E3 enzyme transfers ubiquitin to the target protein.[3] Ubiquitin itself can be internally modified by the conjugation of additional ubiquitin monomers to form poly–ubiquitin chains which when attached via lysine-48 or lysine-11of ubiquitin target the substrate for degradation by the 26S proteasome (Scheme 14.1B and C).[4]

Approaches to targeting enzymes in this cascade will be reviewed below. The ubiquitination cascade is antagonized by a distinct group of enzymes called deubiquitinating enzymes (DUBs). A review of DUB inhibitors has been published[5] and this approach will not be discussed here. There are many ubiquitin–interacting proteins in the human genome and some of these act as binding partners for substrates in order to chaperone them to the 26S proteasome (Scheme 14.1B and C). These ubiquitin–interacting proteins provide an extra layer of complexity to the UPS and, in theory, could provide novel drug targets.[6]

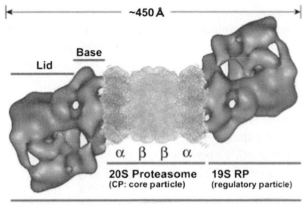

Scheme 14.2 Composition of the 26S proteasome (used with permission).[8] The 26S consist of the 20S core particle and two 19S regulatory particles. The numbers refer to protein subunit size as determined by centrifugation. (See color plate.)

The 26S proteasome is a 2.5 MDa multiprotein assembly spanning over 450 Å[7] that degrades ubiquitinated proteins. The proteolytic activity is found in the 20S core particle (20S CP) and attached on either end of this hollow tube are two 19S regulatory particles (19S RP) (Scheme 14.2). The 20S CP consists of four stacked seven-membered rings—two structural alpha rings (subunits α1–7) and two catalytic beta rings (subunits β1–7) in the arrangement $\alpha\beta\beta\alpha$.

In eukaryotes, three beta subunits β1, β2, and β5 in the 20S CP contain the peptidase activities which cleave substrates into short peptides of 8–15 amino acids in length. The 19S RP consists of a hexameric ring of ATPase subunits (*regulatory particle atpase, rpt* subunits) which forms the base of the 19S RP. These ATPase subunits use the energy of ATP hydrolysis to unwind and translocate substrates into the chamber of the 20S core particle wherein peptide hydrolysis ensues (Schemes 14.1C and 14.2). A number of non-ATPase subunits (*regulatory particle non-atpase, rpn* subunits) form the lid. The Rpn subunits contain ubiquitin-binding proteins (Rpn10 and 13) and a novel metalloenzyme, Rpn11, which is responsible for deubiquitinating substrates just prior to their cleavage by the 20S peptidases.

2. INHIBITORS OF 20S CP PEPTIDASES

The three beta subunits responsible for the peptidase activity in the 20S CP have different cleavage preferences and are inhibited to different extents by various classes of inhibitors. Several groups have used peptide-based approaches to develop selective proteasome inhibitors and two of these (bortezomib and carfilzomib) are now approved for the treatment of patients with multiple myeloma, with several others also in clinical development.[9,10] These agents primarily inhibit the β5 subunit of the 20S CP responsible for the chymotrypsin–like peptidase activity of the proteasome. Three chemical classes of β5 inhibitors have entered clinical trials or been approved as drugs: peptide boronates (bortezomib (**1**), MLN9708 (**2**), and CEP-18770 (**4**)), peptide epoxyketones (carfilzomib (**5**) and ONX 0912 (**6**)), and β-lactones (NPI-0052 (**8**) and PS-519 (**9**)).

2.1. Peptide boronates

Bortezomib (**1**)[11] was initially approved by the FDA in 2003 as a treatment for relapsed and refractory multiple myeloma and was approved as first-line treatment in 2008. The boronic acid portion of **1** binds covalently with the γ-OH group of the N-terminal threonine of the β5 subunit of the 20S CP to reversibly form tetrahedral intermediates (Scheme 14.3).[12] Furthermore, **1** was found to inhibit serine proteases such as HtrA2, which protects neurons from apoptosis. Inhibition of HtrA2 is proposed to be the main cause of peripheral neuropathy, the major dose–limiting toxicity of **1** in patients.[13] Compound **1** manifests its antitumor activity via multiple mechanisms[10] including disruption of cell adhesion- and cytokine–dependent survival pathways (e.g., NF-κB signaling pathway), inhibition of angiogenesis, activation of a misfolded protein stress response (or ER stress), and upregulation of proapoptotic or downregulation of antiapoptotic genes.

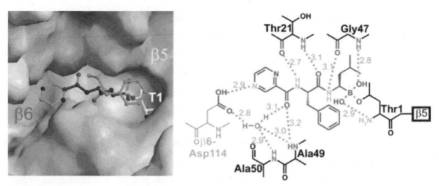

Scheme 14.3 Bortezomib bound to the β5 subunit of the yeast 20S proteasome (used with permission).[12] X-ray structure and schematic representation of covalent binding of **1** to the β5 subunit. (See color plate.)

The tetrahedral adduct that is formed between **1** and the β5 subunit results in a slow dissociation of **1** from its target. The slow release of **1** from red blood cell proteasomes is believed to contribute to the limited effect of **1** in the treatment of solid tumors.[14]

Intensive medicinal chemistry efforts led to the development of a second-generation boronate, MLN9708 (**2**).[14] This compound is a dipeptide boronic ester which is rapidly hydrolyzed in aqueous solutions and plasma to the biologically active species, MLN2238 (**3**). Like bortezomib (**1**), compound **2** selectively inhibits the β5 subunits of the proteasome and possesses a similar *in vitro* inhibitory profile, however unlike **1**, compound **2** is orally bioavailable and does not inhibit HtrA2.[13] It has demonstrated synergistic antimultiple myeloma activity when combined with **1**, histone deacetylase (HDAC) inhibitors, lenalidomide or dexamethasone.[15,16] **2** is also active in bortezomib-resistant multiple myeloma (MM) cell lines.

The most striking difference between compounds **1** and **2** is that the proteasome dissociation half-life of **2** is significantly shorter than **1**.[14] Thus, **2** has a sixfold larger volume of distribution than **1**. This is probably a result of the rapid dissociation of the drug from red blood cell proteasomes resulting in improved tissue distribution, pharmacokinetics, and pharmacodynamics. Upon intravenous administration of **2** into mice, it can achieve greater inhibition of proteasome activity in tumor tissue relative to **1**, and the rate of recovery of proteasome activity in blood was more rapid as compared to **1**.[15,16] Furthermore, compound **2** has a faster off-rate in blood—and perhaps other normal tissues as well—which may contribute partly to more than

10-fold greater maximum tolerated dose in mice as compared to **1**. There was no significant difference in on-rate between the two compounds. Consequently, **2** demonstrated a better antitumor activity in numerous xenograft models including solid tumors. Clinical trials of **2** in MM started in 2009 and many trials of this agent either alone or in combination with other standard of care agents have been initiated in a variety of cancers.

CEP-18770 (**4**, also called delanzomib) is another dipeptide boronic acid-based proteasome inhibitor currently in clinical trials. Though compounds **4** and **1** shared a similar profile of *in vitro* potency, selectivity, and mechanism of action, compound **4** is water soluble and orally bioavailable.[17] It abrogates the production of vascular endothelial growth factor (VEGF) in multiple myeloma cells, which consequently inhibits cell migration and vasculogenesis from endothelial progenitors. Furthermore, the role of **4** in angiogenesis is corroborated by its direct inhibitory effect on endothelial cell proliferation, survival, and capillary tubular morphogenesis. Compound **4** has also been shown to promote apoptosis in human multiple myeloma cell lines.[18] It is more effective in combination with either **1** or melphalan in animal tumor models as compared to its activity as a single agent.[19]

Compound **4** appears to have a significantly reduced toxicity toward human bone marrow progenitors, bone marrow stromal cells, and normal human intestinal cells as compared to **1**. Oral administration of **4** in mice induced an equivalent level of proteasome inhibition in blood and tissues as intravenous administration of **1**.[20] **4** produces greater and more sustained proteasome inhibition in tumor xenograft models. Currently, compound **4** is in two clinical trials in MM utilizing intravenous administration and a weekly administration schedule.

2.2. Peptide α,β-epoxyketones

Peptide epoxyketones react with both the γ-OH and the α-amino groups of the active site threonine in $\beta5$ subunits of the 20S CP and irreversibly form six-membered morpholine rings (Scheme 14.4).[21] The catalytic active sites of serine and cysteine proteases do not possess such a close juxtaposition of the catalytic hydroxyl and the primary amino groups to form such an adduct and this may be important for the observed selectivity of this class of molecules.[22,23]

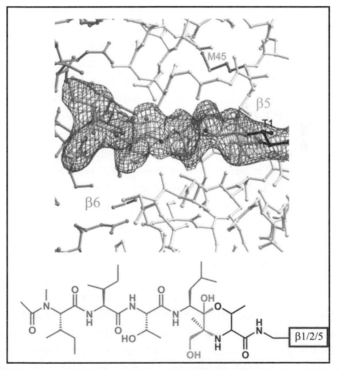

Scheme 14.4 Interaction of an epoxyketone with the β5 subunit of the yeast 20S proteasome (used with permission).[21] The X-ray structure of epoxomicin (**7**) bound to the yeast β5 subunit and the schematic representation. (See color plate.)

Carfilzomib (also known as PR-171) (**5**) is a tetrapeptide epoxyketone and a derivative of the natural product epoxomicin (**7**). An extensive medicinal chemistry effort led to improvement in its inhibitory potency, selectivity, and pharmaceutical properties including solubility by incorporation of a morpholino moiety into the N-cap.[24,25] In 2012, the FDA granted accelerated approval to **5** (Kyprolis™) administered by injection for the treatment of patients with multiple myeloma who have received at least two prior therapies.[26]

Compound **5** irreversibly and selectively inhibits the β5 activity of the 20S proteasome. In distinction to **1**, **5** has shown less inhibitory activity at the other two proteolytic subunits (β1 and β2) of the 20S CP.[24,25] Furthermore, **5** shows minimal reactivity with other protease classes such as serine proteases (e.g., HtrA2).[13] **5** inhibits cell proliferation, and induces apoptosis which is associated with the activation of c-Jun N-terminal kinase (JNK), depolarization of mitochondrial membrane, and release of cytochrome c as well as activation of caspases.[25,27]

Though **5** has demonstrated efficacy in the treatment of multiple myeloma and other malignant diseases, it has to be administrated intravenously owing to its low oral bioavailability. A systematic SAR optimization led to the development of an orally bioavailable tripeptidyl epoxyketone, ONX-0912 (also called PR-047) (**6**), having oral bioavailability of up to 39% in the dog.[28] Compound **6** possesses significantly improved solubility and metabolic stability in gastric fluid, intestinal fluid, liver microsomes, and hepatocytes as well as sensitivity to the multidrug resistance protein 1. Compounds **6** and **5** share similar potency and selectivity profiles for the β5 subunit.[28,29] Compound **6** was well tolerated with repeated oral administration at doses resulting in more than 80% proteasome inhibition in most tissues and elicited an antitumor response or prolonged survival either equivalent or better to intravenously administered **5** in multiple human tumor xenograft and mouse syngeneic models.[28,30] Compound **6** inhibits growth and induces apoptosis in MM cells resistant to conventional and bortezomib therapies. Its anti-MM activity is associated with the activation of caspase-8, caspase-9, caspase-3, and poly(ADP-ribose) polymerase, as well as inhibition of migration of MM cells and angiogenesis.[31] Compound **6** also enhances anti-MM activity in combination with other standard of care agents[32] and is currently being studied in two trials in hematologic malignancies and in a Phase I study in patients with recurrent or refractory solid tumors.

2.3. β-Lactones

The β-lactone class is comprised of lactacystin-related proteasome inhibitors. Among them, NPI-0052 (also called as marizomib, salinosporamide A) (**8**) and PS-519 (**9**) have been intensively studied. Both **8** and **9** have entered into clinical trials, in fact, **9** was the first proteasome inhibitor to enter clinical trials as a treatment of reperfusion injury, inflammation, and ischemia.[33] This class of compounds is not as specific as epoxyketones, and their inhibition of other serine proteases and other cellular proteases have been reported.[24]

Compound **8** was first isolated from the fermentation broth of *Salinispora tropica* strain CNB392.[34,35] Like **9** and other β-lactones, **8** inactivates 20S proteasomes by esterifying the catalytic threonine hydroxyl. Kinetic studies and co-crystal structures with the 20S proteasome show that enzyme inhibition occurs in a two-step mechanism. The first step involves the opening of the β-lactone ring by the threonine hydroxyl followed by rate-limiting formation of a covalent enzyme–inhibitor complex; a tetrahydrofuran ring as the result of nucleophilic displacement of the chloride atom of the inhibitor.[36–38] Consequently, **8** irreversibly inhibits the proteasome, and it also is distinct from compounds **1** and **4** in terms of its inhibitory activity against the three major enzymatic activities of the 20S CP.

Compound **8** induced apoptosis and possessed antitumor activity predominantly via caspase-8 induction and reactive oxygen species–dependent pathways.[39,40] At the maximum tolerated dose without apparent toxicity, **8** produces higher (as high as 90%) proteasome inhibition as compared to 70% inhibition by **1** and 80% inhibition by **5**. The blood proteasome inhibition caused by **8** increased progressively over 24 h, and remained essentially unchanged over a prolonged period, compared to the relatively short duration of **1** and **5**. Furthermore, short exposure of **8** caused cell death, and it has shown effectiveness in multiple myeloma cell lines that are resistant to **1**.[41] Compound **8** is efficacious in multiple animal efficacy models both alone and in combination with bortezomib, lenalidomide, and various HDAC inhibitors[42,43] and is presently undergoing Phase I or II clinical trials for the treatment of patients with multiple myeloma, lymphomas, and solid tumors.[44,45]

Reversible inhibitors may offer advantages in selectivity and allow for greater tissue distribution and perhaps enhanced efficacy in solid tumors.[46] But to date none of the reversible inhibitors have been as efficacious in preclinical models as bortezomib.[47,48]

3. INHIBITORS OF UBIQUITIN CONJUGATION

3.1. Ubiquitin-activating and -conjugating enzymes (E1 and E2)

Two groups have reported 2-nitro-furan-based inhibitors of E1 enzymes.[49,50] PYZD-4409 (10) inhibited Uba1 with an IC_{50} of 20 μM in a biochemical assay. In a cellular assay, 10 caused significant reduction in E1 ubiquitination at 50 μM and complete inhibition at 150 μM concentration. It exhibited cytotoxicity against tumor cells and delayed tumor growth in a mouse model of leukemia.

The E2s or ubiquitin conjugating enzymes contain an active site cysteine residue that accepts an activated ubiquitin from E1~Ub and, in a process requiring an E3, transfers the ubiquitin either directly or indirectly to the substrate lysine residue that is targeted for ubiquitination. Efforts to develop specific E2 inhibitors have met with little success, probably owing to the fact that the active site is quite shallow and contains a reactive cysteine residue. However, recently CC0651 (11) was discovered to be a specific allosteric inhibitor of the E2 enzyme Cdc34.[51] Compound 11 appears to inhibit the transfer of ubiquitin to substrates in biochemical and cellular assays but only had modest effects *in vivo* perhaps owing to poor pharmacokinetics.

10

11

3.2. Ubiquitin ligases (E3s)

The E3s or ubiquitin ligases comprise a family of 500 diverse proteins, which generally supply the substrate specificity function of the ubiquitination cascade. From a biological standpoint, the E3 ubiquitin ligases are attractive as drug targets as they provide the specificity in the ubiquitination cascade.[52] However, the majority of E3 ligases have no inherent catalytic activity or any recognizable "active site" so need to be attacked via interrupting protein–protein interactions—an approach that is much more difficult than developing classical small molecular weight enzyme inhibitors. Several groups are continuing to attempt "to drug" E3s and there has been some success with the E3, HDM2 which is an important regulator of the proapoptotic protein p53.

The transcription factor p53 plays an important role in tumor suppression by affecting cell cycle, DNA repair, and apoptosis.[53] The murine double minute 2 (MDM2 in mice and HDM2 in humans) protein suppresses p53 function in many tumor types. HDM2 has E3 ligase activity which ubiquitinates p53 targeting it for destruction in the 26S proteasome.[54] Therefore, the disruption of HDM2–p53 binding is thought to be a way to increase p53 levels and tumor-suppressing activity. The discovery and early development of a number of small molecules that can effectively inhibit HDM2–p53 interactions has been reviewed[55,56] and a number of these agents have entered early stages of clinical development.

Nutlin-3a (**12**) is the most potent member of the first group of small molecules reported as potent inhibitors of the MDM2–p53 protein–protein interaction.[57] Compound **12** is active against cancer cell lines *in vitro* provided they have no p53 inactivating mutation and in tumor xenograft models.[58] Further investigations around compound **12** resulted in RG-7112 (**13**)[55] which has been taken into Phase I clinical trials for advanced solid tumors, and for refractory acute and chronic leukemia. Elevated p53 levels were detected in tumor tissues of patients treated with **13**. Clinical trials with another molecule in this class, RO5503781 (structure currently unknown), have also been listed.

12   13   14

15   16   17

X-ray crystallography studies of a related molecule (**14**) revealed that these molecules occupy a deep binding pocket in the N-terminal region of MDM2 that recognizes p53.[57] This deep pocket motif is important for this interaction and is thought to be why the MDM2–p53 interaction is more susceptible to inhibition by small molecules than other protein–protein interactions.

Other groups have utilized structural-based approaches to discover novel HDM2 inhibitors. A combination of rational design and systematic optimization produced piperidinone **15**, which had potent affinity for HDM2 and activity against cancer cells *in vitro* and in tumor xenograft models.[59] A structure-based approach resulted in the discovery of the spirooxindole **16** as a potent and selective inhibitor of the HDM2–p53 interaction.[60] Compound **16** increases cellular p53 levels and is active in tumor xenograft models. A compound from this class of molecules entered clinical development in 2012.[55] Published patents and applications have disclosed molecules such as **17**,[61] **18**,[62] and **19**[63] to be potent HDM2 inhibitors and assignees of these patents are each currently conducting Phase I clinical trials with HDM2 inhibitors.

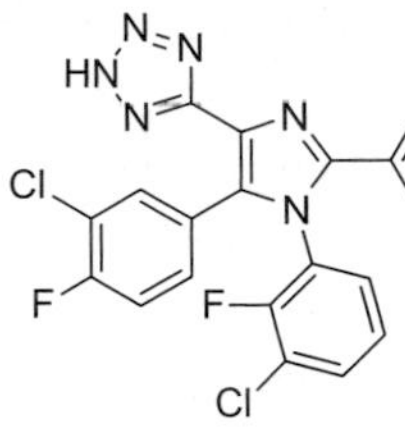

18

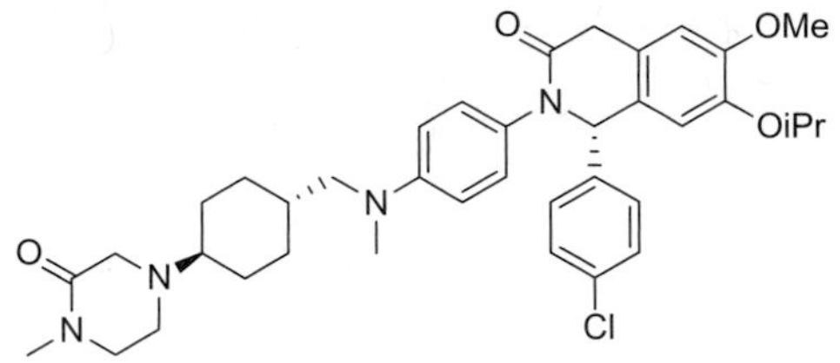

19

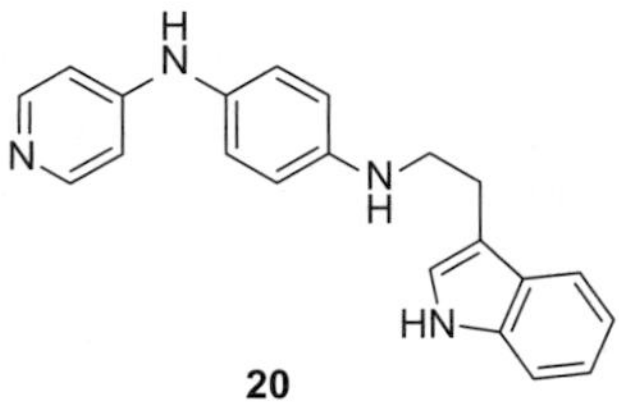

20

Serdemetan (JNJ-26854165, **20**) induces p53 levels in tumors by inter-fering with its ubiquitination and potently inhibits ovarian, lung, and pros-tate cancer cells with IC$_{50}$s ranging from 60 nM to 7.7 μM.[64,65] It binds to the RING domain of HDM2 responsible for its E3 ligase function preventing the ubiquitination and degradation of p53. Compound **20** cau-sed induction of p53 in U87 glioblastoma xenograft tumors and was active in a variety of tumor xenograft models.[64] **20** was tolerated in Phase I trials and dose and plasma concentration–dependent induction of p53 was observed.[66]

The Imids thalidomide (**21**), pomalidomide (**22**), and lenalidomide (**23**) are approved for the treatment of multiple myeloma, but the teratogenicity of these compounds is well documented. In an effort to identify the Imid–protein interactions responsible for these effects, the Imid analog FR 259625 (**24**) was covalently attached to ferrite-glycidyl methacrylate (FG) affinity beads as depicted in Scheme 14.5.[68] HeLa cell extracts were passed over

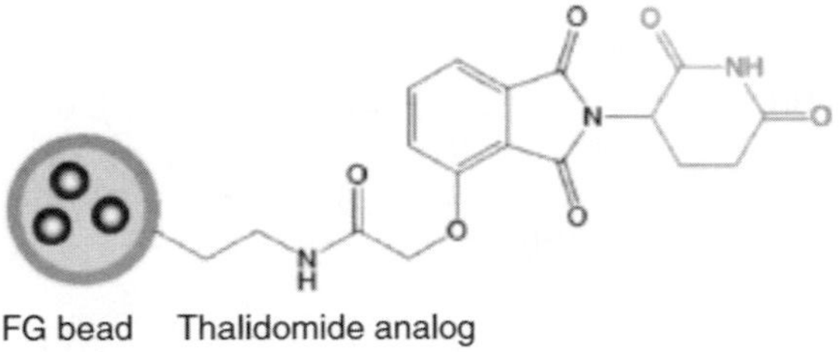

FG bead   Thalidomide analog

Scheme 14.5 Schematic of bead-based probe used to capture the E3 ligase cereblon (used with permission).[67] (For color version of this figure, the reader is referred to the online version of this chapter.)

these FG beads which retained cereblon (CRBN) and identified it as the thalidomide-binding protein. This interaction is responsible for both the teratogenic effects of **21**[68] (and presumably other Imids) as well as the immunomodulatory and antiproliferative effects of **21**, **22**, and **23**.[67] CRBN forms a complex with the DNA damage-binding protein and the E3 ubiquitin ligase Cul4B. Imid binding to CRBN prevents its ubiquitination by this complex. Reduction in cellular CRBN levels results in loss of sensitivity to the antiproliferative effects of the Imids.

The molecular interactions of CRBN with Imid analogs have been investigated by evaluating their ability to disrupt the binding of CRBN to the affinity column.[67] The glutarimide functionality of **21** was found to elute CRBN from the column while the phthalimide function did not. Using similar methodology, it was determined that the interaction of **22** and **23** with CRBN appears to be more potent than **21**. The stereospecificity of the interaction of Imids with CRBN was studied using the methyl substituted enantiomers of **22**, compounds **25** and **26**. Incubation of cellular extracts with the S-isomer **25** significantly inhibited both the ability of CRBN to be retained by affinity chromatography while the R-isomer **26** did not have this effect. Isomer **25** was able to increase IL-2 expression while isomer **26** did not. Further investigations of the structure activity relationship of simple analogs of **23** identified compounds **27** and **28** as having potent effects of modulating cytokine expression, including increasing IL-2 levels and antiproliferative activity against the Namalwa lymphoma cell line.[69]

21

22 X = O
23 X = H$_2$

24

25

26

27 R = CH$_3$
28 R = Cl

3.3. Inhibitors of nedd8-activating enzyme

The cullin ring ligases form a major subgroup of E3 ligases and are attractive as drug discovery targets as they control the turnover of many important

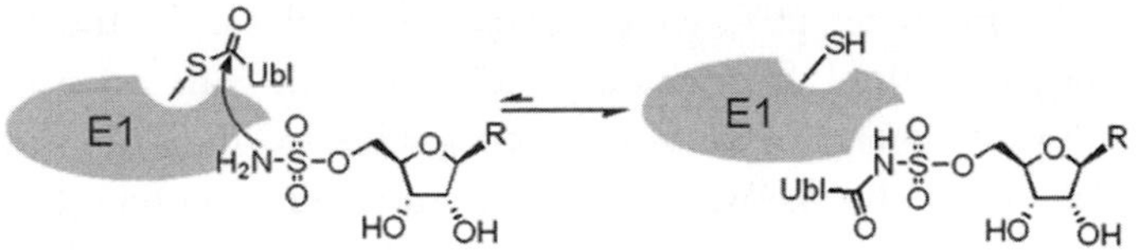

Scheme 14.6 Substrate-assisted inhibition of ubiquitin-activating enzymes by adenosine sulfamate analog (used with permission).[74]

regulatory proteins.[70] In order for the complex to achieve full enzymatic activation the cullin subunit needs to be posttranslationally modified on a specific lysine residue by the small ubiquitin-like protein, nedd8. This so-called neddylation reaction is catalyzed by a distinct E1-like enzyme (nedd8-activating enzyme—NAE) coupled with a specific E2 called ubc12. A potent and specific inhibitor of NAE has been developed. MLN4924 (**29**) exhibited preclinical activity in a wide variety of solid tumor and hematological malignancies and is currently in Phase I trials in cancer patients.[71,72] Compound **29** has a unique mechanism of action—the actual inhibitory moiety is a covalent adduct formed between the C-terminal glycine of nedd8 and the sulfamate of **29** which is generated inside cells by the action of NAE itself.[73,74] In this substrate-assisted inhibition of ubiquitin-activating enzymes, the sulfamate moiety of **29** or related molecules attacks the E1–ubiquitin complexed protein to form the inhibitor (Scheme 14.6).[74]

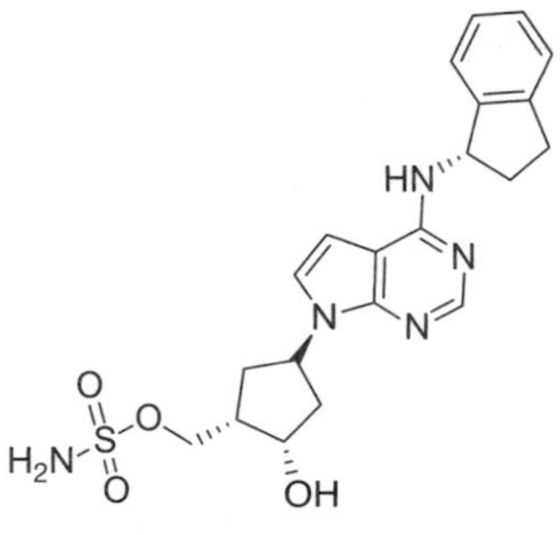

29

4. UBIQUITIN-INTERACTING PROTEINS AND CHAPERONES

There are a large number of ubiquitin-interacting proteins and ubiquitin-binding proteins in the human genome. To interfere with their function would require the disruption of a broad protein–protein interaction surface. An alternative approach being taken by some groups is to target the ATPase function of the molecular machine, p97, which could indirectly affect the numerous functions of p97 in the UPS including the association with its cofactors that mediate the chaperoning of a number of substrates to the 26S proteasome. The ATPase p97 (also known as valosin-containing protein (VCP) and Cdc48) has been shown to be critical for efficient degradation of ubiquitinated proteins as well as other cellular functions.[75] It is a member of a group called the ATPases Associated with diverse cellular Activity (AAA+) family.[76] Small molecules which inhibit the ATPase function of p97 could prevent the mechanical action of p97 assemblies and therefore inhibit the UPS.

A mass screen of p97 ATPase activity identified compound **30** (DBeQ). Compound **30** reversibly inhibits the ATPase function of p97 in an ATP competitive manner.[77] The molecule has shown selectivity for p97 over other ATPases including NSF and the ATPase function of the 26S CP. In cellular assays **30** inhibited the proteasomal degradation of TCRα, caused inhibition of the autophagy pathway and was cytotoxic to a number of cancer cell lines inducing cell death by rapid induction of caspase 3/7.

SAR studies to improve the potency of **30** led to the discovery of quinazolines **31** (ML241) and **32** (ML240) with improved p97 inhibitory potency compared to **30**.[78] Compound **32** had improved solubility compared to **32** and like **30** caused increases in caspase levels leading to cell death. Compound **32** had good selectivity across a panel of other ATPases and kinases.

A series of 2-anilino-thiazole analogs were discovered by high throughput screening (HTS) which were optimized for p97 ATPase activity.[79] Compounds **33**, **34**, and **35** had sub-µM p97 inhibitory potency. Compounds **34** and **35** caused stabilization of a luciferase tagged reporter (UbG76V-luciferase) in HeLa cells with IC_{50} values of 0.09 and 0.12 µM, respectively. Selectivity with this series of compounds may be an issue as compound **33** has also been reported to be a sub-µM inhibitor of sphingosine kinase-1.[80]

A series of substituted thio-triazoles have been discovered that appear to be allosteric p97 inhibitors. The initial hit **36** had µM inhibitory potency against p97 ATPase activity.[81] The apparent IC_{50} of **36** did not shift with increasing ATP concentration suggesting allosteric inhibition of the p97 ATPase activity. Further elaboration of the structure resulted in molecule **37** with improved biochemical potency. Introduction of a methyl group next to a phenyl substituent in **37** resulted in compound **38** which had similar biochemical potency to **37** but had an approximate 10-fold improvement in cell-based potency. Compound **38** did not show measurable activity in a panel of five ATPases and related heat shock proteins or in a panel of 50 kinases. *In vitro* and *in vivo* ADME studies suggested that **38** had poor PK properties owing to rapid metabolic turnover in liver tissue.

5. CONCLUSIONS AND FUTURE DIRECTIONS

The most advanced inhibitors of the UPS exclusively inhibited the 20S CP of the proteasome and have yielded two approved treatments for hematologic malignancies. Now, emerging reports of successful early stage clinical and preclinical studies indicate that agents affecting the ligase and chaperone functions of the UPS show promise as anticancer targets. Compounds which target proteins associated with the 19S regulatory particle of the 26S proteasome could offer an attractive alternative way to develop a whole new class of proteasome inhibitors.[82–84]

REFERENCES

1. Edelmann, M. J.; Nicholson, B.; Kessler, B. M. *Expert Rev. Mol. Med.* **2011**, *13*, e35.
2. Choi, A. M.; Ryter, S. W.; Levine, B. *N. Engl. J. Med.* **2013**, *368*, 651.
3. Kaiser, P.; Huang, L. *Genome Biol.* **2005**, *6*, 233.
4. Gallastegui, N.; Groll, M. *Trends Biochem. Sci.* **2010**, *35*, 634.
5. Lim, K. H.; Baek, K. H. *Curr. Pharm. Des.* **2013**, *19*, 4039.

6. Fu, Q. S.; Song, A. X.; Hu, H. Y. *Curr. Protein Pept. Sci.* **2012**, *13*, 482.
7. Sledz, P.; Forster, F.; Baumeister, W. *J. Mol. Biol.* **2013**, *425*, 1415.
8. Targeted Proteins Research Program of MEXT, Japan. www.tanpaku.org/e_icsg2008/07_01.php.
9. Kisselev, A. F.; van der Linden, W. A.; Overkleeft, H. S. *Chem. Biol. (Oxford, U. K.)* **2012**, *19*, 99.
10. Frankland-Searby, S.; Bhaumik, S. R. *Biochim. Biophys. Acta, Rev. Cancer* **2012**, *1825*, 64.
11. Adams, J. *Curr. Opin. Chem. Biol.* **2002**, *6*, 493.
12. Groll, M.; Berkers, C. R.; Ploegh, H. L.; Ovaa, H. *Structure (London, England: 1993)* **2006**, *14*, 451.
13. Arastu-Kapur, S.; Anderl, J. L.; Kraus, M.; Parlati, F.; Shenk, K. D.; Lee, S. J.; Muchamuel, T.; Bennett, M. K.; Driessen, C.; Ball, A. J.; Kirk, C. J. *Clin. Cancer Res.* **2011**, *17*, 2734.
14. Chauhan, D.; Tian, Z.; Zhou, B.; Kuhn, D.; Orlowski, R.; Raje, N.; Richardson, P.; Anderson, K. C. *Clin. Cancer Res.* **2011**, *17*, 5311.
15. Kupperman, E.; Lee, E. C.; Cao, Y.; Bannerman, B.; Fitzgerald, M.; Berger, A.; Yu, J.; Yang, Y.; Bruzzese, F.; Liu, J.; Blank, J.; Garcia, K.; Tsu, C.; Dick, L.; Fleming, P.; Yu, L.; Manfredi, M.; Rolfe, M.; Bolen, J. *Cancer Res.* **2010**, *70*, 1970.
16. Lee, E. C.; Fitzgerald, M.; Bannerman, B.; Donelan, J.; Bano, K.; Terkelsen, J.; Bradley, D. P.; Subakan, O.; Silva, M. D.; Liu, R.; Pickard, M.; Li, Z.; Tayber, O.; Li, P.; Hales, P.; Carsillo, M.; Neppalli, V. T.; Berger, A. J.; Kupperman, E.; Manfredi, M.; Bolen, J. B.; Van, N. B.; Janz, S. *Clin. Cancer Res.* **2011**, *17*, 7313.
17. Dorsey, B. D.; Iqbal, M.; Chatterjee, S.; Menta, E.; Bernardini, R.; Bernareggi, A.; Cassara, P. G.; D'Arasmo, G.; Ferretti, E.; De, M. S.; Oliva, A.; Pezzoni, G.; Allievi, C.; Strepponi, I.; Ruggeri, B.; Ator, M. A.; Williams, M.; Mallamo, J. P. *J. Med. Chem.* **2008**, *51*, 1068.
18. Piva, R.; Ruggeri, B.; Williams, M.; Costa, G.; Tamagno, I.; Ferrero, D.; Giai, V.; Coscia, M.; Peola, S.; Massaia, M.; Pezzoni, G.; Allievi, C.; Pescalli, N.; Cassin, M.; di, G. S.; Nicoli, P.; de, F. P.; Strepponi, I.; Roato, I.; Ferracini, R.; Bussolati, B.; Camussi, G.; Jones-Bolin, S.; Hunter, K.; Zhao, H.; Neri, A.; Palumbo, A.; Berkers, C.; Ovaa, H.; Bernareggi, A.; Inghirami, G. *Blood* **2008**, *111*, 2765.
19. Sanchez, E.; Li, M.; Steinberg, J. A.; Wang, C.; Shen, J.; Bonavida, B.; Li, Z.-W.; Chen, H.; Berenson, J. R. *Br. J. Haematol.* **2010**, *148*, 569.
20. Gallerani, E.; Zucchetti, M.; Brunelli, D.; Marangon, E.; Noberasco, C.; Hess, D.; Delmonte, A.; Martinelli, G.; Bohm, S.; Driessen, C.; De, B. F.; Marsoni, S.; Cereda, R.; Sala, F.; D'Incalci, M.; Sessa, C. *Eur. J. Cancer* **2013**, *49*, 290.
21. Borissenko, L.; Groll, M. *Chem. Rev.* **2007**, *107*, 687.
22. Kim, K. B.; Myung, J.; Sin, N.; Crews, C. M. *Bioorg. Med. Chem. Lett.* **1999**, *9*, 3335.
23. Groll, M.; Koguchi, Y.; Huber, R.; Kohno, J. *J. Mol. Biol.* **2001**, *311*, 543.
24. Demo, S. D.; Kirk, C. J.; Aujay, M. A.; Buchholz, T. J.; Dajee, M.; Ho, M. N.; Jiang, J.; Laidig, G. J.; Lewis, E. R.; Parlati, F.; Shenk, K. D.; Smyth, M. S.; Sun, C. M.; Vallone, M. K.; Woo, T. M.; Molineaux, C. J.; Bennett, M. K. *Cancer Res.* **2007**, *67*, 6383.
25. Kuhn, D. J.; Chen, Q.; Voorhees, P. M.; Strader, J. S.; Shenk, K. D.; Sun, C. M.; Demo, S. D.; Bennett, M. K.; van Leeuwen, F. W.; Chanan-Khan, A. A.; Orlowski, R. Z. *Blood* **2007**, *110*, 3281.
26. Thompson, J. L. *Ann. Pharmacother.* **2013**, *47*, 56.
27. Sacco, A.; Aujay, M.; Morgan, B.; Azab, A. K.; Maiso, P.; Liu, Y.; Zhang, Y.; Azab, F.; Ngo, H. T.; Issa, G. C.; Quang, P.; Roccaro, A. M.; Ghobrial, I. M. *Clin. Cancer Res.* **2011**, *17*, 1753.

28. Zhou, H.-J.; Aujay, M. A.; Bennett, M. K.; Dajee, M.; Demo, S. D.; Fang, Y.; Ho, M. N.; Jiang, J.; Kirk, C. J.; Laidig, G. J.; Lewis, E. R.; Lu, Y.; Muchamuel, T.; Parlati, F.; Ring, E.; Shenk, K. D.; Shields, J.; Shwonek, P. J.; Stanton, T.; Sun, C. M.; Sylvain, C.; Woo, T. M.; Yang, J. J. *Med. Chem.* **2009**, *52*, 3028.

29. Roccaro, A. M.; Sacco, A.; Aujay, M.; Ngo, H. T.; Azab, A. K.; Azab, F.; Quang, P.; Maiso, P.; Runnels, J.; Anderson, K. C.; Demo, S.; Ghobrial, I. M. *Blood* **2010**, *115*, 4051.

30. Zang, Y.; Thomas, S. M.; Chan, E. T.; Kirk, C. J.; Freilino, M. L.; DeLancey, H. M.; Grandis, J. R.; Li, C.; Johnson, D. E. *Clin. Cancer Res.* **2012**, *18*, 5639.

31. Chauhan, D.; Singh, A. V.; Aujay, M.; Kirk, C. J.; Bandi, M.; Ciccarelli, B.; Raje, N.; Richardson, P.; Anderson, K. C. *Blood* **2010**, *116*, 4906.

32. Verbrugge, S. E.; Assaraf, Y. G.; Dijkmans, B. A.; Scheffer, G. L.; Al, M.; den Uyl, D.; Oerlemans, R.; Chan, E. T.; Kirk, C. J.; van der Heijden, J. W.; De Gruijl, T. D.; Scheper, R. J.; Jansen, G. *J. Pharmacol. Exp. Ther.* **2012**, *341*, 174.

33. Shah, I. M.; Lees, K. R.; Pien, C. P.; Elliott, P. J. *Br. J. Clin. Pharmacol.* **2002**, *54*, 269.

34. Lam, K. S.; Lloyd, G. K.; Neuteboom, S. T. C.; Palladino, M. A.; Sethna, K. M.; Spear, M. A.; Potts, B. C. *Natural Product Chemistry for Drug Discovery*; Neidle S.; Buss, A.; Butler, M., Eds. Royal Society of Chemistry: Cambridge, 2010; p 355.

35. Kisselev, A. F. *Chem. Biol. (Cambridge, MA, U. S.)* **2008**, *15*, 419.

36. Manam, R. R.; McArthur, K. A.; Chao, T.-H.; Weiss, J.; Ali, J. A.; Palombella, V. J.; Groll, M.; Lloyd, G. K.; Palladino, M. A.; Neuteboom, S. T. C.; Macherla, V. R.; Potts, B. C. M. *J. Med. Chem.* **2008**, *51*, 6711.

37. Groll, M.; Huber, R.; Potts, B. C. M. *J. Am. Chem. Soc.* **2006**, *128*, 5136.

38. Macherla, V. R.; Mitchell, S. S.; Manam, R. R.; Reed, K. A.; Chao, T.-H.; Nicholson, B.; Deyanat-Yazdi, G.; Mai, B.; Jensen, P. R.; Fenical, W. F.; Neuteboom, S. T. C.; Lam, K. S.; Palladino, M. A.; Potts, B. C. M. *J. Med. Chem.* **2005**, *48*, 3684.

39. Chauhan, D.; Catley, L.; Li, G.; Podar, K.; Hideshima, T.; Velankar, M.; Mitsiades, C.; Mitsiades, N.; Yasui, H.; Letai, A.; Ovaa, H.; Berkers, C.; Nicholson, B.; Chao, T.-H.; Neuteboom, S. T. C.; Richardson, P.; Palladino, M. A.; Anderson, K. C. *Cancer Cell* **2005**, *8*, 407.

40. Groll, M.; Potts, B. C. *Curr. Top. Med. Chem. (Sharjah, United Arab Emirates)* **2011**, *11*, 2850.

41. Chauhan, D.; Hideshima, T.; Anderson, K. C. *Br. J. Cancer* **2006**, *95*, 961.

42. Potts, B. C.; Albitar, M. X.; Anderson, K. C.; Baritaki, S.; Berkers, C.; Bonavida, B.; Chandra, J.; Chauhan, D.; Cusack, J. C., Jr.; Fenical, W.; Ghobrial, I. M.; Groll, M.; Jensen, P. R.; Lam, K. S.; Lloyd, G. K.; McBride, W.; McConkey, D. J.; Miller, C. P.; Neuteboom, S. T. C.; Oki, Y.; Ovaa, H.; Pajonk, F.; Richardson, P. G.; Roccaro, A. M.; Sloss, C. M.; Spear, M. A.; Valashi, E.; Younes, A.; Palladino, M. A. *Curr. Cancer Drug Targets* **2011**, *11*, 254.

43. Chauhan, D.; Singh, A. V.; Ciccarelli, B.; Richardson, P. G.; Palladino, M. A.; Anderson, K. C. *Blood* **2010**, *115*, 834.

44. Millward, M.; Price, T.; Townsend, A.; Sweeney, C.; Spencer, A.; Sukumaran, S.; Longenecker, A.; Lee, L.; Lay, A.; Sharma, G.; Gemmill, R. M.; Drabkin, H. A.; Lloyd, G. K.; Neuteboom, S. T.; McConkey, D. J.; Palladino, M. A.; Spear, M. A. *Invest. New Drugs* **2012**, *30*, 2303.

45. Richardson, P. G.; Spencer, A.; Cannell, P.; et al. *ASH Ann. Meeting Abs.* **2011**, *118*, 302.

46. Beck, P.; Dubiella, C.; Groll, M. *Biol. Chem.* **2012**, *393*, 1101.

47. Blackburn, C.; Barrett, C.; Blank, J. L.; Bruzzese, F. J.; Bump, N.; Dick, L. R.; Fleming, P.; Garcia, K.; Hales, P.; Hu, Z.; Jones, M.; Liu, J. X.; Sappal, D. S.; Sintchak, M. D.; Tsu, C.; Gigstad, K. M. *Bioorg. Med. Chem. Lett.* **2010**, *20*, 6581.
48. Blackburn, C.; Gigstad, K. M.; Hales, P.; Garcia, K.; Jones, M.; Bruzzese, F. J.; Barrett, C.; Liu, J. X.; Soucy, T. A.; Sappal, D. S.; Bump, N.; Olhava, E. J.; Fleming, P.; Dick, L. R.; Tsu, C.; Sintchak, M. D.; Blank, J. L. *Biochem. J.* **2010**, *430*, 461.
49. Yang, Y.; Kitagaki, J.; Dai, R. M.; Tsai, Y. C.; Lorick, K. L.; Ludwig, R. L.; Pierre, S. A.; Jensen, J. P.; Davydov, I. V.; Oberoi, P.; Li, C. C.; Kenten, J. H.; Beutler, J. A.; Vousden, K. H.; Weissman, A. M. *Cancer Res.* **2007**, *67*, 9472.
50. Xu, G. W.; Ali, M.; Wood, T. E.; Wong, D.; Maclean, N.; Wang, X.; Gronda, M.; Skrtic, M.; Li, X.; Hurren, R.; Mao, X.; Venkatesan, M.; Beheshti Zavareh, R.; Ketela, T.; Reed, J. C.; Rose, D.; Moffat, J.; Batey, R. A.; Dhe-Paganon, S.; Schimmer, A. D. *Blood* **2010**, *115*, 2251.
51. Ceccarelli, D. F.; Tang, X.; Pelletier, B.; Orlicky, S.; Xie, W.; Plantevin, V.; Neculai, D.; Chou, Y.-C.; Ogunjimi, A.; Al-Hakim, A.; Varelas, X.; Koszela, J.; Wasney, G. A.; Vedadi, M.; Dhe-Paganon, S.; Cox, S.; Xu, S.; Lopez-Girona, A.; Mercurio, F.; Wrana, J.; Durocher, D.; Meloche, S.; Webb, D. R.; Tyers, M.; Sicheri, F. *Cell (Cambridge, MA, U. S.)* **2011**, *145*, 1075.
52. Jia, L.; Sun, Y. *Curr. Cancer Drug Targets* **2011**, *11*, 347.
53. Teodoro, J. G.; Evans, S. K.; Green, M. R. *J. Mol. Med. (Berlin, Germany)* **2007**, *85*, 1175.
54. Zhang, H. G.; Wang, J.; Yang, X.; Hsu, H. C.; Mountz, J. D. *Oncogene* **2004**, *23*, 2009.
55. Wang, S.; Yujun, Z.; Bernard, D.; Aguilar, A.; Sanjeev, K. *Protein Interactions*; Wendt, M.D., Ed.; Topics in Medicinal Chemistry; Vol. 8, Springer-Verlag, 2012; p 57.
56. Yuan, Y.; Liao, Y. M.; Hsueh, C. T.; Mirshahidi, H. R. *J. Hematol. Oncol.* **2011**, *4*, 16.
57. Vassilev, L. T.; Vu, B. T.; Graves, B.; Carvajal, D.; Podlaski, F.; Filipovic, Z.; Kong, N.; Kammlott, U.; Lukacs, C.; Klein, C.; Fotouhi, N.; Liu, E. A. *Science (New York, N.Y.)* **2004**, *303*, 844.
58. Becker, K.; Marchenko, N. D.; Maurice, M.; Moll, U. M. *Cell Death Differ.* **2007**, *14*, 1350.
59. Rew, Y.; Sun, D.; Gonzalez-Lopez De Turiso, F.; Bartberger, M. D.; Beck, H. P.; Canon, J.; Chen, A.; Chen, D.; Chow, D.; Deignan, J.; Fox, B. M.; Gustin, D.; Huang, X.; Jiang, M.; Jiao, X.; Jin, L.; Kayser, F.; Kopecky, D. J.; Li, Y.; Lo, M.-C.; Long, A. M.; Michelsen, K.; Oliner, J. D.; Osgood, T.; Ragains, M.; Saiki, A. Y.; Schneider, S.; Toteva, M.; Yakowec, P.; Yan, X.; Ye, Q.; Yu, D.; Zhao, X.; Zhou, J.; Medina, J. C.; Olson, S. H. *J. Med. Chem.* **2012**, *55*, 4936.
60. Yu, S.; Qin, D.; Shangary, S.; Chen, J.; Wang, G.; Ding, K.; McEachern, D.; Qiu, S.; Nikolovska-Coleska, Z.; Miller, R.; Kang, S.; Yang, D.; Wang, S. *J. Med. Chem.* **2009**, *52*, 7970.
61. Ma, Y.; Lahue, B. R.; Shipps, G. W.; Wang, Y.; Bogen, S. L.; Voss, M. E.; Nair, L. G.; Tian, Y.; Doll, R. J.; Guo, Z.; Strickland, C. O.; Zhang, R.; McCoy, M. A.; Pan, W.; Siegel, E. M.; Gibeau, C. R. U.S. Patent 7,884,107, 2011.
62. Bold, G.; Furet, P.; Gessier, F.; Kallen, J.; Hergovich, L. J.; Masuya, K.; Vaupel, A. Patent Application WO 2011/023677, 2011.
63. Berghausen, J.; Buschmann, N.; Furet, P.; Gessier, F.; Hergovich, L. J.; Holzer, P.; Jacoby, E.; Kallen, J.; Masuya, K.; Pissot, S. C.; Ren, H.; Stutz, S. Patent Application WO 2011/076786, 2011.
64. Arts, J. *Proc. Am. Assoc. Cancer Res.* **2008**, *49*, 13.
65. Patel, S.; Player, M. *Expert Opin. Investig. Drugs* **2008**, *17*, 1865.
66. Tabernero, J.; Dirix, L.; Schoffski, P.; Cervantes, A.; Lopez-Martin, J. A.; Capdevila, J.; van Beijsterveldt, L.; Platero, S.; Hall, B.; Yuan, Z.; Knoblauch, R.; Zhuang, S. H.; van

Beijsterveldt, L.; Platero, S.; Hall, B.; Yuan, Z.; Knoblauch, R.; Zhuang, S. H. *Clin. Cancer Res.* **2011**, *17*, 6313.

67. Lopez-Girona, A.; Mendy, D.; Ito, T.; Miller, K.; Gandhi, A. K.; Kang, J.; Karasawa, S.; Carmel, G.; Jackson, P.; Abbasian, M.; Mahmoudi, A.; Cathers, B.; Rychak, E.; Gaidarova, S.; Chen, R.; Schafer, P. H.; Handa, H.; Daniel, T. O.; Evans, J. F.; Chopra, R. *Leukemia* **2012**, *26*, 2326.

68. Ito, T.; Ando, H.; Suzuki, T.; Ogura, T.; Hotta, K.; Imamura, Y.; Yamaguchi, Y.; Handa, H. *Science (New York, N.Y.)* **2010**, *327*, 1345.

69. Ruchelman, A. L.; Man, H. W.; Zhang, W.; Chen, R.; Capone, L.; Kang, J.; Parton, A.; Corral, L.; Schafer, P. H.; Babusis, D.; Moghaddam, M. F.; Tang, Y.; Shirley, M. A.; Muller, G. W. *Bioorg. Med. Chem. Lett.* **2013**, *23*, 360.

70. Zhao, Y.; Sun, Y. *Curr. Pharm. Des.* **2013**, *19*, 3215.

71. Soucy, T. A.; Smith, P. G.; Milhollen, M. A.; Berger, A. J.; Gavin, J. M.; Adhikari, S.; Brownell, J. E.; Burke, K. E.; Cardin, D. P.; Critchley, S.; Cullis, C. A.; Doucette, A.; Garnsey, J. J.; Gaulin, J. L.; Gershman, R. E.; Lublinsky, A. R.; McDonald, A.; Mizutani, H.; Narayanan, U.; Olhava, E. J.; Peluso, S.; Rezaei, M.; Sintchak, M. D.; Talreja, T.; Thomas, M. P.; Traore, T.; Vyskocil, S.; Weatherhead, G. S.; Yu, J.; Zhang, J.; Dick, L. R.; Claiborne, C. F.; Rolfe, M.; Bolen, J. B.; Langston, S. P. *Nature (London, U. K.)* **2009**, *458*, 732.

72. Soucy, T. A.; Smith, P. G.; Rolfe, M. *Clin. Cancer Res.* **2009**, *15*, 3912.

73. Brownell, J. E.; Sintchak, M. D.; Gavin, J. M.; Liao, H.; Bruzzese, F. J.; Bump, N. J.; Soucy, T. A.; Milhollen, M. A.; Yang, X.; Burkhardt, A. L.; Ma, J.; Loke, H. K.; Lingaraj, T.; Wu, D.; Hamman, K. B.; Spelman, J. J.; Cullis, C. A.; Langston, S. P.; Vyskocil, S.; Sells, T. B.; Mallender, W. D.; Visiers, I.; Li, P.; Claiborne, C. F.; Rolfe, M.; Bolen, J. B.; Dick, L. R. *Mol. Cell* **2010**, *37*, 102.

74. Chen, J. J.; Tsu, C. A.; Gavin, J. M.; Milhollen, M. A.; Bruzzese, F. J.; Mallender, W. D.; Sintchak, M. D.; Bump, N. J.; Yang, X.; Ma, J.; Loke, H. K.; Xu, Q.; Li, P.; Bence, N. F.; Brownell, J. E.; Dick, L. R. *J. Biol. Chem.* **2011**, *286*, 40867.

75. Meyer, H.; Bug, M.; Bremer, S. *Nat. Cell Biol.* **2012**, *14*, 117.

76. Erzberger, J. P.; Berger, J. M. *Annu. Rev. Biophys. Biomol. Struct.* **2006**, *35*, 93.

77. Chou, T. F.; Brown, S. J.; Minond, D.; Nordin, B. E.; Li, K.; Jones, A. C.; Chase, P.; Porubsky, P. R.; Stoltz, B. M.; Schoenen, F. J.; Patricelli, M. P.; Hodder, P.; Rosen, H.; Deshaies, R. J. *Proc. Natl. Acad. Sci. U.S.A.* **2011**, *108*, 4834.

78. Chou, T. F.; Li, K.; Frankowski, K. J.; Schoenen, F. J.; Deshaies, R. J. *ChemMedChem* **2013**, *8*, 297.

79. Bursavich, M. G.; Parker, D. P.; Willardsen, J. A.; Gao, Z. H.; Davis, T.; Ostanin, K.; Robinson, R.; Peterson, A.; Cimbora, D. M.; Zhu, J. F.; Richards, B. *Bioorg. Med. Chem. Lett.* **2010**, *20*, 1677.

80. French, K. J.; Schrecengost, R. S.; Lee, B. D.; Zhuang, Y.; Smith, S. N.; Eberly, J. L.; Yun, J. K.; Smith, C. D. *Cancer Res.* **2003**, *63*, 5962.

81. Polucci, P.; Magnaghi, P.; Angiolini, M.; Asa, D.; Avanzi, N.; Badari, A.; Bertrand, J. A.; Casale, E.; Cauteruccio, S.; Cirla, A.; Cozzi, L.; Galvani, A.; Jackson, P. K.; Liu, Y.; Magnuson, S.; Malgesini, B.; Nuvoloni, S.; Orrenius, C.; Riccardi Sirtori, F.; Riceputi, L.; Rizzi, S.; Trucchi, B.; O'Brien, T.; Isacchi, A.; Donati, D.; R., D'Alessio *J. Med. Chem.* **2013**, *56*, 437.

82. Yao, T.; Cohen, R. E. *Nature (London, U. K.)* **2002**, *419*, 403.

83. D'Arcy, P.; Brnjic, S.; Olofsson, M. H.; Fryknaes, M.; Lindsten, K.; De, C. M.; Perego, P.; Sadeghi, B.; Hassan, M.; Larsson, R.; Linder, S. *Nat. Med. (New York, NY, U. S.)* **2011**, *17*, 1636.

84. D'Arcy, P.; Linder, S. *Int. J. Biochem. Cell Biol.* **2012**, *44*, 1729.

CHAPTER FIFTEEN

Targeting Protein–Protein Interactions to Treat Cancer—Recent Progress and Future Directions

William Garland*, Robert Benezra†, Jaideep Chaudhary‡
*AngioGenex Inc., New York, USA
†Memorial Sloan-Kettering Cancer Center, New York, USA
‡Clark Atlanta University, Atlanta, Georgia, USA

Contents

1. INTRODUCTION

Protein–protein interactions (PPIs) are a vast, complex network of reactions important to the regulation and execution of most biological processes. Recently, this network was labeled the "interactome."[1] The number of binary relations between proteins may be 200,000 PPIs[2] or greater with only about 8% identified.[3] The interactome is a target-rich but relatively

Annual Reports in Medicinal Chemistry, Volume 48
ISSN 0065-7743
http://dx.doi.org/10.1016/B978-0-12-417150-3.00015-6

unexplored source for new drugs for untreatable diseases or for drugs superior to currently used agents. The activity of many marketed drugs with unknown or ill-defined mechanisms of action (MOA) is also likely the result of affecting PPIs.

This chapter does not address any analytical issues associated with discovering PPIs or detecting inhibition or stabilization of PPIs. The chapter does highlight PPI-specific issues relevant to drug design and recent progress addressing PPIs with a focus on oncologic applications. The evaluation of PPIs associated with the initiation, growth, and spread of cancer is very active[4,5] because of the limitations of current targets, for example, success inhibiting growth promoting protein kinases, enzymes, and G protein-coupled receptors (GPCRs), are limited because of frequent mutations of the target and the heterogeneous nature of tumors. Increased interest in PPIs among medicinal chemists has resulted from recent success addressing PPIs previously considered intractable, and the intense competition with traditional targets, for example, protein kinases and GPCRs. In addition, the therapeutic effect of inhibitors or stabilizers of PPIs that utilize multiple, relatively low-energy interactions with a protein surface to achieve activity is potentially less sensitive to escape of the target protein through mutation. However, to succeed the chemist and team addressing a specific PPI must fully understand the associated biology and protein biochemistry as well as expanding their chemical creativity.

2. PPIs IMPORTANT TO THE GROWTH AND SPREAD OF CANCER

Cancer is a multistep process of cellular transformations leading to unregulated cell growth. The acquired capabilities of cancer cells[6] are listed in Table 15.1 along with selected examples of PPIs associated with each capability.

These selected but broadly pertinent examples illustrate how significant PPI targets are to cancer.

3. CHALLENGES IN TARGETING PPIs

Most PPI inhibitors fit structurally in chemical space not well served by traditional lead generation approaches and compound libraries.[21] In addition, the need to expand the chemical space explored for PPI modulators is amplified because of the greatly expanded population of targets. The surfaces associated with PPIs are featureless and large (contact surface generally

Table 15.1 Hallmarks of cancer and related protein–protein targets

Acquired capability	Example	Example PPI target
Increased proliferation	Overexpression growth-factor receptors, for example, Her2/neu	c-Myc/Max,[7] ESX/Sur-2[8]
Decreased/inactivation of tumor suppressors	Inactivation retinoblastoma protein, inactivation p53	β-catenin/T-cell factor (TCF),[9] RB/Raf-1,[10] MDM2/p53[11]
Evading immune surveillance	Desensitizing T-cells	PD-L1/PD-1[12]
Increased angiogenesis	Induction vascular endothelial growth factor	E-protein/inhibitor of differentiation (Id)[13]
Activating tissue invasion and metastasis	Alteration cell adhesion and motility	Rac/Tiam1,[14] E-protein/Id[15]
Enabling replicative immortality	Activation of telomerase	TPP1/telomerase[16]
Resisting programmed cell death (apoptosis)	Overexpression of anti-apoptotic proteins	XIAP/caspase,[17] Bcl2[18]
Reprogramming of energy metabolism	Increased uptake and conversion of glucose to lactate by cancer cells	PPARγ and PKM2 and/or PGK1[19]
Recruitment of supportive healthy cells, for example, innate immune cells	Infiltration of pro-growth secreting neutrophil/macrophages into tumors	CXCR2/IQGAP1[20]

750–1500 A^2). The energetics of binding are predominantly hydrophobic best suited for orthosteric interaction with large lipophilic molecules which is not the situation with traditional drug targets. Less than 1% of the roughly 30,000 unique protein sequences that comprise the human proteome have been successfully manipulated to date for therapeutic use[22]; that is, all currently approved small-molecule drugs interact with just over 200 protein targets and ~50% of these fall into just four protein classes: GPCRs, nuclear receptors, voltage-gated, and ligand-gated ion channels.[23] The sites for these targets comprise deep clefts with clearly defined binding sites containing the critical amino acids required for ligand interaction. The molecules that interact with these targets emerge from a relatively small number of molecular scaffolds that, unsurprisingly, are also the basis for most current chemical libraries. Furthermore, active-site clefts tend to shield binding sites from water that can otherwise hinder ligand interactions. This shielding is less possible on the typical surface involved in a PPI.

The targeting approach for classical binding sites will likely not be effective with most PPI, but it can be used to target "hot spots" present on some proteins. These shallow binding cavities provide small subset of residues that contribute most of free energy of binding in the protein partner/small-molecule interaction. The classical approach and associated chemical libraries are expected to find success addressing allosteric binding sites which if occupied will disrupt PPIs if the site resembles a classic site. However, libraries containing very different chemicals than those used to address "classical" targets will probably be required to be routinely successful with PPIs. In particular, libraries containing larger molecular weight (MW) molecules are more suitable as PPI inhibitors because of the possibility of multiple interaction sites over a relatively large surface area. Fortunately, the identification of tractable PPIs and which type of molecule to use to address the PPI will be greatly aided by the new tools to explore protein surfaces emerging from structural biology,[24] computational methodologies[25] and molecular dynamics simulations efforts.[26]

4. METHODS FOR DISCOVERING SMALL-MOLECULE INHIBITORS OF PPIs

What makes PPIs such a challenge for small-molecule intervention is the diverse array of observed molecular moieties at the protein–protein interfaces. Fortunately, a PPI often involves only a few key residues that contribute the majority of the binding affinity to the interaction although the interaction itself takes place over a large surface area. These small subsets of protein surface residues that contribute most of the free energy of binding associated with PPI serve as the starting point for the rational design of orthostatic PPI inhibitors. The approaches typically employed to identify PPI modulators are:

* *Traditional HTS.* Conventional screening has been generally unsuccessful at routinely identifying PPI inhibitors, presumably because conventional libraries are optimized for molecules that bind to enzyme-like or GPCR-like cavities. Nonetheless, combinatorial chemistry and HTS have produced useful inhibitors of PPIs and additional advances with this approach are likely.[27] Potency expectations for "hits" also must be recalibrated: low μM instead of nM or pM to reflect lower binding energetics. Traditional HTS remains an attractive approach for discovering allosteric inhibitors that affect a remote binding site that disrupts the

principal PPI interface by promoting a conformational change that modifies the manner in which one of the proteins subsequently behaves.

- *Virtual screening.* *In silico* screening has been employed to identify inhibitors of PPIs.[28,29] For instance, this application to the (H/M)DM2/p53 cancer target led to a large and diverse set of inhibitors whose validity was confirmed by obtaining co-crystal structure of the inhibitors and the protein.[30]

- *Fragment screening.* ABT-737, an inhibitor of Bcl-2 PPIs that induces regression of solid tumors and was at one time in Phase II clinical trials, was identified by screening a collection of traditional organic molecules for binding to the protein using NMR, followed by chemically linking together the best binders or binding regions of the best binders.[31] Recently, inhibitors of the chemokine CXCL12 PPI with the CXCR4 GPCR, a high-priority target because of its involvement in metastatic cancers, were discovered using fragment-based optimization.[32]

- *Hot spot-based design.* Computational docking,[33] a form of virtual screening, and visual inspection are often used to discover and explore sites for PPI inhibition. Systematic mutations of amino acids in the protein pairs are the most compelling and thorough method for identifying hot spots.[34]

Various chemical scaffolds have been suggested as a more systematic approach to affect PPI inhibition than using *ad hoc* organic chemicals.[35]

- *Constrained secondary structures.* Stable peptides with defined secondary structure are theoretically ideal inhibitors of PPIs. However, those with less than around 15 amino acid residues rarely adopt a defined conformation in isolation and so recent research has focused upon generating peptides with constrained configurations to disrupt PPIs. A recent advance has been synthesis of peptides that are covalently constrained using hydrocarbon "stapling." [36]

- *β-Turns.* β-Turns play an important role in inhibiting or stabilizing PPIs. Thus, there is significant interest in identifying stable β-turns scaffolds useful for inhibiting PPIs.

- *Constrained α-helices and β-sheets.* To satisfy hydrogen binding requirements, α-helices and β-sheets form recognition elements in many naturally occurring PPIs. Thus, small-molecule mimetics or constrained peptides that satisfy this hydrogen binding requirement for either protein of the interaction protein pair can disrupt the PPI. Stabilization of α-helical structure in a peptide chain has been achieved using covalent linkages and noncovalent interactions residues, for example, metal–ligand interactions, salt bridges, π–π stacking and cation–π interactions.

- *Foldamers.* Foldamers, discrete artificial oligomers that adopt a secondary structure stabilized by noncovalent interactions, have been proposed as scaffolds suitable for near universal use in designing PPI inhibitors.[37] Foldamers mimic the ability of proteins to fold into well-defined conformations, such as helices and β-sheets.
- *Surface mimetics.* Surface mimetics rely upon multisite presentation of recognition domains capable of binding "hot spots" on a protein surface over a sufficiently large surface area to be able to bind and block interaction with the partner protein.[38]

5. EXAMPLES OF TARGETING PPIs IMPORTANT TO ONCOLOGY

5.1. Small-molecule inhibitors

Several examples of oncology-relevant, clinical stage applications of small molecules to inhibit PPIs are provided below. The examples are representative and not an exhaustive list.

5.1.1 RG-7112

p53—a potent tumor suppressor and "guardian of the genome" with central roles in the regulation of the cell cycle, apoptosis, DNA repair and senescence[39]—is one of the most frequently mutated proteins in human tumors.[40] Inhibition of p53 by the oncoprotein (H/M)DM2 regulates the basal level and activity of p53 in cells.[41] More specifically, p53 binds to a promoter of the (H/M)DM2 gene and transcriptionally induces expression of (H/M)DM2 which inhibits the transactivation activity of p53 resulting in export of p53 from the nucleus with subsequent proteasomal degradation. Several inhibitors of the p53/(H/M)DM2 PPI have advanced into clinical trials, for example, the "nutlins." [42] The current lead clinical candidate is RG-7112/RO5045337 (**I**),[43] an orally active agent being evaluated in both solid and hematologic malignancies.[44] These inhibitors all require active p53, a situation not always present in tumors.

I

5.1.2 Navitoclax/ABT-263

Cancer cells overexpress antiapoptotic proteins like Bcl-2[45] to promote tumor maintenance, progression, and chemoresistance. Agents designed to target the binding grooves of antiapoptotic Bcl-2 proteins are predicted to induce apoptosis in cancer cells by antagonizing their protective effect that cooperate through PPIs to mediate the intrinsic apoptotic pathway.[46–49] Navitoclax (**II**) binds to the BH3-binding groove of one member of the Bcl-2 family, Bcl-2-L1. Navitoclax is a potent inhibitor of various Bcl molecules by disrupting the PPI between proapoptotic Bcl-2 proteins that block apoptosis-enabling cytochrome *c* release. Navitoclax demonstrated clinical activity as a single agent or used in combination with radiation or other chemotherapeutics. This agent is currently in Phase II clinical trial for the treatments of chronic lymphocytic leukemia (CLL).[50] The clinical development of ABT-737, the precursor to Navitoclax, was dropped because of poor oral bioavailability.[51]

II

5.1.3 GDC-0199/ABT-199[52]

Bcl-2 is highly expressed in various lymphomas. The first-generation Bcl-2 inhibitor Navitoclax showed activity in lymphoma, but co-inhibition of Bcl-xL by Navitoclax resulted in dose-limiting thrombocytopenia that hindered the use of drug in lymphomas. ABT-199 (**III**) is an orally bioavailable, second-generation BH3 mimetic that inhibits Bcl-2, but has 500-fold less activity against Bcl-xL. ABT-199 demonstrated antitumor activity against a variety of human cell lines and xenograft models that include B-cell non-Hodgkin lymphoma and myeloid cell leukemia. Enrollment of CLL patients in a clinical trial with ABT-199 was recently suspended after the death of a patient due to tumor lysis syndrome but dosing of already enrolled patients continues.[53]

5.1.4 R-(−)-Gossypol/AT-101

The R-(−)-enantiomer of gossypol (**IV**) mimics the BH3 domains of various BcL-2 molecules including Mcl-1 (myeloid cell leukemia 1 protein), a dominant member of the BcL-2 antiapoptotic protein family.[54,55] AT-101 disrupts heterodimerization of Bcl-2 with proapoptotic family members. The molecule also induces apoptosis *in vitro* through the activation of caspase-9 and is cytotoxic to multiple myeloma and drug-resistant cell lines. AT-101 delays the onset of androgen-independent growth of VCaP prostate cancer xenografts *in vivo*. Akt and inhibitor of apoptosis (XIAP) are down-regulated in the presence of AT-101. Treatment with bicalutamide and AT-101 increased apoptosis by reducing the expression of prosurvival proteins. AT-101 is being evaluated in Phase 2 studies in various cancers.[56]

IV

5.1.5 Obatoclax/GX15-070

Obatoclax (**V**), another BH3 mimetic, is an inhibitor of Bcl-2 with high nM potency *in vitro*.[57] For Mcl-1, Obatoclax displaces BH3 domains by activation of a pocket of the BcL-2 family member.[58] Obatoclax also binds to a broad spectrum of Bcl-2 family members, including Bcl-2 and Bcl-xL. Inhibition of Obatoclax leads to release of apoptosis-inducing cytochrome *c*. Obatoclax enhances the antimyeloma activity of melphalan, dexamethasone or bortezomib. Obatoclax also potentiates TRAIL-mediated apoptosis in cancer cells as well as exhibiting antitumor activity in severe combined immunodeficiency (SCID) mice bearing various human cancer cell lines. The current, major clinical effort with Obatoclax is an evaluation in combination with bortezomib in patients with relapsed or refractory non-Hodgkin lymphoma.[59]

V

5.1.6 GDC-0152

Inhibitors of apoptosis proteins (IAPs) are endogenous caspase inhibitors that bind and inhibit caspases 3, 7, and 9. Eight members of the IAP family are known to date of which the X-linked inhibitor of apoptosis protein (XIAP) is the best characterized. Endogenous inhibition of the IAP family can occur via DIABLO (also known as Smac) protein released from the mitochondria in response to apoptotic stimuli, which directly interacts with XIAP.[47] The finding that the levels of IAP family members were elevated in several tumors suggested the possibility to treat cancer by inactivating the IAP family to induce apoptosis. A series of compounds mimicking the N-terminus of DIABLO (Smac) were designed and synthesized as antagonists of XIAP and related family members. GDC-0152 (**VI**) had the best profile of these compounds.[60] This compound promoted degradation of relevant IAP molecules, induced activation of caspase 3 and 7, decreased viability of breast cancer cells without affecting normal mammary epithelial cells and inhibited tumor growth in a human breast cancer xenograft model. GDC-0152, dosed orally, recently completed a Phase 1 safety/pharmacokinetic evaluation in cancer patients.[61]

5.1.7 PRI-724/ICG-001[62]

The Wnt/β-catenin signaling pathway regulates cell morphology, motility, and proliferation. Aberrant regulation of this pathway leads to neoplastic proliferation and prevention of apoptosis in a number of human cancers. β-catenin/CBP binds to the Wnt-responsive element and activates transcription of a wide range of target genes associated with Wnt/β-catenin signaling. PRI-724 (**VII**) is a potent, antiproliferative small molecule that specifically inhibits the canonical Wnt signaling pathway in cancer stem cells while also providing antineoplastic activity. By blocking the interaction of β-catenin and CBP, its coactivator, PRI-724 prevents gene expression of many proteins necessary for growth. PRI-724 substantially inhibits tumor growth in animal studies. PRI-724 is currently in early clinical testing for treating solid tumors and myeloid malignancies.[63]

5.1.8 Talbtobulin

During cell division, α/β-tubulin polymerizes into dynamic structures called microtubules. Inhibitors of tubulin either target polymerization (vinca alkaloids and colchicine) or depolymerization (taxanes and epothilones). These two functional classes bind to different regions of the α/β-tubulin heterodimer and allosterically regulate tubulin oligomerization. Although small-molecule inhibitors of tubulin have been in clinical use since 1965, the identification of new inhibitors is still an active area of research, especially because current therapeutics are susceptible to a common mechanism of induced drug resistance (efflux by P-glycoprotein). One novel tubulin polymerization inhibitor undergoing clinical evaluation is talbtobulin (**VIII**).[64] Many classes of tubulin inhibitors are known; for example, colchicine buried and at the heterodimer interface and taxanes binding to a shallow groove found on the β-subunit. The presence of these multiple allosteric binding sites makes tubulin particularly amenable to inhibition by small molecules.

VIII

5.2. Small-molecule stabilizers

Although the vast majority of PPI modulators are inhibitors, an increased lifespan for a PPI associated with an anticancer effect is desirable. Some PPI-stabilizing natural products have already found application as important drugs. One of the most intensely studied modulators of microtubules is paclitaxel. Paclitaxel stabilizes the polymerized microtubule structures by allosteric binding with high affinity to a hydrophobic pocket of polymerized tubulin located on the β-subunit thereby strengthening the lateral contacts of neighboring β-subunits. One common mechanism for stabilization observed for direct PPI stabilizers[65] is binding of a small molecule to one of the proteins, thereby creating or modifying the interaction surface for the second protein. This stabilizing effect can be so strong that two proteins can be induced to dimerize that do not bind to each other in the absence of these molecules. This extreme case is observed for the FKBP binding molecules, FK506 and rapamycin. For cancer, a small molecule that stabilizes the interaction between 14-3-3, a family of conserved regulatory molecules and p53 would be therapeutically desirable because the 14-3-3 proteins bind to the regulatory C-terminal domain of p53 and prevent MDM2-dependent degradation, thereby enhancing p53 stability.[66] The feasibility of achieving this result was recently demonstrated with the discovery by systematic screening of a small molecule that inhibits the PPI of p53 with PMA2, a plant H^+-ATPase.[67]

5.3. Other

Phosphorylation-dependent[68] signaling events govern large multiprotein complexes to regulate cell growth, progression of the cell cycle, differentiation, motility, gene expression, and apoptosis. Both phosphorylation and dephosphorylation of protein substrates have critical functions. In many cases, phosphorylated residues on the substrates create binding sites for phospho-protein binding domains that link upstream kinases and downstream effectors to form protein complexes that regulate the activity, binding partners, and localization of the specific protein.

EGFR–HER2 and HER2–HER3 PPIs are important in cancer for regulation of the human epidermal growth factor receptor. Recently, a small molecule that modulates HER2-mediated signaling was discovered that inhibits the phosphorylation of HER2 kinase domain resulting in inhibition of activation-dependent heterodimerization.[69] Molecular modeling was used to design the inhibitor.

6. TWO RECENT SUCCESSES
6.1. Inhibitor of differentiation 1

The Id genes/proteins are critically important during embryonic development but are only active in adults when cancer is present. The Ids function as negative regulators of transcription (see Fig. 15.1), and are expressed in virtually all cancers. Overexpression of Id genes in tumors is associated with an aggressive phenotype and poor clinical outcome.[70–74] This is not surprising since the effects of Id proteins are linked to almost all signaling elements critical to the initiation and progression of cancer.[75–79] Recently, AGX51 (**IX**), a small-molecule anti-Id agent, was disclosed.[80,81] AGX51 was discovered using X-ray crystallography, gel shift assays, Matrigel evaluations, and xenograft studies (Fig. 15.2). As expected, AGX51 completely blocked tumor-associated angiogenesis in nude mice implanted with human breast cancer cells and also treated briefly with a taxane, blocked new blood vessel formation into implanted vascular endothelial growth factor (VEGF)/fibroblast growth factor (FGF)-treated Matrigel plugs, decreased metastases of injected Lewis

Lung cancer cells, etc. AGX51 also restored levels in cancer cells of mediators of cell cycle regulators, such as p16, p21 and p27.

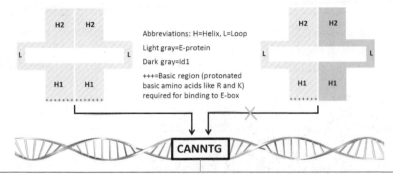

IX

Situation 1: Homodimerization E-protein, for example, E47 with Id1 proteins absent

Situation 2: Heterodimerization E-protein, for example, E47 with Id1

Abbreviations: H=Helix, L=Loop

Light gray=E-protein

Dark gray=Id1

+++=Basic region (protonated basic amino acids like R and K) required for binding to E-box

CANNTG

If E-box occupied, proteins restricting cell-cycle control like p16, p21 and p27 produced. If E-box unoccupied, proteins restricting cell-cycle control **not** produced resulting in a pro-growth environment facilitating, for example neovascularization (angiogenesis). Not shown above is the heterodimerization E-protein like E-47 with another E-protein like c-Myc, Max, HEB, etc. These E-protein/E-protein heterodimers can also bind to the E-box. If Id1 present, E-protein/E-protein heterodimers are not formed because reaction with Id1 is more rapid.

Figure 15.1 Scheme for interaction of E-proteins with E-box on DNA.

6.2. BRD4–histone

In late 2010, Bradner and colleagues reported that JQ1 (**X**) could block the interaction between BRD4 and histones, providing proof-of-principle evidence that PPI inhibition could provide epigenetic control.[82,83] JQ1 was discovered by HTS. A BRD4–NUT translocation has been linked to an aggressive form of human squamous carcinoma, and the BRD4–histone

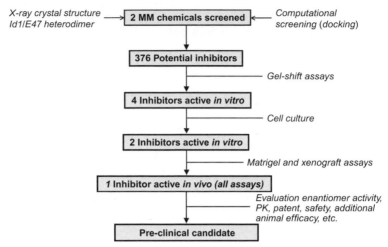

Figure 15.2 Process used to discover small-molecule inhibitors of Id1–E47 interaction.

team was able to generate *in vivo* data demonstrating the potential therapeutic applications of their compound in this setting. Further studies unveiled its potential in other indications, including MYC–associated cancers.[84]

7. FUTURE DEVELOPMENTS

Industry and government scientists have produced improved technologies in several areas important to drug discovery, such as automation, stereo-selective organic syntheses, sophisticated screening assays, analysis of genetic targets, computational strategies and structural biology. However, these advances have not appreciably improved the success of drug discovery in targeting oncogenic PPIs. Most disappointing has been progress with promising oncology targets like Myc.[85] Most of the tractable targets to date, for example, the inhibitors of the p53-(H/M)DM2 PPI or BCL-2 and related molecules have protein interfaces in which a small, linear region of one protein binds into a hydrophobic cleft of the other, not too different than traditional disruption of receptor–substrate fit.

Areas for improvement relevant to discovering new PPI modulators include:

- *PPI-friendly libraries.* The development of chemical libraries is costly and time consuming. However, current chemical libraries are optimized for interaction with classic targets and as a result are consistent with the Lipinski guidelines.[86] However, the interface for PPI is much broader with relatively weak and scattered binding sites (few, or no, highly energetic binding sites). Consistent with this understanding, successful PPI inhibitors possess chemical properties shifted toward higher molecular weight, increased hydrophobicity, and a higher unsaturation index and ring complexity than common drugs.[87] The design of inhibitors that obey the Lipinski guidelines and possess the optimum features for inhibition of PPIs therefore appears to require opposing criteria. Poor oral bioavailability and cell penetration are risks with higher MW agents.

- *PPI-friendly scaffolds.* Because of the need for providing multiple, often widely spaced interactions, scaffolds that can easily be manipulated to move interacting chemical moieties on the scaffold to provide favorable binding are very desirable. The design and production of these flexible scaffolds will likely require new synthetic chemistry approaches and techniques. A similar approach, designated "credit card" libraries, has been proposed.[88] The "credit cards" are defined as flat and rigid small molecules further functionalized to install elements of chemical diversity needed to disrupt a specific PPI.

- *PPI-friendly computational/cheminformatic tools.* Many tools are now available to facilitate the analysis of protein–protein interfaces for small-molecule drug design. Surprisingly, many successful PPI inhibitors have been designed from just visual inspection of the surfaces of the interacting protein without associated computational docking studies. Protein structure determinations have become routine and widespread as have automated docking routines using commercial chemical libraries. Many programs for viewing protein surfaces are now available. However, user-friendly software for seamless integration of all these tools for PPI evaluations is still lacking. In addition, the commercial libraries used for computational docking evaluations have the same previously discussed structural limitations. Besides help in suggesting inhibitors of PPIs, the integrated computational/cheminformatic package could also suggest PPIs whose chemical characteristics are too daunting to attempt manipulation. Such a powerful tool could also be used to rationally design PPI-friendly focused libraries.

- *Covalently bound modulators.* Because of the weak energetics associated with binding to a protein surface associated with a PPI, covalent binding of the inhibitor with the protein should be considered.[89]
- *Mutated proteins and PPIs.* A critical gap in understanding PPIs associated with cancer is the role of protein mutation, which is so important with protein kinases.[90] This possibility is relatively unexplored. However, if new findings suggest mutations play a role in oncologic PPIs, the overall challenge of manipulation PPIs for anticancer activity will become even more complex.

8. CONCLUSIONS

Protein complexes in the interactome provide practical drug targets for oncology drug discovery. Research on a handful of PPIs important to the growth and spread of cancer has produced agents with sufficient potency and cellular activity to become clinical candidates. The list of "tractable protein–protein targets" is growing although still small compared to the list of "considered intractable protein–protein targets." With numerous possible cancer drug targets in the interactome, thoughtful selection of PPI targets that are amenable to inhibition or stabilization by small molecules will become a critical task for researchers. However, the true test of the evolving emphasis on PPI in cancer will be the availability of PPI-based drugs to treat cancer patients.

REFERENCES

1. Vidal, M.; Cusick, M. E.; Barabási, A. L. *Cell* **2011**, *144*, 986–998.
2. Garner, A. L.; Janda, K. D. *Curr. Top. Med. Chem.* **2011**, *11*, 258–280.
3. Venkatesan, K.; Rual, J. F.; Vazquez, A.; Stelzl, U.; Lemmens, I.; Hirozane-Kishikawa, T.; Hao, T.; Zenkner, M.; Xin, X.; Goh, K. I.; Yildirim, M. A.; Simonis, N.; Heinzmann, K.; Gebreab, F.; Sahalie, J. M.; Cevik, S.; Simon, C.; de Smet, A. S.; Dann, E.; Smolyar, A.; Vinayagam, A.; Yu, H.; Szeto, D.; Borick, H.; Dricot, A.; Klitgord, N.; Murray, R. R.; Lin, C.; Lalowski, M.; Timm, J.; Rau, K.; Boone, C.; Braun, P.; Cusick, M. E.; Roth, F. P.; Hill, D.; Tavernier, J.; Wanker, E. E.; Barabási, A. L.; Vidal, M. *Nat. Methods* **2009**, *6*, 83–90.
4. Kamb, A. *Curr. Opin. Pharmacol.* **2010**, *10*, 356–361.
5. Hebar, A.; Valent, P.; Selzer, E. *Expert Rev. Clin. Pharmacol.* **2013**, *6*, 23–34.
6. Hanahan, D.; Weinberg, R. A. *Cell* **2011**, *144*, 646–674.
7. Dang, C. V.; Le, A.; Gao, P. *Clin. Cancer Res.* **2009**, *15*, 6479–6483.
8. Asada, S.; Choi, Y.; Yamada, M.; Wang, S. C.; Hung, M. C.; Qin, J.; Uesugi, M. *Proc. Natl. Acad. Sci. U.S.A.* **2002**, *99*, 12747–12752.

9. Clevers, H.; Nusse, R. *Cell* **2012**, *149*, 1192–1205.
10. Johnson, J. L.; Pillai, S.; Pernazza, D.; Sebti, S. M.; Lawrence, N. J.; Chellappan, S. P. *Cancer Res.* **2012**, *72*, 516–526.
11. Rayburn, E. R.; Ezell, S. J.; Zhang, R. *Anticancer Agents Med Chem.* **2009**, *9*, 882–903.
12. Rozali, E. N.; Hato, S. V.; Robinson, B. W.; Lake, R. A.; Lesterhuis, W. J. *Clin. Dev. Immunol.* **2012**, *2012*, 656340.
13. Lyden, D.; Young, A. Z.; Zagzag, D.; Yan, W.; Gerald, W.; O'Reilly, R.; Bader, B. L.; Hynes, R. O.; Zhuang, Y.; Manova, K.; Benezra, R. *Nature* **1999**, *401*, 670–677.
14. Minard, M. E.; Kim, L. S.; Price, J. E.; Gallick, G. E. *Breast Cancer Res. Treat.* **2004**, *84*, 21–32.
15. Gupta, G. P.; Perk, J.; Acharyya, S.; de Candia, P.; Mittal, V.; Todorova-Manova, K.; Gerald, W. L.; Brogi, E.; Benezra, R.; Massagué, J. *Proc. Natl. Acad. Sci. U.S.A.* **2007**, *104*, 19506–19511.
16. Lue, N. F.; Yu, E. Y.; Lei, M. *Nat. Struct. Mol. Biol.* **2013**, *20*, 10–12.
17. Fiandalo, M. V.; Kyprianou, N. *Exp. Oncol.* **2012**, *34*, 165–175.
18. Vogler, M.; Hamali, H. A.; Sun, X. M.; Bampton, E. T.; Dinsdale, D.; Snowden, R. T.; Dyer, M. J.; Goodall, A. H.; Cohen, G. M. *Blood* **2011**, *117*, 7145–7154.
19. Shashni, B.; Sakharkar, K. R.; Nagasaki, Y.; Sakharkar, M. K. *J. Drug Target.* **2013**, *21*, 161–174.
20. Neel, N. F.; Sai, J.; Ham, A. J.; Sobolik-Delmaire, T.; Mernaugh, R. L.; Richmond, A. *PLoS One* **2011**, *6*, e23813.
21. Bansal, A. T.; Barnes, M. R. *Curr. Opin. Drug Discov. Devel.* **2008**, *11*, 303–310.
22. Whitty, A.; Kumaravel, G. *Nat. Chem. Biol.* **2006**, *2*, 112–118.
23. Bauer, R. A.; Wurst, J. M.; Tan, D. S. *Curr. Opin. Chem. Biol.* **2010**, *14*, 308–314.
24. Jubb, H.; Higueruelo, A. P.; Winter, A.; Blundell, T. L. *Trends Pharmacol. Sci.* **2012**, *33*, 241–248.
25. Morrow, J. K.; Zhang, S. *Curr. Pharm. Des.* **2012**, *18*, 1255–1265.
26. Gago, F. *Future Med. Chem.* **2012**, *4*, 1961–1970.
27. Makley, L. N.; Gestwicki, J. E. *Chem. Biol. Drug Des.* **2013**, *81*, 22–32.
28. Villoutreix, B. O.; Bastard, K.; Sperandio, O.; Fahraeus, R.; Poyet, J. L.; Calvo, F.; Déprez, B.; Miteva, M. A. *Curr. Pharm. Biotechnol.* **2008**, *9*, 103–122.
29. Voet, A.; Zhang, K. Y. *Curr. Pharm. Des.* **2012**, *18*, 4586–4598.
30. Koes, D.; Khoury, K.; Huang, Y.; Wang, W.; Bista, M.; Popowicz, G. M.; Wolf, S.; Holak, T. A.; Dömling, A.; Camacho, C. J. *PLoS One* **2012**, *7*, e32839.
31. Oltersdorf, T.; Elmore, S. W.; Shoemaker, A. R.; Armstrong, R. C.; Augeri, D. J.; Belli, B. A.; Bruncko, M.; Deckwerth, T. L.; Dinges, J.; Hajduk, P. J.; Joseph, M. K.; Kitada, S.; Korsmeyer, S. J.; Kunzer, A. R.; Letai, A.; Li, C.; Mitten, M. J.; Nettesheim, D. G.; Ng, S.; Nimmer, P. M.; O'Connor, J. M.; Oleksijew, A.; Petros, A. M.; Reed, J. C.; Shen, W.; Tahir, S. K.; Thompson, C. B.; Tomaselli, K. J.; Wang, B.; Wendt, M. D.; Zhang, H.; Fesik, S. W.; Rosenberg, S. H. *Nature* **2005**, *435*, 677–681.
32. Ziarek, J. J.; Liu, Y.; Smith, E.; Zhang, G.; Peterson, F. C.; Chen, J.; Yu, Y.; Chen, Y.; Volkman, B. F.; Li, R. *Curr. Top. Med. Chem.* **2012**, *12*, 2727–2740.
33. Grosdidier, S.; Fernández-Recio, J. *Curr. Pharm. Des.* **2012**, *18*, 4607–4618.
34. Schiro, M. M.; Stauber, S. E.; Peterson, T. L.; Krueger, C.; Darnell, S. J.; Satyshur, K. A.; Drinkwater, N. R.; Newton, M. A.; Hoffmann, F. M. *PLoS One* **2011**, *6*, e25021.
35. Azzarito, V.; Long, K.; Murphy, N. S.; Wilson, A. J. *Nat. Chem.* **2013**, *5*, 161–173.
36. Verdine, G. L.; Hilinski, G. J. *Methods Enzymol.* **2012**, *503*, 3–33.
37. Edwards, T. A.; Wilson, A. J. *Amino Acids* **2011**, *41*, 743–754.
38. Tsou, L. K.; Cheng, Y.; Cheng, Y. C. *Curr. Opin. Pharmacol.* **2012**, *12*, 403–407.
39. Wade, M.; Li, Y. C.; Wahl, G. M. *Nat. Rev. Cancer* **2012**, *13*, 83–96.

40. Muller, P. A.; Vousden, K. H. *Nat. Cell Biol.* **2013**, *15*, 2–8.
41. Li, Q.; Lozano, G. *Clin. Cancer Res.* **2013**, *19*, 34–41.
42. Vassilev, L. T. *Trends Mol. Med.* **2007**, *13*, 23–31.
43. Tovar, C.; Graves, B.; Packman, K.; Filipovic, Z.; Xia, B. H.; Tardell, C.; Garrido, R.; Lee, E.; Kolinsky, K.; To, K. H.; Linn, M.; Podlaski, F.; Wovkulich, P.; Vu, B.; Vassilev, L. T. *Cancer Res.* **2013**, *73*, 2587–2597.
44. http://www.clinicaltrials.gov/ct2/results?term=rg-7112&Search=Search.
45. Barillé-Nion, S.; Bah, N.; Véquaud, E.; Juin, P. *Anticancer Res.* **2012**, *32*, 4225–4233.
46. Brown, S. P.; Taygerly, J. P. *Annual Reports Medicinal Chemistry,* Vol. 47; Desai, M., Ed.; Academic Press: UK, 2012 (chapter 17).
47. Condon, S. M. *Annual Reports Medicinal Chemistry,* Vol. 46; Macor, J., Ed.; Academic Press: UK, 2011 (chapter 13).
48. Weyhenmeyer, B.; Murphy, A. C.; Prehn, J. H.; Murphy, B. M. *Exp. Oncol.* **2012**, *34*, 192–199.
49. Thomas, S.; Quinn, B. A.; Das, S. K.; Dash, R.; Emdad, L.; Dasgupta, S.; Wang, X. Y.; Dent, P.; Reed, J. C.; Pellecchia, M.; Sarkar, D.; Fisher, P. B. *Expert Opin. Ther. Targets* **2013**, *17*, 61–75.
50. http://www.clinicaltrials.gov/ct2/results?term=navitoclax+and+cll&Search=Search.
51. Park, C. M.; Bruncko, M.; Adickes, J.; Bauch, J.; Ding, H.; Kunzer, A.; Marsh, K. C.; Nimmer, P.; Shoemaker, A. R.; Song, X.; Tahir, S. K.; Tse, C.; Wang, X.; Wendt, M. D.; Yang, X.; Zhang, H.; Fesik, S. W.; Rosenberg, S. H.; Elmore, S. W. *J. Med. Chem.* **2008**, *51*, 6902–6915.
52. Vandenberg, C. J.; Cory, S. *Blood* **2013**, *121*, 2285–2288.
53. http://www.fiercebiotech.com/story/blog-abbvie-slams-brakes-cancer-drug-trials-after-patient-death/2013-02-15.
54. Liu, Q.; Wang, H. G. *Commun. Integr. Biol.* **2012**, *5*, 557–565.
55. Baggstrom, M. Q.; Qi, Y.; Koczywas, M.; Argiris, A.; Johnson, E. A.; Millward, M. J.; Murphy, S. C.; Erlichman, C.; Rudin, C. M.; Govindan, R. *J. Thorac. Oncol.* **2011**, *6*, 1757–1760.
56. http://www.clinicaltrials.gov/ct2/results?term=AT-101+and+Phase+2&Search=Search.
57. Joudeh, J.; Claxton, D. *Expert Opin. Investig. Drugs* **2012**, *21*, 363–373.
58. Quinn, B. A.; Dash, R.; Azab, B.; Sarkar, S.; Das, S. K.; Kumar, S.; Oyesanya, R. A.; Dasgupta, S.; Dent, P.; Grant, S.; Rahmani, M.; Curiel, D. T.; Dmitriev, I.; Hedvat, M.; Wei, J.; Wu, B.; Stebbins, J. L.; Reed, J. C.; Pellecchia, M.; Sarkar, D.; Fisher, P. B. *Expert Opin. Investig. Drugs* **2011**, *20*, 1397–1411.
59. http://www.clinicaltrials.gov/ct2/results?term=obatoclax&Search=Search.
60. Flygare, J. A.; Beresini, M.; Budha, N.; Chan, H.; Chan, I. T.; Cheeti, S.; Cohen, F.; Deshayes, K.; Doerner, K.; Eckhardt, S. G.; Elliott, L. O.; Feng, B.; Franklin, M. C.; Reisner, S. F.; Gazzard, L.; Halladay, J.; Hymowitz, S. G.; La, H.; LoRusso, P.; Maurer, B.; Murray, L.; Plise, E.; Quan, C.; Stephan, J. P.; Young, S. G.; Tom, J.; Tsui, V.; Um, J.; Varfolomeev, E.; Vucic, D.; Wagner, A. J.; Wallweber, H. J.; Wang, L.; Ware, J.; Wen, Z.; Wong, H.; Wong, J. M.; Wong, M.; Wong, S.; Yu, R.; Zobel, K.; Fairbrother, W. J. *J. Med. Chem.* **2012**, *55*, 4101–4113.
61. http://www.clinicaltrials.gov/ct2/results?term=gdc-0152&Search=Search.
62. Emami, K. H.; Nguyen, C.; Ma, H.; Kim, D. H.; Jeong, K. W.; Eguchi, M.; Moon, R. T.; Teo, J. L.; Kim, H. Y.; Moon, S. H.; Ha, J. R.; Kahn, M. *Proc. Natl. Acad. Sci. U.S.A.* **2004**, *101*, 12682–12687.
63. http://www.clinicaltrials.gov/ct2/results?term=pri-724&Search=Search.
64. For Example: Matsui, Y.; Hadaschik, B. A.; Fazli, L.; Andersen, R. J.; Gleave, M. E.; So, A. I. *Mol. Cancer Ther.* **2009**, *8*, 2402–2411.
65. Thiel, P.; Kaiser, M.; Ottmann, C. *Angew. Chem. Int. Ed. Engl.* **2012**, *51*, 2012–2018.

66. Emami, S. *Clin. Res. Hepatol. Gastroenterol.* **2011**, *35*, 98–104.
67. Richter, A.; Rose, R.; Hedberg, C.; Waldmann, H.; Ottmann, C. *Chemistry* **2012**, *18*, 6520–6527.
68. Watanabe, N.; Osada, H. *Curr. Drug Targets* **2012**, *13*, 1654–1658.
69. Banappagari, S.; Corti, M.; Pincus, S.; Satyanarayanajois, S. *J. Biomol. Struct. Dyn.* **2012**, *30*, 594–606.
70. For Example: Tang, R.; Hirsch, P.; Fava, F.; Lapusan, S.; Marzac, C.; Teyssandier, I.; Pardo, J.; Marie, J. P.; Legrand, O. *Blood* **2009**, *114*, 2993–3000.
71. For Example: Schindl, M.; Oberhuber, G.; Obermair, A.; Schoppmann, S. F.; Karner, B.; Birner, P. *Cancer Res.* **2001**, *61*, 5703–5706.
72. For Example: Sharma, P.; Patel, D.; Chaudhary, J. *Cancer Med.* **2012**, *1*, 187–197.
73. For Example: Castañon, E.; Bosch-Barrera, J.; López, I.; Collado, V.; Moreno, M.; López-Picazo, J. M.; Arbea, L.; Lozano, M. D.; Calvo, A.; Gil-Bazo, I. *J. Transl. Med.* **2013**, *11*, 13.
74. For Example: Luo, K. J.; Wen, J.; Xie, X.; Fu, J. H.; Luo, R. Z.; Wu, Q. L.; Hu, Y. *Ann. Thorac. Surg.* **2012**, *93*, 1682–1688.
75. For Example: Tam, W. F.; Gu, T. L.; Chen, J.; Lee, B. H.; Bullinger, L.; Fröhling, S.; Wang, A.; Monti, S.; Golub, T. R.; Gilliland, D. G. *Blood* **2008**, *112*, 1981–1992.
76. For Example: Minn, A. J.; Gupta, G. P.; Siegel, P. M.; Bos, P. D.; Shu, W.; Giri, D. D.; Viale, A.; Olshen, A. B.; Gerald, W. L.; Massague, J. *Nature* **2005**, *436*, 518–524.
77. For Example: Swarbrick, A.; Roy, E.; Allen, T.; Bishop, J. M. *Proc. Natl. Acad. Sci. U.S.A.* **2008**, *105*, 5402–5407.
78. For Example: Wang, X.; Di, K.; Zhang, X.; Han, H. Y.; Wong, Y. C.; Leung, S. C.; Ling, M. T. *Oncogene* **2008**, *27*, 4456–4466.
79. For Example: Jang, T. J.; Jung, K. H.; Choi, E. A. *Int. J. Cancer* **2006**, *118*, 1356–1363.
80. Garland, W.; Salvador, R.; Chaudhary, J. AACR Meeting, Washington, DC, 2013; Abstract 758/20.
81. Garland, W.; Chaudhary, J. U.S. Patent 8138356, 2012.
82. Filippakopoulos, P.; Qi, J.; Picaud, S.; Shen, Y.; Smith, W. B.; Fedorov, O.; Morse, E. M.; Keates, T.; Hickman, T. T.; Felletar, I.; Philpott, M.; Munro, S.; McKeown, M. R.; Wang, Y.; Christie, A. L.; West, N.; Cameron, M. J.; Schwartz, B.; Heightman, T. D.; La Thangue, N.; French, C. A.; Wiest, O.; Kung, A. L.; Knapp, S.; Bradner, J. E. *Nature* **2010**, *468*, 1067–1073.
83. Also see: Sweis, R.; Michaelides, M. *Annual Reports Medicinal Chemistry*, Vol. 48; Desai, M., Ed.; Academic Press: UK, 2013 (chapter 13).
84. Puissant, A.; Frumm, S. M.; Alexe, G.; Bassil, C. F.; Qi, J.; Chanthery, Y. H.; Nekritz, E. A.; Zeid, R.; Gustafson, W. C.; Greninger, P.; Garnett, M. J.; McDermott, U.; Benes, C. H.; Kung, A. L.; Weiss, W. A.; Bradner, J. E.; Stegmaier, K. *Cancer Discov.* **2013**, *3*, 308–323.
85. Watson, J. *Open Biol.* **2013**, *3*, 120–144.
86. Walters, W. P. *Expert Opin. Drug Discov.* **2012**, *7*, 99–107.
87. Morelli, X.; Bourgeas, R.; Roche, P. *Curr. Opin. Chem. Biol.* **2011**, *15*, 475–481.
88. Xu, Y.; Lu, H.; Kennedy, J. P.; Yan, X.; McAllister, L. A.; Yamamoto, N.; Moss, J. A.; Boldt, G. E.; Jiang, S.; Janda, K. D. *J. Comb. Chem.* **2006**, *8*, 531–539.
89. Singh, J.; Petter, R. C.; Baillie, T. A.; Whitty, A. *Nat. Rev. Drug Discov.* **2011**, *10*, 307–317.
90. Rosell, R. *N. Engl. J. Med.* **2013**, *368*, 1551–1552.

Infectious Diseases

Section Editor: John Primeau
Hancock, Maine

CHAPTER SIXTEEN

Recent Progress in the Discovery of Neuraminidase Inhibitors as Anti-influenza Agents

Aisyah Saad Abdul Rahim*, Mark von Itzstein†
*School of Pharmaceutical Sciences, Universiti Sains Malaysia, Minden, Penang, Malaysia
†Institute for Glycomics, Gold Coast Campus, Griffith University, Queensland, Australia

Contents

1. INTRODUCTION

Influenza remains a serious health threat to mankind and has unleashed devastating socioeconomic burdens in regions afflicted with seasonal and pandemic influenza.[1,2] The pandemic swine influenza virus strain of 2009 had a global impact—more than 600,000 cases and ~17,000 deaths reported around the world. In the United States alone, an estimated 60 million cases were reported.[3-6] Therapeutic options against influenza viruses are limited. Vaccines serve as a prophylactic treatment against seasonal influenza A and B viruses. They are reformulated annually to target specific strains, expected to resurface based on the surveillance and analysis of circulating influenza strains. Vaccine production typically takes several months, thus offering little benefit in containing spontaneous influenza outbreaks or pandemics. Apart from vaccines, antiviral drugs have proved indispensable in the prophylaxis and treatment of influenza infections. The established drug targets against

Annual Reports in Medicinal Chemistry, Volume 48
ISSN 0065-7743
http://dx.doi.org/10.1016/B978-0-12-417150-3.00016-8
249

influenza are the M2-ion channel receptors and neuraminidase (NA) enzyme. M2-ion channel receptor blockers, for example, amantidine, are active only toward influenza A. These agents fell out of use in clinical settings due to severe adverse effects and rapid resistance that consequently reduced their efficacy as anti–influenza agents.[1,7,8]

In contrast, NA inhibitors (NAIs) are effective against both influenza A and B viruses, and show lower frequency of viral resistance *in vitro*.[9] Zanamivir **1**, a highly polar drug, is administered through inhalation, whereas the orally available oseltamivir **2b** has been, to date, the drug of choice in the treatment of influenza viral infections. Stockpiling of oseltamivir **2b** and zanamivir **1** is a part of pandemic preparedness in many countries and has helped mitigate undesirable impacts of influenza pandemics, for instance the threat of the highly pathogenic avian influenza virus H5N1 (2005). The use of oseltamivir, in part, has also led to the emergence of oseltamivir-resistant strains in H5N1 and the pandemic swine influenza H1N1 (pdmH1N1 2009) viruses. However it is believed that oseltamivir has little influence in the spontaneous emergence of the resistant seasonal H1N1 mutant (2007–2008).[7,10,11] To date, zanamivir-resistant strains are rare.

The intravenously administered cyclopentanyl peramivir **3** was approved for a limited time in the United States under Emergency Use Authorization (EUA) for the treatment of hospitalized patients infected with swine influenza (H1N1).[12] The latest addition to the NAI family is laninamivir octanoate **4**. The compound is modified with a methoxy group at C-7 and marketed as a prodrug. Although laninamivir, the active metabolite of the prodrug **4**, is a weaker NAI compared to zanamivir **1** and oseltamivir **2a** *in vitro*, it exhibits superior inhibition of virus replication, in particular *in vivo* models infected with H5N1. The C-7 modification in **4** provides the additional advantage of longer-acting antiviral protection; the compound is inhaled due to its improved pharmacokinetics.[13–15] These anti–influenza therapeutics **3** and **4** were approved in 2010 for clinical use in Japan. More in-depth discussions on NAIs have appeared in other recent literature.[10,13,16–18] This report highlights the progress in the discovery of NAIs that have emerged in the last 6 years.

The invariant key binding moieties in NA are shown in Fig. 16.1A.[13,20] In essence, the NA active site is composed of five subsites (S1–S5). Binding of sialic acid to NA is characterized by polar, charged interactions (S1–S3, S5), and hydrophobic subsite S4 to the acetamido methyl group. Recent X-ray crystallographic studies revealed significant structural flexibility of the 150-loop[21] that is located adjacent to the active site. The discovery of the 150-loop has further distinguished the two phylogenetic groups of influenza A virus NAs: group 1 NAs (N1, N4, N5, and N8) can assume a more "open" form and a "closed" form (Fig. 16.1B and C), whereas group 2 NAs (N2, N3, N6, N7, and N9) typically assume a "closed" form. Design and development of earlier inhibitors have been based on the "closed" form of group 2 and influenza B virus. Given the recent reemergence of pandemic and seasonal influenza A viruses from group 1, the discovery of the 150-cavity presents exciting opportunities for designing NAIs with improved selectivity and potency toward group 1, than group 2.[21–23]

Additionally, changes to the NA active site can arise from mutations caused by occasional genetic reassortments (∼50% difference) and continual antigenic drifts, which are a consequence of selection pressure from host immunity and from drug use.[11,24]

Six point mutations have been recently reported (Fig. 16.1A).[11,13] Of particular note is the mutation His274Tyr (H274Y), which has been found in H5N1 and pandemic swine flu. These mutants exhibit significantly reduced sensitivity toward oseltamivir. In the mutant NA active site, the bulky tyrosine pushes the carboxylate of Glu276 back into the active site. Unfavorable interactions between Glu276 and the hydrophobic C-5 O-pentanyl chain in **2a** lead to the loss of potency for **2a**[25,26] and **3**.[11] The H274Y strain, however, remains susceptible toward zanamivir since the C-8 and C-9 hydroxyl groups on the glycerol side chain maintain contacts with the exposed Glu-276 carboxylate. Other mutant strains, for instance Gln136Lys,[27] Arg292Ser, Glu119Val, and Arg152Lys (Flu B), have also been studied, revealing varying viral "fitness" and enzymatic activities.[11,13,17] These mutations may significantly influence the design of future NAIs.

Central to all NAIs are cyclic scaffolds that are utilized in the presentation of the functionalities that engage with subsites S1–S5. The following sections highlight these flexible and rigid scaffolds as emerging NAIs, including a brief review on natural products as leads for anti-influenza drug development.

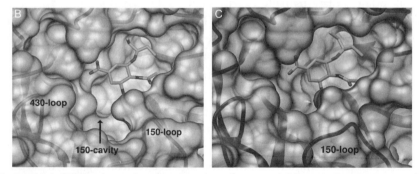

Figure 16.1 (A) A schematic representation of neuraminidase active site in complex with sialic acid (N2 numbering is used). Key binding interactions are shown as dashed lines, while the binding pockets with respective subsites (S1–S5) are highlighted in colors. Observed point mutations in some circulating influenza strains are indicated with an asterisk. (B) Neuraminidase N1-oseltamivir carboxylate complex showing the open 150-loop, 150-cavity, and 430-loop (PDB ID 2HU). (C) Neuraminidase N2-oseltamivir carboxylate complex with a closed 150-loop (PDB ID 2HU4). (B) and (C) were generated using UCSF Chimera package.[19] (See color plate.)

2. INHIBITORS BASED UPON FLEXIBLE SCAFFOLDS

2.1. Pyranose scaffolds

To lock open the 150-loop and explore for additional interactions in the resulting cavity of N1, a series of C-3 analogues based on Neu5Ac2en **5** (K_i 0.67–3.90 μM, N1s and N2) has been synthesized.[22,28] The general trend of these C-3 modified Neu5Ac2en suggests selectivity toward N1s rather than N2. While the unsubstituted **6** showed a modest inhibition against N1 and N2 NAs, the tolyl analogue **7** exhibited up to an ∼200-fold stronger inhibition selective toward the pandemic N1. Tolyl **7** was also active against drug-resistant mutants including the pdmH1N1 2009 H274Y and Q136K mutants. Bulky naphthalene **8**[28] showed low micromolar affinity toward N1, its activity being comparable to **7**. Analogues with non-planar substituents such as cyclohexane, *tert*-Bu, and propyl groups were modest-to-poor inhibitors, as were *p-tert*-Bu–, *p*-COOH, and disubstituted phenyl derivatives. Aided by modeling studies, a solved crystal structure of N8-**7** complex confirmed, for the first time, the role of the C-3 tolyl side chain in **7** in selectively locking open the 150-loop of group 1 NAs.

HO₂C, [structure] OH OH / OH / NHAc / OH **5**

HO₂C, [structure] R OH OH / OH / NHAc / OH
R=
H (allyl), **6**, K_i (μM) 222[a]; 153[b]; 3629[c]
tolyl, **7**, K_i (μM) 7.3[a]; 1.7[b]; 219[c]
napthalene, **8**, K_i (μM) 9.3[a]; 2.8[b]
[a]H5N1 [b]pdm09 H1N1 [c]H3N2

HO₂C, [structure] OH OH / OH / NHAc / HN O / HN / H NH₂O / NH₂ / O **9**

HO₂C, [structure] OH OH / OH / NHAc / HN N / NH₂ / O / N / NH **10**

Several C-4 modified zanamivir analogues have been reported to probe the 150-cavity in N1. A combination of scaffold hopping and bioisostere replacement strategies has led to the design and preparation of a 35-compound series of C-4-substituted zanamivir analogues.[29] Compound **9** with an L-Asn substituent inhibited H3N2 (IC_{50} 0.58 μM) and H5N1 (IC_{50} 2.72 μM) with good anti-influenza activity against H5N1 infection of MDCK cells. The unexpected fourfold stronger potency against N2, rather than N1, was attributed to the L-Asn moiety being better accommodated in N2. Other modifications to the C-4 moiety proved ineffective including substituted urea and thiourea bioisosteres with terminal alkyl, aryl, or amino acids, and various heterocyclic scaffolds. Similar heterocyclic extensions of the internal guanidino group resulted in modest inhibitions against N1, with the exception of the piperazino-substituted **10**.[30] Inhibitor **10** exhibited low micromolar NA activity (IC_{50} 2.15 μM) and submicromolar anti–influenza activity (EC_{50} 0.77 μM) against H1N1 virus. Compound **11** represents a series of novel C-4 triazole substituted zanamivir analogues that showed moderate anti–influenza activities.[31] Triazole **11** was found to be almost equipotent to **1** (**11**, EC_{50} 6.4 μM; **1**, EC_{50} 2.8 μM), even though molecular docking suggested that the triazole group of **11** formed fewer contacts with subsite S2, as compared to the guanidino moiety of **1**. Nevertheless, no additional data against other NAs were reported for C-4 modified zanamivir analogues **9–11** to support the notion of their selectivity toward N1.

Compound **12** exemplifies the remarkable nanomolar inhibition of five different NAs by phosphonate analogues of zanamivir.[32] **12** exhibited approximately three- to sixfold higher NA inhibitory activities compared to **1** and **2a**, including against the H274Y mutant, and has low nanomolar anti–influenza activities with no toxicity to human 294T cells. The nanomolar potencies have been partly attributed to electrostatic interactions of the phosphonate with subsite S1. In another study involving modifications at C-1 and C-4 of **1**, a fluorophenyl **13**[33] emerged as the most potent NAI showing nanomolar inhibition comparable to zanamivir (**1**, IC_{50} 0.0014 μM, H3N2; 0.012 μM, H5N1; 0.001 μM, H1N1). Modifications at C-1 incorporating *para*- and *ortho*-substituted fluorophenyls or other amino acids together with C-4 substituents (guanidino, amino, and hydroxyl groups) afforded modest-to-poor NAIs.

11
EC_{50} 6.4 μM, H5N1

12
IC_{50} 0.3 to 1.0 nM, A/N1s
IC_{50} 4.4 nM, A/N2

13
IC_{50} 0.013 μM, H3N2
IC_{50} 0.001 μM, H5N1
IC_{50} 0.09 μM, H1N1

As a hybrid of Neu5Ac2en **5** and **2a**, compound **14** was designed to explore the hydrophobic interactions at subsite S2 in N2.[34] Superimposition of **14** and **5** revealed that the glucuronic acid scaffold of **14** adopted a dihydropyran ring conformation resembling **5**. Hybrid **14** exhibited a 32-fold stronger activity compared to **5**. Low micromolar affinity was observed with O-methyl **15** or O-acetyl derivatives. Introduction of longer alkyl chains, for instance an O-ethyl in compounds **16** and **18** or O-allyl at C-3 position, dramatically decreased the affinity of this series of analogues toward N2. This observation indicates the intolerance of subsite S2 for bulky hydrophobic functionality.

R =
H, **14**, IC_{50} 0.32 μM^a
CH_3, **15**, IC_{50} 1.46 μM^a
Ethyl, **16**, IC_{50} 28 μM^a
[a]A/N2

R =
OH, **5**, IC_{50} 10.5 μM^a
NH_2, **17**, IC_{50} 0.68 μM^a
OEt, **18**, IC_{50} 479 μM^a
[a]A/N2

Conjugation of the C-7 OH of **1** with anti-inflammatory agents has broadened the potential of NAIs by introducing additional therapeutic value without compromising their nanomolar NA inhibitory activity.[35] This was conceived from the observations of virus-induced hypercytokinemia seen in H5N1-infected patients. Thus, C-7 zanamivir-caffeic acid conjugates **19** and **20** were found to be active against N1, demonstrated good anti-influenza efficacy ($\sim EC_{50}$ 1.4 and 5.1 nM, respectively), and suppressed proinflammatory cytokines. Multivalent presentations of **1**, via L-glutamine polymer, afforded low nanomolar anti-influenza agent **22** showing 15–860 times stronger potency than the monomer **21**.[36]

More recently, two groups[37,38] reported several novel mechanism-based covalent NAIs represented by **23**. Crystallographic studies revealed a covalent linkage formed between the C-2 of **23** with the hydroxyl group of Tyr-406 which is believed to play a critical role in the catalytic activity of influenza NA. Bearing an equatorial fluorine at C-3, the difluoro **23** was active against influenza A and B NAs, including drug-resistant strains, and inhibited viral replications at low nanomolar concentrations. Compound **23** also provided good protection from influenza infection in animal models, with activity comparable to **1**. Generally, the equatorial derivatives showed superior activity compared to the axial epimers.[37]

R=
X=
O, **19** IC_{50} 2.9–7.4 nM, A/N1
NH, **20** IC_{50} 41.3–60.3 nM, A/N1

R = H, **21**
R =

22

23
IC_{50} (nM) 13, H1N1; 44, H1N1 H275Y; 25, H3N2; 2.4, H3N2 E119V; 4.5, Flu B

2.2. Cyclohexene scaffolds

Investigations into the design of noncarbohydrate scaffolds that mimic the transition state intermediate have been extensively carried out. Oseltamivir phosphonates are more active than the corresponding **2a** (IC$_{50}$ 2.6 nM against A/WSN/33, H1N1).[39] Against the H274Y mutant, phosphonates **24** and **25** exhibited over 1000- and 20-fold higher potency than **2a**, respectively. Furthermore, the guanidino derivative **24** inhibited influenza B with an IC$_{50}$ of 2.1 nM (**2a**, 36 nM; **1**, 32.1 nM). These phosphonate–guanidino analogues exhibited anti–influenza activities, including against the H274Y mutant, in cell-based assays. Consistent with their potent NA inhibition and antiviral activity in cell-based assays, oral administration of these analogues, at 1 mg/kg or higher doses, seemed to afford good protection to mice infected with four influenza virus strains (human H1N1, pandemic H1N1, H274Y H1N1 mutant, and avian H5N1). Metabolism, toxicity, and pharmacokinetic studies further showed that **24** and **25** are metabolically stable with low tendency of being protein bound. The low overall bioavailability of these guanidino compounds (less than 12%) requires advanced formulation for oral delivery.

Several C-3 triazoles have been further functionalized with a variety of lipophilic substituents, including a bulky steroidal group, to probe the 150-cavity. Triazole **26** was the most active in the series showing a threefold stronger inhibition toward N1 than N2. The C-3 guanidino analogue **27**, a ring isomer of **2a**, exhibited a nanomolar inhibition against virus-like particles containing N1 (**1**, K_i 0.16 nM).[40] This study also concluded that the position of the C-1 and C-2 double bond has a significant influence in the appropriate orientation of the C-3 substituent toward the 150-loop and has been recently discussed in more detail elsewhere.[13]

In a preliminary screening against the avian influenza A H7N1 and H7N3 viruses, two C-4-substituted 2-propenylamido compounds **28** and

29 were identified as potent NAIs with different selectivity toward NAs.[41] The C-5 amino **28** was active toward N1 (**2a**, IC_{50} 36.1–53.2 nM), whereas the C-5-substituted isopropyl **29** exhibited better inhibitory activity toward N3 (**2a**, IC_{50} 2.9–3.3 nM).

2.3. Cyclopentane and related scaffolds

Five-member ring structures have been investigated as potential scaffolds for the proper presentation of key binding functionalities toward subsites S1–S5. Inspired by **3**, three rigid bicyclic sulfones were found to be competitive inhibitors with K_i of 4.5–700 μM as determined by enzyme kinetic studies using NP40-inactivated influenza A/Brisbane/59/2007 (H1N1) virus.[42] Sulfone **30** (K_i 4.5 μM) is the most active compound in this series. Molecular modeling indicated a good overlay of the carboxylate, guanidino, and acetamido functionalities of **30**. The alignment with key functionalities on peramivir **3** and the promising inhibitory activity of this scaffold hold the potential for the lead development of an anti-influenza drug. Two heterocycles reported recently represent new scaffolds for NAIs. Thiazolidine **31**[43] was reported to inhibit influenza A NA with a submicromolar potency. Modified proline derivatives have also been investigated, affording the low micromolar influenza A virus inhibitor **32**.[44] The virus strains used in the evaluation of **31** and **32** were not described.

30
IC_{50} 32 μM, A/N1

31
IC_{50} 0.14 μM, Flu A

32
IC_{50} 1.56 μM, Flu A

33
IC_{50} 17.7 nM, rH1N1
K_i 7.5 nM, rH1N1

To investigate the role of the guanidino group in peramivir, de-guanidinylated peramivir derivatives were synthesized and evaluated.[45] Inhibitor **33** exhibited a loss in potency that is less than an order of magnitude than peramivir (**3**, IC_{50} 3.4 nM; K_i 0.83 nM) toward recombinant H1N1 (rH1N1). The study suggested that, despite the loss of potency, the predicted improved oral bioavailability of de-guanidinylated **33** could provide an advantage over the low-bioavailable guanidino-containing drugs **1** and **3**. These predictions call for a more in-depth study.

3. INHIBITORS BASED UPON RIGID, PLANAR SCAFFOLDS

Intense medicinal chemistry efforts to harness the potential of aromatic rings as NAIs took place from early 1990 until 2005; nowadays, the effort has dwindled to several groups highlighted below. Viewed as a more economical alternative, aromatic rings and heterocycles tend to have fewer chiral centers and offer the advantage of facile, shorter syntheses compared to the preparations of NAIs based on flexible scaffolds.[8,17,46] Structures **34–39** are some examples of NAIs based on rigid scaffolds.[44,47–50] So far, none has come close to the nanomolar affinities of the current drugs **1–4**.

34
IC$_{50}$ 0.032 μM, N2

35
IC$_{50}$ 2 μM, H3N2; 20 μM, H1N1;
9.1 μM,N9; 180 μM, B/Lee/40

36
IC$_{50}$ 0.20 μM, H3N2; 1.6 μM, H1N1;
3.2 μM,N9; 32 μM, B/Lee/40

37
IC$_{50}$ 26.96 – 27.73 μM
against A/H1N1s

38
IC$_{50}$ 0.06 μM, H1N1

39
IC$_{50}$ 0.05 μM, H1N1

4. NATURAL PRODUCTS AS POTENTIAL ANTI-INFLUENZA DRUG LEADS

Natural products continue to serve as an invaluable source for many current drugs in the market, and for the discovery of novel bioactive compounds and scaffolds.[16,51,52] The quest for such novel, potent bioactive scaffolds as anti-influenza lead structures has attracted considerable effort in screening natural products against NAs; an extensive survey[16] has appeared recently. The majority of natural products reported to inhibit NAs belong to the families of flavonoids and (oligo)stilbenes. This section summarizes this huge field with examples of representative compounds, typically polyphenols having a chalcone backbone as highlighted in **40**.

Chalcones **40** and **41** both were found to not only inhibit NAs noncompetitively, but also synergistically in the presence of **2a**. They enhanced the potency of **2a** by about 3- to 53-fold, depending on the NA subtypes. Speculatively, phenols **40** and **41** may bind to a different binding site or a secondary site[53,54] on NA that brings about improved binding interactions of **2a** to the active site.

40
IC_{50} (μM) 21.46, H1N1; 21.09, H9N2;
10.88, H1N1 WT; 9.10, H1N1 H274Y

41
IC_{50} (μM) 23.21, H1N1; 14.15, H9N2

42
IC_{50} (μM) 20.45, H1N1; 18.57, H9N2;
8.15, H1N1 WT; 3.31, H1N1 H274Y

43
IC_{50} (μM) 0.7, H1N1;
1.1, H3N2; 1.0, H9N2

44
IC_{50} 2.2 μM (rH1N1)

45

Isolated from *Cleistocalyx operculatus*, chalcone **42** emerged as the most active isolate against NAs, and demonstrated antiviral efficacy (EC_{50} 4.90 μM) coupled with low cytotoxicity in MDCK cells (CC_{50} >120 μM).[55] SAR studies[55,56] of flavonoids, exemplified by **43–45**, have revealed the importance of 4'- and 7-hydroxyl groups together with the α,β-unsaturated carbonyl moiety. Weaker activity shown by glycoside flavonoids, compared to aglycones **43–45**, suggests steric hindrance. Both **43**[57] and **44**[58] have been shown to be noncompetitive inhibitors of NAs with low micromolar potency. Recently, several flavonoids were reported to strongly quench the fluorescence of 4-methylumbelliferone, the hydrolysis product in a fluorescent NA inhibition assay, leading to false-positives. Assay results of such compounds should be considered with caution.[59] Screening a library of 2000 structurally diverse compounds has led to the identification of aurintricarboxylic acid **45** as a potent inhibitor of wild-type and H274Y mutant NAs (IC_{50} 8.7 and 18.4 μM, H1N1; IC_{50} 5.4 and 2.3 μM, H5N1).[60] Its NA activities are in agreement with low EC_{50} values for inhibiting viral replication (EC_{50} 4.1 μM, H1N1; 6.3 μM, H3N2; 5.4 μM, H5N1).

Shape-based virtual screening for active molecules from the National Cancer Institute (NCI) database has identified four flavonoids with low micromolar inhibition against H1N1s.[61] The most active compound, artocarpin **46,** exhibited submicromolar NA inhibition, including activity against the oseltamivir-resistant human isolate (A/342/09), presumably through interactions with the active site.

46
IC$_{50}$ (µM) 0.18–0.55, H1N1s

47
IC$_{50}$ 0.59–1.64 µM, H1N1s

48
IC$_{50}$ 4.13 µM, H1N1

Six diarylheptanoids from *Alpinia katsumadai* seeds were reported to show low to modest micromolar inhibition of influenza A/PR/8/34 (H1N1) in a chemiluminescence NA assay.[62] The most potent diarylheptanoid, katsumadain A **47** (IC$_{50}$ 1.05 µM), was chosen for further evaluation. Despite it being a poor inhibitor of the oseltamivir-resistant human isolate, **47** was found to inhibit four swine influenza NA isolates with submicromolar potency and exhibited anti-influenza virus activity in MDCK cells. Modeling studies indicated that **47** seemed well accommodated in the more open conformation of the active site, aided by the flexible 430-loop and 245-loop—similar to the pose assumed by **46**. Interestingly, **48** appeared to retain low micromolar activity against A/PR/8/34 (H1N1). Enzyme kinetic data of **46–48** have not been reported.

5. CONCLUSIONS AND OUTLOOK

Influenza virus always seems two to three steps ahead of us in ensuring its survival, by varying its antigens through genetic reassortment. Thus, even though stockpiling of anti-influenza drugs is essential in countries around the world as short-term measures, over-reliance on current drugs may undermine our efforts in the long run. Sustained focused effort in the discovery and development of novel NAIs is therefore vital.

The deep awareness of how sialic acid or an inhibitor interacts with the active site of NA has culminated in the discovery of **1** and **2a**, and is key toward designing the next-generation NAIs. In view of the widespread H274Y mutants, the significant hurdle remains in improving the pharmaco-kinetic profiles of remaining anti-influenza therapeutics for oral delivery. It is tempting to speculate on the increasing importance of loop flexibility, and perhaps the secondary binding site, in providing exciting opportunities for future inhibitors displaying high group-selectivity. With the recent discovery of the covalent NAIs, further development in these areas holds great promise in producing the next-generation anti-influenza candidates.

REFERENCES

1. von Itzstein, M. *Nat. Rev. Drug Discov.* **2007**, *6*, 967–974.
2. von Itzstein, M., Ed.; *Influenza Virus Sialidase—A Drug Discovery Target*; Springer Basel AG: Basel, 2011.
3. WHO Pandemic (H1N1) 2009. Update 76, Global Alert and Response, 27 November 2009. http://www.who.int/csr/don/2009_11_27a/en/index.html (accessed 13 March 2013).
4. WHO Pandemic (H1N1) 2009. Update 94, Global Alert and Response, 1 April 2010. http://www.who.int/csr/don/2010_04_01/en/index.html (accessed 13 March 2013).
5. Shrestha, S. S.; Swerdlow, D. L.; Borse, R. H.; Prabhu, V. S.; Finelli, L.; Atkins, C. Y.; Owusu-Edusei, K.; Bell, B.; Mead, P. S.; Biggerstaff, M.; Brammer, L.; Davidson, H.; Jernigan, D.; Jhung, M. A.; Kamimoto, L. A.; Merlin, T. L.; Nowell, M.; Redd, S. C.; Reed, C.; Schuchat, A.; Meltzer, M. I. *Clin. Infect. Dis.* **2011**, *52*(Suppl. 1), S75–S82.
6. Thompson, W. W.; Shay, D. K.; Weintraub, E.; Brammer, L.; Cox, N.; Anderson, L. J.; Fukuda, K. *JAMA* **2003**, *289*, 179–186.
7. De Clercq, E. *Nat. Rev. Drug Discov.* **2006**, *5*, 1015–1025.
8. Babu, Y. S.; Chand, P.; Kotian, P. L. *Ann. Rep. Med. Chem.* **2006**, *41*, 287–297.
9. Rameix-Welti, M.-A.; Munier, S.; Naffakh, N. Resistance Development to Influenza Virus Sialidase Inhibitors. In *Influenza Virus Sialidase—A Drug Discovery Target;* von Itzstein, M., Ed.; Springer Basel: Basel, 2011; pp 153–174.
10. von Itzstein, M.; Thomson, R. *Handb. Exp. Pharmacol.* **2009**, *189*, 111–154.
11. Thorlund, K.; Awad, T.; Boivin, G.; Thabane, L. *BMC Infect. Dis.* **2011**, *11*, 134.
12. Bronson, J.; Dhar, M.; Ewing, W. *Ann. Rep. Med. Chem.* **2011**, *46*, 433–502.
13. Thomson, R.; von Itzstein, M. The Development of Carbohydrate-Based Influenza Virus Sialidase Inhibitors. In *Influenza Virus Sialidase—A Drug Discovery Target;* von Itzstein, M., Ed.; Springer Basel: Basel, 2011; pp 77–104.
14. Koyama, K.; Nakai, D.; Takahashi, M.; Nakai, N.; Kobayashi, N.; Imai, T.; Izumi, T. *Drug Metab. Dispos.* **2012**, *41*, 180–187.
15. Chairat, K.; Tarning, J.; White, N. J.; Lindegardh, N. *J. Clin. Pharmacol.* **2012**, *53*, 119–139.
16. Grienke, U.; Schmidtke, M.; Grafenstein, von, S.; Kirchmair, J.; Liedl, K. R.; Rollinger, J. M. *Nat. Prod. Rep.* **2012**, *29*, 11–36.
17. Feng, E.; Ye, D.; Li, J.; Zhang, D.; Wang, J.; Zhao, F.; Hilgenfeld, R.; Zheng, M.; Jiang, H.; Liu, H. *ChemMedChem* **2012**, *7*, 1527–1536.
18. Gong, J.; Xu, W.; Zhang, J. *Curr. Med. Chem.* **2007**, *14*, 113–122.
19. Pettersen, E. F.; Goddard, T. D.; Huang, C. C.; Couch, G. S.; Greenblatt, D. M.; Meng, E. C.; Ferrin, T. E. *J. Comput. Chem.* **2004**, *25*, 1605–1612.
20. Colman, P. M. *Protein Sci.* **1994**, *3*, 1687–1696.
21. Russell, R. J.; Haire, L. F.; Stevens, D. J.; Collins, P. J.; Lin, Y. P.; Blackburn, G. M.; Hay, A. J.; Gamblin, S. J.; Skehel, J. J. *Nature* **2006**, *443*, 45–49.
22. Rudrawar, S.; Dyason, J. C.; Rameix-Welti, M.-A.; Rose, F. J.; Kerry, P. S.; Russell, R. J. M.; van der Werf, S.; Thomson, R. J.; Naffakh, N.; von Itzstein, M. *Nat. Commun.* **2010**, *1*, 113–117.
23. von Itzstein, M. Novel Carbohydrate-Based Inhibitors that Target Influenza A Virus Sialidase. In *Chembiomolecular Science*; Shibasaki, M., Ed.; Springer: New York, 2013; pp 261–267.
24. Govorkova, E. A. *Influenza Other Respi. Viruses* **2012**, *7*, 50–57.

25. Collins, P. J.; Haire, L. F.; Lin, Y. P.; Liu, J.; Russell, R. J.; Walker, P. A.; Skehel, J. J.; Martin, S. R.; Hay, A. J.; Gamblin, S. J. *Nature* **2008**, *453*, 1258–1261.
26. Ives, J. A. L.; Carr, J. A.; Mendel, D. B.; Tai, C. Y.; Lambkin, R.; Kelly, L.; Oxford, J. S.; Hayden, F. G.; Roberts, N. A. *Antiviral Res.* **2002**, *55*, 307–317.
27. Buchy, P. Clinical Experience with Influenza Virus Sialidase Inhibitors. In *Influenza Virus Sialidase—A Drug Discovery Target*; von Itzstein, M., Ed.; Springer Basel: Basel, 2011; pp 131–151.
28. Rudrawar, S.; Kerry, P. S.; Rameix-Welti, M.-A.; Maggioni, A.; Dyason, J. C.; Rose, F. J.; van der Werf, S.; Thomson, R. J.; Naffakh, N.; Russell, R. J. M.; von Itzstein, M. *Org. Biomol. Chem.* **2012**, *10*, 8628–8639.
29. Ye, D.; Shin, W.-J.; Li, N.; Tang, W.; Feng, E.; Li, J.; He, P.-L.; Zuo, J.-P.; Kim, H.; Nam, K.-Y.; Zhu, W.; Seong, B.-L.; Tai No, K.; Jiang, H.; Liu, H. *Eur. J. Med. Chem.* **2012**, *54*, 764–770.
30. Wen, W. H.; Wang, S. Y.; Tsai, K. C.; Cheng, Y. S. E.; Yang, A. S.; Fang, J. M.; Wong, C. H. *Bioorg. Med. Chem.* **2010**, *18*, 4074–4084.
31. Li, J.; Zheng, M.; Tang, W.; He, P.-L.; Zhu, W.; Li, T.; Zuo, J.-P.; Liu, H.; Jiang, H. *Bioorg. Med. Chem. Lett.* **2006**, *16*, 5009–5013.
32. Shie, J.-J.; Fang, J.-M.; Lai, P.-T.; Wen, W.-H.; Wang, S.-Y.; Cheng, Y.-S. E.; Tsai, K.-C.; Yang, A.-S.; Wong, C.-H. *J. Am. Chem. Soc.* **2011**, *133*, 17959–17965.
33. Feng, E.; Shin, W.-J.; Zhu, X.; Li, J.; Ye, D.; Wang, J.; Zheng, M.; Zuo, J.-P.; No, K. T.; Liu, X.; Zhu, W.; Tang, W.; Seong, B.-L.; Jiang, H.; Liu, H. *J. Med. Chem.* **2013**, *56*, 671–684.
34. Bhatt, B.; Böhm, R.; Kerry, P. S.; Dyason, J. C.; Russell, R. J. M.; Thomson, R. J.; von Itzstein, M. *J. Med. Chem.* **2012**, *55*, 8963–8968.
35. Liu, K.-C.; Fang, J.-M.; Jan, J.-T.; Cheng, T.-J. R.; Wang, S.-Y.; Yang, S.-T.; Cheng, Y.-S. E.; Wong, C.-H. *J. Med. Chem.* **2012**, *55*, 8493–8501.
36. Weight, A. K.; Haldar, J.; Álvarez de Cienfuegos, L.; Gubareva, L. V.; Tumpey, T. M.; Chen, J.; Klibanov, A. M. *J. Pharm. Sci.* **2010**, *100*, 831–835.
37. Kim, J. H.; Resende, R.; Wennekes, T.; Chen, H. M.; Bance, N.; Buchini, S.; Watts, A. G.; Pilling, P.; Streltsov, V. A.; Petric, M.; Liggins, R.; Barrett, S.; McKimm-Breschkin, J. L.; Niikura, M.; Withers, S. G. *Science* **2013**, *340*, 71–75.
38. Vavricka, C. J.; Liu, Y.; Kiyota, H.; Sriwilaijaroen, N.; Qi, J.; Tanaka, K.; Wu, Y.; Li, Q.; Li, Y.; Yan, J.; Suzuki, Y.; Gao, G. F. *Nat. Commun.* **2013**, *4*, 1491.
39. Cheng, T.-J. R.; Weinheimer, S.; Tarbet, E. B.; Jan, J.-T.; Cheng, Y.-S. E.; Shie, J.-J.; Chen, C.-L.; Chen, C.-A.; Hsieh, W.-C.; Huang, P.-W.; Lin, W.-H.; Wang, S.-Y.; Fang, J.-M.; Hu, O. Y.-P.; Wong, C.-H. *J. Med. Chem.* **2012**, *55*, 8657–8670.
40. Mohan, S.; McAtamney, S.; Haselhorst, T.; von Itzstein, M.; Pinto, B. M. *J. Med. Chem.* **2010**, *53*, 7377–7391.
41. Kongkamnerd, J.; Cappelletti, L.; Prandi, A.; Seneci, P.; Rungrotmongkol, T.; Jongaroonngamsang, N.; Rojsitthisak, P.; Frecer, V.; Milani, A.; Cattoli, G.; Terregino, C.; Capua, I.; Beneduce, L.; Gallotta, A.; Pengo, P.; Fassina, G.; Miertus, S.; De-Eknamkul, W. *Bioorg. Med. Chem.* **2012**, *20*, 2152–2157.
42. Brant, M. G.; Wulff, J. E. *Org. Lett.* **2012**, *14*, 5876–5879.
43. Liu, Y.; Jing, F.; Xu, Y.; Xie, Y.; Shi, F.; Fang, H.; Li, M.; Xu, W. *Bioorg. Med. Chem.* **2011**, *19*, 2342–2348.
44. Zhang, J.; Wang, Q.; Fang, H.; Xu, W.; Liu, A.; Du, G. *Bioorg. Med. Chem.* **2007**, *15*, 2749–2758.
45. Bromba, C. M.; Mason, J. W.; Brant, M. G.; Chan, T.; Lunke, M. D.; Petric, M.; Boulanger, M. J.; Wulff, J. E. *Bioorg. Med. Chem. Lett.* **2011**, *21*, 7137–7141.
46. Rich, J. R.; Gehle, D.; von Itzstein, M. Design and Synthesis of Sialidase Inhibitors for Influenza Virus Infections. In *Comprehensive Glycoscience*; Kamerling, J. P., Boons,

G.-J., Lee, Y. C., Suzuki, A., Tanigichi, N., Voragen, A. G. J., Eds.; Elsevier: Oxford, 2007; pp 885–922.

47. Venkatramani, L.; Johnson, E. S.; Kolavi, G.; Air, G. M.; Brouillette, W. J.; Mooers, B. H. M. *BMC Struct. Biol.* **2012**, *12*, 7.

48. Li, Y.; Silamkoti, A.; Kolavi, G.; Mou, L.; Gulati, S.; Air, G. M.; Brouillette, W. J. *Bioorg. Med. Chem.* **2012**, *20*, 4582–4589.

49. Li, J.; Zhang, D.; Zhu, X.; He, Z.; Liu, S.; Li, M.; Pang, J.; Lin, Y. *Mar. Drugs* **2011**, *9*, 1887–1901.

50. Sun, C.; Zhang, X.; Huang, H.; Zhou, P. *Bioorg. Med. Chem.* **2006**, *14*, 8574–8581.

51. Marson, C. M. *Chem. Soc. Rev.* **2011**, *40*, 5514–5533.

52. Baker, D. D.; Chu, M.; Oza, U.; Rajgarhia, V. *Nat. Prod. Rep.* **2007**, *24*, 1225–1244.

53. Sung, J. C.; Van Wynsberghe, A. W.; Amaro, R. E.; Li, W. W.; McCammon, J. A. *J. Am. Chem. Soc.* **2010**, *132*, 2883–2885.

54. Lai, J. C. C.; Garcia, J.-M.; Dyason, J. C.; Böhm, R.; Madge, P. D.; Rose, F. J.; Nicholls, J. M.; Peiris, J. S. M.; Haselhorst, T.; von Itzstein, M. *Angew. Chem. Int. Ed.* **2012**, *51*, 2221–2224.

55. Dao, T. T.; Tung, B. T.; Nguyen, P. H.; Thuong, P. T.; Yoo, S. S.; Kim, E. H.; Kim, S. K.; Oh, W. K. *J. Nat. Prod.* **2010**, *73*, 1636–1642.

56. Liu, A.-L.; Wang, H.-D.; Lee, S. M.; Wang, Y.-T.; Du, G.-H. *Bioorg. Med. Chem.* **2008**, *16*, 7141–7147.

57. Jeong, H. J.; Kim, Y. M.; Kim, J. H.; Kim, J. Y.; Park, J.-Y.; Park, S.-J.; Ryu, Y. B.; Lee, W. S. *Biol. Pharm. Bull.* **2012**, *35*, 786–790.

58. Jeong, H. J.; Ryu, Y. B.; Park, S.-J.; Kim, J. H.; Kwon, H.-J.; Kim, J. H.; Park, K. H.; Rho, M.-C.; Lee, W. S. *Bioorg. Med. Chem.* **2009**, *17*, 6816–6823.

59. Kongkamnerd, J.; Milani, A.; Cattoli, G.; Terregino, C.; Capua, I.; Beneduce, L.; Gallotta, A.; Pengo, P.; Fassina, G.; Monthakantirat, O.; Umehara, K.; De-Eknamkul, W.; Miertus, S. *J. Biomol. Screen.* **2011**, *16*, 755–764.

60. Hung, H.-C.; Tseng, C.-P.; Yang, J.-M.; Ju, Y.-W.; Tseng, S.-N.; Chen, Y.-F.; Chao, Y.-S.; Hsieh, H.-P.; Shih, S.-R.; Hsu, J. T. A. *Antiviral Res.* **2009**, *81*, 123–131.

61. Kirchmair, J.; Rollinger, J. M.; Liedl, K. R.; Seidel, N.; Krumbholz, A.; Schmidtke, M. *Future Med. Chem.* **2011**, *3*, 437–450.

62. Grienke, U.; Schmidtke, M.; Kirchmair, J.; Pfarr, K.; Wutzler, P.; Dürrwald, R.; Wolber, G.; Liedl, K. R.; Stuppner, H.; Rollinger, J. M. *J. Med. Chem.* **2010**, *53*, 778–786.

CHAPTER SEVENTEEN

Novel Therapeutics in Discovery and Development for Treatment of Chronic HBV Infection

Yimin Hu[*], Wei Zhu[*], Guozhi Tang[*], Alexander V. Mayweg[*], Guang Yang[†], Jim Z. Wu[†], Hong C. Shen[*]

[*]Department of Medicinal Chemistry, F. Hoffmann-La Roche AG, Pharma Research & Early Development China, Shanghai, China
[†]Department of Virology, F. Hoffmann-La Roche AG, Pharma Research & Early Development China, Shanghai, China

Contents

1. INTRODUCTION OF THE HBV LIFE CYCLE

As a leading cause of cirrhosis and liver cancer, chronic hepatitis B virus (HBV) infection poses a major public health threat. Approximately 2 billion people worldwide carry evidence of prior HBV infection, and 350 million are chronic HBV carriers, resulting in an annual death toll of around

600,000.[1] A major portion of chronic infection occurs through perinatal transmission or early childhood infections.[2,3] While HBV vaccination programs have been implemented extensively, only a modest decline of total chronic HBV cases can be anticipated in the next two decades.[4]

The HBV life cycle is illustrated in Fig. 17.1. It has been shown that HBV particles can be attracted to the host cellular surface by heparin sulfate proteoglycans.[5] A receptor-mediated uptake of viral particles, through the interaction with the pre-S1 region of the virion, enables viral entry into hepatocytes. This process is recently proposed to be facilitated by a multiple transmembrane transporter, sodium taurocholate cotransporting polypeptide, predominantly expressed in the liver.[6] The ensuing endocytosis delivers viral particles to the host cytosol. The decoating of the envelope and the delivery of relaxed circular-DNA (RC-DNA) into the host nucleus proceed prior to the conversion of the RC-DNA to circular covalently closed DNA (cccDNA).[7] The cccDNA is then packed into a stable minichromosome structure together with host and viral proteins. The cccDNA minichromosome subsequently serves as a transcriptional template to synthesize pregenomic RNA (pgRNA) and other viral mRNAs. This process appears to be regulated by the HBV core protein,

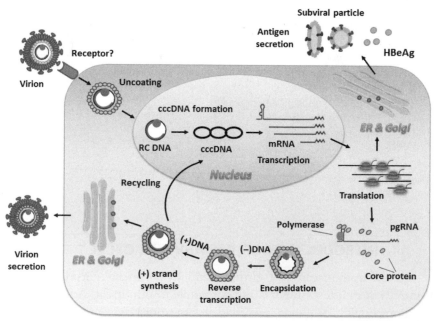

Figure 17.1 Schematic representation of the HBV life cycle. (See color plate.)

x protein (HBx), as well as certain host factors, including histone 3 and 4, histone deacetylases, methyltransferases, and a number of transcriptional factors.[8] The mRNAs are then translated to HBV polymerase, core protein, surface antigen (HBsAg), HBV e-antigen (HBeAg), and HBx proteins. In addition, the pgRNA together with the viral polymerase is packaged into a capsid structure, which is derived from the assembly of core protein dimers, to form a mature nucleocapsid.[9] In the nucleocapsid, viral polymerase catalyzes the reverse transcription of pgRNA to form negative strand DNA and then positive strand DNA, followed by the formation of RC-DNA. The resulting RC-DNA may be shuffled back into the nucleus to amplify the cccDNA pool.[10] Alternatively, the HBV virions may undergo morphogenesis to bud into late endosomes or multivesicular bodies in the endoplasmic reticulum before secretion via the exosome pathway.[9] The released viral particles can then infect other hepatocytes thereby embarking on a new life cycle.

There is only limited number of viral targets available to interfere with the virus due to the small size of HBV genome. Only seven HBV proteins (small, middle, and large surface proteins, core, HBeAg, polymerase, and HBx) can potentially serve as viral protein targets.[11] Host antiviral targets can also be considered, yet potential safety issues associated with chronic treatment aiming at a host target need to be evaluated carefully.

2. CURRENT STANDARD OF CARE THERAPEUTICS FOR HBV INFECTION

Current standard of care treatments for chronic HBV infection include interferon alpha (IFN-α) and pegylated IFN-α targeting the host immune system, as well as five nucleos(t)ide drugs: lamivudine (3TC, LMV), telbivudine, adefovir (ADV), entecavir (ETV), and tenofovir, which target HBV polymerase as a chain terminator.[12] Unfortunately, despite significant reduction of viremia, reflected by achieving undetectable viral DNA in blood, these therapies fail to achieve an acceptable level of clinical cure as determined by HBsAg seroconversion. It is also worth noting that nucleos(t)ide therapies require lifetime treatment in order to prevent rebound of the virus after cessation of the treatment. Compared to nucleos(t)ide therapies, the use of interferons cannot achieve superior response rates regarding serum HBV DNA levels and is associated with issues of parenteral administration and adverse effects.[13,14] Interferon-based treatments show marginally more promise regarding drug efficacy to achieve HBsAg loss. The response rate in HBeAg-positive patients following 4–5 years of treatment ranges from 0% to

10% with nucleos(t)ide therapies, and 8–15% with pegylated IFN-α or pegylated IFN-α plus LMV, whereas the response rate in HBeAg-negative patients ranges from 0% to 5% with nucleos(t)ide therapies, and 8% with pegylated IFN-α or pegylated IFN-α plus LMV.[14] Hence, a tremendous unmet medical need remains for the discovery of drugs which can help achieve persistent HBsAg loss in chronic HBV patients.

Novel compounds targeting various replication steps in the HBV life cycle have recently emerged. Several novel large and small molecules may alter the landscape of future anti-HBV therapy via their unique modes of action. While two excellent reviews reported on small molecule anti-HBV agents up to 2010,[15,16] this review mainly focuses on the most recent progress of small molecule HBV inhibitors in terms of their mechanism of action, biological activities, and structural classes.

3. SELECTED DIRECT ACTING ANTIVIRALS AND IMMUNE MODULATORS UNDER CLINICAL DEVELOPMENT

Selected anti-HBV agents (excluding vaccines) under clinical development are summarized in Table 17.1. Besides nucleos(t)ides and IFNs, therapeutics with new modes of actions have been explored clinically. It is conceivable that the combination of current HBV therapeutics and certain new agents could emerge and redefine the future standard of care for chronic HBV infection.

3.1. Selected nucleos(t)ide HBV polymerase inhibitors

The excellent potency, drug resistance, and safety profiles of the second generation of HBV nucleos(t)ide drugs, especially TDF and ETV, pose a significant challenge to develop new nucleos(t)ide drugs with clear differentiation. However, several new nucleos(t)ide HBV polymerase inhibitors are still in clinical development due to their better activity against resistant HBV variants, PK/PD, or toxicology profiles.[17] For example, LB80380 (**1**, Phase II) showed good efficacy in LMV-resistant chronic hepatitis B patients.[18] Lagociclovir valactate (**2**, MIV-210, Phase I) displayed potent antiviral activity against LMV-, ADV-, and ETV-resistant HBV variants as well as high oral bioavailability in humans.[19] To minimize toxicity due to nucleos(t)ide exposure in nonhepatic tissues such as kidney, pradefovir (**3**, Phase II) was developed as a liver-targeted prodrug of ADV. This resulted in lower exposure in other tissues at an efficacious dose.[20] In addition, other nucleos(t)ides under clinical trials include emtricitabine, valtorcitabine, and amdoxovir.[18]

Table 17.1 Selected anti-HBV agents under clinical development

Mode of action	Examples	Structures	Clinical trial phase
Nucleos(t)ide HBV polymerase inhibitors	LB80380	**1**	II
	Lagociclovir valactate	**2**	I
	Pradefovir	**3**	II
HBsAg release inhibitor	REP 9AC′	Oligonucleotide-based polymer	II
Viral entry inhibitor	Myrcludex-B	Synthetic lipopeptide	I
Immunomodulator	Peg-IFN lambda 1a	Type III interferon	III
TLR-7 agonist	GS-9620	**4**	I

3.2. Modified nucleic acid

REP 9AC′, a second generation nucleic acid-based amphipathic polymer (NAP), inhibited HBsAg release from HBV-infected hepatocytes.[21] NAPs can block the assembly of subviral particles and retain HBsAg in the perinuclear space, thereby preventing its secretion. In an ongoing Phase II study, REP 9AC′ led to rapid clearance of serum HBsAg at 500 mg (i.v.) once weekly in HBV patients.

3.3. Peptide entry inhibitor

Myrcludex-B, a synthetic lipopeptide derived from the pre-S1 domain of the HBV large envelope protein (L-protein), was developed as an HBV entry inhibitor targeting viral entry machinery on the hepatocyte surface.[22] In HepaRG cells, this peptide had no effect on viral replication in infected cells, while demonstrating efficacy blocking HBV infection ($IC_{50} = 8$ nM). A recent study shows that myrcludex-B can block intrahepatic virus spreading in humanized mice infected with HBV and hinder amplification of the cccDNA pool in initially infected hepatocytes.[23]

3.4. Lambda interferon

Type III lambda interferons (IFN-λ) have similar antiviral activity to type I IFNs (IFN-α).[24] However, the receptor distribution for IFN-λ is more restricted to hepatocytes. Pegylated IFN-λ 1a has drawn considerable attention due to the concept that liver-selective distribution may offer a good therapeutic index.[25] PK/PD studies in cynomolgus monkeys demonstrated the absence of pegylated IFN-λ pharmacologic activity in leukocytes, which was consistent with its low receptor expression in blood.[26] On the other hand, strong induction of interferon-stimulated gene expression occurred in cultured hepatocytes and liver biopsies. In a recent 24-week Phase II study, pegylated IFN-λ treatment appeared to result in faster clearance of serum HBV DNA, greater reduction in HBsAg, and better tolerance, compared to the pegylated IFN-α treatment.[27]

3.5. Toll-like receptor-7 agonist

A selective oral Toll-like receptor-7 (TLR-7) agonist GS-9620 (4) has been reported recently.[28] This compound displayed potent induction of IFN-α, other immunomodulatory cytokines, and chemokines, as well as activation of lymphocyte subpopulations *in vitro*.[29] Recently, GS-9620 has been demonstrated to decrease serum and liver HBV DNA, HBsAg, and HBeAg in infected chimpanzees.[30] Elevation of IFN-α, ISGs in PBMCs and liver, and activation of lymphocyte subsets were also observed. It was suggested that the TLR-7 signaling pathway was activated in immune cells leading to an induced clearance of HBV-infected cells. As such, GS-9620 was advanced to Phase I clinical trial in healthy volunteers, where this compound was well tolerated in oral single ascending dose up to 12 mg and its PD effects began to emerge at 2 mg.[31] The Phase I clinical trial in chronic hepatitis B patients is currently ongoing.

4. NONNUCLEOS(T)IDE SMALL MOLECULE HBV INHIBITORS IN DISCOVERY STAGE

4.1. Compounds affecting capsid assembly and maturation

Several small molecule agents were found to modulate capsid assembly or to interfere with the encapsidation of HBV pgRNA. Bay 41-4109 (**5**) was an early example of the heteroaryldihydropyrimidine (HAP) series, which are capsid assembly effectors that prevent the formation of normal nucleocapsids.[32] The IC_{50} of Bay 41-4109 against HBV genome replication in HepG2.2.15 cells was 0.05 μM, and the CC_{50} was 7 μM. Further mechanistic studies revealed that HAP compounds misdirected capsid assembly from core protein to form aberrant products.[33]

5: Bay 41-4109   **6**: GLS-4

Recently, GLS-4 (**6**), which is structurally related to Bay 41-4109, showed improved potency ($IC_{50}=0.012$ μM) and selectivity index ($SI=CC_{50}/IC_{50}$, 6122) in hepG2.2.15 cells.[34] This compound was also active against HBV mutants resistant to ADV. GLS-4's ability to misdirect core protein assembly was studied using truncated core proteins (Cp149), which could spontaneously assemble into capsids under certain conditions.[35] GLS-4 was found to inhibit HBV replication by causing the misassembly of capsid from core proteins. It has also been suggested that core proteins may play multiple roles in the HBV life cycle, especially in viral encapsidation and the regulation of cccDNA function.[36] As such, this series of compounds provides a unique anti-HBV approach by targeting HBV core proteins.

HBV pgRNA is the template for reverse transcription. Inhibition of pgRNA encapsidation by core proteins can block HBV DNA synthesis, which thereby may represent another new therapeutic approach. Phenylpropenamide derivatives have been reported to inhibit HBV DNA replication through this mechanism.[37] In HepAD38 cells, AT-130 (7) demonstrated anti-HBV activity ($EC_{50} = 0.13$ μM, $CC_{50}/EC_{50} > 469$).[38] It was originally reported that AT-130 possessed an E-configuration. A recent paper clarified that most of the HBV inhibition activity was indeed derived from the Z-configuration isomers.[39] In the HepAD38 assay, compound 8 exhibited potent activity ($EC_{50} = 0.39$ μM) against HBV DNA synthesis with no cytotoxicity up to 10 μM.

7: AT-130   8

Oxymatrine (9), a natural product extracted from *Sophora japonica*, has been used to treat hepatitis B patients in China for decades. Oxymatrine inhibited the replication of HBV by interfering with the packing process of pgRNA into the nucleocapsid and/or by inhibiting the viral reverse transcription via destabilizing Hsc70 mRNA.[40,41] SAR studies identified compounds 10 and 11 with better potency in suppressing Hsc70 mRNA levels in HepG2.2.15 cells.[42,43] Anti-HBV activity of compound 10, measured by the inhibition of intracellular HBV DNA replication, is comparable to that of LMV. Compound 11 reduced cellular HBV DNA in HepG2.2.15 cells ($EC_{50} = 10$ μM, SI = 50) and was active against LMV-resistant HBV strains in Huh–7.5 cells. *In vivo* evaluation of **11** in mice for 7 days did not lead to adverse effects up to 750 mg/kg.

9: Oxymatrine   **10**   **11**

Sulfamoylbenzamides were recently disclosed as pgRNA encapsidation inhibitors.[44] A representative compound (DVR–23, **12**) achieved a similar efficacy to LMV (EC$_{50}$ = 0.39 μM, CC$_{50}$ > 50 μM) in AML12HBV10 cells. Mechanistic studies revealed that DVR–23 dose-dependently reduced the levels of encapsidated pgRNA and HBV DNA without affecting the capsid formation.

12

4.2. HBV cccDNA transcriptional modulators

The HBV cccDNA minichromosome has four overlapping open reading frames and its transcription to pgRNA and viral mRNAs is regulated by the core, S1, S2, and X promoters, as well as enhancers I and II in HepA2 cells.[45] Modulation of cccDNA transcription would be an effective strategy to reduce intracellular viral RNA and the production of viral antigens (HBeAg and HBsAg).

Helioxanthin (**13**) and some lactam derivatives demonstrated potent anti–HBV activities in HepG2.2.15 cells. Recently, a close helioxanthin derivative (**14**) was reported to suppress HBsAg and HBeAg in HepG2 cells with EC$_{50}$ values of 0.06 and 0.14 μM, respectively.[46] Compound **14** was also active against the LMV–resistant rtL515M/M539V strain. Compound (**14**) shared the same mechanism as helioxanthin in that it suppressed the activities of S1, S2, core promoters, and enhancer I. In another study, derivative **15** elicited strong suppression of duck HBV RNA production ($EC_{50} = 0.1$ μM) and resulted in dose-dependent reduction of duck HBV core protein and DNA.[47]

Quinolin-2-one **16** was identified as an HBsAg production inhibitor with moderate activity and a narrow selectivity index ($IC_{50} = 86$ μM, SI = 2.6) in HepG2.2.15 cells. To improve the anti-HBV potency and reduce cytotoxicity, analog **17** was synthesized, and it inhibited HBsAg ($EC_{50} = 10$ μM, SI > 135), HBeAg ($EC_{50} = 26$ μM, SI > 51), and DNA replication ($EC_{50} = 45$ μM).[48] Preliminary mechanistic studies revealed that **17** modulated the transcriptional activity of HBV enhancers I and II in HepG2 cells.

Caudatin (**18**), isolated from *Cynanchum bungei Decne*, was reported to be active against hepatocellular carcinoma (HCC).[49] Structural modifications as shown in **19** significantly improved activities against the secretion of HBsAg, HBeAg, and HBV DNA replication with EC_{50} values of 5.5, 5.5, and 2.4 μM, respectively, and a good selectivity index (SI > 330) in the HepG2.2.15 cell line.[50] Compound **19** is known to inhibit HBV X promoter activity and to interfere with enhancer I in a luciferase reporter gene assay.

18: Caudatin

19

4.3. cccDNA biosynthesis inhibitors

The persistent presence of the cccDNA episome in infected hepatocytes is the underlying cause of HBV chronicity and viral rebound after termination of treatment.[51] The formation and maintenance of cccDNA involve many viral intermediates and host factors, some of which may provide opportunities for molecular intervention. For example, sulfonamides CCC-0975 (**20**) and CCC-0346 (**21**) are low micromolar cccDNA biosynthesis inhibitors identified from screening in HepDES19 cells using HBeAg as a surrogate marker of cccDNA. In HepDES19 cells and DHBV–infected primary duck hepatocytes, both compounds blocked the conversion of RC-DNA to deproteinized RC-DNA (DP-RC-DNA), a putative precursor of cccDNA. While the intracellular levels of HBV RNA and RC-DNA remained unchanged at an early time point of the treatment by **20** or **21**, the reduction of DP-RC-DNA and cccDNA was dose-proportional after prolonged treatment at noncytotoxic concentrations.

20   **21**

4.4. Subviral particle secretion inhibitors

HBV viral antigens play an important role in suppressing human immune responses. Lack of efficacy toward abrogation of serum HBsAg is one of the key challenges for current HBV therapies, which could be addressed in part by inhibition of subviral particle secretion. One such example is the substituted tetrahydro-tetrazolo-pyrimidine compound (HBF-0529, **22**), a specific inhibitor of HBsAg secretion, with an EC_{50} of 1.5 µM in HepDE19 cells and 11 µM in HepG2.2.15 cells.[52] No cytotoxicity was observed up to 50 µM in both cell lines. HBF-0529 directly targeted the secretion of particles bearing HBV structural glycoproteins and had no effect on viral DNA synthesis or HBeAg secretion.

22: HBF-0529   **23**: PBHBV-2-15

Further optimization of compounds based on HBF-0529 identified triazolo-pyrimidines such as PBHBV-2-15 (**23**), which inhibited HBsAg in HepG2.2.15 cells with an improved EC_{50} (1.4 µM) and CC_{50} (50 µM).[53] Both compounds **22** and **23** provided comparable potency against HBsAg secretion in HepG2 cells transfected with HBV mutants that are resistant to ADV, LMV/telbivudine, and ETV, respectively. At a 25 mg/kg dose, no adverse effects were observed in either acute (C57BL mice, i.v.) or 14-day (HBV-transgenic mice/strain 1.3.32, p.o., q.d.) toxicity studies. In male Sprague–Dawley rats, compound **23** had a bioavailability of 42% at 25 mg/kg.

4.5. Compounds with undefined mechanism of action

Nitazoxanide (NTZ, **24**) is marketed in the United States for diarrhea and enteritis caused by *Cryptosporidium* spp. or *Giardia lamblia*. NTZ and its active metabolite tizoxanide (**25**) demonstrated a broad-spectrum antiviral activity and are currently under clinical evaluation for HCV and rotavirus.[54] In HepG2.2.15 cells, NTZ inhibited intracellular HBV replication ($EC_{50} = 0.59$ µM) and extracellular virus production ($EC_{50} = 0.12$ µM) with no cytotoxicity up to 100 µM. The drug also inhibited the production of extracellular HBsAg ($EC_{50} = 0.22$ µM) and HBeAg ($EC_{50} = 0.26$ µM), as well as the intracellular HBV core antigen ($EC_{50} = 1.1$ µM). The nitro group in NTZ can be replaced by chloride, as in analog RM-4848 (**26**), which inhibited both intracellular HBV replication ($EC_{50} = 1.0$ µM) and extracellular virus production ($EC_{50} = 0.37$ µM) in HepG2.2.15 cells.[55] While the precise mechanism of thiazolide antiviral activity has not been fully elucidated, current evidence suggests a host-mediated mechanism at the posttranslational stage.

24: R = Ac, Nitazoxanide (NTZ)   **26**: R = H, RM-4848
25: R = H, Tizoxanide (TIZ)

A class of pyridinones were reported to reduce extracellular HBV DNA, HBsAg, and HBeAg in HepG2.2.15 cells.[56] For example, CH04522 (**27**) inhibited extracellular HBV DNA replication, with an $EC_{50} = 1.5$ µM (SI = 151). CH04107 (**28**) was active against HBV DNA ($EC_{50} = 11$ µM, SI > 11), HBsAg ($EC_{50} = 18$ µM, SI > 7), and HBeAg ($EC_{50} = 0.1$ µM,

SI > 1200). In the Peking duck model, compound **28** gave similar potency at 20 and 40 mg/kg to ADV at 20 mg/kg.

27: R = H, CH04522
28: R = Me, CH04107

Arbidol (**29**), a flu drug used in Russia and China, exhibited broad-spectrum antiviral activities mainly through inhibition of viral membrane fusion.[57] A scaffold hopping exercise generated improved compounds containing quinolone,[58] benzimidazole,[59] and imidazo[1,2-*a*]pyridine[60] as core structures. In HepG2.2.15 cells, the IC_{50} values of compounds **30**, **31**, and **32** for inhibition of HBV DNA replication were 3.5 μM (SI = 38), 7.8 μM (SI = 13), and 2.6 μM (SI = 24), respectively. In addition, compound **30** suppressed HBsAg production with modest potency (IC_{50} = 11 μM, SI = 12).

29: Arbidol   **30**   **31**   **32**

Two other series of compounds bearing similar fused bicyclic core structures to arbidol were identified recently. Representative compounds include tryptamine derivative **33** and thiazolylbenzimidazole derivative **34**.[61,62] In HepG2.2.15 cells, both **33** and **34** demonstrated good potency against extra-cellular DNA production (IC_{50} = 0.4 and 1.1 μM, respectively) and favorable cytotoxicity (CC_{50} = 41 and >100 μM, respectively). However, the anti–HBV mechanism of each series of compounds was not revealed.

33   **34**

It was reported recently that naphthoquinone trimer **35** had an EC_{50} of 0.009 μM for the reduction of HBV DNA and a CC_{50} of 280 μM in

HepG2.2.15 cells.[63] Interestingly, the monomer (**36**) was also active in both a transfection assay ($EC_{50} = 2.5-5$ μM, $CC_{50} > 25$ μM in Huh7 cells), and a LMV–resistant anti–HBV cellular assay ($EC_{50} = 1.9$ μM, $CC_{50} > 25$ μM in HepG2 B1/HepG2 D88 cell cells). Unfortunately, compound **36** was inactive in duck and woodchuck models.

35   36

Herbs containing natural products used for hepatitis treatment have provided new opportunities for HBV research. For example, swerilactones H (**37**) and its congeners, isolated from the Chinese herb *Swertia mileensis*, were used for treating viral hepatitis. These compounds have IC_{50} values ranging from 1.5 to 5.3 μM against HBV DNA replication in HepG2.2.15 cells and no cytotoxicity.[64] Magnolol (**38**) and 9-β-xylopyranosylisolariciresinol (**39**) were found in *S. asper*, a remedy for HBV in Southern China.[65] In HepG2.2.15 cells, magnolol and 9-β-xylopyranosylisolariciresinol inhibited the secretion of HBsAg ($IC_{50} = 2.0$ and 6.6 μM, respectively) and HBeAg ($IC_{50} = 3.8$ and 25 μM, respectively). In the same cell line, the IC_{50} values of natural products nirtetralin A (**40**) and B (**41**)[66] against HBsAg secretion were 9.5 and 17 μM, respectively. Compound **41** ($CC_{50} = 1200$ μM) had a better cytotoxicity margin than **40** ($CC_{50} = 70$ μM).

37: Swerilactone H   38: Magnolol   39
                                                  Xyl = Xylosyl

40: 5*S*-, Nirtetralin A
41: 5*R*-, Nirtetralin B

5. CONCLUSION AND OUTLOOK

The goal of chronic HBV infection treatment is to achieve clinical cure and to prevent progression of disease to cirrhosis, liver failure, and

HCC. Most anti–HBV drugs on the market can effectively suppress viral replication, but none of them can achieve the eradication of subviral particles and cccDNA in the majority of patients. Investigation of anti–HBV compounds with novel mechanisms of actions will hopefully help to address this challenge. In addition, combination therapies derived by adding new anti-HBV drugs to the current standard of care may bring additional benefits to patients. With a better understanding of the disease biology, we envision that intense research efforts toward developing novel therapy against HBV are on the horizon and will ultimately lead to improved clinical cure rates.

ACKNOWLEDGMENT

The authors would like to acknowledge Dr. Liping Shi for preparing Fig. 17.1.

REFERENCES

1. WHO. Hepatitis B factsheet: www.who.int/mediacentre/factsheets/fs204/en.
2. WHO, *Weekly Epidemiol. Record* **2009**, *84*, 405.
3. Goldstein, S. T.; Zhou, F.; Hadler, S. C.; Bell, B. P.; Mast, E. E.; Margolis, H. S. *Int. J. Epidemiol.* **2005**, *34*, 1329.
4. Ioannou, G. N. *Ann. Intern. Med.* **2011**, *154*, 319.
5. Schulze, A.; Gripon, P.; Urban, S. *Hepatology* **2007**, *46*, 1759.
6. Yan, H.; Zhong, G.; Xu, G.; He, W.; Jing, Z.; Goa, Z.; Huang, Y.; Qi, Y.; Peng, B.; Wang, H.; Fu, L.; Song, M.; Chen, P.; Gao, W.; Ren, B.; Sun, Y.; Cai, T.; Feng, X.; Sui, J.; Li, W. *eLife* **2012**, *1*, e00049.
7. Rabe, B.; Vlachou, A.; Pante, N.; Helenius, A.; Kann, M. *Proc. Natl. Acad. Sci. U.S.A.* **2003**, *100*, 9849.
8. Quasdorff, M.; Protzer, U. *J. Viral Hepat.* **2010**, *17*, 527.
9. Patient, R.; Hourioux, C.; Roingeard, P. *Cell. Microbiol.* **2009**, *11*, 1561.
10. Zoulim, F.; Locarnini, S. *Gastroenterology* **2009**, *137*, 1593.
11. Lucifora, J.; Zoulim, B. *Future Virol.* **2011**, *6*, 599.
12. Michailidis, E.; Kirby, K. A.; Hachiya, A.; Yoo, W.; Hong, S. P.; Kim, S.-O.; Folk, W. R.; Sarafianos, S. G. *Int. J. Biochem. Cell Biol.* **2012**, *44*, 1060, and references therein.
13. Karayiannis, P. *J. Antimicrob. Chemother.* **2003**, *51*, 761.
14. Kwon, H.; Lok, A. S. *Nat. Rev. Gastroenterol. Hepatol.* **2011**, *8*, 275.
15. Zhan, P.; Jiang, X.; Liu, X. *Mini Rev. Med. Chem.* **2010**, *10*, 162.
16. Kim, K.-H.; Kim, N. D.; Seong, B.-L. *Molecules* **2010**, *15*, 5878.
17. Cox, N.; Tillmann, H. *Expert Opin. Emerging Drugs* **2011**, *16*, 713.
18. Yuen, M.-F.; Han, K.-H.; Um, S.-H.; Yoon, S. K.; Kim, H.-R.; Kim, J.; Kim, C. R.; Lai, C.-L. *Hepatology* **2010**, *51*, 767.
19. Michalak, T. I.; Zhang, H.; Churchill, N. D.; Larsson, T.; Johansson, N. G.; Oberg, B. *Antimicrob. Agents Chemother.* **2009**, *53*, 3803.
20. Tillmann, H. L. *Curr. Opin. Investig. Drugs* **2007**, *8*, 682.
21. Mahtab, M. A.; Bazinet, M.; Vaillant, A. In: *The 63rd Annual Meeting of the American Association for the Study of Liver Diseases, Boston, MA*, 2012, Poster 424.
22. Gripon, P.; Cannie, I.; Urban, S. *J. Virol.* **2005**, *79*, 1613.
23. Volz, T.; Allweiss, L.; M′Barek, M. B.; Warlich, M.; Lohse, A. W.; Pollok, J. M.; Alexandrov, A.; Urban, S.; Petersen, J.; Lütgehetmann, M.; Dandri, M. *J. Hepatol.* **2013**, *58*, 861.

24. Kotenko, S. V.; Gallagher, G.; Baurin, V. V.; Lewis-Antes, A.; Shen, M.; Shah, N. K.; Langer, J. A.; Sheikh, F.; Dickensheets, H.; Donnelly, R. P. *Nat. Immunol.* **2003**, *4*, 69.
25. Sommereyns, C.; Paul, S.; Staeheli, P.; Michiels, T. *PLoS Pathog.* **2008**, *4*, e1000017.
26. Byrnes-Blake, K. A.; Pederson, S.; Klucher, K. M.; Anderson-Haley, M.; Miller, D. M.; Lopez-Talavera, J. C.; Freeman, J. A. *J. Interferon Cytokine Res.* **2012**, *32*, 198.
27. Chan, H. L.; Ahn, S. H.; Chang, T.-T.; Peng, C.-Y.; Wong, D. K.; Coffin, C. S.; Lim, S. G.; Chen, P.-J.; Janssen, H. L.; Marcellin, P.; Serfaty, L.; Zeuzem, S.; Hu, W.; Critelli, L.; Lopez-Talavera, J. C.; Cooney, E. L. In: *The 63rd Annual Meeting of the American Association for the Study of Liver Diseases, Boston, MA*, 2012, Poster LB-14.
28. Desai, M. C.; Halcomb, R. L.; Hrvatin, P.; Hui, H. C.; Mc Fadden, R. Patent Application WO 2010/077613 A1, 2010.
29. Tumas, D.; Zheng, X.; Lu, B.; Rhodes, G.; Duatschek, P.; Hesselgesser, J.; Frey, C.; Henne, I.; Fosdick, A.; Halcomb, R.; Wolfgang, G. In: *46th Annual Meeting of the European Association for the Study of the Liver (EASL 2011), Berlin, March 30–April 3*, 2011, Abstract 1007.
30. Lanford, R. E.; Guerra, B.; Chavez, D.; Giavedoni, L.; Hodara, V. L.; Brasky, K. M.; Fosdick, A.; Frey, C. R.; Zheng, J.; Wolfgang, G.; Halcomb, R. L.; Tumas, D. B. *Gastroenterology* **2013**, *144*, 1508.
31. Lopatin, U.; Wolfgang, G.; Kimberlin, R.; Tumas, D.; Cornprost, M.; Chittick, G.; Frey, C.; Findlay, J.; Ohmstede, C.; Kearney, B.; Barnes, C.; Hirsch, K.; McHutchison, J. In: *46th Annual Meeting of the European Association for the Study of the Liver (EASL 2011), Berlin, March 30–April 3*, 2011, Abstract 614.
32. Deres, K.; SchrÖder, C. H.; Paessens, A.; Goldmann, S.; Hacker, H. J.; Weber, O.; Krämer, T.; NiewÖhner, U.; Pleiss, U.; Stoltefuss, J.; Graef, E.; Koletzki, D.; Masantschek, R. N.; Reimann, A.; Jaeger, R.; Gross, R.; Beckermann, B.; Schlemer, K. H.; Haebich, D.; Rübsamen-Waigmann, H. *Science* **2003**, *299*, 893.
33. Stray, S. J.; Zlotnick, A. *J. Mol. Recognit.* **2006**, *19*, 542.
34. Wang, X.-Y.; Wei, Z.-M.; Wu, G.-Y.; Wang, J.-H.; Zhang, Y.-J.; Li, J.; Zhang, H.-H.; Xie, X.-W.; Wang, X.; Wang, Z.-H.; Wei, L.; Wang, Y.; Chen, H.-S. *Antiviral Ther.* **2012**, *17*, 793.
35. Wynne, S. A.; Crowther, R. A.; Leslie, A. G. *Mol. Cell* **1999**, *3*, 771.
36. Bock, C. T.; Schwinn, S.; Locarnini, S.; Fyfe, J.; Manns, M. P.; Trautwein, C.; Zentgraf, H. *J. Mol. Biol.* **2001**, *307*, 183.
37. Feld, J. J.; Colledge, D.; Sozzi, V.; Edwards, R.; Littlejohn, M.; Locarnini, S. A. *Antiviral Res.* **2007**, *76*, 168.
38. Perni, R. B.; Conway, S. C.; Ladner, S. K.; Zaifert, K.; Otto, M. J.; King, R. W. *Bioorg. Med. Chem. Lett.* **2000**, *10*, 2687.
39. Wang, P.; Naduthambi, D.; Mosley, R. T.; Niu, C.; Furman, P. A.; Otto, M. J.; Sofia, M. J. *Bioorg. Med. Chem. Lett.* **2011**, *21*, 4642.
40. Xu, W.-S.; Zhao, K.-K.; Miao, X.-H.; Ni, W.; Cai, X.; Zhang, R.-Q.; Wang, J.-X. *World J. Gastroenterol.* **2010**, *16*, 2028.
41. Wang, Y.-P.; Liu, F.; He, H.-W.; Han, Y.-X.; Peng, Z.-G.; Li, B.-W.; You, X.-F.; Song, D.-Q.; Li, Z.-R.; Yu, L.-Y.; Cen, S.; Hong, B.; Sun, C.-H.; Zhao, L.-X.; Kreiswirth, B.; Perlin, D.; Shao, R.-G.; Jiang, J.-D. *Antimicrob. Agents Chemother.* **2010**, *54*, 2070.
42. Du, N.-N.; Li, X.; Wang, Y.-P.; Liu, F.; Liu, Y.-X.; Li, C.-X.; Peng, Z.-G.; Gao, L.-M.; Jiang, J.-D.; Song, D.-Q. *Bioorg. Med. Chem. Lett.* **2011**, *21*, 4732.
43. Gao, L.-M.; Han, Y.-X.; Wang, Y.-P.; Li, Y.-H.; Shan, Y.-Q.; Li, X.; Peng, Z.-G.; Bi, C.-W.; Zhang, T.; Du, N.-N.; Jiang, J.-D.; Song, D.-Q. *J. Med. Chem.* **2011**, *54*, 869.
44. Guo, J.-T.; Xu, X.; Block, T. M. Patent Application WO 2013/006394, 2013.
45. Choi, B. H.; Park, G. T.; Rho, H. M. *J. Biol. Chem.* **1999**, *274*, 2858.

46. Janmanchi, D.; Tseng, Y. P.; Wang, K.-C.; Huang, R. L.; Lin, C. H.; Yeh, S. F. *Bioorg. Med. Chem.* **2010**, *18*, 1213.
47. Ying, C.; Tan, S.; Cheng, Y.-C. *Antiviral Chem. Chemother.* **2010**, *21*, 97.
48. Guo, R.-H.; Zhang, Q.; Ma, Y.-B.; Luo, J.; Geng, C.-A.; Wang, L.-J.; Zhang, X.-M.; Zhou, J.; Jiang, Z.-Y.; Chen, J.-J. *Eur. J. Med. Chem.* **2011**, *46*, 307.
49. Fei, H. R.; Chen, H. L.; Xiao, T.; Chen, G.; Wang, F. Z. *Mol. Biol. Rep.* **2012**, *39*, 131.
50. Wang, L.-J.; Geng, C.-A.; Ma, Y.-B.; Luo, J.; Huang, X.-Y.; Chen, H.; Zhou, N.-J.; Zhang, X.-M.; Chen, J.-J. *Eur. J. Med. Chem.* **2012**, *54*, 352.
51. Cai, D.; Mills, C.; Yu, W.; Yan, R.; Aldrich, C. E.; Saputelli, J. R.; Mason, W. S.; Xu, X.; Guo, J.-T.; Block, T. M.; Cuconati, A.; Guo, H. *Antimicrob. Agents Chemother.* **2012**, *56*, 4277.
52. Dougherty, A. M.; Guo, H.; Westby, G.; Liu, Y.; Simsek, E.; Guo, J.-T.; Mehta, A.; Norton, P.; Gu, B.; Block, T.; Cuconati, A. *Antimicrob. Agents Chemother.* **2007**, *51*, 4427.
53. Yu, W.; Goddard, C.; Clearfield, E.; Mills, C.; Xiao, T.; Guo, H.; Morrey, J. D.; Motter, N. E.; Zhao, K.; Block, T. M.; Cuconati, A.; Xu, X. *J. Med. Chem.* **2011**, *54*, 5660.
54. Andrew Hemphill, A.; Müller, N.; Müller, J. *Anti-Infect. Agents* **2013**, *11*, 22.
55. Stachulski, A. V.; Pidathala, C.; Row, E. C.; Sharma, R.; Berry, N. G.; Iqbal, M.; Bentley, J.; Allman, S. A.; Edwards, G.; Helm, A.; Hellier, J.; Korba, B. E.; Semple, J. E.; Rossignol, J.-F. *J. Med. Chem.* **2011**, *54*, 4119.
56. Chen, J.; Zhang, M. Patent Application CN 102584690 A, 2012.
57. Teissier, E.; Zandomeneghi, G.; Loquet, A.; Lavillette, D.; Lavergne, J.-P.; Montserret, R.; Cosset, F.-L.; Böckmann, A.; Meier, B. H.; Penin, F.; Pécheur, E.-I. *PLoS One* **2011**, *6*(1), e15874.
58. Liu, Y.; Zhao, Y.; Zhai, X.; Feng, X.; Wang, J.; Gong, P. *Bioorg. Med. Chem.* **2008**, *16*, 6522.
59. Zhao, Y.; Liu, Y.; Chen, D.; Wei, Z.; Liu, W.; Gong, P. *Bioorg. Med. Chem. Lett.* **2010**, *20*, 7230.
60. Chen, D.; Liu, Y.; Zhang, S.; Guo, D.; Liu, C.; Li, S.; Gong, P. *Arch. Pharm. Chem. Life Sci.* **2011**, *11*, 158.
61. Qu, S.-J.; Wang, G.-F.; Duan, W.-H.; Yao, S.-Y.; Zuo, J.-P.; Tan, C.-H.; Zhu, D.-Y. *Bioorg. Med. Chem.* **2011**, *19*, 3120.
62. Luo, Y.; Yao, J.-P.; Yang, L.; Feng, C.-L.; Tang, W.; Wang, G.-F.; Zuo, J.-P.; Lu, W. *Arch. Pharm. Chem. Life Sci.* **2011**, *2*, 78.
63. Crosby, I. T.; Bourke, D. G.; Jones, E. D.; Jeynes, T. P.; Cox, S.; Coates, J. A. V.; Robertson, A. D. *Bioorg. Med. Chem. Lett.* **2011**, *21*, 1644.
64. Geng, C.-A.; Wang, L.-J.; Zhang, X.-M.; Ma, Y.-B.; Huang, X.-Y.; Luo, J.; Guo, R.-H.; Zhou, J.; Shen, Y.; Zuo, A.-X.; Jiang, Z.-Y.; Chen, J.-J. *Chem. Eur. J.* **2011**, *17*, 3893.
65. Li, J.; Huang, Y.; Guan, X.-L.; Li, J.; Deng, S.-P.; Wu, Q.; Zhang, Y.-J.; Su, X.-J.; Yang, R.-Y. *Phytochemistry* **2012**, *82*, 100.
66. Wei, W.; Li, X.; Wang, K.; Zheng, Z.; Zhou, M. *Phytother. Res.* **2012**, *26*, 964.

CHAPTER EIGHTEEN

Special Challenges to the Rational Design of Antibacterial Agents

Kirk E. Hevener*, Shuyi Cao†, Tian Zhu†, Pin-Chih Su†, Shahila Mehboob†, Michael E. Johnson†

*Department of Biomedical and Pharmaceutical Sciences, Idaho State University, Meridian, Idaho, USA
†Center for Pharmaceutical Biotechnology, University of Illinois at Chicago, Chicago, Illinois, USA

Contents

1. INTRODUCTION

The rational design of pharmaceutical agents with activity against bacterial targets presents several unique challenges due to the significant differences in the target bacterial cells when compared to the eukaryotic cells of their mammalian hosts. The structural features and cellular components commonly targeted in drug design programs are often unique to bacteria. While this provides an excellent opportunity in terms of selectivity and decreased toxicities, there are also special factors that must be considered,

Annual Reports in Medicinal Chemistry, Volume 48
ISSN 0065-7743
http://dx.doi.org/10.1016/B978-0-12-417150-3.00018-1

including distribution to the target, bacterial cell penetration, efflux, metabolism and elimination, and the rapid emergence of bacterial resistance. These factors can play a key role in the design of compounds intended for use against bacterial targets and the application of traditional and nontraditional screening strategies aimed at identifying such compounds. This report will discuss these special issues pertaining to antibacterial drug discovery, present practical approaches to overcoming these challenges, and highlight some recent examples of their application.

2. SPECIFIC ISSUES IN ANTIBACTERIAL DRUG DESIGN

2.1. Bacterial cell penetration

Cell penetration is perhaps the most challenging issue in modern antibacterial drug discovery.[1–3] One of the most significant differences between bacterial cells and the cells of their human hosts is the presence of a cell wall. Aside from their functional purpose of maintaining cell stability and structure, the cell wall presents a significant barrier to the penetration of antibacterial compounds into the cell. The compounds that target Gram-positive bacteria are often very different in terms of structure and physical properties from those that target Gram-negative bacteria.[4] Gram-positive cell walls do not contain the porin channels that are found in Gram-negative cells, necessitating the passive diffusion of compounds targeting these bacteria across the cell wall. Thus, Gram-positive agents are usually more lipophilic than antibacterial compounds that target Gram-negative bacteria. Agents targeting Gram-negative bacteria typically enter the cell by crossing through the porins, followed by diffusion across the cytoplasmic membrane. To facilitate this passage, Gram-negative active compounds are more hydrophilic and carry a practical molecular weight limitation of 600 Da.[4] Often these compounds possess both charged and uncharged species at physiological pH, which facilitates the subsequent diffusion across the cytoplasmic membrane.[1] Additionally, compounds that can carry a positive charge are favored as the negatively charged bacterial membrane can repel a negative charge.

2.2. Special pharmacokinetic issues

Very often antibacterial agents possess molecular weights and lipophilicity which fall outside of the normally accepted range for "acceptable" oral drug candidates.[5] There are two reasons for this: the first is that many of these classes of drugs have been derived from natural products, which tend to yield compounds with higher molecular weight. The second reason has to do with the unique cell penetration requirements of antibacterial agents. The trend toward higher molecular weight and decreased lipophilicity has resulted in special pharmacokinetic issues that must be considered. First, oral absorption of the classes with very high MW and low clogP is significantly decreased, resulting in many agents that can only be given by the intravenous route. The route and mechanism of elimination for these compounds is also affected by their molecular weight and lipophilicity. Compounds with high solubility are primarily eliminated by the kidneys without first being metabolized, while compounds with low solubility are primarily metabolized prior to elimination. The distribution of these agents to the target tissue is also affected by their unique physicochemical properties. The combination of poor oral absorption and low distribution for several antibacterial drug classes makes necessary the use of large doses, often on the gram scale, in order to achieve the required therapeutic concentrations for efficacy.

2.3. Resistance development

The speed at which bacterial organisms can acquire resistance to promising new antibacterial agents is frightening and presents a significant challenge to antibacterial drug discovery.[6–8] Reasons for this rapid emergence of bacterial resistance include selective pressure, rapid replication rates, and antibiotic misuse or overuse. Bacteria have developed a variety of mechanisms to survive exposure to antibacterial agents. Some common mechanisms of bacterial resistance include deactivation of the antibacterial agent by enzymatic modification, decreased permeability of the bacterial cell by altering the cell wall or decreasing porin expression, export of the antibacterial agents by efflux pumps, alteration of the target binding site, protection of the target by producing biomolecules that interfere with binding, and utilization of alternate pathways that bypass the inhibited process or pathway. A common strategy to overcoming bacterial resistance is the use of multiple agents in combination. The additional agent can act at a subsequent step in a common pathway (sequential blocking), by inhibition of an inactivating enzyme, or in a synergistic manner against a distinct target.

3. PRACTICAL APPROACHES

3.1. Strategies to enhance cell penetration

There are a number of strategies that can be used to enhance bacterial cell penetration.[9] The "Trojan horse" strategy employs the use of chemical coupling of an antibacterial compound to another compound that is known to be actively transported into the bacterial cell.[10] Another strategy, known as "self-promoted uptake," involves the disruption of the outer membrane of Gram-negative bacteria by cationic compounds to increase outer membrane permeability, typically via interaction with sites where divalent cations cross-link anionic lipopolysaccharide chains.[11,12] A third strategy for enhancing cell penetration is known as "ion trapping," which takes advantage of the pH gradient between the periplasmic and cytoplasmic space, where the former is slightly acidic (pH ~6.6).[13,14]

A recent review has highlighted the implications of bacterial efflux in drug discovery.[2] There are two principle approaches to preventing bacterial efflux. The first involves the modification or optimization of chemical scaffolds to reduce compound recognition by the efflux pumps; a strategy that is still in its infancy. The second is the use of agents that are directly able to bind and inhibit the actions of bacterial efflux pumps, with the intent of coadministration with antibacterial agents that would otherwise be susceptible. Addressing bacterial efflux, either by inhibitor design or scaffold modification, can be particularly difficult due to the large variety of efflux transporters in the first case and the potential for attenuation of target affinity in the latter.

3.2. Screening library design for antimicrobial targets

As discussed above, antibacterial compounds possess physicochemical properties that are distinct from agents used against eukaryotic targets. These differences have been proposed to be responsible for past failures of large high-throughput screening (HTS) campaigns against bacterial targets using traditional, drug-like compound libraries.[15] To address these difficulties, approaches in antibacterial screening are beginning to involve the design of ID-focused screening libraries, expanded diversity libraries, and specialized libraries including fragment and natural products.[16] Adjusted filters in library design can be used, including a higher molecular weight cutoff, such as 650 Da, and lower clogP cutoff, such as 3.5, when building antibacterial screening libraries (both experimental and virtual). These adjusted filters can be coupled with expanded diversity methods, such as diversity-oriented synthesis (DOS).[17] Unlike traditional combinatorial methods, DOS libraries are smaller but consist of more complex, larger structures that allow for exploration of a greater chemical space with a relatively small library.

A complementary strategy is to develop focused libraries specifically tailored to the discovery of compounds that can kill bacterial cells through effective penetration and inhibition of bacteria-specific targets. Typically these are designed by the application of computational similarity and clustering methods using the structures and scaffolds of agents with known activity. The limitation to diversity using this approach can be decreased by the application of various diversity metrics, within the focusing constraints. Natural products have been deemphasized as HTS resources in the last decade, in part because of difficulties in obtaining high quality natural products screening libraries and difficulties in further synthetic optimization.[18] An alternative strategy is to develop synthesis-friendly and natural-product-like scaffold libraries. Lastly, fragment-based screening has emerged as an important paradigm in early drug discovery as well as to guide late-stage optimization. This approach is believed to more efficiently sample chemical space using fewer compounds and have a more rapid hit-to-lead optimization guided by structural biology.[19]

3.3. ADMET predictions and physicochemical profiling

The molecular structure of compounds influences not only their cell penetration and susceptibility to efflux, but also their biological behavior *in vivo*, including absorption, distribution, metabolism, elimination, and toxicity (ADMET). It is well known that poor ADMET properties are responsible for a large portion of the clinical failures of drug candidates. For this reason, early ADMET studies are gaining popularity, including both *in vitro* methods and *in silico* methods for ADMET prediction and physicochemical property profiling in lead optimization.[20–22] The advantage of the use of *in vitro* methods to predict *in vivo* ADMET properties is that they are well established with a high degree of reliability in most cases. Disadvantages include the requirement for the physical compounds to be on hand and the expense of testing. *In silico* methods offer an advantage of relatively low cost and the ability to make predictions prior to compound synthesis or acquisition, which has significant appeal. However, these methods are significantly less reliable than *in vitro* methods and can require the use of experimental data to train the predictive models.

In silico models that are able to make ADMET property predictions exist in a variety of commercially and publically available software packages.[23–25] These predictive models often employ the use of quantitative structure–property relationships (QSPR) and typically rely on calculated molecular descriptors.[26–28] The methods report generally favorable results; however, it has been argued that QSPR models which only rely on simple physiochemical properties are limited due to inaccuracies in the prediction of the physiochemical properties, limited chemical diversity, and low quality experimental data in the training set.[22,29] New strategies that offer hope to address these concerns include substructure pattern recognition methods, which achieved success in predicting human intestinal absorption,[30] and structure-based ADMET profiling,[29] which utilizes molecular docking against the steadily increasing number of ADMET-related crystal structures. Structure-based studies have focused on a variety of ADMET targets relevant to antibacterial drug discovery, such as CYP enzymes, nuclear receptors regulating metabolizing enzymes, blood plasma proteins, toxicity-associated ion channels, and efflux pumps.[29]

3.4. Strategies to bypass or overcome bacterial resistance

There are a number of rational design strategies to overcome or prevent resistance that can be applied in antibacterial drug discovery, including the backbone-binding concept,[31] the substrate-envelope hypothesis,[32] adaptive flexibility,[33] and the multitargeted design approach.[34] These methods primarily address resistance caused by modification of the antibacterial target by genetic mutation. The first three strategies have yet to be reported in use against bacterial targets, though the concepts are valid for use in any proliferative disease, including bacterial infection, where resistance can emerge to therapeutic intervention. The multitargeted design approach has been successfully employed in the design of antibacterial compounds that are less sensitive to bacterial resistance. This approach, often structure-based, involves the design of compounds that have activity against multiple enzymes in an essential metabolic pathway. This strategy seeks to discover single compounds with affinity for two or more enzyme steps in bacterial biosynthetic pathways by identifying key active site residues conserved (or class-specific) across targets and then targeting them using structure-based design approaches. In addition to lowering susceptibility to resistance, multitarget ligands also have the potential to avoid PK/PD issues, such as drug–drug interactions, as opposed to combination therapies involving multiple drugs. Most of the biological targets that have been targeted using this approach share similar features in the active site, such as key residues, cofactors, or similar pharmacophore models.

4. REVIEW OF RECENT APPLICATIONS

4.1. Enhanced cell penetration or decreased efflux

Using the Trojan horse strategy, researchers have synthesized a variety of siderophore-linked antibacterial conjugates. A recent report discussed the investigation of aminopenicillin conjugates.[10] These conjugates (**1**) were able to achieve excellent antimicrobial activity, with MICs ranging from 0.05 to 0.39 µM against *Pseudomonas aeruginosa*, whereas the unconjugated aminopenicillins showed no activity (MIC > 100 µM). Additional studies reporting the use of alternative siderophores with a variety of antibacterial classes have been reported.[35,36] Although these agents were not able to improve upon the activity of the parent antibacterial, the retention of activity demonstrated that the siderophore conjugation was tolerated at the site of antibacterial action. The authors hypothesized that conjugation of these siderophore moieties to the antibacterial compound masked a recognition site on the siderophore molecule required for active cellular uptake. Studies investigating different modes of linking the compounds are ongoing.

1

Most recent reports utilizing the self-promoted uptake strategy for enhancing penetration involve the use of cationic antimicrobial peptides and polycationic lipopeptide polymyxins.[37] Recently a group reported modifications to the aminoglycoside neamine that resulted in enhanced membrane permeabilizing effect.[38,39] These authors synthesized a 3′,4′,6-tri-2-naphthylmethylene derivative (**2**) of the aminoglycoside neamine in an attempt to improve the ribosomal binding affinity of neamine. They demonstrated activity against resistant *P. aeruginosa*, with the MIC improving from >128 to 8 µg/mL. Follow-up studies showed the antibacterial effect to be primarily due to increased membrane permeabilization and membrane disruption.

2

Another report used a combination of approaches, including the ion-trapping strategy, to increase the cell penetration of a boronic acid based β-lactamase inhibitor.[40] These authors applied physical models of outer membrane permeation in combination with structure-based design to optimize a hit benzo[*b*]thiophene-2-ylboronic acid (3) with an MIC of 359 μM to a triazole-substituted derivative (4) with an MIC of 117 μM. Interestingly, the K_i for β-lactamase inhibition was higher in the optimized compound (0.027 vs. 0.170 μM), which suggests a significant improvement in cell penetration. The cationic property of (4) is likely to have contributed to self-promoted uptake, while the measured pK_a of the boronic acid group of (4) is in the range favorable for ion trapping.

3   4

A classic example of the scaffold modification approach to decreasing efflux susceptibility is tigecycline (5).[41] In this case, the addition of the *t*-butylglyclamido group to the minocycline scaffold allows the compound to evade efflux by the narrow spectrum Tet(A–E) and Tet(K) pumps. The fluoroquinolone, DS-8587 (6), is a recent example of a scaffold modification that reduces efflux pump recognition.[42] This compound has been tested against wild-type and resistant strains of *Acinetobacter baumannii*. In addition to improved antibacterial activity over ciprofloxacin and norfloxacin against the wild type, 6 showed decreased susceptibility to strains with mutations known to confer quinolone resistance and was less affected by the efflux pump systems adeABC and adeM.

5

6

There are numbers of chemotherapeutic examples to minimizing efflux resistance based upon the strategy of efflux pump inhibition.[43] The Gram-positive associated efflux pump inhibitors, pyrazolo[4,3-c][1,2]ben-zothiazine 5,5-dioxide analogs (**7**) and capsaicin (**8**), have recently been reported as *Staphylococcus aureus* NorA efflux pump inhibitors.[44,45] Among recently reported Gram-negative efflux pump inhibitors, the indole deriv-ative (**9**) has been reported as a novel AcrAB–TolC efflux pump inhibitor.[46] The alkylaminoquinazoline analog (**10**) may block both AcrAB–TolC and MexAB–OprM efflux pumps since this derivative can rescue the activity of substrates of both efflux pumps.[35]

7

8

9

10

4.2. HTS and screening library design

A very recent and interesting example of fragment library design combined the features of three techniques: DOS, natural-product-like, and fragment-like strategies.[36,47] Libraries composed of fragments of stereochemically rich scaffolds derived from natural products were used in this study. Over 180,000 natural product structures were decomposed into 2000 clusters of natural-product-derived, sp3-configured, diverse fragments. The authors were able to produce a library that covered areas of chemical space not typically occupied by synthetic libraries; however, they were able to identify commercial sources for representatives from close to half of their clusters. Applying a similar strategy, researchers have developed an "inverse medchem" approach that utilizes synthesis-friendly, natural-product-like scaffolds and modifies the scaffolds with hydrophilic and amphiphilic systems.[48] This method allows for clogP and solubility control by using hydrophilic scaffolds attaching various lipophilic or hydrophilic substituents. The compound library developed using this approach recently generated a 2% hit rate when screened against a panel of bacterial pathogens.[48]

A strategy for designing DOS libraries of structurally unique and diverse macrocyclic, peptidomimetic compounds has recently been proposed.[49] In a proof-of-concept study, the researchers synthesized a library of 14 compounds. Computational analyses demonstrated that the library covered both explored and unexplored chemical spaces. Preliminary biological assays showed that several compounds possessed significant antibacterial activity against *S. aureus*. A "sparse matrix strategy" has also been implemented for the synthesis of DOS libraries.[50] This method allows for the selection of diverse, synthetically efficient members within a user-defined range of physicochemical properties. This approach was demonstrated by the synthesis of a stereochemically diverse 8000-member library. In this strategy, adjustable filters on compound properties can be modified according to the target's biology, such as antibacterial targets. Libraries developed using the DOS method comprising 100,000 customized compounds are currently being investigated to identify new chemical compounds targeting bacterial and viral infections by the library developers in cooperation with industry.[51]

Multiple bacterial targets have been explored using fragment-based screening libraries including bacterial RNAP, FtsZ, and AmpC β-lactamase.[9,52,53] One representative example is the picomolar AmpC β-lactamase inhibitors that were generated by fragment linking.[53] Compound **12** was previously identified as a micromolar inhibitor of the target. Linking the tetrazole of fragment **11–12** led to a low nanomolar inhibitor

(13), and modifying the benzene to a pyridine (14) improved the potency to 800 pM. When 14 was tested in mice infected with cefotaxime-resistant, β-lactamase-expressing *Escherichia coli*, 65% were cleared of infection when treated with a combination of cefotaxime and 14.

11 K_i = 3000 μM
LE = 0.25

12 K_i = 0.21 μM
LE = 0.61

13 K_i = 0.0012 μM
LE = 0.64

14 K_i = 0.0008 μM
LE = 0.66
Active *in vivo*

4.3. ADMET predictions

There are a number of excellent examples in the literature of the use of QSPR/QSAR models in the prediction of ADMET properties and PK optimization.[54,55] One recent study pertaining to antibacterial use involved the development of an *in silico* permeability model to predict *in vivo* corneal permeability of the fluoroquinolone antibacterials.[56] A report related to the prediction of toxicity involved the simultaneous prediction of the affinity of antimicrobial peptides for bacterial cells over erythrocytes.[57] In this study, the authors utilized machine-learning techniques to build classification rules, which showed 80% accuracy on training and external validation sets. In another example, regression modeling using physicochemical parameters is used in the MycPermCheck tool for predicting the *Mycobacterium tuberculosis* permeability of small molecules.[28] In this work, statistical analyses were performed against a training set of known antimycobacterial compounds and a generic dataset of drug-like compounds. A regression model was generated based upon the identified physicochemical differences between the two sets and tested on multiple evaluation datasets. The resulting enrichment analyses showed the model to be highly predictive. It is now available as an online tool.

4.4. Bacterial resistance

There have been several recent reports related to the use of the multitargeted approach to overcoming bacterial resistance. The bacterial peptidoglycan biosynthetic pathway has recently been discussed as an ideal pathway for the multitargeted approach.[34] An interesting report related to overcoming vancomycin resistance discusses a semisynthetic modification of the vancomycin scaffold.[58] Replacement of a key amide with an amidine group improved binding to the D–Ala–D–Lac ligand produced by resistant bacteria by 600-fold, while only decreasing the binding affinity for the typical D–Ala–D–Ala ligand by twofold. Although this details a small molecule target rather than a protein target, the rational synthetic design of this vancomycin analog is noteworthy. Enzymes involved in bacterial DNA replication are also excellent candidates for the multitargeted approach, for example the bacterial type IIA topoisomerases.[59–62] Subunits of DNA gyrase (GyrA/GyrB) and topoisomerase IV (ParC/ParE) share significant structural homology and the conserved ATP-binding pockets between the GyrB and ParE make them attractive targets for development. A series of promising pyrrolopyrimidine inhibitors have been recently reported.[61,63] Compound (15) showed low to subnanomolar activity against GyrB and ParE from both *E. coli* and *Francisella tularensis*, and promising antibacterial activity. Dual-targeted inhibitors of the bacterial DNA polymerase enzymes, pol IIIC and IIIE, have also been reported with potential for broad-spectrum activity and low susceptibility to resistance.[64,65] Compound (16), a substituted (3,4-dichlorobenzyl) guanine, showed subnanomolar inhibition of pol IIIC and IIIE from the Gram-positive *Bacillus subtilis*, low micromolar activity against *E. coli* pol IIIE, and promising antibacterial activity against *B. subtilis*, *S. aureus*, and MRSA.

15

16

A final study that deserves mention dealt with guanidine-substituted aminoglycosides that yielded activity against resistant Gram-positive and Gram-negative organisms, including *P. aeruginosa* (**17**).[66] In this study, the authors attempted to improve affinity for the ribosomal decoding rRNA site by minimally modifying a series of aminoglycosides by substituting a guanidine group at amine or hydroxy functionalities. Several of the compounds did show improved affinity for the target, and interestingly, several analogs showed improved antibacterial activity against resistant strains. The authors hypothesized that the resulting analogs had decreased susceptibility to the aminoglycoside-modifying enzymes.

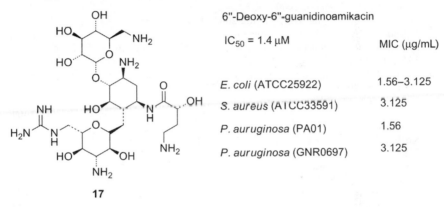

6"-Deoxy-6"-guanidinoamikacin

$IC_{50} = 1.4\ \mu M$

	MIC (µg/mL)
E. coli (ATCC25922)	1.56–3.125
S. aureus (ATCC33591)	3.125
P. auruginosa (PA01)	1.56
P. auruginosa (GNR0697)	3.125

17

5. CONCLUSIONS

In this report we have discussed the unique challenges to the rational design of antibacterial compounds, including bacterial cell penetration, special pharmacokinetic concerns, and the development of bacterial resistance. A variety of practical approaches, related to cheminformatic and structure-guided strategies, have been presented as methods of addressing these challenges and several recent applications of these methods have been presented. With the advent of infectious disease related structural biology initiatives, expanding knowledge related to both the mechanism of bacterial resistance and the structures of various resistance conferring proteins (including compound-modifying enzymes and efflux pumps), it is almost certain that there will be increased utilization of these and other strategies in the rational design of antibacterials addressing these issues.

REFERENCES

1. Silver, L. L. *Clin. Microbiol. Rev.* **2011**, *24*, 71–109.
2. Schweizer, H. P. *Expert Opin. Drug Discovery* **2012**, *7*, 633–642.
3. Gwynn, M. N.; Portnoy, A.; Rittenhouse, S. F.; Payne, D. J. *Ann. N. Y. Acad. Sci.* **2010**, *1213*, 5–19.
4. O'Shea, R.; Moser, H. E. *J. Med. Chem.* **2008**, *51*, 2871–2878.
5. Lipinski, C. A. *J. Pharmacol. Toxicol. Methods* **2000**, *44*, 235–249.
6. Livermore, D. M. *J. Antimicrob. Chemother.* **2011**, *66*, 1941–1944.
7. Wise, R. *J. Antimicrob. Chemother.* **2011**, *66*, 1939–1940.
8. Rennie, R. P. *Handb. Exp. Pharmacol.* **2012**, *211*, 45–65.
9. Chopra, I. *J. Antimicrob. Chemother.* **2012**, *68*, 496–505.
10. Ji, C.; Miller, P. A.; Miller, M. J. *J. Am. Chem. Soc.* **2012**, *134*, 9898–9901.
11. Minnock, A.; Vernon, D. I.; Schofield, J.; Griffiths, J.; Parish, J. H.; Brown, S. B. *Antimicrob. Agents Chemother.* **2000**, *44*, 522–527.
12. Delcour, A. H. *Biochim. Biophys. Acta* **2009**, *1794*, 808–816.
13. Nikaido, H.; Thanassi, D. G. *Antimicrob. Agents Chemother.* **1993**, *37*, 1393–1399.
14. Manchester, J. I.; Buurman, E. T.; Bisacchi, G. S.; McLaughlin, R. E. *J. Med. Chem.* **2012**, *55*, 2532–2537.
15. Payne, D. J.; Gwynn, M. N.; Holmes, D. J.; Pompliano, D. L. *Nat. Rev. Drug Discovery* **2007**, *6*, 29–40.
16. Jones, D. *Nat. Rev. Drug Discovery* **2010**, *9*, 751–752.
17. Galloway, W. R.; Bender, A.; Welch, M.; Spring, D. R. *Chem. Commun. (Camb.)* **2009**, 2446–2462.
18. Koehn, F. E. *Prog. Drug Res.* **2008**, *65, 175*, 177–210.
19. Baker, M. *Nat. Rev. Drug Discovery* **2012**, *12*, 5–7.
20. Wenlock, M. C.; Barton, P. *Mol. Pharm.* **2013**, *10*, 1224–1235.
21. Gertrudes, J. C.; Maltarollo, V. G.; Silva, R. A.; Oliveira, P. R.; Honorio, K. M.; da Silva, A. B. *Curr. Med. Chem.* **2012**, *19*, 4289–4297.
22. Gleeson, M. P.; Hersey, A.; Hannongbua, S. *Curr. Top. Med. Chem.* **2011**, *11*, 358–381.
23. van de Waterbeemd, H.; Gifford, E. *Nat. Rev. Drug Discovery* **2003**, *2*, 192–204.
24. Lagorce, D.; Sperandio, O.; Galons, H.; Miteva, M. A.; Villoutreix, B. O. *BMC Bioinformatics* **2008**, *9*, 396.
25. Lagorce, D.; Maupetit, J.; Baell, J.; Sperandio, O.; Tuffery, P.; Miteva, M. A.; Galons, H.; Villoutreix, B. O. *Bioinformatics* **2011**, *27*, 2018–2020.
26. Gozalbes, R.; Jacewicz, M.; Annand, R.; Tsaioun, K.; Pineda-Lucena, A. *Biorg. Med. Chem.* **2011**, *19*, 2615–2624.
27. Khakar, P. S. *Curr. Top. Med. Chem.* **2010**, *10*, 116–126.
28. Merget, B.; Zilian, D.; Muller, T.; Sotriffer, C. A. *Bioinformatics* **2013**, *29*, 62–68.
29. Moroy, G.; Martiny, V. Y.; Vayer, P.; Villoutreix, B. O.; Miteva, M. A. *Drug Discov. Today* **2012**, *17*, 44–55.
30. Shen, J.; Cheng, F.; Xu, Y.; Li, W.; Tang, Y. *J. Chem. Inf. Model.* **2010**, *50*, 1034–1041.
31. Ghosh, A. K.; Chapsal, B. D.; Weber, I. T.; Mitsuya, H. *Acc. Chem. Res.* **2008**, *41*, 78–86.
32. Kairys, V.; Gilson, M. K.; Lather, V.; Schiffer, C. A.; Fernandes, M. X. *Chem. Biol. Drug Des.* **2009**, *74*, 234–245.
33. Ohtaka, H.; Freire, E. *Prog. Biophys. Mol. Biol.* **2005**, *88*, 193–208.
34. Anusuya, S.; Natarajan, J. *Infect. Genet. Evol.* **2012**, *12*, 1899–1910.
35. Mahamoud, A.; Chevalier, J.; Baitiche, M.; Adam, E.; Pages, J. M. *Microbiology* **2011**, *157*, 566–571.

36. Over, B.; Wetzel, S.; Grutter, C.; Nakai, Y.; Renner, S.; Rauh, D.; Waldmann, H. *Nat. Chem.* **2013**, *5*, 21–28.

37. Ma, Q. Q.; Lv, Y. F.; Gu, Y.; Dong, N.; Li, D. S.; Shan, A. S. *Amino Acids* **2013**, *44*, 1215–1224.

38. Baussanne, I.; Bussiere, A.; Halder, S.; Ganem-Elbaz, C.; Ouberai, M.; Riou, M.; Paris, J. M.; Ennifar, E.; Mingeot-Leclercq, M. P.; Decout, J. L. *J. Med. Chem.* **2010**, *53*, 119–127.

39. Ouberai, M.; El Garch, F.; Bussiere, A.; Riou, M.; Alsteens, D.; Lins, L.; Baussanne, I.; Dufrene, Y. F.; Brasseur, R.; Decout, J. L.; Mingeot-Leclercq, M. P. *Biochim. Biophys. Acta* **2011**, *1808*, 1716–1727.

40. Venturelli, A.; Tondi, D.; Cancian, L.; Morandi, F.; Cannazza, G.; Segatore, B.; Prati, F.; Amicosante, G.; Shoichet, B. K.; Costi, M. P. *J. Med. Chem.* **2007**, *50*, 5644–5654.

41. Petersen, P. J.; Jacobus, N. V.; Weiss, W. J.; Sum, P. E.; Testa, R. T. *Antimicrob. Agents Chemother.* **1999**, *43*, 738–744.

42. Higuchi, S.; Onodera, Y.; Chiba, M.; Hoshino, K.; Gotoh, N. *Antimicrob. Agents Chemother.* **2013**, *57*, 1978–1981.

43. Zechini, B.; Versace, I. *Recent Pat. Antiinfect. Drug Discov.* **2009**, *4*, 37–50.

44. Sabatini, S.; Gosetto, F.; Serritella, S.; Manfroni, G.; Tabarrini, O.; Iraci, N.; Brincat, J. P.; Carosati, E.; Villarini, M.; Kaatz, G. W.; Cecchetti, V. *J. Med. Chem.* **2012**, *55*, 3568–3572.

45. Kalia, N. P.; Mahajan, P.; Mehra, R.; Nargotra, A.; Sharma, J. P.; Koul, S.; Khan, I. A. *J. Antimicrob. Chemother.* **2012**, *67*, 2401–2408.

46. Zeng, B.; Wang, H.; Zou, L.; Zhang, A.; Yang, X.; Guan, Z. *Biosci. Biotechnol. Biochem.* **2010**, *74*, 2237–2241.

47. Shoichet, B. K. *Nat. Chem.* **2013**, *5*, 9–10.

48. McKenna, N. *Genet. Eng. Biotechnol. News* [Online] **2012**, *32*. http://www.gen engnews.com/gen-articles/new-approaches-redefine-small-molecule-discovery/4139/ (Accessed 3/11/2013).

49. Isidro-Llobet, A.; Murillo, T.; Bello, P.; Cilibrizzi, A.; Hodgkinson, J. T.; Galloway, W. R.; Bender, A.; Welch, M.; Spring, D. R. *Proc. Natl. Acad. Sci. U. S. A.* **2011**, *108*, 6793–6798.

50. Akella, L. B.; Marcaurelle, L. A. *ACS Comb. Sci.* **2011**, *13*, 357–364.

51. http://www.broadinstitute.org/news/4331 (Accessed 3/15/2013).

52. Haydon, D. J.; Stokes, N. R.; Ure, R.; Galbraith, G.; Bennett, J. M.; Brown, D. R.; Baker, P. J.; Barynin, V. V.; Rice, D. W.; Sedelnikova, S. E.; Heal, J. R.; Sheridan, J. M.; Aiwale, S. T.; Chauhan, P. K.; Srivastava, A.; Taneja, A.; Collins, I.; Errington, J.; Czaplewski, L. G. *Science* **2008**, *321*, 1673–1675.

53. Eidam, O.; Romagnoli, C.; Dalmasso, G.; Barelier, S.; Caselli, E.; Bonnet, R.; Shoichet, B. K.; Prati, F. *Proc. Natl. Acad. Sci. U. S. A.* **2012**, *109*, 17448–17453.

54. Choi, E.; Lee, C.; Cho, M.; Seo, J. J.; Yang, J. S.; Oh, S. J.; Lee, K.; Park, S. K.; Kim, H. M.; Kwon, H. J.; Han, G. *J. Med. Chem.* **2012**, *55*, 10766–10770.

55. Khan, M. T. *Curr. Drug Metab.* **2010**, *11*, 285–295.

56. Sharma, C.; Velpandian, T.; Biswas, N. R.; Nayak, N.; Vajpayee, R. B.; Ghose, S. *J. Biomed. Biotechnol.* **2011**, *2011*, 483869.

57. Cruz-Monteagudo, M.; Borges, F.; Cordeiro, M. N. *J. Chem. Inf. Model.* **2011**, *51*, 3060–3077.

58. James, R. C.; Pierce, J. G.; Okano, A.; Xie, J.; Boger, D. L. *ACS Chem. Biol.* **2012**, *7*, 797–804.

59. Manchester, J. I.; Dussault, D. D.; Rose, J. A.; Boriack-Sjodin, P. A.; Uria-Nickelsen, M.; Ioannidis, G.; Bist, S.; Fleming, P.; Hull, K. G. *Bioorg. Med. Chem. Lett.* **2012**, *22*, 5150–5156.

60. Nieto, M. J.; Pierini, A. B.; Singh, N.; McCurdy, C. R.; Manzo, R. H.; Mazzieri, M. R. *Med. Chem.* **2012**, *8*, 349–360.
61. Tari, L. W.; Trzoss, M.; Bensen, D. C.; Li, X.; Chen, Z.; Lam, T.; Zhang, J.; Creighton, C. J.; Cunningham, M. L.; Kwan, B.; Stidham, M.; Shaw, K. J.; Lightstone, F. C.; Wong, S. E.; Nguyen, T. B.; Nix, J.; Finn, J. *Bioorg. Med. Chem. Lett.* **2013**, *23*, 1529–1536.
62. Dale, A. G.; Hinds, J.; Mann, J.; Taylor, P. W.; Neidle, S. *Biochemistry* **2012**, *51*, 5860–5871.
63. Trzoss, M.; Bensen, D. C.; Li, X.; Chen, Z.; Lam, T.; Zhang, J.; Creighton, C. J.; Cunningham, M. L.; Kwan, B.; Stidham, M.; Nelson, K.; Brown-Driver, V.; Castellano, A.; Shaw, K. J.; Lightstone, F. C.; Wong, S. E.; Nguyen, T. B.; Finn, J.; Tari, L. W. *Bioorg. Med. Chem. Lett.* **2013**, *23*, 1537–1543.
64. Wright, G. E.; Brown, N. C.; Xu, W. C.; Long, Z. Y.; Zhi, C.; Gambino, J. J.; Barnes, M. H.; Butler, M. M. *Bioorg. Med. Chem. Lett.* **2005**, *15*, 729–732.
65. Xu, W. C.; Wright, G. E.; Brown, N. C.; Long, Z. Y.; Zhi, C. X.; Dvoskin, S.; Gambino, J. J.; Barnes, M. H.; Butler, M. M. *Bioorg. Med. Chem. Lett.* **2011**, *21*, 4197–4202.
66. Fair, R. J.; Hensler, M. E.; Thienphrapa, W.; Dam, Q. N.; Nizet, V.; Tor, Y. *ChemMedChem* **2012**, *7*, 1237–1244.

SECTION 6

Topics in Biology

Section Editor: John Lowe
JL3Pharma LLC, Stonington, Connecticut

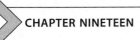

CHAPTER NINETEEN

Recent Advances in Small Molecule Target Identification Methods

Rohan E.J. Beckwith, Rishi K. Jain

Global Discovery Chemistry, Novartis Institutes for Biomedical Research, Cambridge, Massachusetts, USA

Contents

ABBREVIATIONS

ALG7 UDP-*N*-acetylglucosamine–dolichyl-phosphate-*N*-acetylglucosamine phosphotransferase
CRBN cereblon
DDB1 damage-specific DNA-binding protein 1
ECFP extended connectivity fingerprint

Annual Reports in Medicinal Chemistry, Volume 48
ISSN 0065-7743
http://dx.doi.org/10.1016/B978-0-12-417150-3.00019-3

HIP haploinsufficiency profiling
HOP Homozygous profiling
HTS–FP high throughput screening fingerprint
MFSD2a Major f superfamily domain-containing protein 2
NF1 neurofibromatosis type 1
PSMB5 proteasome subunit beta type 5
PTPN1 tyrosine-protein phosphatase nonreceptor type 1
SEA similarity ensemble approach
Y3H yeast three hybrid

1. INTRODUCTION

Phenotypic screening has become increasingly popular for identifying bioactive molecules capable of modulating disease-relevant processes.[1] It holds particular promise against indications where attractive drug targets have been difficult to identify through other means.[2] Even though precise knowledge of the biological target is not required for advancing such molecules into clinical trials,[3] shedding light into this mechanistic black box holds tremendous value when conducting phenotypic drug discovery. For example, if a phenotypically derived lead reaches a dead end during lead optimization with unexpected and insurmountable liabilities, knowledge of the biochemical target provides confidence to bring forward alternative scaffolds with different off-target or liability profiles. When inconsistent phenotypes are observed between cell-based and animal models, knowledge of the target may accelerate the determination of the source of this discrepancy. The target will also aid in the discovery of clinically useful pharmacodynamic markers proximal to the site of drug action. Furthermore, phenotypic screening followed by target identification and validation remains a powerful approach in assigning function to poorly understood proteins.[4]

Given the rapid evolution of target identification methods[5], we herein highlight emerging findings in an intuitive manner for the medicinal chemist. In this chapter we discuss four target identification approaches with demonstrative examples from the last few years. In order to define a tractable scope for this chapter, we cover direct methods that provide target hypotheses of bioactive molecules rather than information surrounding their mechanisms of action. We exclude cellular differential profiling techniques such as proteomic, transcriptomic, or cytological methods for target identification; these are powerful techniques and can often provide strong clues or indirectly enable target identification through data integration and correlation analyses.[6,7]

2. AFFINITY-BASED PROTEOMICS

Affinity-based proteomics[8] is one of the most commonly applied techniques for target identification of bioactive probes; it relies on the premise that proteins possessing high affinity to a ligand are likely to affect its biological activity. Prior to engaging this technology, it is important to characterize the probe through cellular and biochemical profiling assays to assess selectivity and to eliminate common targets and undesirable mechanisms of action. It is also worth noting that a biological effect may be the result of several moderate affinity interactions against multiple targets.[9] The site on the probe for derivatization needs careful evaluation through structure-activity relationship (SAR) studies so that linker incorporation does not impede protein–ligand binding. The probe is then immobilized on a solid support and incubated in cell extract to enrich for target proteins that are then identified by mass spectrometry. The advantage of this technique is exposure of the affinity matrix to the proteome under near physiological conditions, from which direct determination of the protein binding profile of the probe can be garnered in an unbiased fashion.[10] This broad survey can potentially reveal polypharmacology and identify off-targets not captured in traditional target-centric approaches.[11,12] Lysate preparation has a big influence on target procurement since cell lysis may disrupt the molecular and conformational integrity of proteins such that certain classes (e.g., integral membrane proteins) are typically not well represented.[10] Effective pull-downs rely not only on the affinity of the probe/target interaction, but also on the relative abundance of that target protein. Through innovative protein labeling experiments the detection and relative quantification of specifically enriched low abundance proteins is now achievable.[10,13] This quantitative approach, coupled with competitive experiments, helps to differentiate nonspecific binding to the affinity matrix from specific interactions, thereby reducing the incidence of false positives. Additional strategies to minimize nonspecific binding encompass the incorporation of reporter tags and cleavable linkers.[5] Of several emerging affinity-based methods, one of the most notable is drug affinity responsive target stability (DARTS) which promises a unique way to identify targets in a label-free manner.[14] Alternatively, photoaffinity cross-linking can enable covalent capture of target proteins,[8] but general applicability has been hampered by poor yield and nonspecific labeling.[15] Ligand–directed protein labeling using proximity–guided transfer of reporter tags offers an alternative means to covalently engage targets.[15,16]

2.1. Thalidomide affinity matrix identifies cereblon as the target for its teratogenicity

The recent target identification of thalidomide, **1**, exemplifies an attractive workflow for affinity-based proteomics experiments (Fig. 19.1).[12] SAR studies identified a suitable region on thalidomide for derivatization while retaining its bioactivity, affording FR259625, **2**, appended with a carboxylic acid linker. Through immobilization onto an amine-terminated solid support *via* amidation, the resultant thalidomide affinity matrix **3** was incubated in HeLa cell lysates. After washing with buffer to remove proteins that bound nonspecifically to the affinity probe, the bound proteins were eluted with excess free thalidomide. These affinity purified proteins were separated by sodium dodecyl sulfate polyacrylamide gel electrophoresis and visualized. In parallel, negative control matrix **4** lacking the probe was also prepared and tested to reveal nonspecifically bound proteins. Simultaneously, a

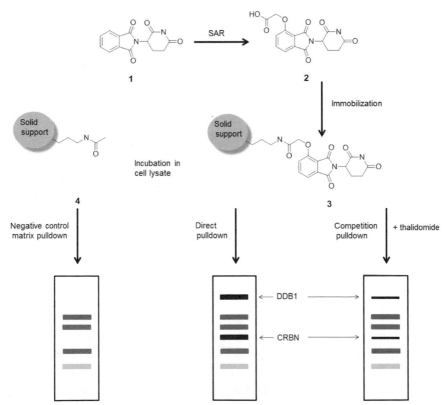

Figure 19.1 Affinity-based proteomics to identify the protein target of thalidomide. (For color version of this figure, the reader is referred to the online version of this chapter.)

competition experiment was undertaken in which excess thalidomide was added to the cell lysate, vying with the affinity matrix for binding to target proteins. Proteins depleted in this competition experiment were likely the most specific and reversible interacting partners with thalidomide and so were prioritized for follow-up. Trypsin digestion and sequencing of the resultant peptides by liquid chromatography–tandem mass spectrometry allowed for protein identification of cereblon (CRBN) and damaged DNA-binding protein 1 (DDB1). Validation experiments revealed that thalidomide initiates its teratogenic effects by binding to CRBN and inhibiting the associated ubiquitin ligase activity. Notably, additional experiments demonstrated that thalidomide indirectly interacts with DDB1 since it participates in a protein–protein complex with CRBN. This illustrates the power of affinity proteomics; both direct and indirect binders and in some cases entire multiprotein complexes can be pulled down, offering valuable insights about the broader molecular mechanism.

2.2. Affinity purification of HDAC megadalton complexes

An immobilized hydroxamate-based probe, "HDAC-bead," was shown to affinity purify megadalton complexes comprised of its target histone deacetylases (HDAC) and associated proteins.[17] Application of this HDAC-bead in competitive binding experiments with an assortment of HDAC inhibitors, coupled to quantitative mass spectrometry, revealed that most inhibitors displayed selectivity for certain HDACs and indeed HDAC-associated protein complexes that had not been realized in assays based solely on purified HDAC catalytic subunits. This emphasizes the need to evaluate drug compounds in a more biological context rather than in reconstituted *in vitro* systems.[11]

3. *IN SILICO* TARGET PREDICTION

This technology offers target hypotheses *en masse*[18] which rely on computer-ascribed similarity metrics such as extended connectivity fingerprints[19] and on increased knowledge of small molecule bioactivity.[20]

3.1. Discovery of adverse drug reaction targets using similarity ensemble approach

A recent study applied novel *in silico* prediction methods on an impressively large scale, leading to the identification of over one hundred newly validated target–ligand pairs starting from 656 marketed drugs.[21] What is most notable is the application of a statistical model referred to as similarity ensemble approach (SEA), where a set of multiple ligands known to bind a target is treated as a reference set rather than relying only on individual pair-wise similarity.[22] SEA also compares the observed overall similarity value to what can be expected by chance. Because of this unique statistical approach a prediction may not necessarily result from high two-dimensional similarity to any particular probe, but rather through the prevalence of similar features to the entire ensemble of ligands hitting the target. Chlorotrianisene, **5**, a nonsteroidal estrogen receptor binder for example, was predicted to hit cyclooxygenase-1 (COX-1) through SEA and validated to be the case with an IC_{50} of 0.16 μM. The ligand which **5** was most similar to in the reference ligand set for COX-1 was indomethacin, **6**. These molecules share little chemical similarity and are likely to be ignored through similarity-based searches as the Tanimoto coefficient (ECFP_4) is only 0.31. Furthermore, the corresponding biological targets, estrogen receptor and COX-1, are entirely unrelated at the primary sequence level, making this a unique, nonobvious prediction both from a chemical and biological perspective. Significantly, COX-1 inhibition by **5** helps explain the clinically observed adverse drug reaction (ADR) of upper abdominal pain for this drug. Several other ADR targets for commonly used drugs were identified illustrating the power of SEA-based *in silico* target identification in tying complex phenotypes such as ADR back to discrete protein targets.

5   6

3.2. Discovery of new targets for PJ34, a heavily utilized PARP probe

Even for probes considered well characterized and heavily utilized in cell biology experiments, opportunities exist to elucidate additional targets. Many small molecules and drugs were not designed to be high quality biological probes devoid of confounding issues such as off-target effects.[23] PJ34, **7**, a widely used chemical tool for probing poly(ADP-ribose) polymerase

(PARP) function was used as a starting point to elucidate additional targets relevant for its reported biological effects.[24] Using a similarity-based prediction strategy,[25] proto-oncogene serine/threonine-protein kinases 1 and 2 (Pim-1 and Pim-2) were predicted to interact with **7** through similarity to **8**, a potent Pim inhibitor. These predictions were experimentally confirmed with IC_{50} values of 3.7 and 16 µM, respectively for **7**.[24] While these potency values are modest, what is striking is that over 60% of 33 publications utilizing this reagent between 2010 and 2011 utilized **7** at concentrations above 5 µM. Although additional experiments need to be conducted to measure target occupancy, it is already revealing that several publications utilizing **7** led to Pim-1- and Pim-2-related phenotypes such as G2/M arrest or other PARP-1 independent effects.[24]

7   8

3.3. HTS fingerprints can identify dissimilar molecules that can modulate the same target

An orthogonal *in silico* target identification approach relies on the concept of biological fingerprinting.[26–28] This method places no reliance on chemical similarity but instead brings together compounds based on similarities between their patterns of biochemical and/or cellular activities across several assays. A recently reported example expands on this approach through utilization of historically accumulated HTS data to assign such fingerprints (HTS-FP).[29] Clustering of roughly 360,000 compounds was accomplished by taking their activities against diverse assay types (phenotypic, biochemical, and biophysics) as inputs. Upon dissection of some of these HTS-FP derived clusters, nonobvious connections emerged. For example, xanthine derivative **9**, established as a selective PDE4c inhibitor, was used as a starting point to recall structurally unrelated PDE4 inhibitors, such as isoquinoline **10** and naphthyridine **11**. Upon closer examination of the actual fingerprint, the relationship between **9** and **10** was primarily driven through shared activities against three kinases, an ion channel, a GPCR, a protease and a pathway assay. It is noteworthy to point out that there was no reliance on PDE4 activity for these two molecules to cluster together. In addition, structural similarity-based methods were unable to tag these molecules, cor-roborating the complementarity of HTS-FP. One limitation of HTS-FP is that newer probes with little to no accumulated data may be unable to reveal relationships to other compounds.

9   **10**   **11**

4. DRUG RESISTANCE AND SEQUENCING-BASED TARGET IDENTIFICATION

Although drug resistance remains a major public health problem for the treatment of infectious agents and cancer, it is a powerful phenomenon for those interested in target identification. In principle, resistant organisms or cells can be obtained by exposure to carefully tuned drug concentrations over a period of time.[30,31] Through genomic sequencing, responsible mutant genes can be identified and validated by reintroduction to the parental model to assess for resistance (Fig. 19.2). Additional validation can come through biochemical characterization of drug affinity against such identified targets. All of this can be accomplished without extensive knowledge of SAR or introduction of chemical modifications, unlike the majority of affinity-based approaches. While conceptually simple and successively applied for organisms with simpler genomes,[30,31] this approach was only

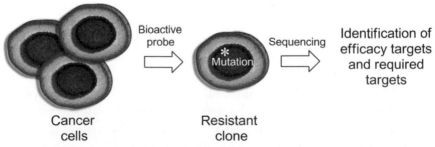

Figure 19.2 Generation of probe resistant cells and elucidation of targets by sequencing. (For color version of this figure, the reader is referred to the online version of this chapter.)

recently reported as an unbiased target identification method for mammalian systems.[32] The redundancy of the mammalian diploid somatic genome makes the selection and identification of resistance conferring mutants often unpredictable and challenging. Fortunately, recent technological strides provide access to genome sequencing with dramatically increased through-put and reduced cost making this an increasingly promising approach.[33]

4.1. Transcriptome sequencing to identify mechanisms of drug action and resistance

A recently reported study demonstrates promising proof of concept of this approach.[32] BI 2536, **12**, a Polo-like-kinase 1 (PLK1) inhibitor currently undergoing clinical trials, was used to select for resistant HCT–116 colon cancer cells. Selection at 10 nM, approximately twofold above the LD_{50} for parental HCT–116 cells, was conducted to obtain six resistant clones for further analysis. Once these clones were analyzed by full transcriptome sequencing,[34] these clones were clustered into four groups based on single nucleotide variations, small insertions or deletions. Upon examination of all mutated genes in these groups, only PLK1 was mutated in more than one group. These mutations mapped back to the binding site of **12** onto PLK1 and when introduced back into an independent cell line, hTERT-RPE1, suppressed toxicity of **12** approximately two to threefold. Resistant clones not carrying PLK1 mutations were characterized and were identified as overexpressing ABCA1, a drug efflux transporter consistent with resistance. To avoid mutations that lead to nonspecific resistance to cytotoxic agents, all clones prior to committing to sequencing can be tested against other cytotoxic agents. Using the same method, resistant colonies against cancer drug bortezomib, **13**, were identified and upon sequencing, two binding site mutants of PSMB5, the known proteosomal subunit target, were discovered. These mutants were introduced into hTERT-RPE1 and demonstrated four to sixfold suppression of toxicity. Finally, two kinesin-5 inhibitors were subjected to the same conditions and not surprisingly the less selective inhibitor led to a lower frequency of mutations in the primary target.

12　　　　　　　　　　　13

4.2. Near haploid genetic screens in human KBM7 leukemia lines

The challenge of obtaining bi-allelic mutants when generating resistance in somatic cells was recently addressed by the application of near-haploid leukemia cells, KBM7.[35] Although this method will not discover the direct target of a bioactive probe, it is a powerful method capable of identifying targets required for activity. This is because the retroviral insertion mutagenesis technology used here results in loss of function mutations; a "gene-trap" cassette inserts into the intron of a gene and attaches a tagged reporter to an upstream exon resulting in a truncated fusion product.[36] If genetic disruption in this manner confers resistance to the probe, genes of surviving colonies can be identified by sequencing of tagged genes. For example, when the BCR-ABL inhibitor imatinib, **14**, was subjected to KBM7 cells known to be sensitive to this drug, gene-trap insertion mutagenesis led to the identification of resistant lines carrying NF1 and PTPN1 insertions. Both genes have been previously shown to be required for CML response to imatinib thus validating this method.[37] In another application of this method, KBM7 cells were treated with tunicamycin, **15**, and subjected to mutagenesis.[38] Even though the target for this ER stress-inducing tri-saccharide derivative is known to be DPAGT1/ALG7, the mode of cell entry is not well understood. This is in spite of **15** being widely used in numerous cell biology experiments since the 1970s. Through this screen, MFSD2a was identified and confirmed as the transporter required for entry and activity of **15**.

14   **15**   **16**

5. YEAST-BASED APPROACHES

Yeast has been extensively used as a model organism to predict the function of orthologous mammalian genes. Owing to their ease of manipulation and complete genome sequencing, several yeast-based strategies have been applied successfully to elucidate the targets of bioactive molecules.[39,40] More recently, the development of a genome-wide deletion mutant collection of the budding yeast *Saccharomyces cerevisiae*, with a

DNA barcode tagging each gene deletion, has enabled the emergence of more systematic approaches for target identification.[41]

5.1. Drug-induced haploinsufficient and homozygous deletion profiling

Diploid yeast carrying a heterozygous deletion has a single copy of the gene and consequently half the concentration of protein relative to the wild type. The drug-induced haploinsufficiency profiling (HIP) assay takes advantage of this concept as the decreased concentration of the target protein can make it hypersensitive to the targeting drug. The deleted genes can be replaced by a unique DNA barcode, enabling the quantitation and deconvolution of a mixture of yeast deletion strains. An early application of this method was the identification of the target of **15** as ALG7 within a mixture of 233 yeast deletion strains.[42] A more recent application of HIP was in the identification of lysyl-RNA synthase as the target of the potent anti-trypanosomal fungal metabolite, cladosporin, **16**.[43]

HIP is an unbiased screen capable of revealing drug targets if they are essential genes for yeast survival. On the other hand, a homozygous deletion profiling (HOP) assay utilizes a collection of strains entirely missing a particular nonessential gene. When analogously conducted to the HIP, HOP can yield information on proteins or pathways that are compensatory to or indirectly affect the action of the drug. HIP and HOP can be integrated along with yeast gene overexpression screens to aid target deconvolution, affording a more complete picture of drug action in cells.[44,45]

5.2. Functional variomics

Functional variomics is built on the theory that genetic variations in drug targets are the major causes of drug resistance. A useful toolbox, referred to as "variomics tool," was created through the incorporation of numerous genetic variants into the *S. cerevisiae* strain, affecting almost all nucleotides of its genome. This exhaustive collection of mutants at the genome-wide level offers a powerful starting point for the identification of compound resistant genes enabling one to elucidate the targets of the drug (*vide supra*, Section 3).[46]

5.3. Yeast three-hybrid system

The transcriptional machinery of *S. cerevisiae* has been exploited as a screening tool to discover small molecule–protein interactions. The yeast three–hybrid system (Y3H) elucidates target–ligand interactions by coupling the assembly of a ternary complex to the expression of a survival gene to enable colony formation followed by sequencing.[47]

The bioactive probe is initially converted to a bifunctional "bait" by conjugation to another small molecule. These typically are ligands with high affinity[47] to, or substrates that can covalently bind[48] to, a genetically encoded enzyme "hook." This hook is fused to a DNA-binding domain allowing association to a specific promoter element upstream of a reporter or survival gene. The bifunctional bait upon entry into the nucleus can associate with the hook, creating a preorganized bait scaffold displayed on the promoter region. To start the "fishing experiment" a cDNA library encoding protein targets fused to a transactivation domain is used to transform the yeast. If the bioactive probe on the bait can bind and recruit its protein target within this library, the assembled "3-hybrid" transcription factor can switch on the reporter or survival gene it regulates, allowing that strain to survive in a medium which it would otherwise not thrive in (Fig. 19.3). Plasmids found in such surviving colonies can be sequenced to identify the targets of interest. Recent application of Y3H led to the identification of the enzyme sepiapterin reductase[48] for the anti-inflammatory drug sulfasalazine, **17**, and PDE6d[49] for glaucoma drug anecortave acetate, **18**. The strength of the Y3H system is that it is an unbiased approach and straightforward to screen multiple protein–ligand interactions in a pool-based library format. The use of cDNA libraries allows normalization of protein concentrations in contrast to methods relying on endogenous levels such as affinity-based

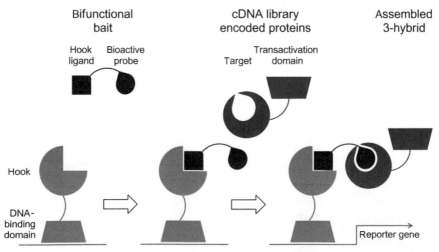

Figure 19.3 Yeast three-hybrid-based target identification.

proteomics, thus giving low abundance proteins improved representation.[50] The reporter gene amplifies the signal ensuring that unstable targets as well as low affinity transient interactions can be detected. A potential limitation of current Y3H is that it requires the interactions to take place within the nucleus of the yeast cell making the identification of integral membrane proteins difficult to investigate. Further, if the target protein requires multimerization to fold properly, it may not be identified.

17   **18**

6. CONCLUSIONS

The reward of target identification is the resolution of the complex pharmacological actions of bioactive probes at the molecular level. Several seminal examples[4,51] demonstrate the impact of small molecule target identification on the discovery of new biological pathways or processes important for therapeutic development.[52] Given the inherent unpredictability of these techniques, it seems logical to use complementary techniques and integrate the results.[53] For example, *in silico* approaches may help gauge novelty and prioritize those that go into full–fledged target identification campaigns. Affinity-based proteomics is the most commonly used approach as it can directly identify targets regardless of the phenotype. It is important, however, to be mindful of potential limitations of such targets of low abundance or those with conformational instability under lysis conditions. Resistance-sequencing methods, though primarily limited to lethal phenotypes with some exceptions,[54] offer orthogonal target identification opportunities as they do not require biochemical purification. Yeast genetics offers a convenient and systematic platform for identifying evolutionary conserved targets. Y3H is well suited for molecules with good uptake and for mammalian targets that express properly within yeast. Finally, given the complexity of polypharmacology, the necessity for stress-testing the biochemical selectivity, stability, and SAR tractability of bioactive probes serves as a reminder of the central role of chemistry in this interdisciplinary field.

REFERENCES

1. Eggert, U. S. *Nat. Chem. Biol.* **2013**, *9*, 206.
2. Swinney, D. C.; Anthony, J. *Nat. Rev. Drug Discov.* **2011**, *10*, 507.
3. Lee, J. A.; Uhlik, M. T.; Moxham, C. M.; Tomandl, D.; Sall, D. J. *J. Med. Chem.* **2012**, *55*, 4527.
4. Schreiber, S. L. *Bioorg. Med. Chem.* **1998**, *6*, 1127.
5. Ziegler, S.; Pries, V.; Hedberg, C.; Waldmann, H. *Angew. Chem. Int. Ed.* **2013**, *52*, 2744.
6. Tashiro, E.; Imoto, M. *Bioorg. Med. Chem.* **2012**, *20*, 1910.
7. Feng, Y.; Mitchison, T. J.; Bender, A.; Young, D. W.; Tallarico, J. A. *Nat. Rev. Drug Discov.* **2009**, *8*, 567.
8. Geoghegan, K. F.; Johnson, D. S. *Annu. Rep. Med. Chem.* **2010**, *45*, 345.
9. Sato, S.-I.; Murata, A.; Shirakawa, T.; Uesugi, M. *Chem. Biol.* **2010**, *17*, 616.
10. Rix, U.; Superti-Furga, G. *Nat. Chem. Biol.* **2009**, *5*, 616.
11. Moellering, R. E.; Cravatt, B. F. *Chem. Biol.* **2012**, *19*, 11.
12. Ito, T.; Ando, H.; Suzuki, T.; Ogura, T.; Hotta, K.; Imamura, Y.; Yamaguchi, Y.; Handa, H. *Science* **2010**, *327*, 1345.
13. Schirle, M.; Bantscheff, M.; Kuster, B. *Chem. Biol.* **2012**, *19*, 72.
14. Lomenick, B.; Olsen, R. W.; Huang, J. *ACS Chem. Biol.* **2011**, *6*, 34.
15. Hughes, C. C.; Yang, Y.-L.; Liu, W.-T.; Dorrestein, P. C.; La Clair, J. J.; Fenical, W. *J. Am. Chem. Soc.* **2009**, *131*, 12094.
16. Tsukiji, S.; Miyagawa, M.; Takaoka, Y.; Tamura, T.; Hamachi, I. *Nat. Chem. Biol.* **2009**, *5*, 341.
17. Bantscheff, M.; Hopf, C.; Savitski, M. M.; Dittmann, A.; Grandi, P.; Michon, A. M.; Schlegl, J.; Abraham, Y.; Becher, I.; Bergamini, G.; Boesche, M.; Delling, M.; Dumpelfeld, B.; Eberhard, D.; Huthmacher, C.; Mathieson, T.; Poeckel, D.; Reader, V.; Strunk, K.; Sweetman, G.; Kruse, U.; Neubauer, G.; Ramsden, N. G.; Drewes, G. *Nat. Biotechnol.* **2011**, *29*, 255.
18. Bender, A.; Young, D. W.; Jenkins, J. L.; Serrano, M.; Mikhailov, D.; Clemons, P. A.; Davies, J. W. *Comb. Chem. High Throughput Screen.* **2007**, *10*, 719.
19. Rogers, D.; Hahn, M. *J. Chem. Inf. Mod.* **2010**, *50*, 742.
20. Olah, M.; Oprea, T. I. Bioactivity Databases. In *Comprehensive Medicinal Chemistry II*; Triggle, D.J., Taylor, J.B., Eds.; Elsevier Ltd: Oxford, 2007; Vol. 3, pp. 219–313.
21. Lounkine, E.; Keiser, M. J.; Whitebread, S.; Mikhailov, D.; Hamon, J.; Jenkins, J. L.; Lavan, P.; Weber, E.; Doak, A. K.; Cote, S.; Shoichet, B. K.; Urban, L. *Nature* **2012**, *486*, 361.
22. Keiser, M. J.; Roth, B. L.; Armbruster, B. N.; Ernsberger, P.; Irwin, J. J.; Shoichet, B. K. *Nat. Biotech.* **2007**, *25*, 197.
23. Workman, P.; Collins, I. *Chem. Biol.* **2010**, *17*, 561.
24. Antolín, A. A.; Jalencas, X.; Yélamos, J.; Mestres, J. *ACS Chem. Biol.* **2012**, *7*, 1962.
25. Vidal, D.; Mestres, J. *Mol. Inf.* **2010**, *29*, 543.
26. Paul, K. D.; Shoemaker, R. H.; Hodes, L.; Monks, A.; Scudiero, D. A.; Rubinstein, L.; Plowman, J.; Boyd, M. R. *J. Natl. Cancer Inst.* **1989**, *14*, 1088.
27. Kauvar, L. M.; Higgins, D. L.; Villar, H. O.; Sportsman, J. R.; Engqvist-Goldstein, A.; Bukar, R.; Bauer, K. E.; Dilley, H.; Rocke, D. M. *Chem. Biol.* **1995**, *2*, 107.
28. Fliri, A. F.; Loging, W. T.; Thadeio, P. F.; Volkmann, R. A. *Proc. Natl. Acad. Sci. U.S.A.* **2005**, *102*, 261.
29. Petrone, P. M.; Simms, G.; Nigsch, F.; Lounkine, E.; Kutchukian, P.; Cornett, A.; Deng, Z.; Davies, J.; Jenkins, J. L.; Glick, M. *ACS Chem. Biol.* **2012**, *7*, 1399.
30. Gao, M.; Nettles, R. E.; Belema, M.; Snyder, L. B.; Nguyen, V. N.; Fridell, R. A.; Serrano-Wu, M. H.; Langley, D. R.; Sun, J. H.; O'Boyle, D. R., II.; Lemm, J. A.;

Wang, C.; Knipe, J. O.; Chien, C.; Colonno, R. J.; Grasela, D. M.; Meanwell, N. A.; Hamann, L. G. *Nature* **2010**, *465*, 96.

31. Rottmann, M.; McNamara, C.; Yeung, B. K. S.; Lee, M. C. S.; Zou, B.; Russell, B.; Seitz, P.; Plouffe, D. M.; Dharia, N. V.; Tan, J.; Cohen, S. B.; Spencer, K. R.; Gonzalez-Paez, G. E.; Lakshminarayana, S. B.; Goh, A.; Suwanarusk, R.; Jegla, T.; Schmitt, E. K.; Beck, H.-P.; Brun, R.; Nosten, F.; Renia, L.; Dartois, V.; Keller, T. H.; Fidock, D. A.; Winzeler, E. A.; Diagana, T. T. *Science* **2010**, *329*, 1175.

32. Wacker, S. A.; Houghtaling, B. R.; Elemento, O.; Kapoor, T. M. *Nat. Chem. Biol.* **2012**, *8*, 235.

33. Metzker, M. L. *Nat. Rev. Genet.* **2010**, *11*, 31.

34. Marguerat, S.; Bahler, J. *Cell. Mol. Life Sci.* **2010**, *67*, 569.

35. Carette, J. E.; Guimaraes, C. P.; Varadarajan, M.; Park, A. S.; Wuethrich, I.; Godarova, A.; Kotecki, M.; Cochran, B. H.; Spooner, E.; Ploegh, H. L.; Brummelkamp, T. R. *Science* **2009**, *326*, 123.

36. Stanford, W. L.; Cohn, J. B.; Cordes, S. P. *Nat. Rev. Genet.* **2001**, *2*, 756.

37. Luoa, B.; Cheung, H. W.; Subramanian, A.; Sharifnia, T.; Okamoto, M.; Yang, X.; Hinkle, G.; Boehm, J. S.; Beroukhim, R.; Weir, B. A.; Mermel, C.; Barbie, D. A.; Awad, T.; Zhou, X.; Nguyen, T.; Piqani, G.; Li, C.; Golub, T. R.; Meyerson, M.; Hacohen, N.; Hahn, W. C.; Lander, E. S.; Sabatini, D. S.; Root, D. E. *Proc. Natl. Acad. Sci. U.S.A.* **2008**, *105*, 20380.

38. Reiling, J. H.; Clish, C. B.; Carette, J. E.; Varadarajan, M.; Brummelkamp, T. R.; Sabatini, D. M. *Proc. Natl. Acad. Sci. U.S.A.* **2011**, *108*, 11731.

39. Heitman, J.; Movva, N. R.; Hall, M. N. *Science* **1991**, *253*, 905.

40. Roemler, T.; Davies, J.; Giaever, G.; Nislow, C. *Nat. Chem. Biol.* **2012**, *8*, 46.

41. Pierce, S. E.; Davis, R. W.; Nislow, C.; Giaever, G. *Nat. Protoc.* **2007**, *2*, 2958.

42. Giaever, G.; Shoemaker, D. D.; Jones, T. W.; Liang, H.; Winzeler, E. A.; Astromoff, A.; Davis, R. W. *Nat. Genet.* **1999**, *21*, 278.

43. Hoepfner, D.; McNamara, C. W.; Lim, C. S.; Studer, C.; Riedl, R.; Aust, T.; McCormack, S. L.; Plouffe, D. M.; Meister, S.; Schuierer, S.; Plikat, U.; Hartmann, N.; Staedtler, F.; Cotesta, S.; Schmitt, E. K.; Petersen, F.; Supek, F.; Glynne, R. J.; Tallarico, J. A.; Porter, J. A.; Fishman, M. C.; Bodenreider, C.; Diagana, T. T.; Movva, N. R.; Winzeler, E. A. *Cell Host Microbe* **2012**, *11*, 654.

44. Andrusiak, K.; Piotrowski, J.; Boone, C. *Bioorg. Med. Chem.* **2012**, *20*, 1952.

45. Kemmer, D.; McHardy, L. M.; Hoon, S.; Reberioux, D.; Giaever, G.; Nislow, C.; Roskelley, C. D.; Roberge, M. *BMC Microbiol.* **2009**, *9*, 9.

46. Huang, Z.; Chen, K.; Zhang, J.; Li, Y.; Wang, H.; Cui, D.; Tang, J.; Liu, Y.; Shi, X.; Li, W.; Liu, D.; Chen, R.; Sucgang, R. S.; Pan, X. *Cell Rep.* **2013**, *3*, 577.

47. Licitra, E. J.; Liu, J. O. *Proc. Natl. Acad. Sci. U.S.A.* **1996**, *93*, 12817.

48. Chidley, C.; Haruki, H.; Pedersen, M. G.; Muller, E.; Johnsson, K. *Nat. Chem. Biol.* **2011**, *7*, 375.

49. Shepard, A. R.; Conrow, R. E.; Pang, I.; Jacobson, N.; Rezwan, M.; Rutschmann, K.; Auerbach, D.; SriRamaratnam, R.; Cornish, V. W. *ACS Chem. Biol.* **2013**, *8*, 549.

50. Rezwan, M.; Auerbach, D. *Methods* **2012**, *57*, 423.

51. Stockwell, B. R. *Nat. Rev. Genet.* **2000**, *1*, 116.

52. Fishman, M. C.; Porter, J. A. *Nature* **2005**, *437*, 491.

53. Schenone, M.; Dancik, V.; Wagner, B. K.; Clemons, P. A. *Nat. Chem. Biol.* **2013**, *9*, 232.

54. Duncan, L. M.; Timms, R. T.; Zavodsky, E.; Cano, F.; Dougan, G.; Randow, F.; Lehner, P. J. *PLoS One* **2012**, *7*, e39651.

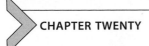

CHAPTER TWENTY

Neuroinflammation in Mood Disorders: Mechanisms and Drug Targets

Allen T. Hopper, Kenneth A. Jones, Brian M. Campbell, Guiying Li
Lundbeck Research USA, Paramus, NJ, USA

Contents

ABBREVIATIONS

3-HK 3-hydroxykynurenine
ATP adenosine triphosphate
CNS central nervous system
ICE interleukin-1beta converting enzyme (Caspase-1)
IDO indolamine 2,3-dioxygenase
IFN-α interferon-alpha
IFN-γ interferon-gamma
IL-1β interleukin-1beta
IL-2 interleukin-2
IL-6 interleukin-6
IL-18 interleukin-18
KMO kynurenine 3-monooxygenase
KYNA kynurenic acid
LPS lipopolysaccharide
MDD major depressive disorder
NLRP3 NOD-like receptor family, pyrin domain containing 3
P2X$_7$ P2X purinoceptor 7

Annual Reports in Medicinal Chemistry, Volume 48
ISSN 0065-7743
http://dx.doi.org/10.1016/B978-0-12-417150-3.00020-X

SNP single nucleotide polymorphism
TDO tryptophan 2,3-dioxygenase
TNF-α tumor necrosis factor alpha

1. INTRODUCTION

Depression and other mood disorders are among the most common diseases worldwide and are associated with a tremendous burden to the people affected as well as high rates of suicide. Despite the fact that there has been substantial effort to develop new drugs, there remains a considerable unmet need for more effective treatments, in particular for therapeutic nonresponders. Most antidepressants function by modulating serotonergic and/or noradrenergic neurotransmission, and while this is an important mechanism therapeutically, its role in disease etiology is yet unclear. The identification of additional pathophysiological changes associated with depressive disorders is therefore important. Growing evidence points to a significant role for elements of the innate immune system to dramatically influence brain function in ways that are relevant to psychiatric and neurological diseases. In this chapter, we review recent advancements in our understanding of how inflammation might cause or exacerbate psychiatric disorders such as depression and schizophrenia.

Three lines of clinical evidence provide the basis for the inflammation hypothesis of depression: the induction of depression with cytokine therapies, increased incidence of depression in the population affected by autoimmune and chronic diseases involving inflammation, and the association of depression with elevated biomarkers of inflammation. Numerous studies point to a causal relationship between certain cytokine-based therapies and depression and these have been the subject of prior reviews.[1–3] Human immunodeficiency virus patients and cancer patients that receive interleukin-2 (IL-2) or interferon-alpha (IFN-α) therapy develop marked cognitive disturbances and neurovegetative symptoms of depression.[4] Approximately 50% of patients treated with cytokine therapy meet criteria for major depressive disorder (MDD) within 3 months of commencing treatment. More than 60% of patients report depressive mood, 30% report anhedonia, and 10% are suicidal. The findings with IFN-α treatment resulted in the change of standard of care for patients, with selective serotonin reuptake inhibitors being administered 3 weeks prior to initiating treatment. Common comorbidities with depression include diseases with clear involvement of the immune system, including cardiovascular disease, type 2 diabetes,[5] and rheumatoid arthritis. Depression in a medically ill population overall occurs 5–10 time more frequently than in the general population.[6–8] The view that depression is secondary to immune activation is supported by observations of elevated plasma cytokines including IL-1β, IL-2, IL-6, tumor necrosis factor alpha (TNF-α) and IFN-α, especially in patients with severe depression.[9,10] Recent meta-analyses have confirmed the association of elevated C-reactive protein, IL-6, and TNF-α, and to a lesser extent IL-1β, with MDD.[11,12]

Although there is substantial support for the association of and, in some instances, a causal role for inflammation in mood disorders, there are currently few options available to clinicians to evaluate alternative treatments that address this mechanism. Anti-inflammatory agents such as celecoxib and omega-3 fatty (eicosapentanoic) acid have been shown to be moderately effective in treating depression in some instances.[13,14] Evidence is beginning to emerge that direct inhibition of TNF-α activity in the periphery by way of neutralizing antibodies or proteins may provide relief of depressive symptoms in patients suffering from chronic inflammation. For example, patients treated with etanercept for psoriasis showed significant improvements in depressive symptoms and fatigue.[15] A recent proof-of-concept study of treatment–resistant depressed patients showed no general effect of TNF-α inhibitor administration on mood scores, but revealed a subset of responders with high baseline TNF-α.[16] Downstream of proinflammatory cytokine release, kynurenine metabolism is also disturbed during inflammation, resulting in metabolism favoring production of quinolinic acid over kynurenic acid (KYNA).[17,18] This imbalance may be neurotoxic and has been reported after both acute and chronic inflammation stimuli to induce mood disturbances in animal models.[19] In these models, symptoms are reversed by inhibition of IDO (indolamine 2,3-dioxygenase) or KMO (kynurenine 3-monooxygenase).[20,21] Thus, direct targeting of key mediators of inflammation may improve depressive symptoms in a subset of patients with high inflammatory biomarkers.

2. INFLAMMASOME PATHWAY AND CYTOKINE MODULATION

The evidence linking elevated cytokines to major depressive disorder has been thoroughly reviewed.[1,22–25] Cytokines are the peptide signaling molecules of the immune system, which when released during a response to injury or infection, elicit an organized set of behaviors that have evolved to optimize survival and recovery.[26] This phenomenon, first defined as "sickness behaviour" by Kent[27] in 1992, is triggered when peripheral proinflammatory cytokines activate a "mirror" brain cytokine-signaling cascade, which acts on neuronal circuits.[1,22] Sickness behaviors are highly conserved across species and include decreased exploration, feeding, grooming, and sexual activity, psychomotor slowing, increased sensitivity to pain, sleep disruption, and altered response to reward stimuli. Patients with elevated plasma biomarkers of inflammation have been reported to exhibit resistance to some antidepressants,[28,29] which suggests that cytokines could trigger pro-depressive mechanisms leading to treatment resistance.

The innate immune system is genetically encoded to stereotypically respond to specific signals derived from pathogens or other danger signals. These signals derive from common pathogens, such as bacterial cell wall components (lipopolysaccharides, LPS), and components of tissue damage, such as adenosine triphosphate (ATP), uric acid, and heat shock proteins. A family of pattern recognition receptors, known as Toll-like receptors, bind specific types of molecular danger signals and induce the synthesis of proinflammatory cytokines including IL-1β, IL-6, and TNF-α.[30,31] IL-6 and TNF-α are immediately released, but IL-1β has an additional point of regulation by a macromolecular complex, the inflammasome.[32] IL-1β is synthesized as the immature pro-IL-1β which is subsequently processed by the NOD-like receptor family, pyrin domain containing 3 (NLRP3) inflammasome complex, which is able to respond to and integrate multiple types of danger signals, including ATP and other damage-associated molecular patterns.[33] One of the components of the NLRP3 inflammasome, caspase-1, is activated when ATP binds the membrane-associated P2X$_7$ (P2X purinoceptor 7) receptor and cleaves pro-IL-1β into active IL-1β. Thus, the P2X$_7$ receptor serves to trigger a complex cascade of events resulting in the release of IL-1β. These two regulators of IL-1β release, caspase-1 and P2X$_7$, are viable points of intervention with small molecules for treating inflammation-associated mood disorders.[34]

2.1. P2X$_7$ receptors

The P2X$_7$ receptor is a member of the family of purinergic receptors that are ion channels gated by ATP and expressed predominantly on cells of the macrophage lineage, including microglia.[35,36] P2X$_7$ is moderately expressed in oligodendrocytes and in reactive astrocytes, and to a lesser extent on neurons, where it is localized to presynaptic glutamatergic terminals in hippocampus and cerebellum.[37] Activation of the receptor by extracellular ATP results in intracellular calcium influx and potassium efflux. In myeloid cells, such as macrophages and microglia, this leads to inflammasome assembly, caspase-1 activation, and ultimately maturation of the inflammatory cytokines IL-1β and IL-18. Prolonged P2X$_7$ activation leads to formation of a large hydrophilic pore by pannexin-1, which is permeable to molecules as large as 900 Da.

Several genetic association studies suggest that the P2X$_7$ gene may be involved in mood disorders,[38] although one large study failed to replicate this effect.[39] An analysis of a French population highlighted a Gln460Arg single nucleotide polymorphism (SNP) of the P2X$_7$ receptor as a potential susceptibility gene for bipolar affective disorder[40] and major depression.[41] Soronen subsequently performed a broader analysis of selected patients from three cohorts and two P2X$_7$ gain-of-function SNPs, His155Tyr and Gln460Arg, were found to be associated with significantly elevated risk for familial mood disorder and increased duration of illness.[42] Another gain-of-function SNP, Ala348Thr,[43] has been associated with increased risk for mania[44] and anxiety.[45] In an examination of plasma markers of inflammation, the expression of P2X$_7$ receptor transcripts along with those of IL-1β and IL-6 showed the highest association among a panel of 20 genes with patients diagnosed with MDD or bipolar depression.[46]

Convergent observations from animal studies point to a role of P2X$_7$ in mood disorders. First, P2X$_7$ antagonists reverse peripheral LPS- and ATP-induced changes in cyclooxygenase-2, cfos, and cortisol, suggesting effects on hypothalamic pituitary-adrenal axis function.[47] Second, hippocampal neurogenesis, which is reduced following chronic mild stress, is normalized by treatment with either antidepressants[48] or a P2X$_7$ antagonist.[49] Finally, the behavioral profile of P2X$_7$ knockout mice shows an "antidepressant-like" phenotype together with a higher responsiveness to a subefficacious dose of the antidepressant imipramine.[50] In addition, P2X$_7$ knockout mice perform better in repeated forced swim testing as compared to wild-type mice,[51] suggesting that P2X$_7$ function is important for stress-induced coping mechanisms. In summary, there is a growing body of data supporting the hypothesis that P2X$_7$ gain of function leads to elevated risk for symptoms of mood disorders and that antagonizing P2X$_7$ in animal models produces an "antidepressant" phenotype.

P2X$_7$ chemistry has been recently reviewed, see Refs. 52, 53 and references therein. While several compounds proceeded to Ph2 clinical development for peripheral inflammatory disorders such as rheumatoid arthritis, they have subsequently been discontinued. Phase 1 results for GSK1482160 (**1**), a P2X$_7$ negative allosteric modulator and the first in development that readily penetrates the central nervous system (CNS), were recently reported.[54] This compound was discontinued post Ph1 after PK/ PD modeling, based on *ex vivo* inhibition of LPS-primed and ATP-stimulated IL-1β release from human whole blood, suggested sufficient plasma concentrations could not be attained to appropriately test the disease hypothesis. Owing to substantial species cross-reactivity issues, there are very few reported compounds with both sufficient rodent potency and adequate CNS bioavailability to probe the pharmacology of this target as relates to mood disorders. The recent report of nicotinamide **2**, a brain penetrant compound with good selectivity, and rodent potency may help address this issue.[55] In a BzATP-induced Ca^{2+} flux, assay **2** inhibited recombinant rat and mouse P2X$_7$ with pIC$_{50}$s of 7.4 and 7.6, respectively. The brain-to-plasma ratio in rat is 0.4, and a 2.5 mg/kg, sc dose produced brain levels 60 min post-dose of 580 ng/mL, which corresponds to 50% receptor occupancy as measured by *ex vivo* autoradiography. Based on these properties, it is suggested that this may be a useful tool to probe neuroinflammatory and neuropsychiatric disorders.[55] New chemical matter reported in the patent literature includes **3** and **4** with IC$_{50}$s of 4.7 and 10 nM, respectively, in a YoPro uptake assay against recombinant human P2X$_7$.[56,57]

Microdialysis methods for measuring brain levels of IL-1β, a potential biomarker for functional P2X$_7$ and caspase-1 activity, have been reported.[58,59] Quantitative assessment of the mature form of IL-1β in homogenized rodent brain, an important mediator of neuroinflammation, had been a challenge as ELISA methods are not able to differentiate the pro and mature forms of the cytokine. Nicotinamide **2** reportedly inhibits BzATP-induced IL-1β release in the hippocampus of freely moving rats in a dose-dependent fashion with maximal effect at 100 mg/kg, sc.[58]

2.2. Caspase-1

The proinflammatory enzyme caspase-1, also known as interleukin converting enzyme, is the key regulator of the conversion of the inactive pro-IL-1β into the mature active IL-1β. A hypothesis has been recently put forward that the inflammasome is a central mediator by which psychological and physical stressors can contribute to the development of depression. It is therefore proposed that the inflammasome, and in particular caspase-1, may be an appropriate target for the development of new treatments for depression.[34]

Selective CNS-penetrating caspase-1 inhibitor tool compounds are desired to further probe this hypothesis. Medicinal chemistry of caspase inhibitors has been reviewed in Ref. 60 and references therein. The ethyl acylal prodrug VX-765 (**5**), currently in phase 2B clinical trials for epilepsy, is readily hydrolyzed to active aspartyl aldehyde **6** (VRT-043198), which has a caspase-1 IC$_{50}$ of 200 pM (Table 20.1) as measured in similar fashion across a panel of caspases.[61] This isosteric aspartic acid replacement in the P1 site is important for activity and most reversible inhibitors possess either an aspartyl aldehyde or ketone in this position. These compounds form a reversible thioacetal covalent bond with the active site cysteine residue. Advances in this area include optimization of the P2–P3 site with succinic acid

Table 20.1 IC$_{50}$ (nM) versus caspase panel.[61]
IC$_{50}$ (nM)

Compound	Casp-1	Casp-3	Casp-4	Casp-5	Casp-8	Casp-9	Casp-10	Casp-14
6	0.20	>10,000	15	11	3.3	5.1	67	59
10	0.023	>10,000	14	3.6	25	2.2	90	800

amides[62] exemplified by prodrug **7** and active analog **8** (caspase-1 IC_{50} 4.8 nM); and aspartyl cyano ester prodrug **9** and active acid **10** (NCGC434, caspase-1 IC_{50} 23 pM).[61] Cyano analog **10** also has an improved caspase selectivity profile over **6** (Table 20.1). While the aspartyl aldehyde and ketone-based analogs possess excellent potency and have afforded clinical development compounds, issues with Pgp efflux leave room for the identification of different chemotypes with improved CNS penetration.[61]

5 (VX-765) R = Et
6 (VRT-043198) R = H

6 (VRT-043198)

7 R = nPr
8 R = H

9 R = Et
10 R = H

3. TRYPTOPHAN METABOLIC PATHWAY

Dysregulation of kynurenine metabolism has been linked to a variety of psychiatric and neurologic disorders with neuroinflammatory components. Mediators of inflammation alter conversion of tryptophan to kynurenine and subsequent metabolism to several neuroactive products. Most notably, proinflammatory cytokines (e.g. IFN-γ, TNF-α) induce transcription and/or activity of the metabolic enzymes in this pathway including IDO, KMO, and kynureninase,[63,17] resulting in an imbalance of metabolism favoring production of quinolinic acid over KYNA (Fig. 20.1). This imbalance is thought to be neurotoxic and may lead to disturbances in mood, thought processes, and motor control.[64] Indeed, it was recently suggested that quinolinic acid may contribute to suicidal tendencies in depression as well as other disorders.[65] A similar correlation was observed in a cohort of schizophrenic patients who possessed

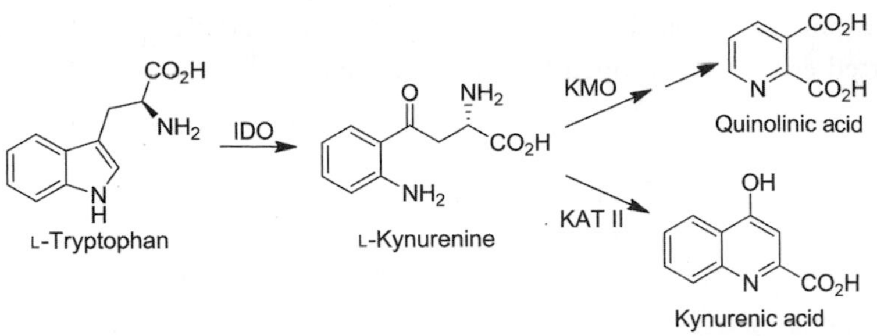

Figure 20.1 Kynurenine pathway. Enzymes regulating kynurenine metabolism in the CNS, including IDO and KMO, are reportedly upregulated in response to inflammation. In each step, multiple enzymes are involved for conversion to the pictured intermediates, but for simplicity only the enzymes and metabolites of interest are shown. For a more detailed view of the pathway, see Ref. 64 Indolamine 2,3-dioxygenase (IDO), kynurenine 3-monooxygenase (KMO), kynurenine aminotransferase II (KAT II).

lower KYNA levels relative to healthy patients[66] further supporting the importance of the quinolinic acid:KYNA ratio in mood disorders.

3.1. Indole 2,3-dioxygenase (IDO1/INDO and IDO2/INDOL1)

Induction of IDO expression and enzyme activity in the brain and periphery by mediators of inflammation is well known.[67] Induction of acute or chronic inflammation by agents such as LPS or Bacillus Calmette-Guerin, respectively, alters the ratio of kynurenine to tryptophan and this effect corresponds to induction of a depression-like phenotype as measured by increased immobility in the forced swim test and tail suspension test, along with anhedonic behavior and disruption of cognitive performance in animal models. Treatment with IDO inhibitors reverses both the disruption in the kynurenine:tryptophan ratio and depression-like symptoms in these animals.[68] Though a variety of IDO inhibitors have been discovered, early compounds including tryptophan analogs and more recently discovered β-carboline inhibitors had micromolar affinity[52] and no reported selectivity between IDO1 and IDO2. A library of known drugs screened for IDO activity identified tenatoprazole (**11**) with an IC_{50} of 1.8 μM for IDO2 and no inhibition of IDO1 or TDO2 (tryptophan 2,3-dioxygenase).[69] Chemical matter reported in patent applications includes novel fused imidazole derivatives **12**, some with IDO1 IC_{50}s < 1 μM,[70] and tryptanthrin derivatives **13** having good *in vitro* potency (recombinant human IDO1 IC_{50} 0.53 μM; whole cell IC_{50} 0.02 μM) and reported efficacy in an open field test for anxiety.[71]

11   12   13

3.2. Kynurenine 3-monooxygenase (KMO)

KMO regulates metabolism of kynurenine to 3-hydroxykynurenine (3-HK) and is the entry point to the quinolinic acid branch of the kynurenine pathway. It is found throughout the brain and body in immune-mediating cells including macrophages, microglia, T cells, and hepatocytes. KMO is induced in experimental models of inflammation, such as LPS administration,[17,18] leading to increased production of 3-HK and quinolinic acid.[19] Inhibition of KMO has been shown to prevent some consequences of neuroinflammation including restoring quinolinic acid/KYNA ratios, reducing neurodegeneration in animal models of Huntington's disease and Alzheimer's disease, and extending life-span in a model of Huntington's disease.[20] However, since current KMO inhibitors have limited ability to enter the brain, the effects of these compounds are attributed to peripheral inhibition of KMO resulting in elevation of L-kynurenine that is transported across the blood–brain barrier and converted to KYNA. Recently, O'Connor et al.[21] showed that local infusion of Ro 61-8048, a KMO inhibitor, into the brain was also able to reverse depression-like symptoms in mice treated with LPS, supporting the hypothesis that KMO inhibitors may have utility in treating mood disorders.

Recent developments in KMO chemistry include two new chemotypes, **14** and **15**, derived from UPF-648, with IC_{50}s of 30 nM.[72–74] These compounds are potent, but have limited brain permeability. Core modification provided pyrimidine **16** and carboxylic acid bioisosteric replacement generated analogs **17** and **18** reported to inhibit KMO 100% at 10 μM.[75–77] However, brain permeability of these analogs is not reported.

In the absence of KMO inhibitors with good brain permeability, an alternate approach for restoring the quinolinic acid/KYNA balance in the brain has been to develop KYNA mimetics that are predicted to antagonize the effects of quinolinic acid on NMDA receptor stimulation. A recent example of this includes KYNA amide derivative **20**, based on **19**, that showed neuroprotective effects in a transgenic mouse model of Huntington's disease[78] and in the four-vessel occlusion model of ischemia.[79]

4. CONCLUDING REMARKS

A variety of clinical observations are pointing to a role of chronic inflammation in the progression of MDD. Attention is therefore shifting away from drugs that regulate monoamines to interventions that normalize an imbalance of inflammatory mediators such as cytokines and metabolites of tryptophan that participate in the CNS response to inflammation.

Regulation of IL–1β biosynthesis is of particular interest since this is the key cytokine for initiating the innate immune response. In this respect, drug discovery efforts are focusing on two important components of the inflammasome: the purinergic receptor P2X₇ and the cysteine protease caspase-1. The identification of $P2X_7$ antagonists with both good CNS penetrability and rodent potency is an important advance, while the reported microdialysis methods for measuring brain IL-1β should also aid in establishing these mechanisms *in vivo*.

Downstream of proinflammatory cytokine release, kynurenine metabolism is also disturbed during inflammation, resulting in metabolism favoring production of quinolinic acid over KYNA. This imbalance may be neurotoxic and has been reported after both acute and chronic inflammation stimuli to induce mood disturbances in animal models. In these models, symptoms are reversed by inhibition of IDO or KMO. Modest improvements in IDO inhibitors over the past 2 years have resulted in the discovery of analogs with selectivity for IDO2 over IDO1. Advancements in KMO chemistry include expanded SAR studies leading to improvements in drug-like properties. Nonetheless, KMO inhibitors with good brain permeability have yet to be reported.

REFERENCES

1. Dantzer, R.; O'Connor, J. C.; Freund, G. G.; Johnson, R. W.; Kelley, K. W. *Nat. Rev. Neurosci.* **2008**, *9*, 46–56.
2. Raison, C. L.; Capuron, L.; Miller, A. H. *Trends Immunol.* **2006**, *27*, 24–31.
3. Miller, A. H.; Maletic, V.; Raison, C. L. *Biol. Psychiatry* **2009**, *65*, 732–741.
4. Anisman, H.; Poulter, M. O.; Gandhi, R.; Merali, Z.; Hayley, S. *J. Neuroimmunol.* **2007**, *186*, 45–53.
5. de Groot, M.; Anderson, R.; Freedland, K. E.; Clouse, R. E.; Lustman, P. J. *Psychosom. Med.* **2001**, *63*, 619–630.
6. Capuron, L.; Pagnoni, G.; Demetrashvili, M.; Woolwine, B. J.; Nemeroff, C. B.; Berns, G. S.; Miller, A. H. *Biol. Psychiatry* **2005**, *58*, 190–196.
7. Johnson, A. K.; Grippo, A. J. *J. Physiol. Pharmacol.* **2006**, *57*(Suppl 11), 5–29.
8. Godha, D.; Shi, L.; Mavronicolas, H. *Curr. Med. Res. Opin.* **2010**, *26*, 1685–1690.
9. Anisman, H.; Ravindran, A. V.; Griffiths, J.; Merali, Z. *Mol. Psychiatry* **1999**, *4*, 182–188.
10. Maes, M. *Adv. Exp. Med. Biol.* **1999**, *461*, 25–46.
11. Dowlati, Y.; Herrmann, N.; Swardfager, W.; Liu, H.; Sham, L.; Reim, E. K.; Lanctot, K. L. *Biol. Psychiatry* **2010**, *67*, 446–457.
12. Howren, M. B.; Lamkin, D. M.; Suls, J. *Psychosom. Med.* **2009**, *71*, 171–186.
13. Song, C.; Zhao, S. *Expert Opin. Investig. Drugs* **2007**, *16*, 1627–1638.
14. Akhondzadeh, S.; Jafari, S.; Raisi, F.; Nasehi, A. A.; Ghoreishi, A.; Salehi, B.; Mohebbi-Rasa, S.; Raznahan, M.; Kamalipour, A. *Depress. Anxiety* **2009**, *26*, 607–611.
15. Tyring, S.; Gottlieb, A.; Papp, K.; Gordon, K.; Leonardi, C.; Wang, A.; Lalla, D.; Woolley, M.; Jahreis, A.; Zitnik, R.; Cella, D.; Krishnan, R. *Lancet* **2006**, *367*, 29–35.
16. Raison, C. L.; Rutherford, R. E.; Woolwine, B. J.; Shuo, C.; Schettler, P.; Drake, D. F.; Haroon, E.; Miller, A. H. *JAMA Psychiatry* **2013**, *70*, 31–41.
17. Connor, T. J.; Starr, N.; O'Sullivan, J. B.; Harkin, A. *Neurosci. Lett.* **2008**, *441*, 29–34.
18. Heisler, J. M.; Doyle, A.; Parrott, J.; O'Connor, J. C. *Program#/Poster#*: 587.10/UU14; Society for Neuroscience: New Orleans, LA, 2012.

19. Bisulco, S.; Lee, A. W.; Budac, D. P.; Lawson, M. A.; Poon, P. A.; Cajina, M.; Jimenez, H.; Doller, D.; Smith, K. E.; Campbell, B.; Dantzer, R. *Program#/Poster#*: 660.24/R18; Society for Neuroscienece: New Orleans, LA, 2012.

20. Zwilling, D.; Huang, S.-Y. ; Sathyasaikumar, K. V.; Notarangelo, F. M.; Guidetti, P.; Wu, H.-Q. ; Lee, J.; Truong, J.; Andrews-Zwilling, Y.; Hsieh, E. W.; Louie, J. Y.; Wu, T.; Scearce-Levie, K.; Patrick, C.; Adame, A.; Giorgini, F.; Moussaoui, S.; Laue, G.; Rassoulpour, A.; Flik, G.; Huang, Y.; Muchowski, J. M.; Masliah, E.; Schwarcz, R.; Muchowski, P. J. *Cell (Cambridge, MA, US)* **2011**, *145*, 863–874.

21. O'Connor, J. C.; Salazar, J. M.; Parrott, J. M.; Redus, L. *Prgram#/Poster#*: 466.21/CC29; Society for Neurocience: Washington, DC, 2011.

22. Dantzer, R.; Capuron, L.; Irwin, M. R.; Miller, A. H.; Ollat, H.; Perry, V. H.; Rousey, S.; Yirmiya, R. *Psychoneuroendocrinology* **2008**, *33*, 18–29.

23. Moreau, M.; Andre, C.; O'Connor, J. C.; Dumich, S. A.; Woods, J. A.; Kelley, K. W.; Dantzer, R.; Lestage, J.; Castanon, N. *Brain Behav. Immun.* **2008**, *22*, 1087–1095.

24. Haroon, E.; Raison, C. L.; Miller, A. H. *Neuropsychopharmacology* **2012**, *37*, 137–162.

25. Pace, T. W. W.; Miller, A. H. *Ann. N. Y. Acad. Sci.* **2009**, *1179*, 86–105.

26. Hart, B. L. *Neurosci. Biobehav. Rev.* **1988**, *12*, 123–137.

27. Kent, S.; Bluthe, R. M.; Kelley, K. W.; Dantzer, R. *Trends Pharmacol. Sci.* **1992**, *13*, 24–28.

28. Maes, M.; Bosmans, E.; De, J. R.; Kenis, G.; Vandoolaeghe, E.; Neels, H. *Cytokine* **1997**, *9*, 853–858.

29. Yoshimura, R.; Hori, H.; Ikenouchi-Sugita, A.; Umene-Nakano, W.; Ueda, N.; Nakamura, J. *Prog. Neuropsychopharmacol. Biol. Psychiatry* **2009**, *33*, 722–726.

30. Gay, N. J.; Keith, F. J. *Nature* **1991**, *351*, 355–356.

31. Tschopp, J.; Martinon, F.; Burns, K. *Nat. Rev. Mol. Cell Biol.* **2003**, *4*, 95–104.

32. Franchi, L.; Eigenbrod, T.; Munoz-Planillo, R.; Nunez, G. *Nat. Immunol.* **2009**, *10*, 241–247.

33. Di Virgilio, F. *Trends Pharmacol. Sci.* **2007**, *28*, 465–472.

34. Iwata, M.; Ota, K. T.; Duman, R. S. *Brain Behav. Immun.* **2013**, *31*, 105–114.

35. North, R. A. *Physiol. Rev.* **2002**, *82*, 1013–1067.

36. Sperlagh, B.; Vizi, E. S.; Wirkner, K.; Illes, P. *Prog. Neurobiol. (Amsterdam, Neth.)* **2006**, *78*, 327–346.

37. Miras-Portugal, M. T.; Diaz-Hernandez, M.; Giraldez, L.; Hervas, C.; Gomez-Villafuertes, R.; Sen, R. P.; Gualix, J.; Pintor, J. *Neurochem. Res.* **2003**, *28*, 1597–1605.

38. Sluyter, R.; Stokes, L. *Recent Pat. DNA Gene Seq.* **2011**, *5*, 41–54.

39. Grigoroiu-Serbanescu, M.; Herms, S.; Muhleisen, T. W.; Georgi, A.; Diaconu, C. C.; Strohmaier, J.; Czerski, P.; Hauser, J.; Leszczgnska-Rodziewicz, A.; Abou, J. R.; Babadjanova, G.; Tiganov, A.; Krasnov, V.; Kapiletti, S.; Neagu, A. L.; Vollmer, J.; Breuer, R.; Rietschel, M.; Propping, P.; Nothen, M. M.; Cichon, S. *Am. J. Med. Genet. B* **2009**, *150B*, 1017–1021.

40. Barden, N.; Harvey, M.; Gagne, B.; Shink, E.; Tremblay, M.; Raymond, C.; Labbe, M.; Villeneuve, A.; Rochette, D.; Bordeleau, L.; Stadler, H.; Holsboer, F.; Mueller-Myhsok, B. *Am. J. Med. Genet. B* **2006**, *141B*, 374–382.

41. Lucae, S.; Salyakina, D.; Barden, N.; Harvey, M.; Gagne, B.; Labbe, M.; Binder, E. B.; Uhr, M.; Paez-Pereda, M.; Sillaber, I.; Ising, M.; Brueckl, T.; Lieb, R.; Holsboer, F.; Mueller-Myhsok, B. *Hum. Mol. Genet.* **2006**, *15*, 2438–2445.

42. Soronen, P.; Mantere, O.; Melartin, T.; Suominen, K.; Vuorilehto, M.; Rytsala, H.; Arvilommi, P.; Holma, I.; Holma, M.; Jylha, P.; Valtonen, H. M.; Haukka, J.; Isometsa, E.; Paunio, T. *Am. J. Med. Genet. Part B* **2011**, *156*, 435–447.

43. Bradley, H. J.; Baldwin, J. M.; Goli, G. R.; Johnson, B.; Zou, J.; Sivaprasadarao, A.; Baldwin, S. A.; Jiang, L.-H. *J. Biol. Chem.* **2011**, *286*, 8176–8187.

44. Backlund, L.; Nikamo, P.; Hukic, D. S.; Ek, I. R.; Traeskman-Bendz, L.; Landen, M.; Edman, G.; Schalling, M.; Frisen, L.; Oesby, U. *Bipolar Disord.* **2011**, *13*, 500–508.
45. Erhardt, A.; Lucae, S.; Unschuld, P. G.; Ising, M.; Kern, N.; Salyakina, D.; Lieb, R.; Uhr, M.; Binder, E. B.; Keck, M. E.; Mueller-Myhsok, B.; Holsboer, F. *J. Affective Disord.* **2007**, *101*, 159–168.
46. Larsen, J. *Eur. Neuropsychopharmacol.* **2011**, *21*(Suppl 3), S623.
47. Pineda, E. A.; Hensler, J. G.; Sankar, R.; Shin, D.; Burke, T. F.; Mazarati, A. M. *Neurotherapeutics* **2012**, *9*, 477–485.
48. Warner-Schmidt, J. L.; Duman, R. S. *Hippocampus* **2006**, *16*, 239–249.
49. Iwata, M.; Li, X.-Y.; Li, N.; Duman, R. S. *Program#/Poster#*: 900.06/HH30; Society for Neuroscience: Washington, DC, Nov 12–16, 2011.
50. Basso, A. M.; Bratcher, N. A.; Harris, R. R.; Jarvis, M. F.; Decker, M. W.; Rueter, L. E. *Behav. Brain Res.* **2009**, *198*, 83–90.
51. Boucher, A. A.; Arnold, J. C.; Hunt, G. E.; Spiro, A.; Spencer, J.; Brown, C.; McGregor, I. S.; Bennett, M. R.; Kassiou, M. *Neuroscience (Amsterdam, Neth.)* **2011**, *189*, 170–177.
52. Hopper, A. T.; Campbell, B. M.; Kao, H.; Pintchovski, S. A.; Staal, R. G. W. *ARMC* **2012**, *47*, 37.
53. Friedle, S. A.; Curet, M. A.; Watters, J. J. *Recent Pat. CNS Drug Discov.* **2010**, *5*, 35–45.
54. Ali, Z.; Laurijssens, B.; Ostenfeld, T.; McHugh, S.; Stylianou, A.; Scott-Stevens, P.; Hosking, L.; Dewit, O.; Richardson, J. C.; Chen, C. *Br. J. Clin. Pharmacol.* **2013**, *75*, 197–207.
55. Letavic, M. A.; Lord, B.; Bischoff, F.; Hawryluk, N. A.; Pieters, S.; Rech, J. C.; Sales, Z.; Velter, A. I.; Ao, H.; Bonaventure, P.; Contreras, V.; Jiang, X.; Morton, K. L.; Scott, B.; Wang, Q.; Wickenden, A. D.; Carruthers, N. I.; Bhattacharya, A. *ACS Med. Chem. Lett.* **2013**, *4*, 419–422.
56. Hilpert, K.; Hubler, F.; Kimmerlin, T.; Murphy, M.; Renneberg, D.; Stamm, S. Patent Application. WO2013014587A1, 2013.
57. Hilpert, K.; Hubler, F.; Murphy, M.; Renneberg, D. Patent Application. WO2012114268A1, 2012.
58. Aluisio, L.; Fraser, I.; Lord, B.; Bhattacharya, A.; Letavic, M.; Carruthers, N.; Lovenberg, T.; Bonaventure, P. In: *Abstract 090213, Monitoring Molecules in Neuroscience; 14th International Conference, London, UK, Sept 16–20, 2012,* 2012.
59. Hopper, A.; Smagin, G.; Song, D.; Zhou, H.; Moller, T. In: *Abstract O4-2-2, Purines 2012 Conference, Fukuoka, Japan, May 31 to June 2, 2012,* 2012.
60. Linton, S. D. *Curr. Top. Med. Chem. (Sharjah, United Arab Emirates)* **2005**, *5*, 1697–1716.
61. Boxer, M. B.; Quinn, A. M.; Shen, M.; Jadhav, A.; Leister, W.; Simeonov, A.; Auld, D. S.; Thomas, C. J. *ChemMedChem* **2010**, *5*, 730–738.
62. Galatsis, P.; Caprathe, B.; Gilmore, J.; Thomas, A.; Linn, K.; Sheehan, S.; Harter, W.; Kostlan, C.; Lunney, E.; Stankovic, C.; Rubin, J.; Brady, K.; Allen, H.; Talanian, R. *Bioorg. Med. Chem. Lett.* **2010**, *20*, 5184–5190.
63. Alberati-Giani, D.; Ricciardi-Castagnoli, P.; Koehler, C.; Cesura, A. M. *J. Neurochem.* **1996**, *66*, 996–1004.
64. Schwarcz, R.; Bruno, J. P.; Muchowski, P. J.; Wu, H.-Q. *Nat. Rev. Neurosci.* **2012**, *13*, 465–477.
65. Erhardt, S.; Lim, C. K.; Linderholm, K. R.; Janelidze, S.; Lindqvist, D.; Samuelsson, M.; Lundberg, K.; Postolache, T. T.; Traskman-Bendz, L.; Guillemin, G. J.; Brundin, L. *Neuropsychopharmacology* **2013**, *38*, 743–752.
66. Carlborg, A.; Jokinen, J.; Joensson, E. G.; Erhardt, S.; Nordstroem, P. *Psychiatry Res.* **2013**, *205*, 165–167.
67. Dantzer, R.; O'Connor, J. C.; Lawson, M. A.; Kelley, K. W. *Psychoneuroendocrinology* **2011**, *36*, 426–436.

68. Corona, A. W.; Norden, D. M.; Skendelas, J. P.; Huang, Y.; O'Connor, J. C.; Lawson, M.; Dantzer, R.; Kelley, K. W.; Godbout, J. P. *Brain Behav. Immun.* **2013**, *31*, 134–142.

69. Bakmiwewa, S. M.; Fatokun, A. A.; Tran, A.; Payne, R. J.; Hunt, N. H.; Ball, H. J. *Bioorg. Med. Chem. Lett.* **2012**, *22*, 7641–7646.

70. Mautino, M.; Kumar, S.; Waldo, J.; Jaipuri, F.; Kesharwani, T. Patent Application. WO2012142237A1, 2012.

71. Kuang, C.; Wang, Y.; Wang, S.; Chen, B.; Guan, Y.; Liu, Y.; Tian, N. Patent Application. CN102579452A, 2012.

72. Prime, M.; Winkler, D.; Beconi, M.; Brookfield, F.; Brown, C.; Courtney, S.; Ebneth, A.; Grigg, R.; Hamelin-Flegg, E.; Johnson, P.; Mack, V.; Marston, R.; Mitchell, W.; Pena, P.; Reed, L.; Suganthan, S.; Munoz-Sanjuan, I.; Schaeffer, E.; Toledo-Sherman, L.; Weddell, D.; Went, N.; Winkler, C.; Wityak, J.; Yarnold, C.; Dominguez, C. In: MEDI-330 Abstracts of Papers, 242nd ACS National Meeting & Exposition, Denver, CO, August 28–September 1, 2011.

73. Wityak, J.; Toledo-Sherman, L. M.; Leeds, J.; Dominguez, C.; Courtney, S. M.; Yarnold, C. J.; De, A. P. P. C.; Scheel, A.; Winkler, D. Patent Application. WO2010011302A1, 2010.

74. Wityak, J.; Toledo-Sherman, L. M.; Leeds, J.; Dominguez, C.; Courtney, S. M.; Yarnold, C. J.; De, A. P. P. C.; Scheel, A.; Winkler, D. Patent Application. WO2010017132A1, 2010.

75. Wityak, J.; Toledo-Sherman, L. M.; Leeds, J.; Dominguez, C.; Courtney, S. M.; Yarnold, C. J.; De, A. P. P. C.; Scheel, A.; Winkler, D. Patent Application. WO2010017179A1, 2010.

76. Dominguez, C.; Toledo-Sherman, L. M.; Winkler, D.; Brookfield, F.; De, A. P. P. C. Patent Application. WO2011091153A1, 2011.

77. Dominguez, C.; Toledo-Sherman, L. M.; Courtney, S. M.; Prime, M.; Mitchell, W.; Brown, C. J.; De, A. P. P. C.; Johnson, P. Patent Application. WO2013016488A1, 2013.

78. Zadori, D.; Nyiri, G.; Szonyi, A.; Szatmari, I.; Fueloep, F.; Toldi, J.; Freund, T. F.; Vecsei, L.; Klivenyi, P. *J. Neural Transm.* **2011**, *118*, 865–875.

79. Gellert, L.; Fuzik, J.; Goebloes, A.; Sarkoezi, K.; Marosi, M.; Kis, Z.; Farkas, T.; Szatmari, I.; Fueloep, F.; Vecsei, L.; Toldi, J. *Eur. J. Pharmacol.* **2011**, *667*, 182–187.

Topics in Drug Design and Discovery

Section Editor: Peter R. Bernstein
PhaRmaB LLC, Rose Valley, Pennsylvania

CHAPTER TWENTY-ONE

Inhibitors of hERG Channel Trafficking: A Cryptic Mechanism for QT Prolongation

Kap-Sun Yeung, Nicholas A. Meanwell

Department of Medicinal Chemistry, Bristol-Myers Squibb Research and Development, Wallingford, CT, USA

Contents

1. INTRODUCTION

The original observation of polymorphic ventricular tachycardia and arrhythmia associated with QT prolongation with a noncardiac drug was in a patient taking the antipsychotic agent thioridazine (**1**).[1,2] However, it was the occurrence of a similar syndrome in a patient taking the recommended dose of the antihistamine terfenadine (**2**), a prodrug of the active species fexofenadine (**3**), concomitantly with a cytochrome P450-inhibiting antifungal agent that brought the issue of drug-induced cardiac side effects into a clear focus for regulatory authorities and the pharmaceutical industry.[2,3] The drug-induced arrhythmia observed in these patients is characterized by a unique shape of the QRS axis in the electrocardiogram which is twisted at the points, an observation that led to the descriptive of *torsade de pointes* (TdP).[4] The underlying cause of QT lengthening with terfenadine is an

Annual Reports in Medicinal Chemistry, Volume 48
ISSN 0065-7743
http://dx.doi.org/10.1016/B978-0-12-417150-3.00021-1

335

1

2: R = CH$_3$
3: R = CO$_2$H

increase in the ventricular action potential duration (APD) which results in a reduction in the rate of phase 3 ventricular repolarization, an event that is dependent on the complex choreography of several interacting ion channels, including the slow and rapid delayed rectifier K$^+$ currents I_{Ks} and I_{Kr}, respectively. Although the incidence of drug-induced QT prolongation may be rare, the ventricular arrhythmias associated with TdP can precipitate sudden cardiac death and this phenomenon has since been linked to the use of a number of drugs, some of which have been removed from the market or have acquired black box warnings as a consequence.[4,5] Subsequent studies revealed that these drugs extended the QT interval by inhibiting the rapid component of the delayed rectifier K$^+$ current (I_{Kr}) that is largely but not completely recapitulated by the *human ether-a-go-go related gene* (hERG) product, also referred to as Kv11.1, the pore-forming subunit of the channel encoded by the *KCNH2* gene.[6–9] The hERG channel is expressed in the heart, smooth muscle cells, several regions of the brain, endocrine cells, and numerous tumor cell lines.[2] The cardiac effects of thioridazine and ter-fenadine are attributed to direct blockade of the cardiac hERG channel and the observed adverse clinical events ultimately led to regulatory guidance that drug candidates be screened for the potential to inhibit hERG prior to clinical evaluation. Safety profiling typically involves an *in vitro* assay that assesses the effect of a test substance on current through either a native or expressed form of the hERG protein, with additional studies conducted to assess the potential to cause QT prolongation *in vivo*.[6–17] However, analysis of a drug candidate using an *in vitro* patch–clamp assay, displacement of [^{3}H]-dofetilide binding from the pore of the channel, or monitoring changes in membrane potential using fluorescence-based dyes assesses only the potential of a molecule to interact directly with the hERG channel protein, while *in vivo* studies typically evaluate the effects of a molecule in the time period immediately following dosing or at C_{max}, the peak plasma concentration.[6–16] It is becoming apparent that while these assays have some predictive value, regulation of hERG channel synthesis and function is complex and drug interference with channel activity can extend well beyond that of simply binding to the channel pore and blocking the K^+-mediated current.[18–24] Consequently, the assays that are typically employed to evaluate this liability may underestimate the potential for problems to occur *in vivo* and some of the alternative mechanisms of drug-induced QT prolongation have begun to be appreciated only with more recent experiences.[25–28] Studies seeking to understand the effect of congenital mutations in the *KCNH2* gene have established that the majority are associated with impaired trafficking of the channel protein from the endoplasmic reticulum (ER) that leads to tagging of the protein for destruction and reduced cell surface expression.[8,29] Initial observations with small molecules that encompass a wide range of structures and biological activities have demonstrated an effect on hERG channel cell surface expression with diverse and complex underlying mechanisms. There has been some initial focus on examining the effects of direct hERG channel blockers on protein expression and several have been shown to possess dual properties, exhibiting both direct channel blockade and trafficking inhibition, while some hERG pore channel blockers promote channel trafficking and are capable of rescuing select genetic variants (hERG trafficking promoters). However, hERG channel surface expression is subject to regulation by a range of mechanisms, many poorly understood, and small molecule interference with these aspects of channel regulation is another source of potential problems. In this chapter, we summarize the biochemical pharmacology of hERG channel expression and discuss the molecules that have been characterized as inhibitors of cell surface expression of this important cardiac ion channel.

2. BIOCHEMICAL PHARMACOLOGY OF THE hERG K^+ CHANNEL

The prevalence of inherited long QT syndrome has been estimated to range from 0.02% to 0.05% of the population and has been associated with mutations in 12 genes that encode for ion channels and associated proteins that most typically lead to a loss–of–function phenotype.[29] The vast majority of cases are accounted for by mutations in just two genes: *KCNQ1* (LQT1, 40–55% of the prevalence) and *KCNH2* (LQT2, 35–45%), which encode for Kv7.1, the α- subunit of I_{Ks}, and Kv11.1, the α- subunit of I_{Kr} (hERG), respectively, and are associated with loss of function, while a third gene, SCN5A (LQT3), which encodes for Nav1.5, the α-subunit of I_{Na}, is estimated to account for between 2% and 8% of the prevalence and expresses a gain-of-function phenotype.[29] The majority (estimated at up to 90%) of the >450 mutations that have been characterized in *KCNH2* have been found to affect hERG protein trafficking in a fashion that leads to retention of the channel protein in the ER and subsequent degradation, although some produce channels that are either nonfunctional or exhibit defective properties.[2,29]

The hERG1a channel is synthesized in the ER as a 1159-residue protein with 6 transmembrane segments, the core of which is glycosylated to bring its apparent molecular weight to 135 kDa and which can assemble as a tetrameric species mediated by structural elements in the amino terminus.[2,30] The hERG1b isoform is missing the first 373 amino acids of the amino terminus of the protein but incorporates a unique 36-residue sequence that is involved in mediating a cotranslational interaction with the hERG1a channel that promotes the preferential formation of a homotetrameric channel. However, the precise ratio of hERG1a and hERG1b proteins that are expressed in the human heart and, particularly, in each of the four chambers is unknown, although there do appear to be some differences in channel distribution.[30] Estimates suggest that the amount of hERG1b expressed ranges from 3% to 30% of the total channel protein and that as little as 3% of this isoform is sufficient to affect channel deactivation kinetics in cardiac tissue.[30] Moreover, the ratio of hERG isoforms and channel composition differs between individuals and varies during the stages of development. The homo- or heterotetrameric channel is trafficked to the Golgi apparatus where additional glycosylation occurs to afford the fully mature, 155 kDa form of the protein that is exported to the plasma membrane.[2,30] Glycosylation is a critical determinant of hERG channel trafficking and surface expression and although hERG contains two potential sites for *N*-glycosylation, Asn598 and Asn629, only Asn598 is required. Incompletely glycosylated hERG protein can be transported to the membrane but exhibits reduced stability, ultimately leading to lower cell surface expression of the protein, while mutation of Asn629 leads to the protein being retained in the ER.[2,31]

Correct folding of the *KCNH2* protein and trafficking to the cell surface are assisted by chaperone and co-chaperone proteins that reside in both the ER and the cytoplasm, a repertoire that includes the heat-shock proteins Hsp70 and Hsp90α, the Hsp-organizing protein (hop), the Golgi-associated protein GM130, the 38 kDa FK506 binding protein FKBP38, the small GTPases Sar1 and Rab11B, and the sigma 1 receptor.[2] The proteins that have been shown to interact with misfolded forms of hERG and facilitate its degradation include the glucose-regulated proteins Grp78 and Grp94, calnexin, calreticulin, the BCL-associated anthanogene Bag-2, the ubiquitin ligases Hsc70-interacting protein CHIP and Nedd4-2, HSP40, DnaJA-1, DnaJA-2, DnaJA-4, Vps24, and EDEM, while low K^+ concentrations also lead to channel degradation.[2] Missense mutations that cause retention of the hERG protein in the ER or Golgi apparatus and tagging for destruction include K28E, F29L, G53R, T65P, C66G, L86R, T421M, A422T, F463L, N470D, G572R, I593R, G601S, Y611H, E637K, R752W, F805C, and V822M.[2,29] Similarly, channels that have deletions, insertions, or frameshifts in the CO_2H terminus are also associated with a trafficking defective phenotype.[2] Because of the tetrameric nature of the hERG channel, a dominant-negative phenotype can also be operative in the channel complex of heterozygotes that can lead to the targeting of the wild-type channels for degradation by the mutant protein or produce a dysfunctional channel.[32–35] The A561V mutant provides an illustrative example of this phenomenon.[32]

The hERG channel is also subject to posttranslational modifications that can modulate both function and cell surface expression, the latter via effects on channel synthesis and trafficking or endocytosis.[2] hERG contains several consensus sequences predicted as potential sites for phosphorylation by cyclic AMP-dependent protein kinase (PKA), protein kinase C (PKC), and tyrosine kinases. PKA-mediated phosphorylation of hERG in the ER is associated with a two to fourfold enhancement of protein abundance measured over a 24 h period, an effect that is not due to increased transcription. The effects of PKA on the phosphorylation of hERG can be influenced in a positive or negative sense by accessory proteins that include the protein kinase A adaptor protein Yotiao and 14-3-3$_\Sigma$.[2] PKA-mediated phosphorylation also exerts direct effects on channel function, rapidly reducing current amplitude and accelerating voltage-dependent deactivation that is accompanied by a depolarizing shift in voltage-dependent activation.[2] cAMP can also bind directly to the hERG protein and exert an opposing effect that is muted when the channel is associated with the accessory proteins MiRP1 or minK.[2] hERG channel current in HEK293 cells is increased when co-expressed with protein kinase B (PKB) in its constitutively active form or with glucocorticoid-inducible kinase-3 (SGK3), both of which appear to activate the channel by indirect means rather than by directly phosphorylating the channel.[2] PKB and SGK3 are, in turn, activated by phosphoinositol- and phosphoinositide-dependent kinases, providing another level of complexity to hERG channel regulation.[2] Activation of

PKC leads to an acute diminution in hERG current in oocytes and isolated guinea pig ventricular myocytes, an effect mediated by a combination of gating changes and endocytosis of the channel protein.[2] However, despite the fact that there may be as many as 18 consensus PKC phosphorylation sites on the hERG protein, it is not clear if this kinase affects hERG function directly by phosphorylation or if the effects are of an indirect nature.[2]

Activation of Src tyrosine kinase leads to increased hERG current in rat microglial cells and these two proteins can be coimmunoprecipitated, suggesting that regulation may depend on complex formation.[2] Commensurate with this observation, inhibitors of Src kinases reduce hERG while inhibitors of the endothelial growth factor receptor kinase also exhibit a similar effect.[2] The tyrosine kinase inhibitor imatinib (4), which exhibits selectivity for the Abelson proto-oncogene (Abl), c-Kit, and platelet-derived growth factor receptor kinases, has been shown to downregulate hERG1 mRNA and protein levels in K562 and bone marrow mononuclear cells collected from chronic myelogenous leukemia (CML) patients.[36] Indeed, the cell proliferative effects of imatinib in K562 cells were augmented by the hERG1 channel blocker E4031 (5), which exhibited only moderate inhibition by itself, although the effect was muted in primary CML cells.

The effects of tyrosine kinase inhibitors on hERG function may also be due to reduced activation of phosphoinositide 3-kinase (PI3K).[37] Nilotinib (6), which shares inhibition of Bcr-Abl and c-Kit with imatinib but inhibits several additional kinases, reduced PI3K activity in isolated cardiac cells and increased the APD. A PI3K inhibitor also increased the APD, while adding the second messenger produced by PI3K action restored the APD to the normal range and isolated mouse hearts lacking a PI3K subunit exhibited a lengthened QT interval on the electrocardiogram. Nilotinib increased the QT interval in WT hearts but not in those deficient in the PI3K subunit; however, both PI3K and nilotinib affect multiple cardiac ion channels, including those conducting Ca^{2+} and Na^+, in addition to hERG, adding to the complexity of the interpretation.[37]

6

The hERG channel can be downregulated by endocytosis into vesicles where it is derivatized by ubiquitination which targets the protein for lysosomal degradation.[2] PKC activation and the sphingolipid ceramide have been shown to stimulate this process in HEK293 cells, while hypokalemia has emerged as a potentially physiologically relevant regulatory mechanism that stimulates this pathway.[2]

3. INHIBITORS OF hERG TRAFFICKING

3.1. Profile of hERG trafficking inhibitors

It has been demonstrated that the hERG channel relies on the cytosolic molecular chaperone heat-shock protein Hsp90 to develop into the functional fully glycosylated 155 kDa mature form that is capable of trafficking from the Golgi to the cell surface.[21] The non–isoform selective Hsp90 inhibitor geldanamycin (**7**) disrupted formation of the complex between the 133 kDa immature hERG protein and Hsp90, leading to increased ubiquitination of the channel protein. This resulted in reduced channel expression on the surface of HEK293 (hERG-HEK293) cells over a 24-h period, detected by a dramatic reduction in I_{Kr} current amplitude, an effect also observed in guinea pig cardiomyocytes after overnight application of geldanamycin at a concentration of 1.8 μM.[21] Of the two cytosolic Hsp90 isoforms, the hERG channel depends only on the α-isoform rather than the β-isoform for maturation and trafficking.[38] The poison arsenic trioxide (As$_2$O$_3$), which is used to treat relapsed or refractory acute promyelocytic leukemia, has been shown to inhibit hERG channel cell surface expression by interfering with the hERG–Hsp90 interaction.[39] After incubation in hERG-HEK293 cells for 24 h, As$_2$O$_3$ inhibited the I_{Kr} current with a clinically relevant IC$_{50}$ of 1.5 μM. However, the effects of As$_2$O$_3$ on cardiac ion channels are quite complex, since it has also been shown to induce acute inhibition of both hERG and minK channels while a 24-h exposure leads to activation of cardiac Kv1.5 and Ca^{2+} channels.[39,40]

Pentamidine (**8**) is an antiprotozoal agent used to treat *Pneumocystis carinii* pneumonia that is commonly administered by i.v. injection. This compound, which shows only a weak direct inhibition of hERG channel current in HEK293 cells, prolongs the APD in guinea pig ventricular myocytes at high (250 µM) concentrations.[41] However, after incubating hERG-HEK293 cells with pentamidine for 16–20 h, the I_{Kr} current was more potently inhibited, $IC_{50} = 5.1$ µM, and expression of mature channels on the cell surface was inhibited with an IC_{50} of 7.8 µM, as determined by Western blot image density.[42] Extending the period of incubation to 48 h led to a further increase in the inhibitory activity of pentamidine with a 1 µM concentration of the drug reducing both the I_{Kr} current and channel expression levels to about 35% of control. In *ex vivo* guinea pig ventricular myocytes, the I_{Kr} tail current was reduced by 35% and the APD_{90} was prolonged by ~135% after an overnight incubation with 10 µM pentamidine. The cholesterol-lowering drug probucol (**9**) demonstrates a hERG channel inhibition profile that is similar to pentamidine, exerting no acute effect on I_{Kr} current in neonatal rat cardiomyocytes even at concentrations as high as 100 µM. However, after 48 h of incubation with probucol, the hERG-mediated current was reduced with an IC_{50} of 20.6 µM and prolongation of the APD_{90} was observed in these myocytes after 24 h of exposure to the drug, producing an IC_{50} of 30 µM.[43] Both pentamidine and probucol have been known since the 1980s to frequently cause LQTS and TdP in the clinic, adverse cardiovascular events that may be due to the fact that the mean therapeutic free plasma concentrations of pentamidine (>0.5 µM) and probucol (35–75 µM) are sufficient to induce chronic inhibition of hERG channel cell surface expression during long-term therapy.[41,43] The *in vitro* incubation results also reflect the delayed onset of LQTS that is typically seen clinically with the effect on channel expression occurring beyond 12 days following the initiation of pentamidine therapy and a recovery that occurs over 7 days after withdrawal of the drug.[44]

7   8   9

The naturally occurring digitalis cardiac glycosides digoxin (**10**) and digitoxin (**11**), which are Na^+/K^+-ATPase inhibitors used to treat atrial fibrillation and congestive heart failure, are potent hERG trafficking inhibitors. Clinical administration of cardiac glycosides does lead to QT prolongation and TdP, although the incidence is rare. A set of 11 digitalis derivatives exhibited significant potency as inhibitors of hERG channel expression in HEK293 cells after an overnight incubation using an antibody-based chemiluminescent assay (hERG-Lite®, see Section 4).[45] Under these conditions, the IC_{50} values for digoxin, digitoxin, and ouabain (**12**) are 73, 14, and 9.8 nM, respectively, although they were somewhat less active in guinea pig ventricular myocytes. The triterpene natural product celastrol (**13**) is an anti-inflammatory and antioxidant agent that significantly reduced cell surface expression of a transfected hERG channel in hERG-HEK293 cells and the native channel in human neuroblastoma SH-SY5Y cells at concentrations of 200–400 nM following an overnight incubation.[46]

10

11

12

13

Results based on an analysis of the relative cell surface expression level of hERG channel protein in HEK293 cells measured using the hERG-Lite® antibody-based chemiluminescent assay have revealed that hERG trafficking inhibitors are more common than anticipated and that 20 of the 50 known direct hERG channel blockers also exhibit indirect channel inhibition.[47] These dual-mode inhibitors of the hERG channel include terfenadine, diphenhydramine (**14**), desipramine (**15**), tamoxifen (**16**), chlorpromazine (**17**), amiodarone (**18**), lovastatin (**19**), and the commonly prescribed azole antifungal ketoconazole (**20**) as well as the antidepressant fluoxetine (**21**), a selective serotonin reuptake inhibitor.

In a patch–clamp experiment using hERG-HEK293 cells, ketoconazole exhibited direct acute inhibition of the I_{Kr} current with an IC_{50} of 1.92 μM via binding to the Tyr652/Phe656 binding site. However, independent of this direct blockade, hERG channel cell surface expression was reduced by over 50% following incubation with 30 μM ketoconazole, an effect that persisted for up to 48 h after drug washout by which time ~74% of channel expression was restored.[48] Fluoxetine is another direct-acting hERG channel blocker, $IC_{50} = 0.7$ μM, that also suppresses cell surface channel expression with an IC_{50} of 2.7 μM (24-h incubation). In these assays, the fluoxetine metabolite norfluoxetine (**22**) was about two- to threefold less active.[49] Similar to the observations with ketoconazole, the inhibition of cell surface channel expression and the recovery period following exposure to 30 μM fluoxetine took place over hours, with 100% inhibition and restoration of channel function achieved by ~24 h. A similar activity profile of direct channel blockade and indirect inhibition after overnight drug treatment was also displayed by the tricyclic antidepressant amoxapine (**23**) which exhibited IC_{50}'s of 5.1 and 15.3 µM, respectively, in these assays.[50] Another tricyclic antidepressant desipramine also inhibited the I_{Kr} current in hERG-HEK293 cells with an IC_{50} of 11.9 µM and cell surface expression in a chemiluminescent assay with an IC_{50} of 17.3 µM upon 24-h incubation.[51] The longer term inhibition of hERG by desipramine was also manifested in a reduction of native I_{Kr} current and cell surface levels of the mature 155 kDa hERG channel in murine HL-1 cardiomyocytes.[51] More interestingly, desipramine and amoxapine showed a significant (40%) reduction in hERG channel cell surface density within 1–2 h of exposure, an effect that is more pronounced than that observed, for example, with pentamidine. QT prolongation related to ketoconazole use is rare while the clinical incidence with fluoxetine or amoxapine therapy is sporadic and mostly associated with drug overdose. The maximum unbound plasma concentration of each of these drugs obtained at therapeutic doses is <60 nM, which is over 40-fold lower than their *in vitro* IC_{50} values for both the direct and indirect effects on hERG channel function, providing a safety window that would appear to be sufficient to avoid both the acute and chronic cardiac effects.[48–50,52]

3.2. Mechanisms of hERG trafficking inhibition

The mode of action of drugs that share the phenotype of reducing the cell surface expression of the hERG channel has begun to be understood only recently. Early studies in this direction revealed a complex and diverse underlying biochemical pharmacology that included interference with hERG channel protein synthesis and processing, ultimately affecting trafficking or promoting internalization and degradation of the membrane-bound channel. Other than inhibition of the association of the hERG protein and the chaperone Hsp90, as described for geldanamycin, it appears that multiple mechanisms can be expressed by trafficking inhibitors and that different mechanisms may be utilized by different inhibitors. Since the density of the core-glycosylated, 135 kDa nascent hERG channel protein synthesized in the ER was not affected by these inhibitors, it has been suggested that inhibition is unlikely to be occurring at the level of protein synthesis. Experiments with desipramine, ketoconazole, fluoxetine, and amoxapine using Tyr652/Phe656 mutant hERG channels also indicated that binding to the Tyr652/Phe656 binding site in the S6 pore domain occupied by direct blockers is not involved in their inhibitory effects on hERG trafficking.[48–51] Mechanistic studies using hERG–HEK293 cells exposed to 30 μM pentamidine showed that this compound inhibits hERG forward trafficking from the ER, resulting in the retention of the 135 kDa form of the hERG channel protein in the ER.[53] The inhibition of hERG trafficking mediated by pentamidine could be rescued by the antihistamine astemizole (24), a potent direct hERG channel blocker that has been shown to mitigate conformational defects in misfolded, trafficking-deficient inherited LQTS2 hERG mutants.[54] Pentamidine did not disrupt the interaction between the hERG channel protein and the chaperones Hsp/c70 and Hsp90, or enhance ubiquitination of the channel. More intriguingly, mutational analysis showed that trafficking inhibition by pentamidine was eliminated by a mutation at the pore Phe656 residue. Taken together, these observations led to the proposal that pentamidine stabilized an immature, trafficking-incompatible conformation of the hERG channel protein via binding to the transmembrane S6 pore domain.[53]

In contrast, probucol appears to act by a different mechanism. Results from incubation with 100 μM of probucol indicated that it increased channel degradation via increasing turnover of hERG-associated caveolin-1 which is expressed in HEK293 cells and is a cholesterol-binding integral membrane protein that is incorporated into membrane lipid rafts.[55] This effect presumably resulted from the cholesterol-lowering action of probucol as low density lipoprotein prevented the probucol-induced reduction in the I_{Kr} current in hERG–HEK293 cells as well as in the native I_{Kr} current in cultured neonatal rat cardiomyocytes. In cardiomyocytes, the caveolin-3 isoform rather than caveolin-1 is expressed. Indeed, it was shown by using hERG–HEK293 cells overexpressing caveolin-3 and studies in rat or rabbit ventricular myocytes that the endocytotic degradation of the hERG channel protein is regulated by caveolin-3 through the ubiquitin-ligase Nedd4-2 that targets the hERG channel for ubiquitination, most likely via the formation of a complex between the three proteins.[56]

24   25

Based on studies with progesterone (**25**), it was further proposed that interference of intracellular cholesterol homeostasis may lead to a reduction in hERG channel cell surface expression. In hERG-HEK293 cells, 5 µM of progesterone, a concentration that is relevant to those experiencing pregnancy, reduced the level of the mature 155 kDa hERG channel at the cell surface and led to retention of the immature 135 kDa protein in the ER in a time-dependent manner, exhibiting the typical delayed onset and recovery phenomenon.[57] However, the decrease in hERG cell surface expression could be rescued by the application of 5 µM of the potent, direct-acting hERG channel inhibitor E4031, a class-III antiarrhythmic agent. The mature channel levels and I_{Kr} current in rat neonatal cardiac myocytes were also significantly reduced by 5 µM of progesterone after a 24-h incubation period, inhibitory effects that were linked to the accumulation of free cholesterol in the cytosol by progesterone. The lowering of cholesterol levels by 10 µM simvastatin (**26**) or increasing the levels by addition of 20 µg/ml of exogenous cholesterol also induced the inhibition of hERG channel cell surface expression, further suggesting the importance of maintaining careful control of cholesterol homeostasis.[57]

26   **27**

A combination of two mechanisms for the indirect inhibition of the hERG channel was determined for desipramine.[58] It was demonstrated that at a concentration of 30 µM, desipramine inhibited hERG forward trafficking from the ER in a fashion that did not involve incorrectly folded hERG protein or the interaction with Hsp chaperones. Interestingly, desipramine also increased channel endocytosis from the membrane by inducing a rapid, time-dependent increase in ubiquitination, as shown by immunoblotting experiments using hemagglutinin (HA)-tagged ubiquitin, that led to subsequent lysosomal degradation of the hERG protein in hERG-HEK293 cells.[58] These findings are consistent with the effect of desipramine on reducing hERG cell surface expression occurring within 1–2 h, which is less than the 8–11-h half-life of the hERG proteins in HEK293 cells. The precise molecular determinants of these two pathways utilized by desipramine are not well understood; however, it is most likely not related to binding of the compound to the hERG channel pore domain based on mutational analysis. Not surprisingly, amoxapine shows a mode of action similar to desipramine while amitriptyline (**27**), another tricyclic antidepressant that also reduces hERG cell surface expression, was shown to increase ubiquitination of the hERG channel protein.[58]

4. SCREENING ASSAYS FOR hERG TRAFFICKING INHIBITION

To date, only non–electrophysiological assay methods have been reported for screening for compounds that inhibit hERG channel trafficking. A 96-well chemiluminescent assay, designated HERG-Lite®, utilizes hERG-HEK293 cells tagged with a HA epitope at the extracellular loop connecting the transmembrane voltage-sensing domains S1 and S2.[47] The electrophysiological properties and trafficking behavior of the HA-hERG channel were shown to be similar to that of the wild-type channel. After a 16-h overnight incubation with test compounds, the primary antibody rat anti-HA and a secondary antibody cocktail are employed to measure the relative cell surface expression levels of the hERG channel protein compared to a drug-free control. As noted above, this assay was used to determine the inhibitory activity of several cardiac glycosides and desipramine on hERG channel expression and is of utility for the broader screening of hERG trafficking inhibitors. A related ELISA-based method, designated as CHAN-Lite®, has also been developed.[59]

An interesting hERG trafficking assay based on a microfluidics array platform that employs polymer microchannels has been developed to study the inhibition of hERG channel protein expression on the cell surface.[60] hERG-HEK293 cells were found to readily proliferate in the microchannel environment and stably express a transfected hERG channel protein that maintains normal electrophysiological properties. After a 22-h incubation with test compound, anti-hERG antibody followed by IR dye-conjugated donkey anti-rabbit IgG antibody is utilized to quantify the levels of cell surface expression of the hERG channel by live-cell Western analysis using an infrared scanning technique.[60]

More recently, a high-throughput thallium flux assay that uses hERG-HEK293 cells has been developed by modification of the FluxOR potassium ion channel assay platform that is amenable to a 384-well format.[61] hERG trafficking inhibitors can be distinguished from acute inhibitors by extending the drug incubation time out to 16 h. In this assay, the cardiac glycoside ouabain showed a time–dependent increase in inhibition of hERG channel expression, with the maximal effect reached by 90 min with an IC_{50} of 360 nM. However, the inhibitory potency of ouabain was enhanced with a prolonged incubation time with an IC_{50} of 18 nM, similar to that obtained from the HERG-Lite® assay, measured after 16 h of exposure.[45] Pentamidine also exhibited hERG inhibition after overnight incubation in this assay with an IC_{50} of 11.3 µM, which is consistent with the previously reported values for the effect of this compound on channel trafficking.[42] The observation that maximal inhibition of hERG by ouabain was achieved in less than the 11-h half-life of the hERG protein in HEK293 cells may suggest that promotion of hERG channel internalization and degradation underlie the effects of ouabain, as determined for desipramine.

5. CONCLUSION

Inhibition of hERG channel expression on the cell surface represents an alternative, indirect inhibitory mechanism for drug-induced LQTS that, to date, has been under-appreciated when compared to the more conventional direct channel blockade that has been the focus of safety pharmacology testing and medicinal chemistry studies seeking to design away from this side effect.[12,13,19,62–65] The trafficking inhibitors reported in the literature to date are structurally diverse and mitigation of the inhibition of cell surface hERG channel expression may be more challenging since multiple molecular mechanisms appear to operate. Unlike the direct hERG channel blockers that share a common binding site in the S6 domain, trafficking inhibitors can exert their inhibitory activity by either binding to the hERG channel (e.g., pentamidine) or other proteins in the hERG channel anteretrograde and retrograde trafficking pathways (e.g., geldanamycin). Although the IC_{50}s exhibited by hERG channel trafficking inhibitors are often in the micromolar range, the chronic nature of the inhibition may increase the cardiac safety risk associated with the use of such inhibitors and may render them difficult to detect in safety monitoring as currently configured. Indeed, this alternative mechanism for drug-induced LQTS is not effectively captured by the assays currently employed for compound profiling during preclinical development or in a secondary cardiovascular *in vivo* safety screen. Recently developed screening assays demonstrate the capacity to assess the effect of test molecules on hERG trafficking and cell surface expression and safety screening for the potential for cardiotoxicity that would result from this type of chronic inhibitory pathway should be afforded a higher level of consideration.

REFERENCES

1. Schoonmaker, F. W.; Osteen, R. T.; Greenfield, J. C., Jr. *Ann. Intern. Med.* **1966**, *65*, 1076.
2. Vandenberg, J. I.; Perry, M. D.; Perrin, M. J.; Mann, S. A.; Ke, Y.; Hill, A. P. *Physiol. Rev.* **2012**, *92*, 1393.
3. Monahan, B. P.; Ferguson, C. L.; Killeavy, E. S.; Lloyd, B. K.; Troy, J.; Cantilena, L. R., Jr. *J. Am. Med. Assoc.* **1990**, *264*, 2788.
4. Roden, D. M. *N. Engl. J. Med.* **2004**, *350*, 1013.
5. A compilation of drugs with the potential to induce QT prolongation can be found at www.torsades.org.
6. Sanguinetti, M. C.; Jurkiewicz, N. K. *J. Gen. Physiol.* **1990**, *96*, 195.
7. Warmke, J. W.; Ganetzky, B. *Proc. Natl. Acad. Sci. U. S. A.* **1994**, *91*, 3438.
8. Curran, M. E.; Splawski, I.; Timothy, K. W.; Vincent, G. M.; Green, E. D.; Keating, M. T. *Cell* **1995**, *80*, 795.
9. Sanguinetti, M. C.; Jiang, C.; Curran, M. E.; Keating, M. T. *Cell* **1995**, *81*, 299.
10. Pugsley, M. K.; Authier, S.; Curtis, M. J. *Br. J. Pharmacol.* **2008**, *154*, 1382.
11. FDA guidance. *S7B Nonclinical Evaluation of the Potential for Delayed Ventricular Repolarization (QT Interval Prolongation) by Human Pharmaceuticals* can be found at http://www.fda.gov/downloads/Drugs/GuidanceComplianceRegulatoryInformation/Guidances/ucm074963.pdf, October, 2005.
12. Fermini, B.; Fossa, A. A. *Nat. Rev. Drug Discov.* **2003**, *2*, 439.
13. Fermini, B.; Fossa, A. A. *Annu. Rep. Med. Chem.* **2004**, *39*, 323.
14. Bowlby, M. R.; Peri, R.; Zhang, H.; Dunlop, J. *Curr. Drug Metab.* **2008**, *9*, 965.
15. Diaz, G. J.; Daniell, K.; Leitza, S. T.; Martin, R. L.; Su, Z.; McDermott, J. S.; Cox, B. F.; Gintant, G. A. *J. Pharmacol. Toxicol. Methods* **2004**, *50*, 187.
16. Murphy, S. M.; Palmer, M.; Fontilla Poole, M.; Padegimas, L.; Hunady, K.; Danzig, J.; Gill, S.; Gill, R.; Ting, A.; Sherf, B.; Brunden, K.; Stricker-Krongrad, A. *J. Pharmacol. Toxicol. Methods* **2006**, *54*, 42.
17. Deacon, M.; Singleton, D.; Szalkai, N.; Pasieczny, R.; Peacock, C.; Price, D.; Boyd, J.; Boyd, H.; Steidl-Nichols, J. V.; Williams, C. *J. Pharmacol. Toxicol. Methods* **2007**, *55*, 255.
18. Pollard, C. E.; Abi Gerges, N.; Bridgland-Taylor, M. H.; Easter, A.; Hammond, T. G.; Valentin, J.-P. *Br. J. Pharmacol.* **2010**, *159*, 12.
19. Jamieson, C.; Moir, E. M.; Rankovic, Z.; Wishart, G. *J. Med. Chem.* **2006**, *49*, 5029.
20. Eckhardt, L. L.; Rajamani, S.; January, C. T. *Br. J. Pharmacol.* **2005**, *145*, 3.
21. Dennis, A.; Wang, L.; Wan, X.; Ficker, E. *Biochem. Soc. Trans.* **2007**, *35*, 1060.
22. Yeung, K.-S.; Meanwell, N. A. *ChemMedChem* **2008**, *3*, 1501.
23. van der Heyden, M. A. G.; Smits, M. E.; Vos, M. A. *Br. J. Pharmacol.* **2008**, *153*, 406.
24. Staudacher, I.; Schweizer, P. A.; Katus, H. A.; Thomas, D. *Curr. Opin. Drug Discov. Devel.* **2010**, *13*, 23.
25. Guth, B. D.; Rast, G. *Br. J. Pharmacol.* **2010**, *159*, 22.
26. Valentin, J.-P.; Pollard, C.; Lainée, P.; Hammond, T. *Br. J. Pharmacol.* **2010**, *159*, 25.
27. Salvi, V.; Karnad, D. R.; Panicker, G. K.; Kothari, S. *Br. J. Pharmacol.* **2010**, *159*, 34.
28. Hoffmann, P.; Warner, B. *J. Pharmacol. Toxicol. Methods* **2006**, *53*, 87.
29. Hedley, P. L.; Jorgensen, P.; Schlamowitz, S.; Wangari, R.; Moolman-Smook, J.; Brink, P. A.; Kanters, J. K.; Corfield, V. A.; Christiansen, M. *Hum. Mutat.* **2009**, *30*, 1486.
30. Jonsson, M. K. B.; van der Heyden, M. A. G.; van Veen, T. A. B. *J. Mol. Cell. Cardiol.* **2012**, *53*, 369.
31. Gong, Q.; Anderson, C. L.; January, C. T.; Zhou, Z. *Am. J. Physiol. Heart Circ. Physiol.* **2002**, *283*, H77.

32. Ficker, E.; Dennis, A. T.; Obejero-Paz, C. A.; Castaldo, P.; Taglialatela, M.; Brown, A. M. *J. Mol. Cell. Cardiol.* **2000**, *32*, 2327.
33. Huo, J.; Zhang, Y.; Huang, N.; Liu, P.; Huang, C.; Guo, X.; Jiang, W.; Zhou, N.; Grace, A.; Huang, C. L. H.; Ma, A. *Pflugers Arch. – Eur. J. Physiol.* **2008**, *456*, 917.
34. Zhao, J. T.; Hill, A. P.; Varghese, A.; Cooper, A. A.; Swan, H.; Laitinen-Forsblom, P. J.; Rees, M. I.; Skinner, J. R.; Campbell, T. J.; Vandenberg, J. I. *J. Cardiovasc. Electrophysiol.* **2009**, *20*, 923.
35. Zhang, Y.; Wang, H.; Wang, J.; Han, H.; Nattel, S.; Wang, Z. *FEBS Lett.* **2003**, *534*, 125.
36. Zheng, F.; Li, H.; Liang, K.; Du, Y.; Guo, D.; Huang, S. *Mol. Oncol.* **2012**, *29*, 2127.
37. Lu, Z.; Wu, C.-Y.; Jiang, Y.-P.; Ballou, L. M.; Clausen, C.; Cohen, I. S.; Lin, R. Z. *Sci. Transl. Med.* **2012**, *4*, 131ra50.
38. Peterson, L. B.; Eskew, J. D.; Vilhauer, G. A.; Blagg, B. S. J. *Mol. Pharm.* **2012**, *9*, 1841.
39. Ficker, E.; Kuryshev, Y. A.; Dennis, A. T.; Obejero-Paz, C.; Wang, L.; Hawryluk, P.; Wible, B. A.; Brown, A. M. *Mol. Pharmacol.* **2004**, *66*, 33.
40. Drolet, B.; Simard, C.; Roden, D. M. *Circulation* **2004**, *109*, 26.
41. Cordes, J. S.; Sun, Z.; Lloyd, D. B.; Bradley, J. A.; Opsahl, A. C.; Tengowski, M. W.; Chen, X.; Zhou, J. *Br. J. Pharmacol.* **2005**, *145*, 15.
42. Kuryshev, Y. A.; Ficker, E.; Wang, L.; Hawryluk, P.; Dennis, A. T.; Wible, B. A.; Brown, M. A.; Kang, J.; Chen, X.-L.; Sawamura, K.; Reynolds, W.; Rampe, D. *J. Pharmacol. Exp. Ther.* **2005**, *312*, 316.
43. Guo, J.; Massaeli, H.; Li, W.; Xu, J.; Luo, T.; Shaw, J.; Kirshenbaum, L. A.; Zhang, S. *J. Pharmacol. Exp. Ther.* **2007**, *321*, 911.
44. Quadrel, M. A.; Atkin, S. H.; Jaker, M. A. *Am. Heart J.* **1992**, *123*, 1377.
45. Wang, L.; Wible, B. A.; Wan, X.; Ficker, E. *J. Pharmacol. Exp. Ther.* **2007**, *320*, 525.
46. Sun, H.; Liu, X.; Xiong, Q.; Shikano, S.; Li, M. *J. Biol. Chem.* **2006**, *281*, 5877.
47. Wible, B. A.; Hawryluk, P.; Ficker, E.; Kuryshev, Y. A.; Kirsch, G.; Brown, A. M. *J. Pharmacol. Toxicol. Methods* **2005**, *52*, 136.
48. Takemasa, H.; Nagatomo, T.; Abe, H.; Kawakami, K.; Igarashi, T.; Tsurugi, T.; Kabashima, N.; Tamura, M.; Okazaki, M.; Delisle, B. P.; January, C. T.; Qtsuji, Y. *Br. J. Pharmacol.* **2008**, *153*, 439.
49. Rajamani, S.; Eckhardt, L. L.; Valdivia, C. R.; Klemens, C. A.; Gillman, B. M.; Anderson, C. L.; Holzem, K. M.; Delisle, B. P.; Anson, B. D.; Makielski, J. C.; January, C. T. *Br. J. Pharmacol.* **2006**, *149*, 481.
50. Obers, S.; Staudacher, I.; Ficker, E.; Dennis, A.; Koschny, R.; Erdal, H.; Bloehs, R.; Kisselbach, J.; Karle, C. A.; Schweizer, P. A.; Katus, H. A.; Thomas, D. *Naunyn-Schmied Arch. Pharmacol.* **2010**, *381*, 385.
51. Staudacher, I.; Wang, L.; Wan, X.; Obers, S.; Wenzel, W.; Tristram, F.; Koschny, R.; Staudacher, K.; Kisselbach, J.; Koelsch, P.; Schweizer, P. A.; Katus, H. A.; Ficker, E.; Thomas, D. *Naunyn-Schmied Arch. Pharmacol.* **2011**, *383*, 119.
52. Redfern, S. W.; Carlsson, L.; Davis, A. S.; Lynch, W. G.; MacKenzie, I.; Palethorpe, S.; Sieglf, P. K. S.; Stranga, I.; Sullivang, A. T.; Wallish, R.; Cammi, A. J.; Hammond, T. G. *Cardiovasc. Res.* **2003**, *58*, 32.
53. Dennis, A. T.; Wang, L.; Wan, H.; Nassal, D.; Deschenes, I.; Ficker, E. *Mol. Pharmacol.* **2012**, *81*, 198.
54. Ficker, E.; Obejero-Paz, C. A.; Zhao, S.; Brown, A. M. *J. Biol. Chem.* **2002**, *277*, 4989.
55. Guo, J.; Li, X.; Shallow, H.; Xu, J.; Yang, T.; Massaeli, H.; Li, W.; Sun, T.; Pierce, G. N.; Zhang, S. *Mol. Pharmacol.* **2011**, *79*, 806.
56. Guo, J.; Wang, T.; Li, X.; Shallow, H.; Yang, T.; Li, W.; Xu, J.; Fridman, M. D.; Yang, X.; Zhang, S. *J. Biol. Chem.* **2012**, *287*, 33132.
57. Wu, Z.-Y.; Yu, D.-J.; Soong, T. W.; Dawe, G. S.; Bian, J.-S. *J. Biol. Chem.* **2011**, *286*, 22186.

58. Dennis, A. T.; Nassal, D.; Deschenes, I.; Thomas, D.; Ficker, E. *J. Biol. Chem.* **2011**, *286*, 34413.
59. ChanTest web-site http://www.chantest.com accessed on Jan 9, 2013.
60. Su, X.; Young, E. W. K.; Underkofler, H. A. S.; Kamp, T. J.; January, C. T.; Beebe, D. J. *J. Biomol. Screen.* **2011**, *16*, 101.
61. Huang, X.-P.; Mangano, T.; Hufeisen, S.; Setola, V.; Roth, B. L. *Assay Drug Dev. Technol.* **2010**, *8*, 727.
62. Vaz, R. J.; Li, Y.; Rampe, D. *Prog. Med. Chem.* **2005**, *43*, 1.
63. Diller, D. J. *Curr. Comput. Aided Drug Des.* **2009**, *5*, 106.
64. Lagrutta, A. A.; Trepakova, E. S.; Salata, J. J. *Curr. Top. Med. Chem.* **2008**, *8*, 1102.
65. Bell, I. M.; Bilodeau, M. T. *Curr. Top. Med. Chem.* **2008**, *8*, 1128.

> CHAPTER TWENTY-TWO

Recent Progress in Small-Molecule Agents Against Age-Related Macular Degeneration

Muneto Mogi, Christopher M. Adams, Nan Ji, Nello Mainolfi

Global Discovery Chemistry, Novartis Institutes for Biomedical Research, Cambridge, Massachusetts, USA

Contents

> ## 1. INTRODUCTION

Age-related macular degeneration (AMD) is the leading cause of vision loss and blindness in industrialized countries.[1] Drusen and hyper- or hypopigmentation of the retinal pigment epithelium (RPE) are the characteristics of early AMD (also known as dry AMD).[2] This asymptomatic abnormality of the RPE advances to two forms of AMD with a permanent loss of vision: geographic atrophy (GA) and/or choroidal neovascularization

Annual Reports in Medicinal Chemistry, Volume 48
ISSN 0065-7743
http://dx.doi.org/10.1016/B978-0-12-417150-3.00022-3
353

(CNV or also known as wet AMD). Although significant progress has been made in recent years for the treatment of wet AMD using intravitreal (IVT) injections of anti–VEGF biologics such as pegaptanib, ranibizumab, and aflibercept, there is still a need for less invasive treatment options. Furthermore, unlike for wet AMD, there is currently no treatment available for dry AMD or GA. This review will cover three emerging therapeutic target classes of low molecular weight agents for AMD: anti-angiogenesis agents, visual cycle inhibitors, and complement pathway inhibitors. Other modalities such as antibody, siRNA, stem cell, or gene therapies will be out of scope of this review.

2. ANTI-ANGIOGENESIS AGENTS

The hallmark of wet AMD is pathological neovascularization, and anti-angiogenics have long been postulated as a potential therapy.[3] The clinical success of the anti–VEGF biologics such as pegaptanib, ranibizumab, and aflibercept have validated this concept, and prompted research into small molecule anti-angiogenic agents. A key challenge associated with developing anti-angiogenics for clinical use is overcoming the risk of exaggerated pharmacology or "on-target" toxicity such as hypertension, hemorrhage, and thromboembolism.[4] Although topical ocular administration of drugs has proven to be a successful strategy to mitigate systemic side effects for ocular diseases associated with the front of the eye (e.g., glaucoma therapies), there are currently no Food and Drug Administration (FDA) approved topical therapies for AMD. This is due in large part to anatomical and physiological barriers that have evolved to protect the back of the eye from exogenous agents.[5]

2.1. VEGF receptor 2 inhibitors

1, Vatalanib   **2**, Sorafenib   **3**, Pazopanib   **4**, GW771806

5, TG100801   6   7, Linifanib

Inhibitors of VEGF receptor 2 have been the most thoroughly explored class of anti-angiogenic agents for wet AMD. Vatalanib (**1**) was one of the earliest VEGFR2 inhibitors used for wet AMD.[6,7] When vatalanib was dosed at 50 mg/kg once a day for 2 weeks in a mouse CNV model, there was a significant (>70%) reduction in lesion area.[7] Subsequently, it was reported that oral vatalanib was being evaluated for safety and efficacy in patients with subfoveal CNV at doses of 500 and 100 mg/day.[8] Supporting this approach are case reports employing oral dosing of sorafenib (**2**) which hinted at the possibility of RTK inhibitors as a useful therapy.[9] Pazopanib (**3**) has also been explored as both an oral and topical therapy.[10] Initial investigations of topically dosed pazopanib indicated reduced lesion size in laser mediated CNV rodent models.[11] However, neither topical nor oral dosing of pazopanib afforded efficacy in a rabbit model of retinal neovascularization.[11] Further confusing the issue, a more recent report stated that topical pazopanib was ineffective in a rat laser mediated CNV model, but efficacious when dosed orally.[12] The report also explored the ocular pharmacokinetics of pazopanib and the impact of ocular melanin on efficacy. Pazopanib binds melanin leading to a depot of the drug in ocular tissues. Pazopanib was retained in the eye for greater than 35 days after a single oral dose in pigmented animals, while it was cleared after 3 days in non-pigmented animals. Binding to melanin was hypothesized to underlie the efficacy of pazopanib as a single oral dose in a laser CNV pigmented rat model. Despite the conflicting preclinical results, pazobanib was evaluated both orally and topically in patients with neovascular AMD.[13,14] Initially, 15 patients were given a relatively low oral dose of 15 mg/day for 28 days. Although 6 out of 15 patients required anti-VEGF rescue therapy, the remaining nine patients showed a statistically significant increase in best corrected visual acuity (BCVA) (8.0 ETDRS letters) and a decrease in central retinal thickness (CRT) ($-50.3\ \mu m$) on Day 29.[13] Topical studies involved 19 patients and employed pazopanib drops 4 times per day (QID) for over 12 weeks.[14] However, there was no significant change from baseline in CNV, CRT, nor BCVA in these patients. Promisingly, a recently disclosed analog of pazopanib, GW771806 (4) has exhibited similar melanin binding properties and ocular half-life as pazopanib.[12,15] However, 4 achieved 2.7-fold higher exposure in the RPE/choroid in pigmented rats after topical dosing. It is speculated that the greater exposure relative to pazopanib may be due to the enhanced solubility of 4. GW771806 was also efficacious with 30 μL drops at 5 mg/mL, two times per day (BID), affording regression of CNV lesions.[12]

Unlike the aforementioned VEGFR2 inhibitors which were originally developed for oncology indications, TG100801 (5) was designed with the specific aim of a topical therapy for wet AMD.[16] The development of 5 initially focused on identifying compounds having inhibitory activity against VEGFR, Src, and YES kinases. Src and YES have both been implicated as relevant targets in angiogenesis.[16] Ultimately, this activity led to identification of phenol 6, which displayed IC_{50} values of 1.8, 2.2, and 0.3 nM against Src, VEGFR2, and YES kinases, respectively. Phenol 6 showed exposure in the retina and choroid/sclera in mice after topical dosing with C_{Max} values of 3.6 and 23 μM, respectively with a 10 μL drop of a 0.7% (w/v) solution. Phenol 6 also exhibited limited exposure in plasma, which potentially could mitigate the risk of systemic toxicity. However, the half-life in ocular tissues was relatively short ($T_{1/2} = 0.5$ h) and frequent administration resulted in ocular irritation. To address the short ocular half-life and to increase ocular exposure, several ester prodrugs were examined with the aim of slow hydrolysis to afford the active drug. Ultimately, TG100801 was identified, which furnished an 11-fold increase in retinal AUC_{0-24h} of 6 compared to dosing 6 directly in mice. TG100801 was also evaluated in the mouse CNV model; after 14 days of three times per day (TID) treatment, there was a reduction in CNV size relative to vehicle control of $39 \pm 8\%$ or $26 \pm 10\%$ with 10 μL drops of 0.61% or 1.83% (w/v) TG100801, respectively. TG100801 was reported to have entered the clinic for a pilot study in patients with AMD,[17] however there has been no report of the outcome to date.

VEGFR inhibitors have also been administered by IVT injections. Linifanib[18] (7) was shown to be efficacious in a mouse CNV model after IVT injection.[19] It was reported that IVT linifanib was evaluated in both "prevention" and "regression" modes. In the prevention paradigm, dosing was initiated the same day the lesions were induced and continued for 14 days. 5 μL IVT doses of a 0.3 wt%, 1 wt%, or 3 wt% formulation were employed. In the regression mode, dosing was initiated 7 days after the initial laser induced lesion employing 5 μL IVT doses of 1% and 3% formulation. The animals treated in the prevention mode exhibited 84.1% and 83.0% reduction in CNV lesion size relative to untreated controls at the 1% and 3% doses, respectively. In the regression mode 1% and 3% formulations afforded a 45.4% and 41.0% regression in CNV compared to controls, respectively.

2.2. mTOR/TORC inhibitors[20]

Inhibitors of the mammalian target of rapamycin (mTOR) have also been evaluated as potential anti-angiogenic therapies for wet AMD. Systemic administration of rapamycin (**8**), which inhibits mTOR through binding to FKB12, was shown to be efficacious in both a mouse CNV model as well as a mouse oxygen induced retinopathy (OIR) model.[21] Mice receiving intraperitoneal injections of rapamycin (2 or 4 mg/kg/day) for 1–2 weeks showed a statistically significant decrease in CNV area of 29.8% and 40% relative to untreated controls for the low and high dose groups, respectively. In the OIR model, there was a decrease in neovascular tufts into the vitreous of 72.4% and 81.1% for the low and high dose groups, respectively. Rapamycin, along with the immunosuppressive agents daclizumab and infliximab, was evaluated in a small number of wet AMD patients.[22] Thirteen patients were randomized between the three drugs with three patients receiving rapamycin orally, 2 mg every other day for 6 months. The primary goal was to determine if any of the therapies would reduce the number of anti–VEGF injections needed during the 6-month period of the study. At the completion of the study, patients receiving the rapamycin treatment required anti-VEGF IVT injections at rates of 0.34 per month, while the rates for daclizumab and infliximab were 0.42 and 0.83 per month, respectively. The control group had an injection rate of 0.83 per month. Although none of the three drugs achieved statistically significant reduction in injection rates, a trend was seen with both daclizumab and rapamycin suggesting that the immunosuppressive effects of rapamycin may also play a role in AMD progression. More recently, it was reported that local delivery of rapamycin was being evaluated in the clinic in combination with ranibizumab. In this study, rapamycin was injected subconjuctively at doses of 440 and 1320 μg.[23] There have been no reports to date on the outcome of this study.

Everolimus (**9**), an analog of rapamycin with improved physicochemical properties suitable as oral agent, was evaluated in wet AMD patients.[24] In this study, everolimus was provided daily (5 mg dose) as a monotherapy to six patients. A separate six patient cohort received a combination of everolimus and a single IVT injection of ranibizumab. One patient received ranibizumab alone. Patients were evaluated after 4 weeks, and the patient receiving ranibizumab had an increase in BCVA of 2.0 letters, patients on combination therapy exhibited an increase of 4.4 ± 5.3 letters and achieved a 41 ± 56.2 µm decrease in CRT and patients receiving everolimus alone exhibited a decrease of 3 ± 7.59 letters with an increase in CRT of 37.3 ± 67.3 µm.[25]

An alternative approach to inhibiting the function of mTOR is to cause dissociation of the transducer of regulated CREB activity (TORC1 and TORC2) from the mTOR complex.[26] It has recently been reported that Palomid-529 (**10**) is an allosteric inhibitor of the PI3K/Akt/mTOR pathway functioning by causing the dissociation of TORC1 and TORC2.[27] It has been stated that Palomid-529 also inhibits retinal and subretinal vascularization in a rodent OIR model and retinal fibrosis in a rabbit model of retinal detachment after intraocular injection.[27] Interestingly, it is reported that this small molecule can persist in a rabbit eye for as long as 6 months at levels expected to show pharmacological activity after IVT injection. These results have led to the initiation of a clinical trial exploring both IVT and subconjuctival dosing of patients with advanced neovascular AMD.[28]

8, Rapamycin (R=H)
9, Everolimus (R=-CH₂CH₂OH)

10, Palomid-529

11, JSM6427 scaffold

12, Squalamine

13, Cortistatin A

14, Cortistatin analog

2.3. Other anti-angiogenic agents

Although to date receptor tyrosine kinase and mTOR/TORC inhibitors have been the primary focus of preclinical and clinical efforts, other anti-angiogenic compounds have been explored with diverse or unknown mechanisms of action. Most notable are antagonists of integrin $\alpha5\beta1$, such as JSM6427 (**11**),[29] and the natural products squalamine (**12**),[30] and cortistatin A (**13**).[31] Cortistatin A and squalamine do not yet have a proven target linked to their observed anti-angiogenic properties.[32,33] JSM6427, an antagonist of integrin $\alpha5\beta1$, exhibited efficacy in rabbit and cynomolgus monkey models of CNV when administered as weekly IVT injections.[34] In the monkey CNV model, 2 weeks after laser inducement, 42% of the generated lesions in the vehicle controls groups were graded as clinically significant (grade 4/5). Animals dosed with 100, 300, and 1000 µg of JSM6427 afforded 25%, 19%, and 10% reductions in grade 4/5 lesions, respectively. In a rabbit model where the animals were implanted with VEGF/bFGF containing pellets to cause CNV there was a dose dependent suppression of choroidal vascular leakage after 3 weeks of treatment. These results led to initiation of a Phase 1 clinical trial employing IVT JSM6427 at doses of 0.15–1.5 mg per eye.[35]

Squalamine was investigated in the clinic as an IV infusion in patients with wet AMD.[36] Squalamine was infused once a week for 4 weeks, with the high dose group (40 mg/infusion) affording a mean change in BCVA of +3.8 letters after the second dose and +5.8 letters 1 week after the final dose. Subsequent to this result it was reported that a Phase 2 study was initiated with a topical eye drop formulation of squalamine.[37]

The anti-angiogenic property of the cortistatin family of natural products has received considerable attention in recent years.[32] A paper in 2009 describes the synthesis of simplified analogs of cortistatin A (**13**) and highlights the discovery of compound **14** which was shown to inhibit retinal vessel formation in newborn mice after a single IVT injection of 500 pmol.[38]

3. VISUAL CYCLE INHIBITORS

The visual or retinoid cycle is a sequence of enzymatic reactions involved in the photo-transduction cascade where 11-*cis*-retinal undergoes rhodopsin-catalyzed photo-isomerization, and the generated all-*trans*-retinal is recycled to form 11-*cis*-retinal.[39] As the efficiency of visual system decreases with aging, all-*trans*-retinal is not effectively processed by the visual cycle and part of it reacts with phosphatidylethanolamine (PE) to form by-products such as N-retinide-N-retinyl ethanolamine (A2E) that is reported to be toxic.[40] Furthermore, disabling mutations in genes encoding visual cycle proteins (e.g., ABCA4) can speed up the accumulation of such toxic products. A2E can produce toxic effects on RPE cells by a number of mechanisms: destabilizing membranes, inhibiting respiration in mitochondria, increasing blue-light photodamage, impairing lysosomal acidification, and impairing degradation of phospholipids from phagocytosed outer segments.[41] More recently, it has also been shown that A2E can potently inhibit isomerohydrolase (RPE65) in the visual cycle.[41] Diseases of retinal degeneration such as AMD and Stargardt disease have been associated with accumulation of A2E in the eye. As a consequence, modulation of the visual cycle to reduce the formation of toxic by-products such as A2E has emerged as a potential approach for the treatment of diseases of retinal degeneration.

15, Fenretinide   **16, A1120**   **17, ACU-4429**

18, 11-trans-retinyl palmitate   **21, Ret-NH$_2$**

19, TDH   **22, PBN**

20, TDT   **23, C20-D3-vitamin A**

3.1. Retinol binding protein (RBP) blockers

While being evaluated in patients for the treatment of multiple basal cell carcinomas, fenretinide (**15**), a retinoid analogue, was found to cause abnormalities of rod photoreceptor function which rapidly reversed upon cessation of the therapy.[42] It was hypothesized that fenretinide may have interfered with the binding or the transport of vitamin A. A later study showed that chronic administration of fenretinide produced significant reduction of visual cycle retinoids and arrested accumulation of A2E.[43] A1120 (**16**) is a nonretinoid RBP inhibitor that was originally identified from a screen to identify compounds to improve insulin sensitivity.[44] A1120 decreases RBP level in serum, partially depletes certain visual cycle retinoids, and reduces the level of lipofuscin accumulation in the Abcr4$^{-/-}$ mouse model.[45] When A1120 was dosed orally in Abcr4$^{-/-}$ mice at 30 mg/kg/day, it showed a 64% decrease in serum RBP level and about 50% reduction in lipofuscin bisretinoid level compared to a vehicle control.

3.2. RPE65 isomerase inhibitor, antagonists, and spin traps

ACU-4429 (**17**) is an inhibitor of RPE65 isomerase, which catalyzes the *trans* to *cis* isomerization process in the visual cycle.[46] AMD patients who received 5, 10, and 20 mg daily of ACU-4429 for 3 months showed dose dependent suppression (maximum of 50% suppression with the 10 mg cohort) of ERG b-wave rod recovery.[47] All-*trans*-retinyl palmitate (**18**) is an endogenous substrate of RPE65,[48] and its analogs TDH (**19**) and TDT (**20**) were shown to bind to mouse RPE65 with K_ds of 96 and 58 nM, respectively, comparable to the affinity of **18** (47 nM).[49] A single i.p. injection of TDT (50 mg/kg) in rats inhibited 11-*cis*-retinal synthesis after bleaching by 79%. Chronic treatment of TDT in Abcr$^{-/-}$ mice at 50 mg/kg i.p. twice a week for 2 months reduced A2E levels in eye cups by 85% compared to control. However, after reanalysis of the HPLC data, the reduction of 11-*cis*-retinal synthesis following single i.p. injection of TDT (50 mg/kg) was revised to be less than 20%.[50] Developed as a transition state analog for the isomerization reaction catalyzed by RPE65, Ret-NH$_2$ (**21**) was shown to inhibit 11-*cis*-retinol production.[51] Ret-NH$_2$ binds to RPE65 with an average K_d of 80 nM and its *in vitro* potency compares favorably to visual cycle modulators, such as fenretinide and TDH.[52] The effect of the chronic treatment of Abcr$^{-/-}$ mice with Ret-NH$_2$ on A2E accumulation was evaluated. One-month-old Abcr$^{-/-}$ mice were given Ret-NH$_2$ (1 mg every 2 weeks) orally for 2 months. The level of A2E in the treatment group was significantly reduced compared to the control group. RPE65 catalyzes the isomerization from all-*trans*-retinyl ester to 11-*cis*-retinyl ester through a radical cation mechanism.[53] It was then postulated that spin traps will inhibit the activity of RPE65 by stabilizing the radical intermediates.[54] In a cell-based assay, N-*tert*-butyl-α-phenylnitrone (PBN) **22**, a known lipophilic spin trap, demonstrated an inhibitory effect on RPE65 activity with an IC$_{50}$ of $\sim$100 μM. Under the assay conditions, levels of RPE65 protein expression and all-*trans*-retinyl ester were not reduced. A lower IC$_{50}$ for PBN in the assay was observed when the concentration of substrate was increased, indicating a noncompetitive mechanism of inhibition. Taken together, these observations suggested that a PBN–retinyl ester spin adduct directly binds to RPE65 and inhibits its activity.

3.3. Other retinoid analogs

13-*cis*-Retinoic acid (RA), along with its analog fenretinide, has been used for decades in the treatment of acne and cancer.[55] 13-*cis*-RA was found to inhibit 11-*cis*-retinoldehydrogenase,[56] another important enzyme in the visual cycle, and also bind RPE65.[57] In an *in vitro* isomerization activity assay, 13-*cis*-RA did not show any significant inhibitory activity at up to 100 μM whereas Ret-NH$_2$ (**21**) had an IC$_{50}$ of 650 nM. Similarly, a single i.p. injection of 13-*cis*-RA (0.3 mg) in mice inhibited the ability of 11-*cis*-retinal synthesis after bleaching by 20% whereas 0.3 mg of Ret-NH$_2$ (**21**) provided 90% reduction.[54] C20-D3-vitamin A (**23**) is vitamin A enriched with the stable isotope deuterium at C20.[58] It was shown that the dimerization process of C20-D3-vitamin A was considerably slower than that of vitamin A *in vitro*. In order to evaluate the impact of C20-D3-vitamin A on the biosynthesis of A2E *in vivo*, two groups of 8-week-old ICR mice were administered (i.p.) 1.5 mg of either all-*trans*-retinal or C20-D3-all-*trans*-retinal twice a week for 6 weeks. At the end of the study, mice administered C20-D3-vitamin A had 68% less A2E relative to mice administered vitamin A ($p = 0.006$). These data also corroborated the proposed mechanism of A2E biosynthesis.[40]

4. COMPLEMENT PATHWAY INHIBITORS

The complement system is a complex network of over 40 proteins and regulators found mostly in the systemic circulation that helps or "complements" the ability of antibodies and phagocytic cells to clear pathogens from an organism.[59] Evidence of complement involvement in one class of sub-RPE deposits (basal laminar deposits) began to emerge[60] and it became evident that a large number of complement activators, complement components, and complement regulatory proteins are molecular constituents of drusen, the hallmark extracellular deposits associated with early stage AMD.[61] Identification of this compositional profile formed the basis for a new paradigm of AMD pathogenesis in which drusen is regarded as a byproduct of chronic, local inflammatory events near Bruch's membrane. Multiple genetic studies have revealed highly significant associations between AMD and sequence variants of several complement pathway associated genes including complement factor H (CFH), complement factor B (CFB), complement factor I (CFI), complement component 2 (C2), and 3 (C3).[59] Whilst there are no small molecule complement inhibitors in AMD clinical trials to date, a number of biologics (that are out of scope for this review) are being developed to target specific components of the complement cascade, many of which are thought to have potential impact on early stage of the disease, preventing progression to late AMD.[62]

24, Compstatin   **25**   **26**

27 R = H, K-76
28 R = OH, K-76 COOH

29 R = Me
30 R = H

31   **32**

33   **34**

IC50 = 13 nM (fetal complement hemolytic assay) C9 deposition
IC50 = 3.68 mg/mL (classical pathway)
IC50 = 4.46 mg/mL (lectin pathway)
IC50 = 3.59 mg/mL (alternative pathway)

4.1. C3 and C1s inhibitors

Compstatin (**24**), a highly potent and selective cyclic tridecapeptide C3 inhibitor, was discovered more than 15 years ago by screening phage-display libraries in the search for C3b-binding peptides.[63] It inhibits both the classical and the alternative complement pathway with IC_{50} values of 63 and 12 μM, respectively. Reduction and alkylation of the intramolecular disulfide bond between cysteines 2 and 12 resulted in loss of function, indicating the importance of the cyclic structure.[63] The cyclic peptide, which was later named compstatin, binds to C3 in a reversible manner and inhibits the convertase mediated cleavage of C3 to C3a and C3b. A series of surface plasmon resonance (SPR) studies confirmed and quantified the selectivity and binding mode of compstatin.[64] While the original compstatin peptide represented a highly selective and potent lead structure itself, the combination of rational, combinatorial, and computational optimization methods was able to drastically improve its activity.[65] POT-4, a derivative of compstatin [structure not disclosed], is a potent C3 inhibitor and is being pursued in the clinic for wet/dry AMD. POT-4 (administered as intravitreal injection) has completed Phase 1 study.[66] Recently, a class of small molecule inhibitors of C1s trypsin-like serine protease was reported.[67] A series of arylsulfonylthiophene-2-carboxamide such as compounds **25** and **26** showed K_i value of 30 and 20 nM, respectively, against the enzyme.

4.2. Filifolinol derivatives

As a result of a large effort involving screening for complement inhibitory activity of over 3000 strains of actinomycetes and fungi, K-76 (**27**) was isolated and characterized from the culture broth of *Stachybotrys complementi*, obtained from soil of the Ishigaki Island in Japan.[68] Oxidation of the natural product yielded K-76 COOH (**28**); the sodium salt of **28** was found to be less toxic and displayed better aqueous solubility.[69] K-76 COOH has been shown to play important roles in several complement mediated diseases in animal models.[70] Synthetic analogs of the natural product have been prepared and tested for their ability to inhibit activation of the classical complement pathway.[71]

Filifolinol (**29**) is a bioactive tricyclic terpenoid with antioxidant, antibacterial, antifungal, and antiviral properties.[72] Filifolinol and the related acid (**30**) carry the grisan (3H-spiro[benzofuran-2,10-cyclohexane]) motif, a privileged structure found in many bioactive natural products including

K-76 and its derivatives. Given the structural analogies between filifolinol and the BCD-ring system of K-76 COOH, the former was used a starting point for the synthesis of several analogs such as **31** and **32** that were shown to be more potent than the natural compound in inhibiting the classical complement pathway in an hemolytic assay.[73,74]

4.3. Factor B inhibitors

A library of peptidic aldehydes designed and synthesized to probe the structure–activity requirements for inhibition of factor B at pH 9.5 lead to the identification of peptide covalent inhibitors such as **33**, which showed an IC_{50} of 250 nM when incubated with factor B (50 nM) in glycine–NaOH buffer.[75] This compound also showed activity ($IC_{50} = 320$ nM) in the membrane attack complex (MAC) formation assay using the Wieslab Complement System Screen based on the principle of ELISA.

4.4. Other complement inhibitors

As one of the downstream targets of the complement pathway, a number of C5a receptor antagonists have been developed and extensively reviewed recently.[76] A series of 1-phenyl-3-(1-phenylethyl)urea derivatives, identified from high-throughput screening using fetal complement hemolytic assay, were reported to show complement inhibition. Optimized compounds, such as **34** inhibit C9 deposition (but not C3 and C4) through the classical, lectin, and alternative pathways.[77] The exact target of this class of compounds is not known.

5. CONCLUSIONS

Although progress has been made in recent years against wet AMD with the introduction of IVT anti-VEGF therapies, development of a less invasive treatment (e.g., topical or oral administration) remains a challenge. Moreover, there is currently no effective therapy available for dry AMD or GA. The emerging molecules and continued research targeting angiogenesis, the visual cycle, and the complement pathway should reveal the underlying disease mechanism and eventually help to address the medical needs.

REFERENCES

1. Hubschman, J. P.; Reddy, S.; Schwartz, S. D. *Clin. Ophthalmol.* **2009**, *3*, 155.
2. Zhang, K.; Zhang, L.; Weinreb, R. N. *Nat. Rev. Drug Discov.* **2012**, *11*, 541.
3. Miller, J. W.; Stinson, W. G.; Folkman, J. *Ophthalmology* **1993**, *100*, 9.
4. Keefe, D.; Bowen, J.; Gibson, R.; Tan, T.; Okera, M.; Stringer, A. *Oncologist* **2011**, *16*, 432.
5. Gaudana, R.; Ananthula, H. K.; Parenky, A.; Mitra, A. K. *AAPS J.* **2010**, *12*, 348.
6. Bold, G.; Altmann, K. H.; Frei, J.; Lang, M.; Manley, P. W.; Traxler, P.; Wietfeld, B.; Bruggen, J.; Buchdunger, E.; Cozens, R.; Ferrari, S.; Furet, P.; Hofmann, F.; Martiny-Baron, G.; Mestan, J.; Rosel, J.; Sills, M.; Stover, D.; Acemoglu, F.; Boss, E.; Emmenegger, R.; Lasser, L.; Masso, E.; Roth, R.; Schlachter, C.; Vetterli, W. *J. Med. Chem.* **2000**, *43*, 2310.
7. Kwak, N.; Okamoto, N.; Wood, J. M.; Campochiaro, P. A. *Invest. Ophthalmol. Vis. Sci.* **2000**, *41*, 3158.
8. ClinicalTrials.gov Identifier: NCT00138632 at http://clinicaltrials.gov/ct2/show/NCT00138632.
9. Diago, T.; Pulido, J. S.; Molina, J. R.; Collet, L. C.; Link, T. P.; Ryan, E. H., Jr. *Mayo Clin. Proc.* **2008**, *83*, 231.
10. Harris, P. A.; Boloor, A.; Cheung, M.; Kumar, R.; Crosby, R. M.; Davis-Ward, R. G.; Epperly, A. H.; Hinkle, K. W.; Hunter, R. N., 3rd.; Johnson, J. H.; Knick, V. B.; Laudeman, C. P.; Luttrell, D. K.; Mook, R. A.; Nolte, R. T.; Rudolph, S. K.; Szewczyk, J. R.; Truesdale, A. T.; Veal, J. M.; Wang, L.; Stafford, J. A. *J. Med. Chem.* **2008**, *51*, 4632.
11. Iwase, T.; Oveson, B. C.; Hashida, N.; Lima e Silva, R.; Shen, J.; Krauss, A. H.; Gale, D. C.; Adamson, P.; Campochiaro, P. A. *Invest. Ophthalmol. Vis. Sci.* **2013**, *54*, 503, and references therein.
12. Robbie, S. J.; Lundh von Leithner, P.; Ju, M.; Lange, C. A.; King, A. G.; Adamson, P.; Lee, D.; Sychterz, C.; Coffey, P.; Ng, Y.-S.; Bainbridge, J. W.; Shima, D. T. *Invest. Ophthalmol. Vis. Sci.* **2013**, *54*, 1490.
13. Slakterm, J. S.; Tolentino, M. J.; Berger, B. B.; Pearlman, J.; Noble, R.; Ye, L.; Xu, C.-F.; Danis, R. P.; McLaughlin, M.; Kim, R. *Invest. Ophthalmol. Vis. Sci.* **2012**, *53*, E-Abstract 2038.
14. Publically accessible GSK clinical study results can be found at the link listed here under study ID: 114987, http://www.gsk-clinicalstudyregister.com/result_comp_list.jsp?phase=Phase+2&studyType=All&population=All&marketing=All&compound=pazopanib.
15. Kumar, R.; Knick, V. B.; Rudolph, S. K.; Johnson, J. H.; Crosby, R. M.; Crouthamel, M. C.; Hopper, T. M.; Miller, C. G.; Harrington, L. E.; Onori, J. A.; Mullin, R. J.; Gilmer, T. M.; Truesdale, A. T.; Epperly, A. H.; Boloor, A.; Stafford, J. A.; Luttrell, D. K.; Cheung, M. *Mol. Cancer Ther.* **2007**, *6*, 2012.
16. Palanki, M. S.; Akiyama, H.; Campochiaro, P.; Cao, J.; Chow, C. P.; Dellamary, L.; Doukas, J.; Fine, R.; Gritzen, C.; Hood, J. D.; Hu, S.; Kachi, S.; Kang, X.; Klebansky, B.; Kousba, A.; Lohse, D.; Mak, C. C.; Martin, M.; McPherson, A.; Pathak, V. P.; Renick, J.; Soll, R.; Umeda, N.; Yee, S.; Yokoi, K.; Zeng, B.; Zhu, H.; Noronha, G. *J. Med. Chem.* **2008**, *51*, 1546.
17. ClinicalTrials.gov Identifier: NCT00509548 at http://www.clinicaltrials.gov/ct/show/NCT00509548?order=1.
18. Dai, Y.; Hartandi, K.; Ji, Z.; Ahmed, A. A.; Albert, D. H.; Bauch, J. L.; Bouska, J. J.; Bousquet, P. F.; Cunha, G. A.; Glaser, K. B.; Harris, C. M.; Hickman, D.; Guo, J.; Li, J.; Marcotte, P. A.; Marsh, K. C.; Moskey, M. D.; Martin, R. L.; Olson, A. M.; Osterling, D. J.; Pease, L. J.; Soni, N. B.; Stewart, K. D.; Stoll, V. S.; Tapang, P.; Reuter, D. R.; Davidsen, S. K.; Michaelides, M. R. *J. Med. Chem.* **2007**, *50*, 1584.

19. Bingaman, D. P.; Liu, C.; Landers, R. A.; Gu, X. *Invest. Ophthalmol. Vis. Sci.* **2005**, *46*, E-Abstract 464.

20. Hudson, C. C.; Liu, M.; Chiang, G. G.; Otterness, D. M.; Loomis, D. C.; Kaper, F.; Giaccia, A. J.; Abraham, R. T. *Mol. Cell. Biol.* **2002**, *22*, 7004.

21. Dejneka, N. S.; Kuroki, A. M.; Fosnot, J.; Tang, W.; Tolentino, M. J.; Bennett, J. *Mol. Vis.* **2004**, *10*, 964.

22. Nussenblatt, R. B.; Byrnes, G.; Sen, H. N.; Yeh, S.; Faia, L.; Meyerle, C.; Wroblewski, K.; Li, Z.; Liu, B.; Chew, E.; Sherry, P. R.; Friedman, P.; Gill, F.; Ferris, F., 3rd. *Retina* **2010**, *30*, 1579.

23. ClinicalTrials.gov Identifier: NCT00766337 at http://www.clinicaltrials.gov/ct2/show/NCT00766337?term=NCT00766337&rank=1.

24. Sedrani, R.; Cottens, S.; Kallen, J.; Schuler, W. *Transplant. Proc.* **1998**, *30*, 2192.

25. ClinicalTrials.gov Identifier: NCT00857259 at http://www.clinicaltrials.gov/ct2/show/results/NCT00857259?term=everolimusþandþAMD&rank=1&sect=X0125#all.

26. Liu, L.; Parent, C. A. *J. Cell Biol.* **2011**, *194*, 815.

27. Sherris, D. In: *240th ACS National Meeting, Boston, MA* 2010, Abstract MEDI-309.

28. ClinicalTrials.gov Identifier: NCT01033721 at http://www.clinicaltrials.gov/ct2/show/NCT01033721?term=palomid+529&rank=2.

29. The complete structure of JSM6427 has not been disclosed, however the general scaffold to which it belongs has been reported in: Stragies, R.; Osterkamp, F.; Zischinisky, G.; Vossmeyer, D.; Kalkhof, H.; Reimer, U.; Zahn, G. *J. Med. Chem.* **2007**, *50*, 3786.

30. Moriarty, R. M.; Tuladhar, S. M.; Guo, L.; Wehrli, S. *Tetrahedron Lett.* **1994**, *35*, 8103.

31. Shenvi, R. A.; Guerrero, C. A.; Shi, J.; Li, C. C.; Baran, P. S. *J. Am. Chem. Soc.* **2008**, *130*, 7241.

32. Flyer, A. N.; Chong, S.; Myers, A. G. *Nat. Chem.* **2010**, *2*, 886, and references therein.

33. Sills, A. K., Jr.; Williams, J. I.; Tyler, B. M.; Epstein, D. S.; Sipos, E. P.; Davis, J. D.; McLane, M. P.; Pitchford, S.; Cheshire, K.; Gannon, F. H.; Kinney, W. A.; Chao, T. L.; Donowitz, M.; Laterra, J.; Zasloff, M.; Brem, H. *Cancer Res.* **1998**, *58*, 2784.

34. Zahn, G.; Vossmeyer, D.; Stragies, R.; Wills, M.; Wong, C. G.; Löffler, K. U.; Adamis, A. P.; Knolle, J. *Arch. Ophthalmol.* **2009**, *127*, 1329.

35. ClinicalTrials.gov Identifier: NCT00536016 at http://www.clinicaltrials.gov/ct2/show/NCT00536016?term=NCT00536016&rank=1.

36. Garcia, C. A.; Connolly, B.; Thomas, E.; Levitt, R.; Desai, A.; Nau, J.; Smith, I.; Regillo, C. *Invest. Ophthalmol. Vis. Sci.* **2005**, *46*, E-Abstract 206.

37. ClinicalTrials.gov Identifier: NCT01678963 at http://clinicaltrials.gov/ct2/show/NCT01678963?term=squalamine&rank=3.

38. Czakó, B.; Laszló, K.; Mammoto, A.; Ingber, D. E.; Corey, E. J. *J. Am. Chem. Soc.* **2009**, *131*, 9014.

39. Travis, G. H.; Golczak, M.; Moise, A. R.; Palczewski, K. *Annu. Rev. Pharmacol. Toxicol.* **2007**, *47*, 469.

40. Parish, C. A.; Hashimoto, M.; Nakanishi, K.; Dillon, J.; Sparrow, J. *Proc. Natl. Acad. Sci. U.S.A.* **1998**, *95*, 14609.

41. Moiseyev, G.; Nikolaeva, O.; Chen, Y.; Farjo, K.; Takahashi, Y.; Ma, J.-X. *Proc. Natl. Acad. Sci. U.S.A.* **2010**, *107*, 17551, and references therein.

42. Kaiser-Kupfer, M. I.; Peck, G. L.; Caruso, R. C.; Jaffe, M. J.; DiGiovanna, J. J.; Gross, E. G. *Arch. Ophthalmol.* **1986**, *104*, 69.

43. Radu, R.; Han, Y.; Bui, T. V.; Nusinowitz, S.; Bok, D.; Lichter, J.; Widder, K.; Travis, G. H.; Mata, N. L. *Invest. Ophthalmol. Vis. Sci.* **2005**, *46*, 4393.

44. Motani, A.; Wang, Z.; Conn, M.; Siegler, K.; Zhang, Y.; Liu, Q.; Johnstone, S.; Xu, H.; Thibault, St.; Wang, Y.; Fan, P.; Connors, R.; Le, H.; Xu, G.; Walker, N.; Shan, B.; Coward, P. *J. Biol. Chem.* **2009**, *284*, 7673.

45. Dobri, N.; Qin, Q.; Kong, J.; Yamamoto, K.; Liu, Z.; Moiseyev, G.; Ma, J.-X.; Allikmets, R.; Sparrow, J. R.; Petrukhin, K. *Invest. Ophthalmol. Vis. Sci.* **2013**, *54*, 85.
46. Kubota, R.; Boman, N. L.; David, R.; Mallikaarjun, S.; Patil, S.; Birch, D. *Retina* **2012**, *32*, 183.
47. Birch, D. G.; Chandler, J. W.; Reaves, S. I. In: Association for Research in Vision and Ophthalmology Annual Meeting, May 07, 2012 (Abs 2044).
48. Mata, N. L.; Moghrabi, W. N.; Lee, J. S.; Bui, T. V.; Radu, R. A.; Horwitz, J.; Travis, G. H. *J. Biol. Chem.* **2004**, *279*, 635.
49. Maiti, P.; Kong, J.; Kim, S. R.; Sparrow, J. R.; Allikmets, R.; Rando, R. R. *Biochemistry* **2006**, *45*, 852.
50. Maiti, P.; Kong, J.; Kim, S. R.; Sparrow, J. R.; Allikmets, R.; Rando, R. R. *Biochemistry* **2007**, *46*, 8700.
51. Golczak, M.; Kuksa, V.; Maeda, T.; Moise, A. R.; Palczewski, K. *Proc. Natl. Acad. Sci. U.S.A.* **2005**, *102*, 8162.
52. Golczak, M.; Maeda, A.; Bereta, G.; Maeda, T.; Kiser, P.; Hunzelmann, S.; Lintig, J.; Blaner, W. S.; Palczewski, K. *J. Biol. Chem.* **2008**, *283*, 9543.
53. Redmond, T. M.; Poliakov, E.; Kuo, S.; Chander, P.; Gentleman, S. *J. Biol. Chem.* **2010**, *285*, 1919.
54. Poliakov, E.; Parikh, R.; Ayele, M.; Kuo, S.; Chander, P.; Gentleman, S.; Redmond, T. M. *Biochemistry* **2011**, *50*, 6739.
55. Conley, B.; O'Shaughnessy, J.; Prindiville, S.; Lawrence, J.; Chow, C.; Jones, E.; Merino, M. J.; Kaiser-Kupfer, M. I.; Caruso, R. C.; Podgor, M.; Goldspiel, B.; Venzon, D.; Danforth, D.; Wu, S.; Noone, M.; Goldstein, J.; Cowan, K. H.; Zujewski, J. *J. Clin. Oncol.* **2000**, *18*, 275.
56. Gamble, M. V.; Mata, N. L.; Tsin, A. T.; Mertz, J. R.; Blaner, W. S. *Biochim. Biophys. Acta* **2000**, *1476*, 3.
57. Gollapalli, D. R.; Rando, R. R. *Proc. Natl. Acad. Sci. U.S.A.* **2004**, *101*, 10030.
58. Kaufman, Y.; Ma, L.; Washington, I. *J. Biol. Chem.* **2011**, *289*, 7958.
59. (a) Khandhadia, S.; Cipriani, V.; Yates, J. R. W.; Lotery, A. J. *Immunobiology* **2012**, *217*, 127, and references therein; (b) Ricklin, D.; Hajishengallis, G.; Yang, K.; Lambris, J. D. *Nat. Immunol.* **2010**, *11*, 785, and references therein.
60. Van der Schaft, T.; Mooy, C.; de Bruijn, W.; de Jong, P. *Br. J. Ophthalmol.* **1993**, 77, 657.
61. (a) Anderson, D. H.; Mullins, R. F.; Hageman, G. S.; Johnson, L. V. *Am. J. Ophthalmol.* **2002**, *134*, 411; (b) Anderson, D. H.; Talaga, K. C.; Rivest, A. J.; Barron, E.; Hageman, G. S.; Johnson, L. V. *Exp. Eye Res.* **2004**, *78*, 243; (c) Crabb, J. W.; Miyagi, M.; Gu, X.; Shadrach, K.; West, K. A.; Sakaguchi, H.; Kamei, M.; Hasan, A.; Yan, L.; Rayborn, M. E.; Salomon, R. G.; Hollyfield, J. G. *Proc. Natl. Acad. Sci. U.S.A.* **2002**, *99*, 14682; (d) Hageman, G. S.; Mullins, R. F. *Mol. Vis.* **1999**, *5*, 28; (e) Johnson, L. V.; Ozaki, S.; Staples, M. K.; Erickson, P. A.; Anderson, D. H. *Exp. Eye Res.* **2000**, *70*, 441; (f) Johnson, L. V.; Leitner, W. P.; Staples, M. K.; Anderson, D. H. *Exp. Eye Res.* **2001**, *73*, 887; (g) Johnson, L. V.; Leitner, W. P.; Rivest, A. J.; Staples, M. K.; Radeke, M. J.; Anderson, D. H. *Proc. Natl. Acad. Sci. U.S.A.* **2002**, *99*, 11830; (h) Johnson, P. T.; Betts, K. E.; Radeke, M. J.; Hageman, G. S.; Anderson, D. H.; Johnson, L. V. *Proc. Natl. Acad. Sci. U.S.A.* **2006**, *103*, 17456.
62. Yehoshua, Z.; Rosenfeld, P. J.; Albini, T. A. *Semin. Ophthalmol.* **2011**, *26*, 167.
63. Sahu, A.; Kay, B. K.; Lambris, J. D. *J. Immunol.* **1996**, *157*, 884.
64. Ricklin, D.; Lambris, J. D. *Adv. Exp. Med. Biol.* **2007**, *598*, 260.
65. Ricklin, D.; Lambris, J. D. *Adv. Exp. Med. Biol.* **2008**, *632*, 273, and references therein.
66. ClinicalTrials.gov Identifier: NCT00473928 at http://clinicaltrials.gov/ct2/show/study/NCT00473928.

67. Travins, J. M.; Ali, F.; Huang, H.; Ballentine, S. K.; Khalil, E.; Hufnagel, H. R.; Pan, W.; Gushue, J.; Leonard, K.; Bone, R. F.; Soll, R. M.; DesJarlais, R. L.; Crysler, C. S.; Ninan, N.; Kirkpatrick, J.; Cummings, M. D.; Huebert, N.; Molloy, C. J.; Gaul, M.; Tomczuk, B. E.; Subasinghe, N. L. *Bioorg. Med. Chem. Lett.* **2008**, *18*, 1603.
68. Miyazaki, W.; Tomaoka, H.; Shinohara, M.; Kaise, H.; Izawa, T.; Nakano, Y.; Kinoshita, T.; Hong, K.; Inoue, K. *Microbiol. Immunol.* **1980**, *24*, 1091.
69. Hong, K.; Kinoshita, T.; Miyazaki, M.; Izawa, T.; Inoue, K. *J. Immunol.* **1979**, *122*, 2418.
70. (a) Joh, T.; Ikai, M.; Oshima, T.; Kurokawa, T.; Seno, K.; Yokoyama, Y.; Okada, N.; Ito, M. *Life Sci.* **2001**, *70*, 109; (b) Roos, A.; Ramwadhdoebé, T. H.; Nauta, A. J.; Hack, E.; Daha, M. R. *Immunobiology* **2002**, *205*, 595; (c) Hudig, D.; Redelman, D.; Minning, L.; Carine, K. *J. Immunol.* **1984**, *133*, 408; (d) Endo, Y.; Yokochi, T.; Matsushita, M.; Fujita, T.; Takada, H. *J. Endotoxin Res.* **2001**, 7, 451.
71. Kaufman, T. S.; Srivastava, R. P.; Sindelar, R. D.; Scesney, S. M.; Marsh, H. C., Jr. *J. Med. Chem.* **1995**, *38*, 1437.
72. (a) Torres, R.; Villarroel, L.; Urzúa, A.; Delle Monache, F.; Delle Monache, G.; Gacs-Baitz, E. *Phytochemistry* **1994**, *36*, 249; (b) Modak, B.; Salina, M.; Rodilla, J.; Torres, R. *Molecules* **2009**, *14*, 4625; (c) Modak, M.; Rojas, M.; Torres, R. *Molecules* **2009**, *14*, 1980; (d) Modak, B.; Sandino, A. M.; Arata, L.; Cárdenas-Jirón, G.; Torres, R. *Vet. Microbiol.* **2010**, *141*, 53.
73. Larghi, E. L.; Operto, M. A.; Torres, R.; Kaufman, T. S. *Bioorg. Med. Chem. Lett.* **2009**, *19*, 6172.
74. Larghi, E. L.; Operto, M. A.; Torres, R.; Kaufman, T. S. *Eur. J. Med. Chem.* **2012**, *55*, 74.
75. Ruiz-Gomez, G.; Lim, J.; Halili, M. A.; Le, G. T.; Madala, P. K.; Abbenante, G.; Fairlie, D. P. *J. Med. Chem.* **2009**, *52*, 6042.
76. Powers, J. P.; Dairaghi, G. J.; Jaen, J. C. *Annu. Rep. Med. Chem.* **2011**, *46*, 171.
77. Zhang, M.; Yang, X.-Y.; Tang, W.; Groeneveld, T. L.; He, P.-L.; Zhu, F.-H.; Li, J.; Lu, W.; Blom, A. M.; Zuo, J.-P.; Nan, F.-J. *ACS Med. Chem. Lett.* **2012**, *3*, 317.

CHAPTER TWENTY-THREE

Synthetic Macrocycles in Small-Molecule Drug Discovery

Sandrine Vendeville*, **Maxwell D. Cummings**†
*Janssen Research & Development, LLC, Beerse, Belgium
†Janssen Research & Development, LLC, Spring House, Pennsylvania, USA

Contents

1. INTRODUCTION

Macrocycles, molecules containing 12-membered or larger rings, are receiving increased attention in small-molecule drug discovery. The reasons are several, including providing access to novel chemical space, challenging new protein targets, improved pharmacokinetics for relatively large small-molecules, and for peptides. The considerable benefit humanity has reaped from the pharmaceutical exploitation of macrocyclic natural products is undeniable.

Annual Reports in Medicinal Chemistry, Volume 48
ISSN 0065-7743
http://dx.doi.org/10.1016/B978-0-12-417150-3.00023-5

Macrocyclic small–molecule drugs are not a new idea; the well-known macrolide antibiotic erythromycin (**1**) was launched commercially in 1952,[1] and since then many other macrocyclic drugs (e.g., rifampicin (**2**), vancomycin (**3**)) have been discovered.[2–4] Recent reviews provide some perspective on natural products as a resource for both past and future drug discovery, and highlight technological advances that facilitate natural product-based drug discovery.[5,6] Natural product-derived principles have been established to guide the synthesis of new natural product-like molecules.[7,8] Advances in synthetic chemistry are facilitating broader application of macrocyclic strategies in synthetic small-molecule drug discovery.[3,9,10]

1   **2**   **3**

Recent analyses have contemplated the state of pharmaceutical discovery, concluding that productivity is low and exploring possible causes. The contrast between the more recent approach of mining vast high throughput screening (HTS) collections and the more historical medicinal chemistry-based approach to "hit" discovery has been noted, along with the challenge of representing the vastness of chemical space in a screening collection.[11] An analysis of recently approved first-in-class small-molecule drugs indicated that phenotypic screening significantly outperformed target-based discovery in driving drug discovery, and concluded that target-based screening, while popular, is contributing to high attrition and low productivity.[12] Concepts of drug-likeness are widely employed, but can be quite simplistic and can overlook many known drugs.[13,14] Some consensus has emerged that synthetic practices have biased discovery molecules in undesirable ways, and that a return to the "lost art" of more difficult and lower throughput medicinal chemistry is called for.[14,15]

In recent years pharmaceutical industry drug hunters have moved beyond "synthetic small-molecules only" to (re-)consider a broader palette of molecular starting points for drug discovery. In addition to the factors noted above, another motivation is the exploration of new target classes. For example, protein–protein interactions are an emerging source of new targets for drug action; however, their (generally) relatively large and "non-active site-like" (i.e., "non-druggable") binding interfaces present a challenge to typical discovery approaches (e.g., HTS, substrate or physiological ligand-based design), and researchers have cast about for alternative approaches for these non-traditional targets. Re-consideration of natural products as a source for drugs and/or leads is also driving this trend. It's somewhat ironic, perhaps, that in this context longstanding bioactives such as peptides and macrocycles are often presented as "new" or "non-traditional."

Increasingly, macrocycles are being incorporated into ongoing small-molecule drug discovery efforts. Here we briefly describe a few representative examples of recent studies in which macrocyclization has had significant impact on PK/PD and/or potency and/or target selectivity. We conclude with a few relevant issues and several studies aimed at broader understanding in this area.

2. HEPATITIS C VIRUS NS3/4A PROTEASE AND NS5B POLYMERASE

The quest for new hepatitis C drugs that act specifically against viral target proteins has been intense over the past 20 years, with most drug discovery efforts aimed at the NS3/4A protease, the NS5b RNA polymerase and the NS5A protein.[16–19] A significant milestone was reached in 2011, with the U.S. Food and Drug Administration (FDA)/European Medicines Evaluation Agency (EMEA) approvals of two NS3/4A protease inhibitor drugs.[20] Several groups, including our own, have pursued macrocyclic hepatitis C virus (HCV) protease inhibitors,[21] and multiple representatives are in late stage clinical evaluation. The NS5b polymerase has also been well-studied as a drug target,[19] and our own group has recently described macrocycle work in this area that has yielded a clinical candidate.[22–24]

2.1. HCV NS3/4A protease inhibitors: BILN 2061

The discovery of the macrocyclic NS3/4A protease inhibitor BILN 2061 (4) represents one important milestone along the path to the development of new and improved HCV drugs. The relevant discovery work has been thoroughly described,[25] and is worthwhile reading for those interested in reviewing a relatively modern-day peptide-based drug discovery story. Initial reports described substrate-based studies including alanine scanning and truncation,[26,27] progressing to more sophisticated NMR studies that suggested (macro)cyclization between the P1 and P3 sidechains of the peptide substrate,[28] the pursuit of which ultimately yielded the clinical candidate **4**. In the progression toward **4** macrocyclization led to dramatically improved cell-based antiviral activity, in an effort that had previously yielded high affinity inhibitors hampered by relatively poor cellular potencies.[28] Protease selectivity, whilst very good, was also good for non-macrocyclic predecessors. **4** entered clinical evaluation, but unfortunately was withdrawn due to cardiotoxicity observed in primates.[29] As the discovery effort around **4** proceeded, many groups began working on macrocyclic NS3/4A inhibitors.

4   **5**   **6**

2.2. HCV NS3/4A protease inhibitors: simeprevir, vaniprevir, and others

As noted above, many groups have based HCV drug discovery efforts on macrocyclic NS3/4A inhibitors.[21] Two relatively advanced molecules in this class are **5** (TMC435, simeprevir, from our own group)[30] and **6** (MK-7009, vaniprevir),[31] both of which have progressed to Phase 3 clinical study. From an early drug discovery perspective, these molecules both show sub-nanomolar binding affinities for the genotype 1b enzyme, acceptable selectivities with respect to human proteases, good activities in replicon (cell-based) screens and promising pharmacokinetic profiles. While both **5** and **6** feature a cyclopropylacylsulfonamide head group and retain some features of their peptidic parentage, they are distinctly different in their macrocyclization pattern, with **5** following a P1–P3 cyclization scheme and **6** employing the distinct P2–P4 cyclization pattern, the design of which has been described.[32]

2.3. HCV NS5b RNA-dependent RNA polymerase

Structure-based macrocyclization was used to design HCV NS5b polymerase inhibitors with improved binding affinities and pharmacokinetic properties compared to their acyclic analogues.[23,25,26] NS5b has multiple druggable binding sites, and the different NS5b inhibitors have been reviewed recently.[19] Here we will refer only to inhibitors binding to the thumb 1 site (i.e., finger-loop domain, NNI-1 site). These inhibitors have been pursued by several groups, and their structures present a common benzimidazole- or indole-carboxylate chemotype, initially discovered through HTS; **7** (MK-3281) is one example that reached clinical trials.[33] Our group used structure-based molecular modeling to design new macrocyclic indole-based inhibitors, incorporating a carboxylate bioisostere and a solvent-exposed linker that serves to rigidify the structure and enhance PK properties.[22] The resulting 17-membered ring macrocycle **8**, in which the linker, acid bioisostere and macrocycle ring size were optimized to favor the bioactive conformation and enhance drug–like properties, has been selected for clinical evaluation based on its *in vitro* antiviral potency ($EC_{50} = 82$ nM), good oral bioavailability in rats and dogs: >66% and 87%, respectively; high distribution to the liver (target organ for HCV therapy), and acceptable safety profile.[24,26] Noteworthy in this evolution is the observation that the basic nitrogen in **7** was not required with the macrocycles.[33] Furthermore, a 3D complex structure obtained with a close analogue of **8** revealed a slightly different binding mode, with additional protein residues being resolved.[22] **8** recently entered phase 2 clinical trials, after showing robust viral load reduction in patients in a phase 1b study.[34]

7   8

3. KINASE INHIBITORS WITH IMPROVED SELECTIVITY

Kinases are enzymes involved in cellular signal transmission, and their deregulation can lead to diverse pathologies including inflammation, immune disorders, and cancers. They represent attractive therapeutic targets, and ~20 kinase inhibitors have been approved by the FDA. Adenosine triphosphate (ATP)–competitive kinase inhibitors have been most commonly pursued, with a general challenge being a lack of kinase specificity due to the conserved nature of the ATP binding site across kinases and the number of human kinases. Rational macrocyclization of linear kinase inhibitors as well as HTS of macrocycle-based libraries have both yielded inhibitors with improved selectivity, with the HTS-based approach also leading to a new class of bi-substrate ATP/peptide binding site inhibitors.

3.1. JAK2/FLT3 inhibitors: from SB1518 to SB1578

Janus kinases (JAK, comprising JAK1, JAK2, JAK3, and TYK2) are critical in the pathogenesis of immunological and inflammatory disorders and cancers.[35,36] Several ATP-competitive inhibitors with diverse JAK specificities have shown efficacy in clinical trials, for example ruxolitinib, a dual JAK1/JAK2 inhibitor recently approved by the FDA for the treatment of intermediate or high-risk myelofibrosis,[37] and tofacitinib, a JAK1/JAK2/JAK3 inhibitor approved for the treatment of rheumatoid arthritis (RA).[38] In the search for dual JAK2/FLT3 (a class III receptor tyrosine kinase involved in RA) inhibitors, devoid of JAK3 inhibition to avoid immunosuppressive effects, a macrocyclization strategy was applied to **9**, a relatively nonselective HTS hit.[39] Conformational restriction imposed by a linker between the two phenyl rings was anticipated to affect potency and selectivity by favoring the bioactive conformation. An eight atom dibenzylic ether linker, incorporating a *trans* olefin, offered a good starting point, with inhibitor(s) **10** showing some specificity for JAK2 (IC$_{50}$ JAK3/IC$_{50}$ JAK2 ~ 20) and submicromolar JAK2 inhibition. Further optimization guided by docking studies involved the introduction of a solubilizing group on the aniline moiety (i.e., R_1 and R_2 groups) and small substituents on the other two aromatic rings. Pacratinib (SB1518, **11**) was identified as a promising molecule, combining potent inhibition of JAK2/FLT3 (IC$_{50}$ = 48 and 22 nM, respectively) and good selectivity versus JAK3 (17-fold), with favorable *in vitro* and *in vivo* ADME (absorption, distribution, metabolism and excretion) profiles. This first-generation inhibitor is currently in phase 2 clinical study for the treatment of myeloid and lymphoid malignancies.[40] Second-generation inhibitors originating from this conformationally constrained macrocyclic series have also been developed, with improvements in selectivity, solubility, and therapeutic window.[41] Replacement of the upper (see **9–12**) phenyl ring by heterocycles, combined with previously described beneficial modifications (dibenzyl ether linker, pyrrolidinylethoxy solubilizing group), led to the identification of SB1578, **12**, which is now being evaluated in a phase 1 clinical study for RA. Although not the most active of the series against JAK2 and FLT3, **12** represents the best compromise between potency (IC_{50} JAK2 = 46 nM, IC_{50} FLT3 = 60 nM), selectivity (ratio IC_{50} JAK3/JAK2 = 93), ADME, and toxicological profile.

9 **10** (R_1 = OMe or OH, R_2 = H) **12**
11 (R_1 = pyrrolidinylethoxy, R_2 = H)

3.2. CDK2/JAK2/FLT3 inhibitors: discovery of SB1317

In an effort to develop compounds with the addition of cyclin–dependent kinase CDK2 inhibitory activity, the previously reported ATP-competitive JAK2/FLT3 macrocyclic scaffold **10** was modified.[42] A basic nitrogen in the linker was fundamental to good CDK2 inhibition, with this effect being pK_a dependent.[42] While variation of the substituent on the basic nitrogen indicated the methyl group was optimal, diverse substitutions on the three aromatic rings revealed no significant improvements. Based on the observed SAR, cellular potency, *in vitro* and *in vivo* ADME profiles and oral dose-dependent efficacy in mouse models, SB1317, **13**, was selected as a lead candidate and is currently undergoing phase 1 clinical evaluation in leukemia patients.

13

3.3. ALK inhibitors: macrocyclic diaminopyrimidines

ATP-competitive small-molecule inhibitors of the analplastic lymphoma kinase (ALK) have shown oncolytic efficacy in ALK-positive lung cancers *in vivo*. Crizotinib, a dual ALK/cMet inhibitor, was approved by the FDA in 2011 to treat non-small cell lung carcinomas.[43] The diaminopyrimidine (DAP) scaffold represents a well-documented kinase platform, and some DAP analogues are potent ALK inhibitors.[44] Structure-based modeling around DAP **14** revealed a "U" shape conformation that could be locked by the formation of a macrocycle.[45] Introduction of an ethylene linker between the two phenyl rings led to macrocycle **15**, with a minimum energy conformation that overlaps well with the minimum energy conformer of **14** as well as a small-molecule crystal structure of a close acyclic analogue. This constrained bioactive conformation led to improved enzymatic potency versus acyclic **14** ($IC_{50} = 3$ nM versus 22 nM), although the cellular potencies were similar ($EC_{50} = 150$ nM (**15**) versus 100 nM (**14**)). Selectivity with respect to the homologous insulin receptor kinase was also increased (IR IC_{50}/ALK $IC_{50} = 150$ for **15** and 3.6 for **14**), further validating this macrocyclization strategy. Incorporation of modifications derived from the acyclic parent series allowed rapid optimization of the macrocycles; with just 13 compounds this structure-based macrocyclic design strategy led to the discovery of **16**, a potent enzymatic and cellular ALK inhibitor ($IC_{50} = 0.51$ nM, $EC_{50} = 10$ nM) with improved selectivity *versus* IR (IR IC_{50}/ALK $IC_{50} = 173$).

14   **15**   **16**

3.4. Src allosteric inhibitors

The structure, function, and biological roles of Src have been reviewed.[46–48] Src inhibitors with different degrees of selectivity have been developed, and various design strategies have been described.[48–50] Dasatinib (BMS-354825) and bosutinib (SKI-606) are dual Src/Bcr-Abl ATP-competitive inhibitors that have recently received FDA approval for the treatment of chronic myelogenous leukemia.[50,51] Achieving high specificity for Src remains challenging because of the high sequence conservation within the ATP-binding pocket across kinases. It is postulated that non-ATP competitive inhibitors have the potential to show higher specificity due to the more diverse nature of their binding sites.[52] Screening of a DNA-templated macrocycle library of >13,000 members allowed the identification of two families of sub-micromolar Src inhibitors with high selectivity.[52,53] Optimization of **17** (an 18-membered macrocycle incorporating a *cis* olefin) and **18** (a 19-membered macrocycle incorporating a *trans* olefin) *via* systematic modification of the three building blocks led to macrocycles **19** and **20**, with a 240-fold improvement in potency (Src $IC_{50} < 0.004$ nM for both compounds) and retaining good selectivity. Crystallography with **19** and biological studies have revealed that these macrocycles are bi-substrate inhibitors, simultaneously occupying the ATP-binding site and the substrate peptide-binding patch, thereby locking the kinase in an inactive conformation. These results also provide insight into the binding mode and specificity for wild-type Src and the Src^{T338I} gate-keeper mutant, a common drug-resistant form of tyrosine kinases, explaining why macrocycle **19**, unlike **20**, loses potency toward Src^{T338I}.

17 (R$_1$: furyl, R$_2$: cyclopropyl, R$_3$: NH$_2$)
19 (R$_1$: phenyl, R$_2$: cyclohexyl, R$_3$: OH)

18 (R$_1$: furyl, R$_2$: cyclopropyl, R$_3$: NH$_2$)
20 (R$_1$: *p*-F-phenyl, R$_2$: cyclohexyl, R$_3$: OH)

4. GHRELIN AGONISTS: DISCOVERY OF TZP-101

Ghrelin is an endogeneous octanoylated 28-mer peptide that binds to GRLN (previously termed the human growth hormone secretagogue receptor, hGHS-R1a), a member of the G-protein coupled receptor (GPCR) superfamily.[54] Amongst its multiple physiological roles (release of growth hormone, regulation of glucose homeostasis, adipogenesis, cardioprotection), its potent gastro-prokinetic effect has led to a new therapeutic approach for gastrointestinal (GI) dysmotility disorders.[54] HTS of a diversity-oriented macrocyclic library of tethered tripeptides yielded new druggable ghrelin agonists.[3] **21** was chosen as a lead based on its potent activity ($EC_{50} = 68$ nM) and good selectivity versus the TOR-M receptor. Optimization aimed at PK property improvement involved stereospecific introduction of methyl groups, replacement of the *n*-propyl sidechain, introduction of substituents on the phenyl rings and reduction of the olefinic moiety, and led to the discovery of **22** (ulimorelin, TZP-101), a low nanomolar agonist ($EC_{50} = 29$ nM) with a suitable PK profile (e.g., 24% oral bioavailability in rats and monkeys). Interestingly, both crystallographic and NMR studies suggest an intramolecular hydrogen bond between the phenoxy oxygen and the protonatable amine; since both the cyclopropyl and methyl (from the alanine moiety) groups are required for improved PK properties, a cooperative PK effect was postulated for **22**, through stabilization of the proposed interaction. In phase 1 clinical trials **22** was safe and well-tolerated, and led to more profound effects on GI activity than on growth hormone release,[55,56] thus differentiating it from non-macrocyclic full GRLN agonists (e.g., anamorelin[57] and ibutamoren[58]). In phase 2 clinical trials, i.v. administration of **22** demonstrated efficacy for acute treatment of GI disorders such as post-operative ileus and diabetic gastroparesis; **22** is currently undergoing phase 3 clinical evaluation. In addition, a second generation analogue (of **22**, structure undisclosed) with improved PK properties has shown oral efficacy in a phase 2 clinical study.[54]

21   **22**

5. STAPLED PEPTIDES

Mimickry of specific protein structural features has long been pursued in small-molecule drug discovery. Early work on structure-based analysis of the nature of protein–protein interactions[59–61] provided initial general overviews and a foundation for more recent analyses.[62–65] The pursuit of general molecular strategies for targeting specific protein secondary structure elements has broad application to drug discovery in the area of protein–protein interactions.

XXX-XXX-XXX-XXX-XXX-XXX-XXX-XXX-XXX-XXX-XXX-XXX-XXX-XXX-XXX-XXX-XXX-XXX

i to *i*+4 linker   *I* to *i*+7 linker

Generic peptide sequence with two helical stapling schemes

Peptide stapling refers to the covalent linkage of two amino acid sidechains to stabilize a helical peptide conformation, a macrocyclization approach aimed at improving the pharmaceutical properties of peptides.[66–68] Such modification should provide benefits in peptide-based drug discovery, with possible effects including stabilization of the bioactive conformation leading to improved binding affinity, greater selectivity through disfavoring one or more undesirable conformations, increased proteolytic stability from biasing toward a helical conformation as well as by steric block, improved cellular penetration and altered physicochemical properties. Realization of these potential benefits could lead to broader use of peptide-based drugs.

In the well-studied HDM2-p53 protein–protein interaction system, p53 binds to HDM2 *via* an α-helix, and small helical peptides (and synthetic small molecules) compete for binding at the HDM2 site. For HDM2-p53, multiple groups have shown that different peptide stapling methods lead to increased helicity of unbound peptides,[66,69] and in some cases this has led to increased binding affinity.[69] Stapling in conjunction with other targeted modifications led to improved cellular penetration, although the effect of stapling alone on this property was unclear.[69] Recent structural studies indicate that the staple itself can add productive binding contacts in this system.[70] In the BIM–BCL-2 protein–protein system, peptide stapling yielded a stabilized α-helical peptide that was protease-resistant, cell permeable, and bound to multiple BCL-2 family members.[71] Double stapling of peptides related to the HIV drug Fuzeon, a 36-mer peptide that acts by inhibiting viral fusion (i.e., targeting HIV entry), gave rise to modified peptides with decreased protease susceptibility, somewhat improved antiviral activity and enhanced systemic exposure when compared to unmodified or singly stapled analogues.[72]

6. CONCLUSION

We have presented examples of synthetic macrocycles in the areas of HCV protease inhibitors, HCV polymerase inhibitors, inhibitors of various kinases, ghrelin agonists and stapled peptides, and in all of these cases macrocyclization has helped progress drug discovery and drug development. It's appealing to imagine that macrocyclization offers a broadly applicable strategy: for example, allowing the systematic bending or breaking of typical physicochemical drug-likeness rules. While it may be attractive to speculate thus, such a principle remains to be established. Furthermore, the data is complex: in the SAR studies reported for the HCV protease inhibitor **6**, macrocycle saturation and substitution influenced measured liver concentrations, and liver uptake for these molecules was shown to be *via* specific liver transporters[31,73]; stapled p53-based 16-mer peptides show improved proteolytic stability, but alteration of charge by amino acid replacement was (also) required to achieve cellular penetration[69]; when comparing the ALK inhibitors **14** and **15**, macrocyclization gave improved binding affinity and binding specificity, but this did not translate to improved cellular potency.[45] Deciphering broad effects and principles are complicated by inconsistency across experiments, as in many fields. Also, since acyclic versions of advanced macrocyclic candidates are unlikely to be synthesized and comprehensively tested and non-cyclized versions of advanced macrocyclic candidates may not be good representatives of optimized molecules in the relevant non-macrocyclic chemotype, extracting the broad impact of macrocyclization on drug discovery has been and will remain challenging.

Nevertheless, studies in this and related areas are starting to shed light. Recent analyses support the relevance of intramolecular hydrogen bonding to drug membrane permeation,[74,75] and indicate that it may be of particular relevance to larger small-molecule drugs that may not adhere to typical drug-likeness criteria.[76] Macrocyclization may be similar to intramolecular hydrogen bonding in that it may effectively "bury" hydrogen bonding partners and may lead to more spherical and less extended molecular shapes. Studies with macrocyclic peptides have also indicated the importance of internal hydrogen bonding, and have been used to develop computational approaches to the prediction of membrane permeation.[77,78] One plausible molecular design strategy may be to aim for ~isoenergetic conformations with/without one or more intramolecular hydrogen bonds, so that a "buried" conformation would be accessible for permeation while an "exposed" conformation could allow intermolecular hydrogen bonding with the drug target.[75,76]

Synthetic macrocycles show some promise as a molecular design strategy in small-molecule drug discovery, and it seems likely that their use will continue to increase. Hopefully, future research will lead to greater understanding of how macrocyclization of small-molecules affects drug-like behavior.

REFERENCES

1. Washington, J. A., 2nd.; Wilson, W. R. *Mayo Clin. Proc.* **1985**, *60*, 271.
2. Driggers, E. M.; Hale, S. P.; Lee, J.; Terrett, N. K. *Nat. Rev. Drug Discov.* **2008**, *7*, 608.
3. Marsault, E.; Peterson, M. L. *J. Med. Chem.* **2011**, *54*, 1961.
4. Wessjohann, L. A.; Ruijter, E.; Garcia-Rivera, D.; Brandt, W. *Mol. Divers.* **2005**, *9*, 171.
5. Koehn, F. E.; Carter, G. T. *Nat. Rev. Drug Discov.* **2005**, *4*, 206.
6. Li, J. W.; Vederas, J. C. *Science* **2009**, *325*, 161.
7. Clardy, J.; Walsh, C. *Nature* **2004**, *432*, 829.
8. Lachance, H.; Wetzel, S.; Kumar, K.; Waldmann, H. *J. Med. Chem.* **2012**, *55*, 5989.
9. Vougioukalakis, G. C.; Grubbs, R. H. *Chem. Rev.* **2010**, *110*, 1746.
10. White, C. J.; Yudin, A. K. *Nat. Chem.* **2011**, *3*, 509.
11. Scannell, J. W.; Blanckley, A.; Boldon, H.; Warrington, B. *Nat. Rev. Drug Discov.* **2012**, *11*, 191.
12. Swinney, D. C.; Anthony, J. *Nat. Rev. Drug Discov.* **2011**, *10*, 507.
13. Kenny, P. W.; Montanari, C. A. *J. Comput. Aided Mol. Des.* **2012**, *27*, 1.
14. Walters, W. P.; Green, J.; Weiss, J. R.; Murcko, M. A. *J. Med. Chem.* **2011**, *54*, 6405.
15. Lowe, D. B. *ACS Med. Chem. Lett.* **2012**, *3*, 3.
16. Calle Serrano, B.; Manns, M. P. *Antivir. Ther.* **2012**, *17*, 1133.
17. Gao, M.; Nettles, R. E.; Belema, M.; Snyder, L. B.; Nguyen, V. N.; Fridell, R. A.; Serrano-Wu, M. H.; Langley, D. R.; Sun, J. H.; O'Boyle, D. R., 2nd.; Lemm, J. A.; Wang, C.; Knipe, J. O.; Chien, C.; Colonno, R. J.; Grasela, D. M.; Meanwell, N. A.; Hamann, L. G. *Nature* **2010**, *465*, 96.
18. Kwong, A. D.; McNair, L.; Jacobson, I.; George, S. *Curr. Opin. Pharmacol.* **2008**, *8*, 522.
19. Sofia, M. J.; Chang, W.; Furman, P. A.; Mosley, R. T.; Ross, B. S. *J. Med. Chem.* **2012**, *55*, 2481.
20. Marks, K. M.; Jacobson, I. M. *Antivir. Ther.* **2012**, *17*, 1119.
21. Venkatraman, S.; Njoroge, F. G. *Expert Opin. Ther. Pat.* **2009**, *19*, 1277.
22. Cummings, M. D.; Lin, T. I.; Hu, L.; Tahri, A.; McGowan, D.; Amssoms, K.; Last, S.; Devogelaere, B.; Rouan, M. C.; Vijgen, L.; Berke, J. M.; Dehertogh, P.; Fransen, E.; Cleiren, E.; van der Helm, L.; Fanning, G.; Van Emelen, K.; Nyanguile, O.; Simmen, K.; Raboisson, P.; Vendeville, S. *Angew. Chem. Int. Ed. Engl.* **2012**, *51*, 4637.
23. McGowan, D.; Vendeville, S.; Lin, T. I.; Tahri, A.; Hu, L.; Cummings, M. D.; Amssoms, K.; Berke, J. M.; Canard, M.; Cleiren, E.; Dehertogh, P.; Last, S.; Fransen, E.; Van Der Helm, E.; Van den Steen, I.; Vijgen, L.; Rouan, M. C.; Fanning, G.; Nyanguile, O.; Van Emelen, K.; Simmen, K.; Raboisson, P. *Bioorg. Med. Chem. Lett.* **2012**, *22*, 4431.
24. Vendeville, S.; Lin, T. I.; Hu, L.; Tahri, A.; McGowan, D.; Cummings, M. D.; Amssoms, K.; Canard, M.; Last, S.; Van den Steen, I.; Devogelaere, B.; Rouan, M. C.; Vijgen, L.; Berke, J. M.; Dehertogh, P.; Fransen, E.; Cleiren, E.; van der Helm, L.; Fanning, G.; Van Emelen, K.; Nyanguile, O.; Simmen, K.; Raboisson, P. *Bioorg. Med. Chem. Lett.* **2012**, *22*, 4437.
25. Lamarre, D.; Anderson, P. C.; Bailey, M.; Beaulieu, P.; Bolger, G.; Bonneau, P.; Bos, M.; Cameron, D. R.; Cartier, M.; Cordingley, M. G.; Faucher, A. M.; Goudreau, N.; Kawai, S. H.; Kukolj, G.; Lagace, L.; LaPlante, S. R.; Narjes, H.; Poupart, M. A.; Rancourt, J.; Sentjens, R. E.; St George, R.; Simoneau, B.; Steinmann, G.; Thibeault, D.; Tsantrizos, Y. S.; Weldon, S. M.; Yong, C. L.; Llinas-Brunet, M. *Nature* **2003**, *426*, 186.
26. Llinas-Brunet, M.; Bailey, M.; Deziel, R.; Fazal, G.; Gorys, V.; Goulet, S.; Halmos, T.; Maurice, R.; Poirier, M.; Poupart, M. A.; Rancourt, J.; Thibeault, D.; Wernic, D.; Lamarre, D. *Bioorg. Med. Chem. Lett.* **1998**, *8*, 2719.

27. Llinas-Brunet, M.; Bailey, M.; Fazal, G.; Goulet, S.; Halmos, T.; Laplante, S.; Maurice, R.; Poirier, M.; Poupart, M. A.; Thibeault, D.; Wernic, D.; Lamarre, D. *Bioorg. Med. Chem. Lett.* **1998**, *8*, 1713.
28. Tsantrizos, Y. S.; Bolger, G.; Bonneau, P.; Cameron, D. R.; Goudreau, N.; Kukolj, G.; LaPlante, S. R.; Llinas-Brunet, M.; Nar, H.; Lamarre, D. *Angew. Chem. Int. Ed. Engl.* **2003**, *42*, 1356.
29. Hinrichsen, H.; Benhamou, Y.; Wedemeyer, H.; Reiser, M.; Sentjens, R. E.; Calleja, J. L.; Forns, X.; Erhardt, A.; Cronlein, J.; Chaves, R. L.; Yong, C. L.; Nehmiz, G.; Steinmann, G. G. *Gastroenterology* **2004**, *127*, 1347.
30. Lin, T. I.; Lenz, O.; Fanning, G.; Verbinnen, T.; Delouvroy, F.; Scholliers, A.; Vermeiren, K.; Rosenquist, A.; Edlund, M.; Samuelsson, B.; Vrang, L.; de Kock, H.; Wigerinck, P.; Raboisson, P.; Simmen, K. *Antimicrob. Agents Chemother.* **2009**, *53*, 1377.
31. McCauley, J. A.; McIntyre, C. J.; Rudd, M. T.; Nguyen, K. T.; Romano, J. J.; Butcher, J. W.; Gilbert, K. F.; Bush, K. J.; Holloway, M. K.; Swestock, J.; Wan, B. L.; Carroll, S. S.; DiMuzio, J. M.; Graham, D. J.; Ludmerer, S. W.; Mao, S. S.; Stahlhut, M. W.; Fandozzi, C. M.; Trainor, N.; Olsen, D. B.; Vacca, J. P.; Liverton, N. J. *J. Med. Chem.* **2010**, *53*, 2443.
32. Liverton, N. J.; Holloway, M. K.; McCauley, J. A.; Rudd, M. T.; Butcher, J. W.; Carroll, S. S.; DiMuzio, J.; Fandozzi, C.; Gilbert, K. F.; Mao, S. S.; McIntyre, C. J.; Nguyen, K. T.; Romano, J. J.; Stahlhut, M.; Wan, B. L.; Olsen, D. B.; Vacca, J. P. *J. Am. Chem. Soc.* **2008**, *130*, 4607.
33. Narjes, F.; Crescenzi, B.; Ferrara, M.; Habermann, J.; Colarusso, S.; Ferreira Mdel, R.; Stansfield, I.; Mackay, A. C.; Conte, I.; Ercolani, C.; Zaramella, S.; Palumbi, M. C.; Meuleman, P.; Leroux-Roels, G.; Giuliano, C.; Fiore, F.; Di Marco, S.; Baiocco, P.; Koch, U.; Migliaccio, G.; Altamura, S.; Laufer, R.; De Francesco, R.; Rowley, M. *J. Med. Chem.* **2011**, *54*, 289.
34. Leempoels, J.; Reesink, H. W.; Bourgeois, S.; Vijgen, L.; Rouan, M. C.; Vandebosch, A.; Ispas, G.; Marien, K.; Van Remoortere, P.; Fanning, G.; Simmen, K.; Verloes, R. In: Vol. 54; *62th Annual Meeting of the American Association for the Study of Liver Diseases, San Francisco,* 2011; p 533A.
35. Kontzias, A.; Kotlyar, A.; Laurence, A.; Changelian, P.; O'Shea, J. J. *Curr. Opin. Pharmacol.* **2012**, *12*, 464.
36. Seavey, M. M.; Dobrzanski, P. *Biochem. Pharmacol.* **2012**, *83*, 1136.
37. Mesa, R. A.; Yasothan, U.; Kirkpatrick, P. *Nat. Rev. Drug Discov.* **2012**, *11*, 103.
38. Flanagan, M. E.; Blumenkopf, T. A.; Brissette, W. H.; Brown, M. F.; Casavant, J. M.; Shang-Poa, C.; Doty, J. L.; Elliott, E. A.; Fisher, M. B.; Hines, M.; Kent, C.; Kudlacz, E. M.; Lillie, B. M.; Magnuson, K. S.; McCurdy, S. P.; Munchhof, M. J.; Perry, B. D.; Sawyer, P. S.; Strelevitz, T. J.; Subramanyam, C.; Sun, J.; Whipple, D. A.; Changelian, P. S. *J. Med. Chem.* **2010**, *53*, 8468.
39. William, A. D.; Lee, A. C.; Blanchard, S.; Poulsen, A.; Teo, E. L.; Nagaraj, H.; Tan, E.; Chen, D.; Williams, M.; Sun, E. T.; Goh, K. C.; Ong, W. C.; Goh, S. K.; Hart, S.; Jayaraman, R.; Pasha, M. K.; Ethirajulu, K.; Wood, J. M.; Dymock, B. W. *J. Med. Chem.* **2011**, *54*, 4638.
40. Hart, S.; Goh, K. C.; Novotny-Diermayr, V.; Hu, C. Y.; Hentze, H.; Tan, Y. C.; Madan, B.; Amalini, C.; Loh, Y. K.; Ong, L. C.; William, A. D.; Lee, A.; Poulsen, A.; Jayaraman, R.; Ong, K. H.; Ethirajulu, K.; Dymock, B. W.; Wood, J. W. *Leukemia* **2011**, *25*, 1751.
41. William, A. D.; Lee, A. C.; Poulsen, A.; Goh, K. C.; Madan, B.; Hart, S.; Tan, E.; Wang, H.; Nagaraj, H.; Chen, D.; Lee, C. P.; Sun, E. T.; Jayaraman, R.; Pasha, M. K.; Ethirajulu, K.; Wood, J. M.; Dymock, B. W. *J. Med. Chem.* **2012**, *55*, 2623.

42. William, A. D.; Lee, A. C.; Goh, K. C.; Blanchard, S.; Poulsen, A.; Teo, E. L.; Nagaraj, H.; Lee, C. P.; Wang, H.; Williams, M.; Sun, E. T.; Hu, C.; Jayaraman, R.; Pasha, M. K.; Ethirajulu, K.; Wood, J. M.; Dymock, B. W. *J. Med. Chem.* **2012**, *55*, 169.

43. Cui, J. J.; Tran-Dube, M.; Shen, H.; Nambu, M.; Kung, P. P.; Pairish, M.; Jia, L.; Meng, J.; Funk, L.; Botrous, I.; McTigue, M.; Grodsky, N.; Ryan, K.; Padrique, E.; Alton, G.; Timofeevski, S.; Yamazaki, S.; Li, Q.; Zou, H.; Christensen, J.; Mroczkowski, B.; Bender, S.; Kania, R. S.; Edwards, M. P. *J. Med. Chem.* **2011**, *54*, 6342.

44. Milkiewicz, K. L.; Ott, G. R. *Expert Opin. Ther. Pat.* **2010**, *20*, 1653.

45. Breslin, H. J.; Lane, B. M.; Ott, G. R.; Ghose, A. K.; Angeles, T. S.; Albom, M. S.; Cheng, M.; Wan, W.; Haltiwanger, R. C.; Wells-Knecht, K. J.; Dorsey, B. D. *J. Med. Chem.* **2012**, *55*, 449.

46. Edwards, J. *Expert Opin. Investig. Drugs* **2010**, *19*, 605.

47. Puls, L. N.; Eadens, M.; Messersmith, W. *Oncologist* **2011**, *16*, 566.

48. Sawyer, T.; Boyce, B.; Dalgarno, D.; Iuliucci, J. *Expert Opin. Investig. Drugs* **2001**, *10*, 1327.

49. Brandvold, K. R.; Steffey, M. E.; Fox, C. C.; Soellner, M. B. *ACS Chem. Biol.* **2012**, *7*, 1393.

50. Das, J.; Chen, P.; Norris, D.; Padmanabha, R.; Lin, J.; Moquin, R. V.; Shen, Z.; Cook, L. S.; Doweyko, A. M.; Pitt, S.; Pang, S.; Shen, D. R.; Fang, Q.; de Fex, H. F.; McIntyre, K. W.; Shuster, D. J.; Gillooly, K. M.; Behnia, K.; Schieven, G. L.; Wityak, J.; Barrish, J. C. *J. Med. Chem.* **2006**, *49*, 6819.

51. Cortes, J. E.; Kantarjian, H. M.; Brummendorf, T. H.; Kim, D. W.; Turkina, A. G.; Shen, Z. X.; Pasquini, R.; Khoury, H. J.; Arkin, S.; Volkert, A.; Besson, N.; Abbas, R.; Wang, J.; Leip, E.; Gambacorti-Passerini, C. *Blood* **2011**, *118*, 4567.

52. Georghiou, G.; Kleiner, R. E.; Pulkoski-Gross, M.; Liu, D. R.; Seeliger, M. A. *Nat. Chem. Biol.* **2012**, *8*, 366.

53. Kleiner, R. E.; Dumelin, C. E.; Tiu, G. C.; Sakurai, K.; Liu, D. R. *J. Am. Chem. Soc.* **2010**, *132*, 11779.

54. Tack, J.; Janssen, P. *Expert Opin. Emerg. Drugs* **2011**, *16*, 283.

55. Fraser, G. L.; Hoveyda, H. R.; Tannenbaum, G. S. *Endocrinology* **2008**, *149*, 6280.

56. Lasseter, K. C.; Shaughnessy, L.; Cummings, D.; Pezzullo, J. C.; Wargin, W.; Gagnon, R.; Oliva, J.; Kosutic, G. *J. Clin. Pharmacol.* **2008**, *48*, 193.

57. Garcia, J. M.; Polvino, W. J. *Growth Horm. IGF Res.* **2009**, *19*, 267.

58. Patchett, A. A.; Nargund, R. P.; Tata, J. R.; Chen, M. H.; Barakat, K. J.; Johnston, D. B.; Cheng, K.; Chan, W. W.; Butler, B.; Hickey, G.; Jacks, T.; Schleim, K.; Pong, S.-S.; Chaung, L.-Y. P.; Chen, H. Y.; Frazier, E.; Leung, K. H.; Chiu, S.-H. L.; Smith, R. G. *Proc. Natl. Acad. Sci. U.S.A.* **1995**, *92*, 7001.

59. Chothia, C. *Nature* **1974**, *248*, 338.

60. Janin, J.; Chothia, C. *J. Biol. Chem.* **1990**, *265*, 16027.

61. Jones, S.; Thornton, J. M. *Prog. Biophys. Mol. Biol.* **1995**, *63*, 31.

62. Bullock, B. N.; Jochim, A. L.; Arora, P. S. *J. Am. Chem. Soc.* **2011**, *133*, 14220.

63. Cummings, M. D.; Hart, T. N.; Read, R. J. *Protein Sci.* **1995**, *4*, 2087.

64. Eisenberg, D.; McLachlan, A. D. *Nature* **1986**, *319*, 199.

65. Horton, N.; Lewis, M. *Protein Sci.* **1992**, *1*, 169.

66. Henchey, L. K.; Porter, J. R.; Ghosh, I.; Arora, P. S. *Chembiochem* **2010**, *11*, 2104.

67. Smeenk, L. E.; Dailly, N.; Hiemstra, H.; van Maarseveen, J. H.; Timmerman, P. *Org. Lett.* **2012**, *14*, 1194.

68. Verdine, G. L.; Hilinski, G. J. *Methods Enzymol.* **2012**, *503*, 3.

69. Bernal, F.; Tyler, A. F.; Korsmeyer, S. J.; Walensky, L. D.; Verdine, G. L. *J. Am. Chem. Soc.* **2007**, *129*, 2456.

70. Baek, S.; Kutchukian, P. S.; Verdine, G. L.; Huber, R.; Holak, T. A.; Lee, K. W.; Popowicz, G. M. *J. Am. Chem. Soc.* **2012**, *134*, 103.

71. LaBelle, J. L.; Katz, S. G.; Bird, G. H.; Gavathiotis, E.; Stewart, M. L.; Lawrence, C.; Fisher, J. K.; Godes, M.; Pitter, K.; Kung, A. L.; Walensky, L. D. *J. Clin. Invest.* **2012**, *122*, 2018.

72. Bird, G. H.; Madani, N.; Perry, A. F.; Princiotto, A. M.; Supko, J. G.; He, X.; Gavathiotis, E.; Sodroski, J. G.; Walensky, L. D. *Proc. Natl. Acad. Sci. U.S.A.* **2010**, *107*, 14093.

73. Monteagudo, E.; Fonsi, M.; Chu, X.; Bleasby, K.; Evers, R.; Pucci, V.; Orsale, M. V.; Cianetti, S.; Ferrara, M.; Harper, S.; Laufer, R.; Rowley, M.; Summa, V. *Xenobiotica* **2010**, *40*, 826.

74. Desai, P. V.; Raub, T. J.; Blanco, M. J. *Bioorg. Med. Chem. Lett.* **2012**, *22*, 6540.

75. Kuhn, B.; Mohr, P.; Stahl, M. *J. Med. Chem.* **2010**, *53*, 2601.

76. Alex, A.; Millan, D. S.; Perez, M.; Wakenhut, F.; Whitlock, G. A. *Med. Chem. Commun.* **2011**, *2*, 669.

77. Leung, S. S.; Mijalkovic, J.; Borrelli, K.; Jacobson, M. P. *J. Chem. Inf. Model.* **2012**, *52*, 1621.

78. Rezai, T.; Bock, J. E.; Zhou, M. V.; Kalyanaraman, C.; Lokey, R. S.; Jacobson, M. P. *J. Am. Chem. Soc.* **2006**, *128*, 14073.

INTERNATIONAL UNION OF PURE AND APPLIED CHEMISTRY

CHEMISTRY AND HUMAN HEALTH DIVISION[a]
SUBCOMMITTEE ON MEDICINAL CHEMISTRY AND DRUG
DEVELOPMENT

CHAPTER TWENTY-FOUR

Glossary of Terms Used in Medicinal Chemistry Part II (IUPAC Recommendations 2013)

Derek R. Buckle*, **Paul W. Erhardt**[†], **C. Robin Ganellin**[‡],
Toshi Kobayashi[§], **Thomas J. Perun**[¶], **John Proudfoot**[||],
Joerg Senn-Bilfinger[#]

*DRB Associates, Redhill, Surrey, United Kingdom
[†]Center for Drug Design & Development, College of Pharmacy, University of Toledo, Toledo, Ohio, USA
[‡]Christopher Ingold Laboratory, Department of Chemistry, University College London, London,
United Kingdom
[§]PhRMA, Minato-ku, Japan
[¶]47731 Old Houston Hwy, Hempstead, Texas, USA
[||]Boehringer Ingelheim Pharmaceuticals Inc., Ridgefield, Connecticut, USA
[#]Altana Pharma AG, Konstanz, Germany

[a] Membership of the Division Committee of the Chemistry and Human Health Division during the
preparation of this report (2008–2011) was as follows:
President: D.M. Templeton (Canada, 2008-2011); *Secretary:* M.S. Chorghade (USA, 2008-2009); M.
Schwenk (Germany, 2010-2011); *Past President:* P.W. Erhardt (USA 2008-2009); *Vice President:* F.
Pontet (France, 2010-2011); *Titular Members:* O. Andersen (Denmark 2008-2011); S.O. Bachurin
(Russia 2010-2011); D.R. Buckle (UK 2010-2011); X. Fuentes-Arderiu (Spain 2008-2011); H.P.
A. Illing (UK 2010-2011); M.N. Liebman (USA 2008-2009); Y.C. Martin (USA 2010-2011); T.
Nagano (Japan 2010-2011); M. Nordberg (Sweden 2008-2009); F. Pontet (France, 2008-2009); F.
Sanz (Spain 2008-2009); G. Tarzia (Italy 2008-2010); *Affiliate Members:* C.R. Ganellin (UK,
2008-2011); T.J. Perun (USA, 2008-2011); J.H. Duffus (UK, 2008-2011).
Active Membership of the Subcommittee on Medicinal Chemistry and Drug Development (2008-
2011) was as follows: C.R. Ganellin (UK, *Chairperson*), J. Proudfoot (USA, *Secretary*), S.O. Bachurin
(Russia), E. Breuer (Israel), D.R. Buckle (UK), M.S. Chorghade (USA), P.W. Erhardt (USA), J.
Fischer (Hungary), A. Ganesan (UK), G. Gaviraghi (Italy), T. Kobayashi (Japan), M.N. Liebman
(USA), P. Lindberg (Sweden), Y. Martin (USA), P. Matyus (Hungary), A. Monge (Spain), T.J. Perun
(USA), F. Sanz (Spain), J. Senn-Bilfinger (Germany), N.J. de Souza (India), G. Tarzia (Italy), H.
Timmerman (Netherlands), M. Varasi (Italy), Yao, Zhu-Jun (PR China).

Annual Reports in Medicinal Chemistry, Volume 48
ISSN 0065-7743
http://dx.doi.org/10.1016/B978-0-12-417150-3.00024-7

Contents

1. INTRODUCTION

Since publication of the first Glossary of Terms used in Medicinal Chemistry over 10 years ago, the practice of medicinal chemistry has undergone a rapid and continuous change. This change, which by necessity has blurred the boundaries between what was traditionally seen as classical medicinal chemistry and its associated scientific disciplines, has resulted in a considerable expansion of related terminology. Medicinal chemists are increasingly required to understand, and interpret, language that was formerly the predominant domain of those involved in a much broader array of biological sciences in which chemistry has an underlying involvement. To reflect this change, the authors have compiled this supplementary Glossary of over 180 additional terms that were not previously collated into a defining document. In compiling this augmented list, a large body of experts actively involved in the practice of medicinal chemistry was consulted, but inevitably with terminology extending into multiple disciplines it is impossible to guarantee that all useful terms have been included. Within an entry, when a term is defined elsewhere in the Glossary, it is italicised.

To avoid a repetition of terms included in the original Glossary, we have chosen to keep this supplement as a separate document and to identify it by the designation Part II. By inference, therefore, the earlier Glossary necessarily becomes Part I. Those searching for specific terminology are advised to refer to both Glossaries. For simplicity, it is recommended that Part I[1] is searched prior to Part II, although where descriptive terms defined earlier have been used these are noted as embedded references in the text.

2. ALPHABETICAL ENTRIES

1. ADMET

Acronym referring to the absorption, distribution, metabolism,[1] excretion, and toxicity profile or processes for a xenobiotic[1] upon its administration *in vivo*.

Note: ADME[1] is also used to delineate these selected parameters within the context of a xenobiotic's pharmacokinetic profile. Because any of the five characteristics may become hurdles during drug development, ADMET behavior is typically studied and optimized among efficacious analogues during the early drug discovery stage by using *in vitro* models that attempt to predict such behaviors in clinical studies.
See also *pharmacokinetics, drug distribution*

2. adverse effect
   adverse event
   adverse drug event
   1. (In medicinal chemistry). Undesirable reaction in response to the administration of a drug or test compound.
      Note: In most instances, such effects result from off-*target* interactions.
   2. (In toxicokinetics). Change in biochemistry, morphology, physiology, growth, development, or lifespan of an organism which results in impairment of functional capacity or impairment of capacity to compensate for additional stress or increase in susceptibility to other environmental influences.[2]

3. allosteric antagonist
   Compound that binds to a receptor[1] at a site separate from, but actively coupled to, that of the endogenous agonist to actively reduce receptor signals.
   Note: The terms allosteric antagonist and *non-competitive antagonist* are often synonymous but not necessarily so.
   See also *non-competitive antagonist*

4. *alpha*-helix
   Secondary three-dimensional structure of a protein or peptide containing a right-handed coiled or spiral conformation, in which every backbone N–H group donates a hydrogen bond to the backbone C=O group of the amino acid four residues earlier.

5. analogue
   analog
   Chemical compound having structural similarity to a reference compound.
   Note: Despite the structural similarity, an analogue may display different chemical and/or biological properties, as is often intentionally the case during design and synthesis to optimize either efficacy or *ADMET* properties within a given series.
   See also *analogue-based drug discovery, congener, follow-on drug*

6. analogue-based drug discovery
   analog-based drug discovery
   Strategy for drug[1] discovery and/or optimization in which structural modification of an existing drug provides a new drug with improved chemical and/or biological properties.
   Note: Within the context of analogue-based drug discovery, three categories of drug analogues are recognized: Compounds possessing structural, chemical, and pharmacological similarities, termed "direct analogues" and sometimes referred to as "*me-too*" *drugs*; compounds possessing structural and often chemical but not pharmacological similarities, termed "structural analogues"; and structurally different compounds displaying similar pharmacological properties, termed "pharmacological analogues."
   See also *ligand-based drug design*[3]

7. ATP-binding cassette protein
   ABC protein
   Large gene family of transporter proteins that bind ATP and use the energy to transport substrates (e.g., sugars, amino acids, metal ions, peptides, proteins, and a large number of hydrophobic compounds and metabolites) across lipid membranes.
   Note: These proteins have an important role in limiting oral absorption and brain penetration of xenobiotics.[1]
   See also *efflux pump, P-glycoprotein*[4]

8. atropisomer
   Stereoisomer resulting from hindered rotation about a single bond in which steric hindrance to rotation is sufficient to allow isolation of individual isomers.[5]

9. attrition rate
   Rate of loss of drug[1] candidates during progression through the optimization and developmental stages while on route to the marketplace.
   Note: It has been estimated that for every 10,000 compounds examined during the early stages of biological testing, just one reaches the market.

10. autoinduction
    Capacity of a drug[1] to induce enzymes that mediate its own metabolism.[1]
    Note: This often results in lower, often subtherapeutic, drug exposure on prolonged or multiple dosing.[6]

11. back–up compound

Molecule selected as a replacement for the lead drug[1] candidate should this subsequently fail during further preclinical evaluation or in clinical studies.

Note: Ideally a back-up should be pharmacologically equivalent to the lead drug but have significant structural differences. Possession of a distinct core *scaffold* is optimal.

12. best–in–class

Drug acting on a specific *molecular target* that provides the best balance between efficacy[1] and adverse effects.

13. *beta*-barrel

Secondary three-dimensional protein structure containing a large *beta-sheet* that twists and coils to form a closed structure in which the first strand hydrogen bonds to the last.

14. *beta*-sheet

Secondary three-dimensional structure of a protein that takes on a flat, pleated appearance.

Note: Other common protein structural motifs are the *alpha-helix* and *beta*-barrel.

15. *beta*-sheet breaker

Compound that on binding to a protein disrupts *beta*-sheet formation.

16. bioinformatics

Discipline encompassing the development and utilization of computational tools to store, analyze, and interpret biological data.

Note: Typically protein or DNA sequence or three-dimensional information.

17. biological agent

Biopolymer-based pharmaceutical, such as a protein, applicable to the prevention, treatment, or cure of diseases or injuries to man.

Note: Biological agents may be any virus, therapeutic serum, toxin, antitoxin, vaccine, blood component or derivative, allergenic product, or analogous products.[7]

18. biomarker

Indicator signaling an event or condition in a biological system or sample and giving a measure of exposure, effect, or susceptibility.[2]

19. blockbuster drug

Drug which generates annual sales of US $1 billion ($10^9$) or more.

20. blood–brain barrier (BBB)

Layer of endothelial cells which line the small blood vessels of the brain.

Note 1: These cells form "tight junctions" which restrict the free exchange of substances between the blood and the brain. Such cells are rich in *P-glycoprotein* which serves to pump substrates back to the peripheral side of the vasculature.

Note 2: Passive diffusion across the BBB is highly dependent on drug lipophilicity and very few orally active agents acting in the central nervous system have a *polar surface area* greater than 0.9 nm^2.[8,9]

21. carcinogen

Agent (chemical, physical, or biological) which is capable of increasing the incidence of malignant neoplasms, thus causing cancer.[10]

22. chemical biology

Application of chemistry to the study of molecular events in biological systems; often using tool compounds.

Note: Distinguished from medicinal chemistry which is focused on the design and optimization of compounds for specific *molecular targets*.[11]

23. chemical database

Specific electronic repository for storage and retrieval of chemical information.

Note 1: Chemical structural information is sometimes stored in string notation such as the *InChI* or *SMILES notations*.

Note 2: Such databases can be searched to retrieve structural information and data on specific or related molecules.

Note 3: A free database of chemical structures of small organic molecules and information on their biological activities is available from PubChem.[12]

24. chemical diversity

See *diversity*

25. chemical library

compound library

compound collection

1. Collection of samples (e.g., chemical compounds, natural products, overexpression library of a microbe) available for biological screening.

2. Set of compounds produced through combinatorial chemistry or other means which expands around a single core structure or *scaffold*.

26. chemical space

Set of all possible stable molecules based on a specific chemical entity which interacts at one or more specific *molecular targets*.

27. cheminformatics

chemoinformatics

Use of computational, mathematical, statistical, and information techniques to address chemistry-related problems.[13]

28. chemogenomics

chemical genomics

Systematic screening of *chemical libraries* of *congeneric* compounds against members of a target family of proteins.[14]

29. chemokine

Member of a superfamily of proteins with the primary function to control leukocyte activity and trafficking through tissues.[15]

30. CLOGP values

Calculated 1-octanol/water partition coefficients.

Note: Frequently used in structure–property correlation or quantitative structure–activity relationship[1] studies.

See also *log P, log D*.[2]

31. cluster

Group of compounds which are related by structural, physicochemical, or biological properties.

Note: Organizing a set of compounds into clusters is often used to assess the *diversity* of those compounds, or to develop structure–activity relationship[1] models.[16–18]

32. codrug

mutual prodrug[1]

Two chemically linked synergistic drugs[1] designed to improve the drug delivery properties of one or both drugs.

Note: The constituent drugs are indicated for the same disease, but may exert different therapeutic effects via disparate mechanisms of action.[19]

33. congener

Substance structurally related to another and linked by origin or function.

Note: Congeners may be *analogues* or vice versa but not necessarily. The term congener, while most often a synonym for homologue, has become somewhat more diffuse in meaning so that the terms congener and *analogue* are frequently used interchangeably in the literature.

See also *analogue, follow-on drug*[2,20,21]

34. constitutive activity
Receptor or enzymatic function displayed in the absence of an agonist[1] or activator.

35. contract research organization (CRO)
Commercial organization that can be engaged to undertake specifically defined chemical, biological, safety, or clinical studies.
Note: Typically such studies are subject to confidentially agreements.

36. covalent drug
Ligand that binds irreversibly to its *molecular target* through the formation of a new chemical bond.

37. cytochrome P450 (CYP450)
Member of a superfamily of heme-containing monooxygenases involved in xenobiotic[1] metabolism, cholesterol biosynthesis, and steroidogenesis, in eukaryotic organisms found mainly in the endoplasmic reticulum and inner mitochondrial membrane of cells.[2]

38. descriptors
See *molecular descriptors*

39. designed multiple ligands
Compounds conceived and synthesized to act on two or more *molecular targets*.

40. diversity
Unrelatedness of a set of molecules (e.g., building blocks or members of a compound library), as measured by properties such as atom connectivity, physical properties, or computationally generated descriptors.
Note: Inverse of *molecular similarity*.

41. diversity-oriented synthesis (DOS)
Efficient production of a range of structures and *templates* with skeletal and stereochemical diversity as opposed to the synthesis of a specific target molecule.

42. drug cocktail
1. (In drug therapy). Administration of two or more distinct pharmacological agents to achieve a combination of their individual effects.
Note 1: The combined effect may be additive, synergistic, or designed to reduce side effects.
Note 2: This term is often used synonymously with that of drug combination but is preferred in order to avoid confusion with medications in which different drugs are included in a single formulation.

2. (In drug testing). Administration of two or more distinct compounds to test simultaneously their individual behaviors (e.g., pharmacological effects in high-throughput screens or drug metabolism[1]).

43. drug combination

See *drug cocktail*

44. drug delivery

Process by which a drug[1] is administered to its intended recipient.

Note: Examples include administration orally, intravenously, or by inhalation.

See also *drug distribution, targeted drug delivery*

45. drug distribution

Measured amounts of an administered compound in various parts of the organism to which it is given.

See also *drug delivery*

46. druggable target

See *druggability*

47. druggability

Capacity of a molecular target to be modulated in a favorable manner by a small molecule drug.[1]

Note: It is estimated that only around 10% of the human genome affords druggable targets.[22,23]

48. drug-like(ness)

Physical and chemical properties in a small molecule that make it likely to perform efficiently as a drug.[1]

49. drug repurposing

drug repositioning

drug reprofiling

Strategy that seeks to discover new applications for an existing drug[1] that were not previously referenced and not currently prescribed or investigated.

Note: Various additional synonymous terms have been used to describe the process of drug repurposing. All appear to be used interchangeably.[24]

50. drug safety

Assessment of the nontolerable biological effects of a drug.[1]

Note: Because the nontolerable effects of a drug are directly related to its concentration or dose, safety is generally ranked relative to the dose required to obtain the desirable effect.

See also *therapeutic index*

51. dual–binding site

Presence of two distinct *ligand* binding sites on the same *molecular target*.

52. effective concentration (EC)

Concentration of a substance that produces a defined magnitude of response in a given system.

Note 1: EC_{50} is the median dose that causes 50% of the maximal response.

Note 2: The term usually refers to an agonist[1] in a receptor system effect and could represent either an increase or a decrease in a biological function.

See also *IC_{50}, effective dose*[10,25]

53. effective dose (ED)

Dose of a substance that causes a defined magnitude of response in a given system.

Note: ED_{50} is the median dose that causes 50% of the maximal response.

See also *IC_{50}, effective concentration*[10]

54. efflux pump

Transporter protein located in the membrane of cells which utilizes active transport[1] to move a compound from the internal to the external environment.

See also *P-glycoprotein, ATP-binding cassette*

55. epigenetic(s)

Phenotypic change(s) in an organism brought about by alteration in the expression of genetic information without any change in the genomic sequence itself.

Note: Common examples include changes in nucleotide base methylation and changes in histone acetylation. Changes of this type may become heritable.[2,26]

56. equilibrium solubility

Analytical composition of a mixture or solution which is saturated with one of the components in the designated mixture or solution.

Note 1: Solubility may be expressed in any units corresponding to quantities that denote relative composition, such as mass, amount concentration, molality, etc.

Note 2: The mixture or solution may involve any physical state: solid, liquid, gas, vapor, supercritical fluid.

Note 3: The term "solubility" is also often used in a more general sense to refer to processes and phenomena related to dissolution.

See also *intrinsic solubility, kinetic solubility, solubility, supersaturated solution*[27–29]

57. fast follower

Compound selected as a rapid successor to a lead drug[1] candidate.

Note: Fast followers usually possess a marked increase in one or more pharmacological/pharmaceutical characteristics such as potency, efficacy,[1] *therapeutic index*, or physicochemical parameters (e.g., *solubility*).

58. fingerprint

Representation of a compound or *chemical library* by attributes (descriptors) such as atom connectivities, 3D structure, or physical properties.[30,31]

59. first–in–class

First drug acting on a hitherto unaddressed molecular target to reach the market.

60. follow-on drug

Drug[1] having a similar mechanism of action to an existing drug.

Note: Compounds may be of the same or different chemical class. A therapeutic advantage over *first-in-class* drugs must be demonstrated for regulatory approval.

See also *analogue, analogue-based drug design, congener*[32,33]

61. fragment

Low molecular weight *ligand* (typically smaller than 200 Da) that binds to a target with low affinity[1] but high *ligand efficiency*.

Note: Typically fragments have affinities in a concentration interval from 0.1 to 1.0 mM.

62. fragment-based lead discovery

Screening libraries of low molecular weight compounds (typically 120–250 Da) using sensitive biophysical techniques capable of detecting weakly binding lead compounds.

Note: X-ray structures are frequently used to drive the optimization of fragment *hits* to *leads*.

See also *fragment, ligand efficiency*[34]

63. frequent hitter

Structural feature that regularly results in a positive response in a variety of *high-throughput* or primary screens.

Note: Such compounds often exert their actions through nonspecific mechanisms and are therefore unreliable leads.[35]

64. genomics

Science of using DNA- and RNA-based technologies to demonstrate alterations in gene expression.[2]

65. good laboratory practice (GLP)

Set of principles that provides a framework within which laboratory studies are planned, performed, monitored, recorded, reported, and archived.

Note: These studies are undertaken to generate data by which the hazards and risks to users, consumers, and third parties, including the environment, can be assessed for pharmaceuticals (only preclinical studies), agrochemicals, cosmetics, food additives, feed additives and contaminants, novel foods, biocides, detergents, etc. GLP helps assure regulatory authorities that the data submitted are a true reflection of the results obtained during the study and can therefore be relied upon when making risk/safety assessments.[36]

66. good manufacturing practice (GMP)

Quality assurance process which ensures that medicinal products are consistently produced and controlled to the standards appropriate to their intended use.

Note: Quality standards are those required under marketing authorization or product specification. GMP is concerned with both production and quality control.[37]

67. G-protein

Guanine-binding protein

Member of a family of membrane-associated proteins which on activation by cellular receptors[1] lead to signal transduction.

See also *G-protein-coupled receptor*

68. G-protein-coupled receptor (GPCR)

Large family of cell surface receptors[1] in which seven portions of the protein cross the cellular membrane and which are linked to internal *G-proteins*.

Note 1: Interaction of these receptors with extracellular ligands activates signal transduction pathways and, ultimately, cellular responses.

Note 2: G-protein-coupled receptors are found only in eukaryotes.

See also *G-protein*

69. green chemistry

Invention, design, and application of chemical products and processes to reduce or eliminate the use and generation of hazardous substances.[38]

70. high-throughput screening (HTS)

Method for the rapid assessment of the activity of samples from large compound collections.

Note 1: Typically these assays are carried out in microplates of at least 96 wells using automated or robotic techniques.

Note 2: The rate of at least 10^5 assays per day has been termed ultra high-throughput screening (UHTS).[25]

71. hit

Molecule that produces reproducible activity above a defined threshold in a biological assay and whose structural identity has been established. Note: Hits typically derive from *high-throughput screening* initiatives or other relatively extensive primary assays and do not become true hits until fully validated.

72. hit expansion

Generation of additional compound sets which contain chemical motifs and *scaffolds* that have activity in the primary screen.

Note: This methodology permits the identification of additional *hits* and new *scaffolds* and develops structure–activity relationships[1] around existing hits.

73. hit-to-lead chemistry

Process by which a proven molecule or series derived from *high-throughput screening* or primary screens is chemically optimized to a viable *lead* or series.

74. homology model

Computational representation of a protein built from the 3D structure of a similar protein or proteins using alignment techniques and homology arguments.

75. hydrophobic fragmental constant

Representation of the lipophilicity[1] contribution of a constituent part of a structure to the total lipophilicity.[39]

76. hydrophobic interaction

Entropically driven favorable interaction between nonpolar substructures or surfaces in aqueous solution.

See Ref.[2] for alternative definition

77. IC$_{50}$ (inhibitory concentration 50)

The concentration of an enzyme inhibitor or receptor antagonist[1] that reduces the enzyme activity or agonist response by 50%.

Note: IC$_{50}$ values are influenced by experimental conditions (e.g.: substrate or agonist[1] concentration—which should be specified).

Related terms: *Inhibition constant, K$_i$*[25]

78. inhibition constant, K$_i$

1. Equilibrium dissociation constant of an enzyme inhibitor complex: $K_i = [E][I]/[EI]$.

2. The equilibrium dissociation constant of a receptor–ligand complex.

Note: This value is usually obtained through competition binding experiments, where K_i is determined after the *IC$_{50}$* obtained in a competition assay performed in the presence of a known concentration of labeled reference ligand.

Related term: IC_{50} [25]

79. *in silico* screening

See *virtual screening*

80. International Chemical Identifier (InChI)

Nonproprietary identifier for chemical substances that can be used in printed and electronic data sources to enable easier linking of diverse data compilations.

Note: This IUPAC notation frequently replaces the earlier *SMILES* notation.[40]

81. intercalation

Thermodynamically favorable, reversible inclusion of a molecule (or group) between two other molecules (or groups).

Examples include DNA intercalation.

82. intrinsic solubility

Equilibrium solubility of the uncharged form of an ionizable compound at a pH where it is fully unionized.

$$HA(s) = HA(aq) \qquad\qquad (24.1)$$

$$HA(aq) = H^+ + A^- \qquad\qquad (24.2)$$

The intrinsic solubility can be determined from the analytical composition at a pH where [HA] is very much greater than $[A^-]$.

See also *equilibrium solubility, kinetic solubility, solubility, supersaturated solution*[27–29]

83. Investigational New Drug (IND)

Drug[1] not yet approved for general use by the national authority, such as the Food and Drug Administration of the United States of America, but undergoing clinical investigation to assess its safety and efficacy.

84. ionotropic receptor

Transmembrane ion channel that opens or closes in response to the binding of a *ligand*.

85. kinase

phosphotransferase

Enzyme[1] that transfers a phosphate group from high-energy donor molecules, such as ATP, to specific target molecules.

86. kinetic solubility
turbidimetric solubility
Composition of a solution with respect to a compound when its induced precipitate first appears.
See also *equilibrium solubility, intrinsic solubility, solubility, supersaturated solution*[41]

87. lead
Compound (or compound series) that satisfies predefined minimum criteria for further structure and activity optimization.
Note: Typically, a lead will demonstrate appropriate activity, selectivity, tractable SAR,[1] and have confirmed activity in a relevant cell-based assay.
See *lead validation*[24]

88. lead validation
Process by which a *lead* compound is authenticated by the confirmation of its expected pharmacological properties.
Note: Usually a cluster of structurally similar compounds showing discernable structure–activity relationships will support the validation process.

89. ligand
Ion or molecule that binds to a *molecular target* to elicit, block, or attenuate a biological response.

90. ligand-based drug design
Method of drug discovery and/or optimization in which the pursuit of new structures and/or structural modifications is based upon one or more *ligands* known to interact with the *molecular target* of interest.
Note: This approach is applicable even when no structural detail of the target is known. In such cases, a series of *analogues* is usually prepared and tested to produce structure–activity relationship data[1] that can be extrapolated to indirectly derive a topographical map of the biological surface.
See also *analogue-based drug discovery, structure-based drug design*

91. ligand efficiency (LE)
Measure of the free energy of binding per heavy atom count (i.e., non-hydrogen) of a molecule.
Note 1: It is used to rank the quality of molecules in drug discovery, particularly in fragment-based lead discovery.
Note 2: A LE value of 1.25 kJ/mol/nonhydrogen atom is the minimum requirement of a good lead or fragment.[34,42]

92. ligand lipophilic efficiency (LLE)
lipophilic efficiency

Parameter used to identify *ligands* with a high degree of specific inter-action toward the desired *molecular target*.

Note 1: The potency of a *ligand* toward a *molecular target* may be dom-inated by nonspecific partitioning from the aqueous phase. It can be advantageous to separate out the nonspecific component of the potency in order to identify more specific interactions; typically using an equation such as:

LLE, symbol E_{LL}, is defined by the logarithm of the potency minus a lipophilicity measure, where a typical example would be:

$$E_{LL} = -\log(IC_{50}) - \log P$$

Note 2: E_{LL} can be regarded as part of a thermodynamic cycle used as a complementary measure to potency in the search for specific target interactions. In this case, the dissociation constant, K_d, is a more appropriate measure than IC_{50} since it refers to the Gibbs Energy of the binding process[43]:

$$E_{LL} = -\log K_d - \log P$$

93. Lipinski's rule
See *rule of five*

94. log *D*
lg *D*
Logarithm of the apparent partition coefficient at a specified pH.
See *CLOGP*, log *P*

95. log *P*
lg *P*
Measure of the lipophilicity[1] of a compound by its partition coeffi-cient between an apolar solvent (e.g., 1-octanol) and an aqueous buffer.
Thus, *P* is the quotient of the concentration of nonionized drug in the solvent divided by the respective concentration in buffer.
See *CLOGP*, log *D*

96. Markush structure
Generalized formula or description for a related set of chemical com-pounds used in patent applications and chemical papers.

97. metabotropic receptor
Receptor[1] that modulates electric potential-gated channels via *G-proteins*.

Note: The interaction can occur entirely within the membrane or by the generation of diffusible second messengers. The involvement of *G-proteins* causes the activation of these receptors to last tens of seconds to minutes, in contrast with the brief effect of *ionotropic receptors*.[44]

98. me-too drug

Term applied to drug analogues that offer no significant advantage over the prototype compound.

Note: Continued use of this term is not recommended.

See *analogue, follow-on drug*[21]

99. microarray

Planar surface where assay reagents and samples are distributed as submicroliter drops.

Note: This screening format is a direct offshoot of genomic microarray technologies and makes use of ultra-low volume miniaturization provided by nanodispensing technologies.[25]

100. microRNA (miRNA)

Small single-stranded RNA molecules that play a significant role in the posttranscriptional regulation of gene expression.

Note: MicroRNA usually comprises approximately 22 nucleotides.[45]

101. molecular descriptor

Parameter that characterizes a specific structural or physicochemical aspect of a molecule.[46]

102. molecular diversity

See *diversity*

103. molecular dynamics

Computational simulation of the motion of atoms in a molecule or of individual atoms or molecules in solids, liquids, and gases, according to Newton's laws of motion.

Note: The forces acting on the atoms, required to simulate their motions, are generally calculated using molecular mechanics force fields.

See Ref.[2] for alternative definition[47]

104. molecular similarity

Measure of the coincidence or overlap between the structural and physicochemical profiles of compounds.

105. molecular target

Protein (e.g., receptor,[1] enzyme,[1] or ion channel), RNA, or DNA that is implicated in a clinical disorder or the propagation of any untoward event.

Note: Usually biochemical, pharmacological, or genomic information supporting the role of such a target in disease will be available.

106. multidrug resistance (MDR)

Characteristic of cells that confers resistance to the effects of several different classes of drugs.

Note: There are several forms of drug resistance. Each is determined by genes that govern how cells will respond to chemical agents. One type of multidrug resistance involves the ability to eject several drugs out of cells (e.g., *efflux pumps* such as *P-glycoprotein*).[48]

107. multi-parameter optimization (MPO)

Drug-likeness penetrability algorithm derived from *CLOGP*, *dogD*, molecular weight, *topological polar surface area*, number of hydrogen bond donors, and pKa.

Note: The MPO desirability score is larger or equal to 4 on a scale of 0–6.[49]

108. multitarget-directed ligand (MTDL)

multitarget drug

Ligand acting on more than one distinct *molecular target*. Targets may be of the same or different mechanistic classes.

109. multitarget drug discovery (MTDD)

Deliberate design of compounds which act on more than one *molecular target*.

110. Murcko assembly

Core *scaffold* of a molecule that remains after all chain substituents that do not terminate in a ring are removed. Single atoms connected by a double bond are typically also retained.[50]

111. neglected disease

Term used to emphasize an imbalance between therapeutic needs and resource allocation.

Examples include tropical infections which are especially endemic in regions of Africa, Asia, Central and South America.

Note 1: Different groups define neglected diseases differently, but typically they include diseases such as schistosomiasis, Chagas disease, tuberculosis, malaria, as well as other parasitic and vector-borne diseases.

Note 2: The terms tropical disease and neglected disease are sometimes indistinguishable, or used interchangeably.

See also *orphan disease*

112. neural network

Statistical analysis procedure based on models of nervous system learning in animals.

Note: Neural networks have the ability to "learn" from a collection of examples to discover patterns and trends. These data-mining techniques can be used in forecasting or prediction.[51]

113. neutral antagonist (in pharmacology)
Ligand that blocks the responses of a receptor[1] to both agonists[1] and inverse agonists[1] with the same intensity. It binds to the receptor without evoking any change of conformation or change to the ratio of activated to inactivated conformations.
Note: Perfect neutral antagonism is difficult to achieve.

114. new chemical entity
Drug[1] that contains no active moiety previously approved for use by the national authority, such as the Food and Drug Administration of the United States of America.
See also *new molecular entity*[52]

115. new molecular entity
Active ingredient that has never before been approved in any form by the national authority, such as the Food and Drug Administration of the United States of America.
See also *new chemical entity*[52]

116. noncompetitive antagonist
Functional antagonist[1] that either binds irreversibly to a receptor[1] or to a site distinct from that of the natural agonist.[1]
See *allosteric antagonist*

117. nuclear hormone receptor
nuclear receptor
Ligand-activated transcription factor that regulates gene expression by interacting with specific DNA sequences upstream of its target gene(s).[53]

118. obviousness
Term associated with intellectual property wherein the latter's *patentability* is assessed relative to the combination of more than one item of "prior art."
Note: To be patentable within the context of medicinal chemistry, a given compound must be: (i) novel, in that its specific arrangement of atoms has never been previously disclosed; (ii) nonobvious, in that its specific arrangement of atoms is not readily suggested to be of benefit by a person having ordinary skill in the art upon considering two or more other, previously disclosed structures; and, (iii) useful, in that it should have some benefit, the disclosure of the latter encompassing a valid "reduction to practice."
See also *patentability*

119. off-target effect

Pharmacological action induced by any molecule at molecular/biological sites distinct from that for which it was designed.

Note: Such effects are dose-dependent and may be beneficial, adverse, or neutral.

120. orphan disease

Disease for which drug research, development, and marketing are economically unfavorable.

Note 1: The poor commercial environment could be due to a lack of economic incentives or a lack of understanding of the diseases or a combination of both.

Note 2: Sometimes the term "rare disease" is used synonymously with orphan disease, although there is a slight difference. For example, a rare disease is so uncommon that there is no drug development effort.

Note 3: Which diseases are classified as orphan depends strongly on the country that classifies it. In the United States, for example, any disease affecting less than 200,000 people is considered an orphan or rare disease. Europe and countries such as Japan, Australia, and Singapore have a different definition.

See also *orphan drug, neglected disease*

121. orphan drug

Pharmaceutical agent that has been approved specifically to treat a rare and commercially unfavorable medical condition.

See also *orphan disease*

122. orphan receptor

Receptor for which an endogenous *ligand* has yet to be identified.

123. parallel synthesis

Simultaneous preparation of sets of discrete compounds in arrays of physically separate reaction vessels or microcompartments without interchange of intermediates during the assembly process.[54]

124. patentability

Set of criteria which must be satisfied in order to achieve commercial exclusivity for an invention.

Note: These criteria are essentially the same in all major countries and include suitability, novelty, inventiveness, utility, and the provision of an adequate description.

See also *obviousness*[55]

125. peptidase
      See *proteinase*
126. peptoid
      Peptide-like oligomer consisting of repeating *N*-substituted glycine units.

      Where R_1 to R_4 are typically alkyl or aryl and may be the same or different.[54]

127. P–glycoprotein (Pgp)
      ATP-binding cassette transporter responsible for the efflux of small molecules from cells.
      Note: P-glycoproteins can play a major role in limiting brain penetration and restricting the intestinal absorption of drugs. Their overexpression in cancer cells becomes a common mechanism of *multidrug resistance*.
      See also *blood–brain barrier, efflux pump*[56]

128. pharmacogenetics
      Study of inherited differences (variation) in drug metabolism[1] and response.
      See *pharmacogenomics* [57]

129. pharmacogenomics
      General study of all of the many different genes that determine drug behavior.
      Note: The distinction between the terms *pharmacogenetics* and *pharmacogenomics* has blurred with time and they are now frequently used interchangeably.
      See *pharmacogenetics*[57]

130. phase 0 clinical studies
      Exploratory investigational new drug
      Exploratory first-in-human trials that involve microdosing of drug[1] to allow the assessment of pharmacokinetic parameters with limited drug exposure.

Note: These trials have no therapeutic or diagnostic intent but are designed to assist decision making by providing bioavailability, metabolism, and other limited data from a small number of patients.
See also *phase I, II, III, IV clinical studies*[52]

131. phase I clinical studies
Initial introduction of an *investigational new drug* into humans.
Note: These studies are designed to determine the metabolic and pharmacologic actions of the drug[1] in humans, the side effects associated with increasing doses, and, if possible, to gain early evidence on effectiveness.
See also *phase 0, II, III, IV clinical studies*[52]

132. phase II clinical studies
Controlled clinical studies conducted in a limited number of individuals to obtain some preliminary data on the effectiveness of an *investigational new drug* for a particular indication or indications in patients with the disease or condition.
Note: Phase II clinical trials have two subclasses, IIa and IIb. Phase IIa trials are essentially pilot clinical trials designed to evaluate efficacy (and safety) in selected populations of patients with the disease or condition to be treated, diagnosed, or prevented. Phase IIb trials extend those of Phase IIa to well-controlled trials that evaluate the same parameters in similar patient populations.
See also *phase 0, I, III, IV clinical studies*[52]

133. phase III clinical studies
Expanded controlled and uncontrolled trials in humans.
Note: These trials are performed after preliminary evidence suggesting effectiveness of the drug[1] has been obtained in *Phase II*, and are intended to gather the additional information about effectiveness and safety that is needed to evaluate the overall benefit–risk relationship of the drug.
See also *phase 0, I, II, IV clinical studies*[52]

134. phase IV clinical studies
Extended postmarketing studies in humans.
Note: These trials are designed to broaden information concerning treatment risks, benefits, and optimal drug use.
See also *phase 0, I, II, III clinical studies*[52]

135. phenotypic screening
Evaluation of compounds (small molecules, peptides, *siRNA*, etc.) in cells, tissues, or organisms for their ability to modify the system in a measurable manner.

Note: Phenotypic screening differs from target-based screens in that it assesses the overall response of the compound under investigation rather than its specific response on a purified molecular target. Advances in molecular biology resulted in a marked shift away from phenotypic screening, although the latter is now regaining popularity.[58]

136. phosphotransferase

See *kinase*

137. pipeline

Discovery and development compound portfolio of a pharmaceutical company or research organization.

138. pivotal study

Experiment that provides strong support for or against the drug[1] or *molecular target* under investigation.

See also *proof of concept*

139. polar surface area

topological polar surface area

Surface area over all polar atoms (usually oxygen and nitrogen), including any attached hydrogen atoms, of a molecule.

Note: Polar surface area is a commonly used metric (cf. *molecular descriptor*) for the optimization of cell permeability. Molecules with a PSA of greater than 1.4 nm^2 are usually poor at permeating cell membranes. For molecules to penetrate the *blood–brain barrier*, the polar surface area should normally be smaller than 0.6 nm^2 although values up to 0.9 nm^2 can be tolerated.

See also *blood–brain barrier*[59]

140. polymorphism

Ability of a compound to exist in more than one crystalline form (polymorph) with each having a different arrangement or conformation of the molecules within the crystal lattice.

Note: Polymorphs generally differ in their melting points, *solubility*, and relative intestinal absorption such that optimal polymorphs can markedly enhance the attractiveness of some drugs.[60]

141. positron emission tomography (PET)

Imaging technique used to visualize small amounts of a compound in biological tissues by the use of radionuclide labels. These radionuclides, such as ^{11}C, ^{18}F, ^{13}N, and ^{15}O, are positron emitters.

142. potential genotoxic impurity

Confirmed or suspected presence of one or more known genotoxic materials in an active pharmaceutical ingredient.

Note: Typically such impurities arise from reactive intermediates involved in the synthetic pathway and have the potential to adversely affect subsequent genotoxicity tests.

143. preclinical candidate (PCC)
safety assessment candidate
Optimized *lead* compound successfully passing key screening, selectivity, and physicochemical criteria sufficient to warrant further detailed pharmacological and pharmacokinetic evaluation in animal models.
Note: Critical studies usually include bioavailability, therapeutic efficacy in an appropriate disease model, and side effect profiling.

144. privileged structure
Substructural feature which confers desirable (often drug-like) properties on compounds containing that feature. They often consist of a semirigid *scaffold* which presents multiple hydrophobic residues without undergoing hydrophobic collapse.
Note 1: For example, diazepam (below) in which the diphenylmethane moiety prevents association of the aromatic rings.

Note 2: Such structures are commonly found to confer activity against different targets belonging to the same receptor[1] family.[61]

145. proof of concept (in pharmacology)
Procedure by which a specific therapeutic mechanism or treatment/diagnostic paradigm is shown to be beneficial.
Note: Similar to, and often simultaneously associated with, that for a new drug[1] candidate. This process usually involves an early supporting step prior to clinical testing and final validation within human studies.
See also *lead validation, pivotal study*

146. proteinase
protease
Enzyme[1] that catalyzes the hydrolysis of proteins.

Note: Usually several proteolytic enzymes are necessary for the complete breakdown of polypeptides to their amino acids.[2]

147. protein data bank (PDB)

Repository for the three-dimensional structural data of large biological molecules including proteins and nucleic acids.

Note: These high-resolution structures, generated predominantly by X-ray or NMR spectroscopic techniques, provide a major resource for structural biology.[62]

148. protein–protein interaction (PPI)

Association of one protein with one or more other proteins to form either homomeric or heteromeric proteins.

Note: Such associations are common in biological systems and are responsible for the regulation of numerous cellular functions in addition to the mediation of disease morphology where aberrant interactions play a significant role.

149. prototype drug

Early compound that has biological properties suitable for *target* validation but may not necessarily be adequate for clinical studies.

150. QTc interval

Time between the start of the Q wave and the end of the T wave in the heart's electrical cycle. When corrected for individual heart rate, it becomes known as the corrected QT, or QTc interval.

Note: Significant prolongation of the QTc interval by pharmaceutical agents can induce life-threatening ventricular arrythmia (Torsades de Pointes), typically by interacting with the hERG channel.[63]

151. retrosynthesis

Process of conceptually deconstructing complex molecules into simpler fragments capable of chemical manipulation to reform the parent compound.

152. RNA interference (RNAi)

Physiological process used by cells to attenuate, or silence, the activity of specific genes.

Note: RNAi offers a means to manipulate gene expression experimentally and to probe gene function on a whole-genome scale.

See also *siRNA*

153. rule of five

Set of *molecular descriptors* used to assess the potential oral bioavailability of a compound.

Note 1: Characterized by a mass of less than 500 Da, less than or equal to 5 hydrogen bond donors, less than or equal to 10 hydrogen bond acceptors (usually using the $N+O$ count as a surrogate for the number of hydrogen bond acceptors), and a *CLOGP* less than or equal to 5.

Note 2: Frequently used to profile a *chemical library* or *virtual chemical library* with respect to the proportion of *drug-like* members which it contains. Often used as a surrogate for *drug-likeness*.

Note 3: While these criteria are frequently referred to as Lipinski's rules, the de-personalized term "rule of five" is preferred.[64]

154. rule of three

Set of *molecular descriptors* used to assess the quality of *hit* or *lead* molecules.

Note 1: Most commonly applied to *fragments* which ideally are characterized by a mass of less than 300 Da, less than or equal to 3 hydrogen bond donors, less than or equal to 3 hydrogen bond acceptors, and a *CLOGP* of less than or equal to 3. In addition, the number of rotatable bonds should average or be less than 3 and the *polar surface area* about $0.6 \, \mathrm{nm}^2$.

Note 2: Often used to distinguish "lead-like" from *drug-like* molecules. See *fragment-based lead discovery*[65]

155. scaffold

template

Core portion of a molecule common to all members of a *chemical library* or compound series.

156. scaffold hopping

Exchange of one *scaffold* for another while maintaining molecular features that are important for biological properties.[66]

157. similarity

See *molecular similarity*

158. siRNA

See *small inhibitory double-stranded RNA*

159. site–directed mutagenesis

Molecular biology technique in which mutations are created at one or more defined sites in a DNA molecule.

Note: Typically used in *molecular target* validation and to determine whether specific amino acids are involved at *ligand* or substrate interaction sites.

160. small inhibitory double-stranded RNA (siRNA)

small interfering double-stranded RNA

Small RNA fragments that can combine with specific genes to silence their expression.

Note: siRNAs are powerful tools that are often used for manipulating gene expression during *molecular target* validation.

See also *RNAi*[67]

161. SMILES notation (Simplified Molecular Input Line Entry System)

String notation used to describe the atom type and connectivity of molecular structures.

Note: Primarily used to input chemical structures into electronic databases and now frequently replaced by the *InChI* notation.[68]

162. solubility

Analytical composition of a mixture or solution which is saturated with one of the components of the mixture or solution, expressed in terms of the proportion of the designated component in the designated mixture or solution.

Note 1: Solubility may be expressed in any units corresponding to quantities that denote relative composition, such as mass, amount concentration, molality, etc.

Note 2: The mixture or solution may involve any physical state: solid, liquid, gas, vapor, supercritical fluid.

Note 3: The term "solubility" is also often used in a more general sense to refer to processes and phenomena related to dissolution.

See *equilibrium solubility, intrinsic solubility, kinetic solubility, supersaturated solution*[29]

163. spare receptor

receptor reserve

Residual binding site still available to an endogenous *ligand* after sufficient sites have already been filled to elicit the maximal response possible for that particular biological system.

Note: Xenobiotics[1] such as drugs[1] may similarly interact with such receptors[1] but by definition, their identification and quantification occurs via use of the natural *ligand*.

164. stem cell

Multipotent cell with mitotic potential that may serve as a precursor for many kinds of differentiated cells.

Note: Unipotent stem cells can differentiate into one mature cell type only.[2]

165. structural alert

Chemical features present in a *hit* or *lead* molecule indicative of potential toxicity.

Note: Typically such features include chemically reactive functionality and components known to metabolize to chemically reactive entities. Examples include anhydrides, aromatic amines, and epoxides.

166. supersaturated solution

Solution that has a greater composition of a solute than one that is in equilibrium with undissolved solute at specified values of temperature and pressure.

See also *equilibrium solubility, intrinsic solubility, kinetic solubility*[29]

167. systems biology

Integration of high-throughput biology measurements with computational models that study the projection of the mechanistic characteristics of metabolic and signaling pathways onto physiological and pathological phenotypes.[69]

168. target

See *molecular target*

169. targeted drug delivery

Approach to target a drug to a specific tissue or *molecular target* using a prodrug[1] or antibody recognition systems.

See site-specific delivery in Ref. [1]

170. target validation

Process by which a protein, RNA, or DNA is implicated in a biological pathway thought to be of relevance to a disease or adverse pathology.

Note: Typically validation will involve location of the *molecular target* in relevant cells, organs, or tissues, evidence for its upregulation/activation in the disorder, and the ability to attenuate adverse responses by agents known to interfere with the *target*.

171. tautomer

Structural isomer that can readily convert to another form which differs only by the attachment position of a hydrogen atom and the location of double bond(s).

Note: In most cases, these isomers are formed by a proton shift to or from heteroatoms such as O, N, or S as typified by the enol and keto forms of carbonyl compounds. Tautomers rapidly interconvert by proton transfer and are usually in equilibrium with one another.

See also *tautomerism*[1,70]

172. tautomerism

Reversible interconversion of two different *tautomers*.

Note: For an expanded definition see Ref. [1]

173. template

See *scaffold*

174. therapeutic index
therapeutic ratio
Ratio of the exposure/concentration of a therapeutic agent that causes beneficial effects to that which causes the first observed adverse effect.
Note: A commonly used measure of therapeutic index is the toxic dose of a drug for 50% of a population divided by the minimum effective dose for 50% of a population.

175. topological polar surface area
See *polar surface area*

176. training set
Specific group of compounds selected for characterization of both the *molecular descriptors* and the measured values of the targeted property.
Note: Statistical methods applied to the set are used to derive a function between the *molecular descriptors* and the targeted property.

177. ultra high-throughput screening (UHTS)
See *high-throughput screening*

178. unmet medical need
Term used for diseases or other disorders for which no optimal therapeutic options exist.

179. virtual chemical library
Collection of chemical structures constructed solely in electronic form or on paper.
Note: The building blocks required for such a library may not exist, and the chemical steps for such a library may not have been tested. These libraries are used in the design and evaluation of possible libraries.
See *virtual screening*[71]

180. virtual screening
in silico screening
Evaluation of compounds using computational methods.
Note: The source of the model could be a macromolecular structure or based on physicochemical parameters or *ligand* structure–activity relationships.

181. volume of distribution (V_d)
Apparent (hypothetical) volume of fluid required to contain the total amount of a substance in the body at the same concentration as that present in the plasma assuming equilibrium has been attained.[10]

181. wild-type receptor
Receptor that occurs naturally in human and other species.

Annex 1: Acronyms Used in Medicinal Chemistry Literature

ADMET Absorption, distribution, metabolism, excretion, toxicology
ABC ATP-binding cassette
BBB Blood–brain barrier
CRO Contract research organization
DOS Diversity-oriented synthesis
EC Effective concentration
ED Effective dose
GLP Good Laboratory Practice
GMP Good Manufacturing Practice
GPCR G-protein-coupled receptor
HTS High-throughput screening
InChI International Chemical identifier
IND Investigational new drug
LE Ligand efficiency
LLE Ligand lipophilic efficiency
MDR Multidrug resistance
MPO Multi-parameter optimization
MTDL Multitarget-directed ligand
MTDD Multitarget drug discovery
PgP P-Glycoprotein
PDB Protein data bank
PPI Protein–protein interaction
PET Positron emission tomography
SAR Structure–activity relationship
siRNA Small inhibitory double-stranded RNA
SMILES Simplified molecular input line entry system
UHTS Ultra high-throughput screening
V_d Volume of distribution

ACKNOWLEDGMENT

The authors are grateful to the following individuals for their valuable support, comments, or suggestions: Koen Augustyns (University of Antwerp, Belgium), John Comer (Sirius, UK), Gunda Georg (University of Minnesota, USA), William J. Greenlee (MedChem Discovery Consulting, USA), Trevor Grinter, (GSK, UK), Philip Jones (Consultant, UK), Hugo Kubinyi (Germany), John Macor (BMS, USA), Carlo Melchiorre (University of Bologna, Italy), J. Richard Morphy (Schering Plough Research, UK), Philip Portoghese (University of Minnesota, USA), Hans Ulrich Stiltz (Sanofi-Aventis, Germany), Hiromitsu Takayama (Chiba University, Japan), Antoni Torrens (Esteve Quimica, Spain), Shaomeng Wang (University of Michigan, USA), and Camille G. Wermuth (Prestwick Chemical, France).

REFERENCES

1. Wermuth, C. G.; Ganellin, C. R.; Lindberg, P.; Mitscher, L. A. *Pure Appl. Chem.* **1998**, *70*, 1129.
2. McNaught, A. D.; Wilkinson, A. *Compendium of Chemical Terminology (the "Gold Book")*, 2nd ed.; Blackwell Scientific: Oxford, 1997;XML on-line corrected version: http://goldbook. iupac.org (2006-) created by M. Nic, J. Jirat, B. Kosata; updates compiled by A. Jenkins.
3. Fischer, J., Ganellin, C.R., Eds.; In *Analogue-Based Drug Discovery*; Wiley-VCH: Weinheim, 2006.
4. Dean, M.; Hamon, Y.; Chimini, G. *J. Lipid Res.* **2001**, *42*, 1007.
5. Bringmann, G.; Mortimer, A. J. P.; Keller, P. A.; Gresser, M. J.; Garner, J.; Breuning, M. *Angew. Chem. Int. Ed. Engl.* **2005**, *44*, 5384.
6. Sinz, M. W. *Annu. Rep. Med. Chem.* **2008**, *43*, 405.
7. Ng, R. *Drugs from Discovery to Approval.* Wiley-Blackwell: Hoboken, NJ, 2009. ISBN 978-0-470-19510-9.
8. Kelder, J.; Grootenhuis, P. D. J.; Bayada, D. M.; Delbressine, L. P. C.; Ploemen, J. P. *Pharm. Res.* **1999**, *16*, 1514.
9. Hitchcock, S. A.; Pennington, L. D. *J. Med. Chem.* **2006**, *49*, 7559.
10. Nordberg, M.; Duffus, J. H.; Templeton, D. M. *Pure Appl. Chem.* **2007**, *79*, 1583.
11. Begley, T. P. *Nat. Chem. Biol.* **2005**, *1*, 236.
12. Pubchem; http://pubchem.ncbi.nlm.nih.gov/.
13. Brown, F. K. *Annu. Rep. Med. Chem.* **1998**, *33*, 375.
14. Kubinyi, H. www.kubinyi.de/schering58-2006.pdf.
15. Carson, K. G.; Jaffee, B. D.; Harriman, G. C. B. *Annu. Rep. Med. Chem.* **2004**, *39*, 149.
16. Downs, G. M.; Willett, P.; Fisanick, W. *J. Chem. Inf. Comput. Sci.* **1994**, *34*, 1094.
17. Brown, R. D.; Martin, Y. C. *J. Chem. Inf. Comput. Sci.* **1996**, *36*, 572.
18. Gorse, D.; Rees, A.; Kaczorek, M.; Lahana, R. *Drug Discov. Today* **1999**, *4*, 257.
19. Lau, W. M.; White, A. W.; Gallagher, S. J.; Donaldson, M.; McNaughton, G.; Heard, C. M. *Curr. Pharm. Design* **2008**, *14*, 794.
20. Schueller, F. W. *Chemobiodynamics and Drug Design*; McGraw-Hill, The Blakiston Division: New York, 1960; p 405.
21. Wermuth, C. G. *Drug Discov. Today* **2006**, *11*, 348.
22. Owens, J. *Nat. Rev. Drug Discov.* **2007**, *6*, 187.
23. Cheng, A. C.; Coleman, R. G.; Smyth, K. T.; Cao, Q.; Soulard, P.; Caffrey, D. R.; Salzberg, A. C.; Huang, E. S. *Nat. Biotechnol.* **2007**, *25*, 71.
24. Doan, T. L.; Pollastri, M.; Walters, M. A.; Georg, G. I. *Annu. Rep. Med. Chem.* **2011**, *46*, 385.
25. Proudfoot, J.; Nosjean, O.; Blanchard, J.; Wang, J.; Besson, D.; Crankshaw, D.; Gauglitz, G.; Hertzberg, R.; Homon, C.; Llewellyn, L.; Neubig, R.; Walker, L.; Villa, P. *Pure Appl. Chem.* **2011**, *83*, 1129.
26. Duffus, J. H.; Nordberg, M.; Templeton, D. M. *Pure Appl. Chem.* **2007**, *79*, 1153.
27. Box, K. J.; Comer, J. E. *Curr. Drug Metab.* **2008**, *9*, 869.
28. Box, K. J.; Comer, J. E.; Gravestock, T.; Stuart, M. *Chem. Biodivers.* **2009**, *6*, 1767.
29. Gamsjäger, H.; Lorimer, J. W.; Scharlin, P.; Shaw, D. G. *Pure Appl. Chem.* **2008**, *80*, 233.
30. Pickett, S. D.; Luttmann, C.; Guerin, V.; Laoui, A.; James, E. *J. Chem. Inf. Comput. Sci.* **1998**, *38*, 144.
31. McGregor, M. J.; Muskal, S. M. *J. Chem. Inf. Comput. Sci.* **1999**, *39*, 569.
32. Zhao, H.; Guo, Z. *Drug Discov. Today* **2009**, *14*, 516.
33. Di Massi, J. A.; Faden, L. A. *Nat. Rev. Drug Discov.* **2011**, *10*, 23.
34. Congreave, M.; Murray, C. W.; Carr, R.; Rees, D. C. *Annu. Rep. Med. Chem.* **2007**, *42*, 431.
35. Baell, J. B.; Holloway, G. A. *J. Med. Chem.* **2010**, *53*, 2719.
36. Medicines and Healthcare Products Regulatory Agency; http://www.mhra.gov.uk/ Howweregulate/Medicines/Inspectionandstandards/GoodLaboratoryPractice/Structure/ index.htm.

37. Medicines and Healthcare Products Regulatory Agency; http://www.mhra.gov.uk/ Howweregulate/Medicines/Inspectionandstandards/GoodManufacturingPractice/ index.htm.
38. IUPAC Green Chemistry Directory; http://www.incaweb.org/transit/iupacgcdir/ overview.htm.
39. Rekker, R. F. *Hydrophobic Fragmental Constant (Pharmacochemistry Library)*; Elsevier Scientific: New York, 1977.
40. The IUPAC International Chemical Identifier (InChI); http://IUPAC.org/inchi/.
41. Solubility Definitions; http://sirius-analytical.com; Application Note 08/12.
42. Hopkins, A. L.; Groom, C. R.; Alex, A. *Drug Discov. Today* **2004**, *9*, 430.
43. Leeson, P. D.; Springthorpe, B. *Nat. Rev. Drug Discov.* **2007**, *6*, 881.
44. Katzung, B. G. *Basic and Clinical Pharmacology*, 9th ed.; Lange Medical Books/McGraw-Hill: New York, 2004.
45. Loomis, K. A.; Brock, G. J. *Annu. Rep. Med. Chem.* **2011**, *46*, 351.
46. van de Waterbeemd, H.; Testa, B. *Adv. Drug Res.* **1987**, *16*, 85.
47. van de Waterbeemd, H.; Carter, R. E.; Grassy, G.; Kubiny, H.; Martin, Y. C.; Tute, M. S.; Willett, P. *Pure Appl. Chem.* **1997**, *69*, 1137.
48. Chronic Lymphocytic Leukemia Research Consortium Glossary; http://cll.ucsd.edu/ glossary/glossary_m.html.
49. Wager, T. T.; Hou, X.; Verhoest, P. R.; Villalobos, A. *ACS Chem. Neurosci.* **2010**, *1*, 435.
50. Bemis, G. W.; Murcko, M. A. *J. Med. Chem.* **1996**, *39*, 2887.
51. Australian Academy of Science; http://www.science.org.au/nova/050/050glo.htm.
52. US Food and Drug Administration; http://www.fda.gov.
53. NIH Center for Macromolecular Modeling and Bioinformatics; http://www.ks.uiuc. edu/Research/pro_DNA/ster_horm_rec/.
54. Maclean, D.; Baldwin, J. J.; Ivanov, V. T.; Kato, Y.; Shaw, A.; Schneider, P.; Gordon, E. M. *Pure Appl. Chem.* **1999**, *71*, 2349.
55. Smith, S. C. In *Analogue-Based Drug Discovery II*; Fischer, J.; Ganellin, C.R., Eds.; Wiley-VCH: Weinheim, 2010; p 83.
56. Egan, W. J. *Annu. Rep. Med. Chem.* **2007**, *42*, 449.
57. National Center for Biotechnology Information. http://www.ncbi.nlm.nih.gov/ About/primer/pharm.html.
58. Kotz, J. *SciBX* **2012**, *5*(15). http://dx.doi.org/10.1038/scibx.2012.380.
59. Kelder, J.; Grootenhuis, P. D. J.; Bayada, D. M.; Delbressine, L. P. C.; Ploemen, J.-P. *Pharmaceutical Res.* **1999**, *16*, 1514.
60. Grant, D. J. W. Theory and Origin of Polymorphism. In *Polymorphism in Pharmaceutical Solids*; Brittain, H.G., Ed.; Marcel Dekker: New York, 1999; pp 1–34.
61. Evans, B. E.; Rittle, K. E.; Bock, M. G.; DiPardo, R. M.; Freidinger, R. M.; Whitter, W. L.; Lundell, G. F.; Veber, D. F.; Anderson, P. S.; Chang, R. S. *J. Med. Chem.* **1988**, *31*, 2235.
62. Worldwide Protein Databank; http://www.wwpdb.org.
63. Fermini, B.; Fossa, A. F. *Annu. Rep. Med. Chem.* **2004**, *39*, 323.
64. Lipinski, C.; Lombardo, F.; Dominy, B. W.; Feeney, P. J. *Adv. Drug. Deliv. Rev.* **1997**, *23*, 3.
65. Rees, D. C.; Congreve, M.; Murray, C. W.; Carr, R. *Nat. Rev. Drug Discov.* **2004**, *3*, 660.
66. See Böhm, H.-J.; Flohr, A.; Stahl, M. *Drug Discov. Today: Technol.* **2004**, *1*, 217.
67. Sharmoon, B.-M.; Reinhard, C. *Annu. Rep. Med. Chem.* **2003**, *38*, 261.
68. Daylight Chemical Information Systems; http://www.daylight.com.
69. Gomes, B.; de Graaf, D. *Annu. Rep. Med. Chem.* **2007**, *42*, 393.
70. Patrick, G. L. In *Organic Chemistry*, Instant Notes Series: Chemistry Series; 2nd ed. Taylor & Francis: Routledge, London, New York, 2004.
71. Sheridan, R. P.; Kearsley, S. K. *J. Chem. Inf. Comput. Sci.* **1995**, *35*, 310.

Case Histories and NCEs

Section Editor: Joanne Bronson
Bristol-Myers Squibb, Wallingford, Connecticut

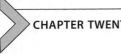

CHAPTER TWENTY-FIVE

Case History: Xalkori™ (Crizotinib), a Potent and Selective Dual Inhibitor of Mesenchymal Epithelial Transition (MET) and Anaplastic Lymphoma Kinase (ALK) for Cancer Treatment

J. Jean Cui, Michele McTigue, Robert Kania, Martin Edwards

La Jolla Laboratories, Pfizer Worldwide Research and Development, San Diego, California, USA

Contents

1. INTRODUCTION

Receptor tyrosine kinases (RTKs) play fundamental roles in cellular processes, including cell proliferation, migration, metabolism, differentiation, and survival.[1] Constitutively enhanced RTK activities resulting from point mutation, amplification or rearrangement of the corresponding genes,

Annual Reports in Medicinal Chemistry, Volume 48
ISSN 0065-7743
http://dx.doi.org/10.1016/B978-0-12-417150-3.00025-9

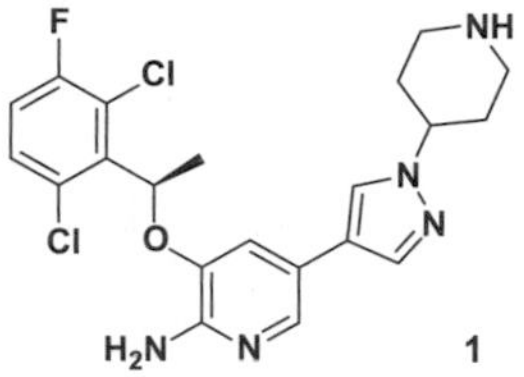

Figure 25.1 Xalkori™ (crizotinib, **PF-02341066**).

as well as aberrant RTK activation through enhanced autocrine or paracrine ligand production have been implicated in the development and progression of many types of cancer.[2] Since the first Abelson (ABL) tyrosine kinase inhibitor imatinib was approved for the treatment of chronic myelogenous leukemia in 2001, inhibitors of a number of RTKs including vascular endothelial growth factor receptor (VEGFR) (axitinib), epidermal growth factor receptor (EGFR) (gefitinib), and mast/stem cell growth factor receptor (c-KIT) (sunitinib) inhibitors have been successfully discovered and developed against aberrant RTK signaling in cancers and approved for the treatments of cancer patients. Xalkori™ (crizotinib, **PF-02341066**, Fig. 25.1), a RTK inhibitor targeting MET (mesenchymal epithelial transition factor) and ALK (anaplastic lymphoma kinase), was granted fast track approval in 2011 to treat certain patients with late-stage (locally advanced or metastatic) non-small cell lung cancer (NLCSC) that expresses the abnormal *ALK* gene along with a companion diagnostic test (Vysis ALK Break Apart FISH Probe Kit).

2. TARGETING MET IN CANCER

MET, along with macrophage stimulating 1 receptor (MST1R, or RON), belongs to a unique subfamily of RTKs, and is mainly produced in cells of epithelial or endothelial origin.[3] Hepatocyte growth factor (HGF), also known as scatter factor (SF), is the only known natural high-affinity ligand of MET, and is mainly expressed in cells of mesenchymal origin.[4] As is common for most RTKs, MET dimerizes and trans-phosphorylates Y1234/Y1235 at the activation loop upon HGF binding to induce kinase activity. Subsequent phosphorylation at Y1349/Y1356 near the carboxyl terminus creates a docking site for intracellular adaptor proteins leading to the activation of the downstream signal transduction pathways, including the mitogen-activated protein kinase (MAPK), phosphoinositide 3-kinase (PI3K)–protein kinase B (AKT), and signal transducer and activator of transcription (STAT) signaling cascades.[5] HGF/MET signaling controls MET-dependent cell proliferation, survival and migration that is critical for invasive growth during embryonic development and postnatal organ regeneration, and only fully active in adults for wound healing and tissue regeneration processes.[5] The HGF/MET axis is frequently upregulated through activating mutation, gene amplification, aberrant paracrine, or autocrine ligand production in many cancers, and is strongly linked with tumorigenesis, invasive growth, and metastasis.[5,6] In addition, the activation of HGF/MET signaling is emerging as an important mechanism in resistance to EGFR and B-type Raf kinase (BRAF) inhibitor treatments *via MET* amplification[7] and/or upregulation of stromal HGF.[8–10] Due to the role of aberrant HGF/MET signaling in human oncogenesis, invasion/metastasis, and resistance, the inhibition of the HGF/MET signaling pathway has great potential in cancer therapy.

3. TARGETING ALK IN CANCER

ALK, along with leukocyte tyrosine kinase (LTK), is grouped within the insulin receptor (IR) superfamily of RTKs. ALK is mainly expressed in the central and peripheral nervous systems suggesting a potential role in normal development and function of the nervous system.[11] Pleiotrophin (PTN) and midkine (MK) have been proposed as ALK ligands in mammals.[11] ALK was first discovered as a fusion protein, NPM (nucleophosmin)-ALK encoded by a fusion gene arising from the t(2;5)(p23;q35) chromosomal translocation in anaplastic large cell lymphoma (ALCL) cell lines in 1994.[12] More than 20 distinct *ALK* translocation partners have been discovered in many cancers, including ALCL (60–90% incidence), inflammatory myofibroblastic tumors (IMTs, 50–60%), non-small cell lung carcinomas (NSCLCs, 3–7%), colorectal cancers (CRCs, 0–2.4%), breast cancers (0–2.4%), and other carcinomas with rare incidence.[13] The ALK-fusion proteins are located in the cytoplasm, and the fusion partners with ALK play a role in dimerization or oligomerization of the fusion proteins through a coil–coil interaction to generate constitutive activation of ALK kinase function.[14] *EML4-ALK*, comprising portions of the echinoderm microtubule associated protein-like 4 (*EML4*) gene and the *ALK* gene, was first discovered in both the archived NSCLC specimens[15] and NSCLC cell lines.[16] EML4-ALK fusion variants are highly oncogenic and caused lung adenocarcinoma in transgenic mice.[17] ALK is highly expressed in subtypes of breast cancer.[18] Oncogenic point mutations of ALK in both familial[19] and sporadic cases[20] of neuroblastoma, and in anaplastic thyroid cancer[21] have been reported. Although the normal physiologic roles of ALK are not completely elucidated, ALK is an attractive molecular target for cancer therapeutic intervention because of the important roles in hematopoietic, solid, and mesenchymal tumors.

4. MEDICINAL CHEMISTRY EFFORTS IN DISCOVERING CRIZOTINIB[22]

4.1. Discovery of the potent and selective MET inhibitor PHA-665752

Crizotinib originates from a drug discovery program aimed at inhibiting the MET RTK. Pyrrole substituted 2-indolinones were discovered as ATP competitive tyrosine kinase inhibitors in the middle of 1990s,[23] and the kinase selectivity was tuned with the substituents on oxindole and pyrrole rings.[24] The potent VEGFR/c-KIT inhibitor sunitinib (**2**, Fig. 25.2) was a weak MET inhibitor. With a benzylsulfonyl substituent at the 5-position, **PHA-665752** (**3**, Fig. 25.2) was a potent and selective inhibitor of MET cellular activity ($IC_{50} = 9$ nM in GTL-16 cell line).[25] With both high molecular weight and high lipophilicity **PHA-665752** had poor pharmaceutical properties (low solubility and high metabolic clearance), which limited its further development as a clinical candidate.[22]

4.2. Cocrystal structure of PHA-665752 with unphosphorylated MET kinase domain (KD)

The cocrystal structure of **PHA-665752** with unphosphorylated MET KD (Fig. 25.3) revealed that the MET KD adopted the unique auto-inhibitory conformation observed previously in crystal structures of the apo-enzyme and a complex with the staurosporine analog K252a.[26] In these MET crystal structures, the beginning of the kinase activation loop (residues 1222–1227)

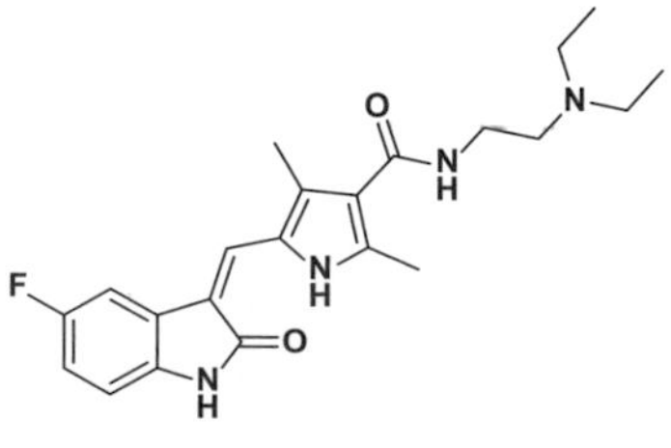

2 (Sunitinib)
MW 398.48; LogD 1.90
MET K_i 5.31 µM
MET Cell IC_{50} 11.7 µM

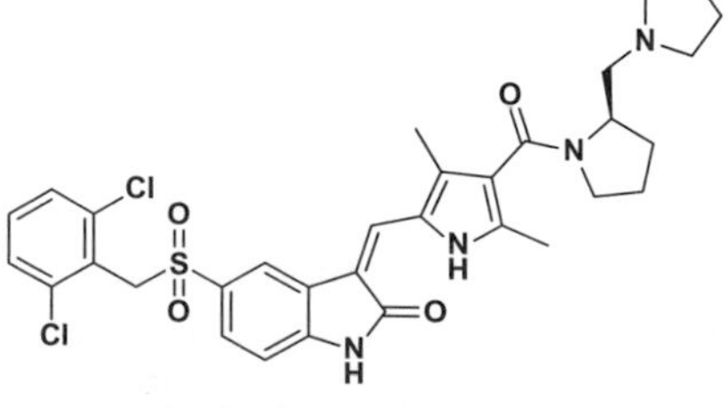

3 (PHA-665752)
MW 641.61; LogD 3.20
MET K_i 0.0005 µM
MET Cell IC_{50} 0.009 µM
MET Cell LE 0.250; LipE 4.85

Figure 25.2 Pyrrole substituted indolin-2-one RTK inhibitors.

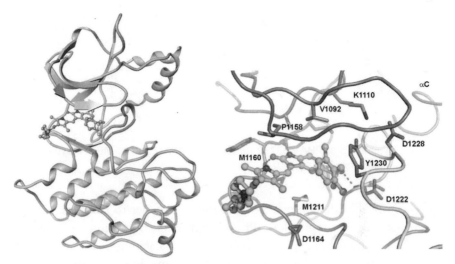

Figure 25.3 Cocrystal structure of **PHA-665752** with unphosphorylated MET KD. (See color plate.)

forms a turn (red color) that wedges between the β-sheet and the αC-helix. Consequently, the activation loop (yellow color) significantly displaces the αC-helix from a catalytically competent position and the downstream activation loop residues (1228–1245) are in a position that interferes with ATP and substrate binding. This unusual kinase activation loop conformation creates a unique inhibitor binding pocket that can enable high kinase selectivity. As expected, the oxindole in **PHA-665752** forms two hydrogen bonds as a mimic of ATP adenine to the kinase hinge backbone residues Pro-1159 and Met-1160, and the pyrrole ring amide substituent extends into the solvent (Fig. 25.3). However, in order to create a space for the benzyl group bending back to the ATP adenine pocket and forming a π–π stacking interaction with Tyr-1230 at A-loop, the plane of the oxindole–pyrrole in **PHA-665752** is tilted away from the glycine-rich-loop significantly compared to the ATP adenine surface (Fig. 25.4). Also, a hydrogen bond between the oxygen of the sulfonyl group and the backbone N–H of Asp-1222 of the aspartic acid-phenylalanine-glycine (DFG) segment is formed in this orientation.

4.3. Design of 5-aryl-3-benzyloxypyridin-2-amine for more efficient interaction with MET

Although the **PHA-665752**/MET complex showed unique inhibitor-kinase binding interactions, it was clear that some of the interactions were not efficient. The pyrrole indolin-2-one core in **PHA-665752** occupies the

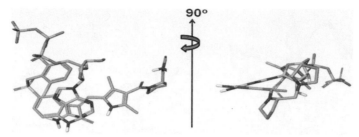

Figure 25.4 Overlap of **PHA-665752** (green) with ATP (cyan) in MET cocrystal structures. (See color plate.)

full width of the adenine pocket from the gatekeeper to the solvent. As a result, it needs to adopt a low tilt and high energy position to allow the 2,6-dichlorophenyl group to achieve a π–π stacking interaction with Tyr-1230 with the sulfone methylene linker making a U-turn. Consequently, a significant portion of the indolinone ring along with the sulfone linker was identified as inefficient scaffolding for the 2,6-dichlorophenyl group. A 5-aryl-3-benzyloxy-pyridin-2-amine scaffold was designed for more efficient interactions with the MET KD. A mono-aromatic pyridin-2-amine was designed to interact with the hinge residues Pro-1158 and Met-1160, and the 3-benzyloxy group is positioned to interact with Tyr-1230 in a direct manner, which will allow the hinge scaffold to be tilted back to the ATP adenine position. Lastly, the 5-aryl group was anticipated to point toward solvent in the same way as oxindole pyrrole substituents, providing a handle to modulate lipophilicity. To test the design, a small set of compounds was prepared (Fig. 25.5), and demonstrated moderate inhibition against MET. Further optimization of MET potency and physical properties were carried out in parallel.

4.4. Medicinal chemistry optimization leading to crizotinib

Optimization of the 3-benzyloxy group substituents led to the discovery of the 2,6-dichloro-3-fluoro-α-methyl-benzyl group as an important component for MET cell potency, as exemplified by compound **7** (Fig. 25.6), which was 30-fold more potent than the initial lead **6**. As expected, a variety of soluble amide substituents added to the 5-phenyl group provided similar potency. Compound **8** was a potent MET inhibitor *in vitro* and *in vivo* leading to complete tumor growth inhibition in nude mice in a human U87 glioblastoma xenograft tumor model.[27] However, **8** was a potent CYP 3A4 inhibitor with an IC_{50} of 0.6 μM.[27] Optimization of the 5-aryl

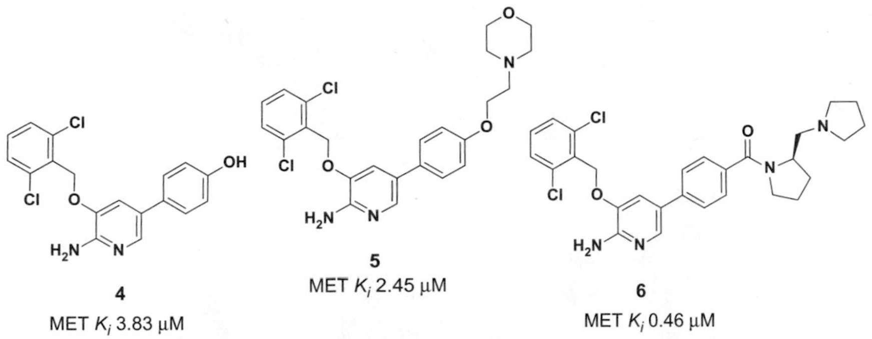

Figure 25.5 5-Aryl-3-benzyloxypyridin-2-amines show MET inhibition.

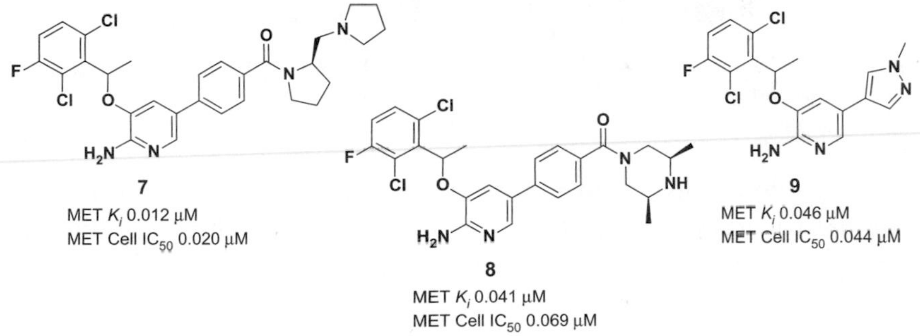

Figure 25.6 Optimization of 5-aryl-3-benzyloxypyridin-2-amines.

substituent with the goal of improving lipophilic efficiency (LipE) to enable highly potent inhibition in less lipophilic compounds that would have reduced CYP 3A4 inhibition as well as permeability, metabolic clearance, and solubility consistent with low dose oral activity led to the discovery of 5-pyrazol-4-yl group as a significantly more efficient (both LipE and LE) substituent than the 5-phenyl analogues **7** and **8**, as exemplified by compound **9**. Optimization of the 5-pyrazol-4-yl N-substituent generated the clinical candidate **PF-02341066**, later named crizotinib with a MET cell IC_{50} of 0.008 µM, and much reduced inhibition of CYP 3A4 (IC_{50} 5 µM).

The co-crystal structure of crizotinib bound to unphosphorylated MET KD confirms the original design hypotheses (Fig. 25.7). Crizotinib shares a similar but significantly more atom efficient binding mode than **PHA-665752** having the key interaction with Tyr-1230 and stabilizing the MET kinase in an inactive autoinhibitory conformation. Indeed, the plane of pyridin-2-amine core is tilted back to the adenine surface of ATP, and the

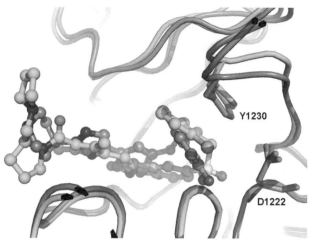

Figure 25.7 Overlay of crizotinib (gray color) and **PHA-665752** (cyan color) bound to unphosphorylated MET kinase domain. (See color plate.)

phenyl group is delivered in a more direct way to interact with Tyr-1230. The alpha-methyl group improves cell potency profoundly. In addition to rigidifying the rotation of benzyl group, the alpha-methyl group is bound in a tight hydrophobic pocket surrounded with the hydrophobic side chains of Ala-1226, Leu-1157, Lys-1110, and Val-1092. Overall, the more efficient binding mode results in much improved cell-based ligand efficacy (LE) and LipE for crizotinib (LE 0.38 and LipE 6.14) in comparison with the early lead **PHA-665752** (LE 0.26 and LipE 4.81), and leads to the desired drug-like properties.

5. KINASE SELECTIVITY OF CRIZOTINIB

The kinase selectivity of crizotinib was evaluated against a panel of more than 120 human kinases from Upstate Inc.[28] There were 13 kinases with a less than 100-fold enzymatic selectivity window over MET kinase. The enzymatic assays used at Upstate measure the inhibition of substrate phosphorylation by a phosphorylated (active) kinase. The binding of an inhibitor to either active or inactive kinase proteins can result in the inhibition of kinase phosphorylation in cells. According to the Cheng–Prusoff equation $(IC_{50} = K_i(1 + [ATP]/K_{m,ATP}))$, ATP competitive inhibitors binding to inactive kinase will produce more effective inhibition of kinase activity in cell because of much higher $K_{m,ATP}$ of inactive kinase.

Table 25.1 Kinase selectivity of crizotinib

Kinase	MET	ALK	ROS	RON	AXL	TIE2	TRKA	TRKB	ABL	IR	LCK
Enzym. IC_{50} (nM)	<1.0	<1.0	<1.0	NA	<1.0	5.0	<1.0	2.0	24	102	<1.0
Cell IC_{50} (nM)	8.0	24	31	80	294	448	580	399	1159	2887	2741

Considering the unique and inactive MET autoinhibitory conformation accessed by crizotinib,[22] and the consequent expectation that biochemical inhibition of other kinases may not translate to potent cell kinase inhibition for these kinases, cell-based autophosphorylation assays were employed to determine a more accurate picture of kinase selectivity in the whole cell context (Table 25.1). Crizotinib demonstrated a potent cell IC_{50} of 20 nM against an oncogenic fusion ALK (NPM-ALK) in a human lymphoma cell line. It also inhibited ROS (a closely related kinase to ALK) with an IC_{50} of 31 nM in a HCC78 cell line, and RON in the transfected NIH 3T3 cell line (IC_{50} 80 nM). In summary, crizotinib is a selective and potent inhibitor against MET/ALK/ROS at the cellular level.

6. PRECLINICAL PHARMACOLOGY OF CRIZOTINIB

Crizotinib is a free base, anhydrous crystalline compound with a pH dependent solubility profile. Crizotinib showed moderate metabolic stability in human hepatocytes, moderate to high plasma protein binding (92–97%) in human and animal species, and low to moderate permeability in the Caco-2 cell assay. Biotransformation mediated by CYP3A4 is the major clearance mechanism. Crizotinib is not a potent inhibitor of CYP1A2, 2C8, 2C9, 2C19, and 2D6 ($IC_{50} > 30$ μM) in human liver microsomes, but CYP3A isozymes are inhibited in a time-dependent manner ($K_I = 3.7$ μM and $k_{inact} = 6.9\,h^{-1}$ in human liver microsomes, and $K_I = 0.89$ μM and $k_{inact} = 0.78\,h^{-1}$ in cryopreserved human hepatocytes).[29,30] Crizotinib demonstrated moderate clearance, long half life, large volume of distribution, and moderate oral availability in different species.[29]

The pharmacologic effects of crizotinib in MET or ALK-driven tumor cell lines and models were evaluated.[31,32] Crizotinib potently inhibited MET phosphorylation and MET-dependent proliferation, migration, or invasion of *MET* amplified GTL-16 tumor cells *in vitro* (IC_{50} values, 5–20 nM). Crizotinib showed marked efficacy *in vivo* at well-tolerated doses in MET-driven tumor models, including GTL-16 gastric, NCI-H441 NSCL, CAKi-1 renal, and PC-3 prostate carcinoma xenograft models. The antitumor efficacy of crizotinib was dose dependent and showed a strong correlation with the inhibition of MET phosphorylation *in vivo*. Near maximal inhibition of MET activity for the full dosing interval was necessary to maximize *in vivo* efficacy. The antitumor mechanisms *in vivo* were associated with the inhibition of MET-dependent signal transduction, tumor cell proliferation (Ki67), induction of apoptosis (caspase-3), and reduction of microvessel density (CD31). Crizotinib was also evaluated *in vitro* and *in vivo* in NPM-ALK-positive ALCL cell lines and related tumor models.[32] Crizotinib potently inhibited NPM-ALK phosphorylation in Karpas299 or SU-DHL-1 ALCL cells (mean IC50 value, 24 nM) leading to the inhibition of NPM-ALK dependent cell proliferation (IC50 values, ~30 nM) and induced apoptosis (IC50 values, 25–50 nM). The observed cytoreductive activity of crizotinib was associated with the inhibition of NPM-ALK dependent cell cycle progression at the G1–S-phase checkpoint, and the induction of cell death pathways. Crizotinib demonstrated dose-dependent antitumor efficacy in severe combined immunodeficient-Beige mice bearing Karpas299 ALCL tumor xenografts. A strong correlation was observed between antitumor response and inhibition of NPM-ALK phosphorylation and induction of apoptosis in tumor tissue. The *in vitro* and *in vivo* preclinical studies indicated that NPM-ALK was a potential oncogene driver for ALCL, and crizotinib effectively blocked the tumor growth.

7. CRIZOTINIB HUMAN CLINICAL EFFICACY

The human clinical Phase I trial of crizotinib was initiated in 2006 in patients with advanced cancer. Crizotinib was first dosed at 50 mg orally once daily and eventually escalated to 300 mg twice a day (BID). The maximum tolerated dose (MTD) and the recommended clinical dose is 250 mg BID.[33] After repeated dosing at 250 mg BID, the plasma concentrations of crizotinib appeared to reach steady state within 15 days with a mean steady-state trough plasma level of 256 ng/mL or 45 nM of free drug.[34] A 3.6-fold increase in the single-dose oral midazolam AUC was observed following 28 days of dosing at 250 mg BID, suggesting that crizotinib is a moderate CYP3A4 inhibitor.[34] The AUC generally increased proportionally with doses over the therapeutic dose range studied, and accumulated by 4.0–5.9-fold after multiple doses with a terminal half-life of 43–51 h.[34]

Crizotinib has demonstrated marked efficacies for patients harboring fusion *ALK*, or *ROS* genes, or having *de novo MET* gene amplification.

Table 25.2 Summary of clinical activities of crizotinib in patients enrolled in the Phase I (A8081001) and Phase II (A8081005) trials

Clinical trial	# of patients	ORR[a] (95% CI[b])	DCR[c] (95% CI)	Estimated PFS[d] (range)	Median duration of response
A8081001	143	60.8 (52.3–68.9)	82.5 at 8 weeks (75.3–88.4)	9.7 months (7.7–12.8)	49.1 weeks
A8081005	255	53 (47–60)	85 at 12 weeks (80–89)	8.5 months (6.2–9.9)	43 weeks

[a]Overall response rate.
[b]Confidence interval.
[c]Disease control rate.
[d]Progression free survival.

The efficacy results from a subset of NSCLC patients harboring the *EML4-ALK* translocation gene in the crizotinib Phase Ib (A8081001)[35] and Phase II studies (A8081005)[36] are summarized in Table 25.2.

Overall, the treatment-related adverse events were mostly grade 1 or 2. The most common adverse events were visual effects, nausea, diarrhea, constipation, vomiting, and peripheral edema. The most common treatment-related grade 3 or 4 adverse events were neutropenia, raised alanine aminotransferase, hypophosphataemia, and lymphopenia.[35,36]

In addition, marked responses to crizotinib were observed in advanced, chemoresistant ALK + ALCL patients,[37,38] ALK mutated children neuroblastoma patients,[38] and ALK-translocated IMT patients.[39] Crizotinib is a multi-target protein kinase inhibitor of MET/ALK/ROS. Approximately 0.88–2% of patients with NSCLC harbor *ROS1* rearrangements.[40] Crizotinib demonstrates marked antitumor activity in patients with advanced NSCLC harboring *ROS1* rearrangements.[41] The mechanisms and efficacy of Met inhibitors in cancer patients are under investigations. One patient with a *de novo* highly *MET*-amplified NSCLC achieved a confirmed partial response with crizotinib.[42]

8. CONCLUSIONS

Crizotinib was identified using structure-based drug design and medicinal chemistry lead optimization. It demonstrated potent *in vitro* and *in vivo* inhibition of MET/ALK/ROS, effective tumor growth inhibition *in vivo*, and good pharmaceutical and safety properties in preclinical studies.

Crizotinib was well tolerated in human Phase I studies at the efficacious doses. The discovery of EML4-ALK as an oncogenic driver in NSCLC accelerated the clinical development of crizotinib in the EML4-ALK-driven patient population. The U.S. Food and Drug Administration granted fast-track approval of crizotinib on August 26, 2011 based on the marked efficacy of crizotinib in patients with ALK-positive advanced NSCLC and good safety profile in Phases I and II trials. Promising antitumor activity has been observed with crizotinib in patients with abnormal *ALK* genes of ALCL, IMT, and neuroblastoma; with abnormal *ROS1* gene in NSCLC; and with *MET* gene amplified NSCLC. The broad antitumor activities of crizotinib based on molecular targets across many cancers indicate the importance of understanding tumor biology to identify the oncogenic driver targets for the stratification of the right patient population.

REFERENCES

1. Lemmon, M. A.; Schlessinger, J. *Cell* **2010**, *141*, 1117.
2. Gschwind, A.; Fischer, O. M.; Ullrich, A. *Nat. Rev. Cancer* **2004**, *4*, 361.
3. Park, M.; Dean, M.; Cooper, C. S.; Schmidt, M.; O'Brien, S. J.; Blair, D. G.; Vande Woude, G. F. *Cell* **1986**, *45*, 895.
4. Bottaro, D. P.; Rubin, J. S.; Faletto, D. L.; Chan, A. M.; Kmiecik, T. E.; Vande Woude, G. F.; Aaronson, S. A. *Science* **1991**, *251*, 802.
5. Trusolino, L.; Bertotti, A.; Comoglio, P. M. *Nat. Rev. Mol. Cell Biol.* **2010**, *11*, 834.
6. Gherardi, E.; Birchmeier, W.; Birchmeier, C.; Vande Woude, G. *Nat. Rev. Cancer* **2012**, *12*, 89.
7. Engelman, J. A.; Zejnullahu, K.; Mitsudomi, T.; Song, Y.; Hyland, C.; Park, J. O.; Lindeman, N.; Gale, C. M.; Zhao, X.; Christensen, J.; Kosaka, T.; Holmes, A. J.; Rogers, A. M.; Cappuzzo, F.; Mok, T.; Lee, C.; Johnson, B. E.; Cantley, L. C.; Jänne, P. A. *Science* **2007**, *316*, 1039.
8. Wilson, T. R.; Fridlyand, J.; Yan, Y.; Penuel, E.; Burton, L.; Chan, E.; Peng, J.; Lin, E.; Wang, Y.; Sosman, J.; Ribas, A.; Li, J.; Moffat, J.; Sutherlin, D. P.; Koeppen, H.; Merchant, M.; Neve, R.; Settleman, J. *Nature* **2012**, *487*, 505.
9. Turke, A. B.; Zejnullahu, K.; Wu, Y. L.; Song, Y.; Dias-Santagata, D.; Lifshits, E.; Toschi, L.; Rogers, A.; Mok, T.; Sequist, L.; Lindeman, N. I.; Murphy, C.; Akhavanfard, S.; Yeap, B. Y.; Xiao, Y.; Capelletti, M.; Iafrate, A. J.; Lee, C.; Christensen, J. G.; Engelman, J. A.; Jänne, P. A. *Cancer Cell* **2010**, *17*, 77.
10. Straussman, R.; Morikawa, T.; Shee, K.; Barzily-Rokni, M.; Qian, Z. R.; Du, J.; Davis, A.; Mongare, M. M.; Gould, J.; Frederick, D. T.; Cooper, Z. A.; Chapman, P. B.; Solit, D. B.; Ribas, A.; Lo, R. S.; Flaherty, K. T.; Ogino, S.; Wargo, J. A.; Golub, T. R. *Nature* **2012**, *487*, 500.
11. Pulford, K.; Lamant, L.; Espinos, E.; Jiang, Q.; Xue, L.; Turturro, F.; Delsol, G.; Morris, S. W. *Cell. Mol. Life Sci.* **2004**, *61*, 2939.
12. Morris, S. W.; Kirstein, M. N.; Valentine, M. B.; Dittmer, K. G.; Shapiro, D. N.; Saltman, D. L.; Look, A. T. *Science* **1994**, *263*, 1281.
13. Grande, E.; Bolós, M.-V.; Arriola, E. *Mol. Cancer Ther.* **2011**, *10*, 569.
14. Bischof, D.; Pulford, K.; Mason, D. Y.; Morris, S. W. *Mol. Cell. Biol.* **1997**, *17*, 2312.
15. Soda, M.; Choi, Y. L.; Enomoto, M.; Takada, S.; Yamashita, Y.; Ishikawa, S.; Fujiwara, S.; Watanabe, H.; Kurashina, K.; Hatanaka, H.; Bando, M.; Ohno, S.;

Ishikawa, Y.; Aburatani, H.; Niki, T.; Sohara, Y.; Sugiyama, Y.; Mano, H. *Nature* **2007**, *448*, 561.

16. Rikova, K.; Guo, A.; Zeng, Q.; Possemato, A.; Yu, J.; Haack, H.; Nardone, J.; Lee, K.; Reeves, C.; Li, Y.; Hu, Y.; Tan, Z.; Stokes, M.; Sullivan, L.; Mitchell, J.; Wetzel, R.; MacNeill, J.; Ren, J. M.; Yuan, J.; Bakalarski, C. E.; Villen, J.; Kornhauser, J. M.; Smith, B.; Li, D.; Zhou, X.; Gygi, S. P.; Gu, T.-L.; Polakiewicz, R. D.; Rush, J.; Comb, M. J. *Cell* **2007**, *131*, 1190.

17. Soda, M.; Takada, S.; Takeuchi, K.; Choi, Y. L.; Enomoto, M.; Ueno, T.; Haruta, H.; Hamada, T.; Yamashita, Y.; Ishikawa, Y.; Sugiyama, Y.; Mano, H. *Proc. Natl. Acad. Sci. U.S.A.* **2008**, *105*, 19893.

18. Perez-Pinera, P.; Chang, Y.; Astudillo, A.; Mortimer, J.; Deuel, T. F. *Biochem. Biophys. Res. Commun.* **2007**, *358*, 399.

19. Mossé, Y. P.; Laudenslager, M.; Longo, L.; Cole, K. A.; Wood, A.; Attiyeh, E. F.; Laquaglia, M. J.; Sennett, R.; Lynch, J. E.; Perri, P.; Laureys, G.; Speleman, F.; Kim, C.; Hou, C.; Hakonarson, H.; Torkamani, A.; Schork, N. J.; Brodeur, G. M.; Tonini, G. P.; Rappaport, E.; Devoto, M.; Maris, J. M. *Nature* **2008**, *455*, 930.

20. Caren, H.; Abel, F.; Kogner, P.; Martinsson, I. *Biochem. J.* **2008**, *416*, 153.

21. Murugan, A. K.; Xing, M. M. *Cancer Res.* **2011**, *71*, 4403.

22. Cui, J. J.; Tran-Dubé, M.; Shen, H.; Nambu, M.; Kung, P. P.; Pairish, M.; Jia, L.; Meng, J.; Funk, L.; Botrous, I.; McTigue, M.; Grodsky, N.; Ryan, K.; Padrique, E.; Alton, G.; Timofeevski, S.; Yamazaki, S.; Li, Q.; Zou, H.; Christensen, J.; Mroczkowski, B.; Bender, S.; Kania, R. S.; Edwards, M. P. *J. Med. Chem.* **2011**, *54*, 6342.

23. Mohammadi, M.; McMahon, G.; Sun, L.; Tang, C.; Hirth, P.; Yeh, B. K.; Hubbard, S. R.; Schlessinger, J. *Science* **1997**, *276*, 955.

24. Sun, L.; Liang, C.; Shirazian, S.; Zhou, Y.; Miller, T.; Cui, J.; Fukuda, J. Y.; Chu, J.-Y.; Nematalla, A.; Wang, X.; Chen, H.; Sistla, S.; Luu, T. L.; Tang, F.; Wei, J.; Cho, T. *J. Med. Chem.* **2003**, *46*, 1116.

25. Christensen, J. G.; Schreck, R.; Burrows, J.; Kuruganti, P.; Chan, E.; Le, P.; Chen, J.; Wang, X.; Ruslim, L.; Blake, R.; Lipson, K. E.; Ramphal, J.; Do, S.; Cui, J. J.; Cherrington, J. M.; Mendel, D. B. *Cancer Res.* **2003**, *63*, 7345.

26. Schiering, N.; Knapp, S.; Marconi, M.; Flocco, M. M.; Cui, J.; Perego, R.; Rusconi, L.; Cristiani, C. *Proc. Natl. Acad. Sci. U.S.A.* **2003**, *100*, 12654.

27. Cui, J. J.; Nambu, M.; Kung, P.-P.; Tran-Dube, M.; Shen, H.; Pairish, M.; Lei, J.; Meng, J.; Lee, F.; McTigue, M.; Yamazaki, S.; Alton, G.; Zou, H.; Christensen, J.; Mroczkowski, B. In: *242nd ACS National Meeting & Exposition, Denver, CO, United States, August 28–September 1, 2011*; 2011, Abstr. MEDI-190.

28. Timofeevski, S. L.; McTigue, M. A.; Ryan, K.; Cui, J.; Zou, H. Y.; Zhu, J. X.; Chau, F.; Alton, G.; Karlicek, S.; Christensen, J. G.; Murray, B. W. *Biochemistry* **2009**, *48*, 5339.

29. Yamazaki, S.; Skaptason, J.; Romero, D.; Vekich, S.; Jones, H. M.; Tan, W.; Wilner, K.; Koudriakova, T. *Drug Metab. Dispos.* **2011**, *39*, 383.

30. Mao, J.; Johnson, T. R.; Shen, Z.; Yamazaki, S. *Drug Metab. Dispos.* **2013**, *41*, 343.

31. Zou, H. Y.; Li, Q.; Lee, J. H.; Arango, M. E.; McDonnell, S. R.; Yamazaki, S.; Koudriakova, T. B.; Alton, G.; Cui, J. J.; Kung, P.-P.; Nambu, M. D.; Los, G.; Bender, B. L.; Mroczkowski, B.; Christensen, J. G. *Cancer Res.* **2007**, *67*, 4408.

32. Christensen, J. G.; Zou, H. Y.; Arango, M. E.; Li, Q.; Lee, J. H.; McDonnell, S. R.; Yamazaki, S.; Alton, G.; Mroczkowski, B.; Los, G. *Mol. Cancer Ther.* **2007**, *6*, 3314.

33. Kwak, E. L.; Camidge, D. R.; Clark, J.; Shapiro, G. I.; Maki, R. G.; Ratain, M. J.; Solomon, B.; Bang, Y.; Ou, S.; Salgia, R. *J. Clin. Oncol.* **2009**, *27* (suppl, abstr. e3509).

34. Tan, W.; Wilner, K. D.; Bang, Y. E.; Kwak, L.; Maki, R. G.; Camidge, D. R.; Solomon, B. J.; Ou, S. I.; Salgia, R.; Clark, J. W. *J. Clin. Oncol.* **2010**, *28* (suppl, abstr. e2596).

35. Camidge, D. R.; Bang, Y.; Kwak, E. L.; Iafrate, A. J.; Varella-Garcia, M.; Fox, S. B.; Riely, G. J.; Solomon, B.; Ou, S. I.; Kim, D.; Salgia, R.; Fidias, P.; Engelman, J. A.; Gandhi, L.; Jänne, P. A.; Costa, D. B.; Shapiro, G. I.; Lorusso, P.; Ruffner, K.; Stephenson, P.; Tang, Y.; Wilner, K.; Clark, J. W.; Shaw, A. T. *Lancet Oncol.* **2012**, *13*, 1011.

36. Kim, D.-W.; Ahn, M.-J.; Shi, Y.; Martino De Pas, T.; Yang, P.-C.; Riely, G. J.; Crinò, L.; Evans, T. L.; Liu, X.; Han, J.-Y.; Salgia, R.; Moro-Sibilot, D.; Ou, S.-H. I.; Gettinger, S. N.; Wu, Y. L.; Lanzalone, S.; Polli, A.; Iyer, S.; Shaw, A. T. *J. Clin. Oncol.* **2012**, *30* (suppl, abstr. e7533).

37. Pogliani, E. M.; Dilda, I.; Villa, F.; Farina, F.; Giudici, G.; Guerra, L.; Di Lelio, A.; Borin, L.; Casaroli, I.; Verga, L.; Gambacorti-Passerini, C. *J. Clin. Oncol.* **2011**, *29* (suppl, abstr. e18507).

38. Mosse, Y. P.; Balis, F. M.; Lim, M. S.; Laliberte, J.; Voss, S. D.; Fox, E.; Bagatell, R.; Weigel, B.; Adamson, P. C.; Ingle, A. M.; Ahern, C. H.; Blaney, S. *J. Clin. Oncol.* **2012**, *30* (suppl, abstr. e9500).

39. Butrynski, J. E.; D'Adamo, D. R.; Hornick, J. L.; Dal Cin, P.; Antonescu, C. R.; Jhanwar, S. C.; Ladanyi, M.; Capelletti, M.; Rodig, S. J.; Ramaiya, N.; Kwak, E. L.; Clark, J. W.; Wilner, K. D.; Christensen, J. G.; Jänne, P. A.; Maki, R. G.; Demetri, G. D.; Shapiro, G. I. *N. Engl. J. Med.* **2010**, *363*, 1727.

40. Takeuchi, K.; Soda, M.; Togashi, Y.; Suzuki, R.; Sakata, S.; Hatano, S.; Asaka, R.; Hamanaka, W.; Ninomiya, H.; Uehara, H.; Choi, Y. L.; Satoh, Y.; Okumura, S.; Nakagawa, K.; Mano, H.; Ishikawa, Y. *Nat. Med.* **2012**, *18*, 378.

41. Shaw, A. T.; Camidge, D. R.; Engelman, J. A.; Solomon, B. J.; Kwak, E. L.; Clark, J. W.; Salgia, R.; Shapiro, G.; Bang, Y.-J.; Tan, W.; Tye, L.; Wilner, K. D.; Stephenson, P.; Varella-Garcia, M.; Bergethon, K.; Iafrate, A. J.; Ou, S.-H. I. *J. Clin. Oncol.* **2012**, *30* (suppl, abstr. e7508).

42. Ou, S.-H. I.; Kwak, E. L.; Siwak-Tapp, C.; Dy, J.; Bergethon, K.; Clark, J. W.; Camidge, D. R.; Solomon, B. J.; Maki, R. G.; Bang, Y.-J.; Kim, D.-W.; Christensen, J.; Tan, W.; Wilner, K. D.; Salgia, R.; Iafrate, A. J. *J. Thorac. Oncol.* **2011**, *6*, 942.

CHAPTER TWENTY-SIX

Case History: Vemurafenib, a Potent, Selective, and First-in-Class Inhibitor of Mutant BRAF for the Treatment of Metastatic Melanoma

Prabha N. Ibrahim, Jiazhong Zhang, Chao Zhang, Gideon Bollag
Plexxikon Inc., Berkeley, California, USA

Contents

1. INTRODUCTION

Melanoma, a malignant tumor originating in melanocytes, is the cause for the majority (75%) of deaths related to skin cancer.[1] Although the exact cause of melanoma is not clear, exposure to ultraviolet (UV) radiation from sunlight or tanning lamps increases the risk of developing melanoma. About 160,000 new cases of melanoma are diagnosed worldwide each year. Melanomas are classified according to four stages based on the thickness, depth of penetration, and the degree of spread. Stages I and II, the early melanomas, are localized while the advanced stages III and IV have metastasized locally or to other parts of the body. Survival rates correlate with the stage of the cancer such that patients with lower stage cancers have a better outlook for a cure or long-term survival.[2]

Treatment for early-stage melanomas usually includes surgery to remove the melanoma. If melanoma has spread beyond the skin, treatment options may include surgical removal of affected lymph nodes, chemotherapy, radiotherapy, biological therapy, and targeted therapy. The opportunity for the targeted therapy was presented with the identification of the role of oncogenic BRAF mutations and their downstream signaling pathways in melanoma.[3]

2. RATIONALE FOR TARGETING MUTANT BRAF

The RAS–RAF–MEK–ERK signaling pathway plays an important role in mediating cellular responses to growth signals (Fig. 26.1). The RAF family of serine–threonine kinases includes three isoforms, ARAF, BRAF, and CRAF, which are regulated by RAS binding. BRAF is the most frequently mutated oncogene, present in a large proportion of patients with metastatic melanoma, and at lower frequency in a wide range of human cancers. Mutations in the BRAF oncogene are present in approximately 6–8% of human cancers, occurring in high frequency in melanoma (30–70%), thyroid cancer (30–50%), colorectal cancer (5–20%), and ovarian cancer (~30%).[4] About 90% of the activating BRAF mutation consists of glutamic acid substitution for valine at position 600 (BRAF^{V600E}). A high frequency

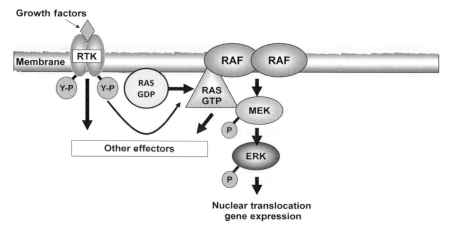

Figure 26.1 RAS–RAF–MEK–ERK signaling pathway. (See color plate.)

of BRAF mutations was found in benign and atypical nevi (skin anomalies such as moles and birthmarks); the link to melanoma is currently being investigated.[5] Thus, a melanocyte specific signaling pathway controlling proliferation and differentiation through BRAF activation presents a new drug target for melanoma.

BRAF^{V600E}, the major driver mutation, has been shown to be 500-fold activated and can constitutively stimulate MEK–ERK signaling in cells[6] and induce melanoma in mice.[7] Inhibition of BRAF^{V600E} blocks cell proliferation and induces apoptosis *in vitro* and xenograft growth *in vivo*.[6] Taken together, these findings validate the BRAF mutation as a therapeutic target in melanoma.

3. MEDICINAL CHEMISTRY EFFORTS CULMINATING IN VEMURAFENIB

3.1. Mutant BRAF program objectives

Therapeutic intervention targeting the inhibition of oncogenic BRAF kinase activity provides an avenue to treat BRAF-mutated tumors. The objective of the program was to identify a novel, potent, and selective inhibitor with a highly favorable balance of efficacy and safety using a scaffold-based drug discovery approach.

3.2. Early preclinical and clinical leads

When our mutant-BRAF program began in 2005, the published RAF inhibitors included GW5074,[8,9] ZM336372,[10] L-779450,[11] and sorafenib, a multikinase inhibitor[12–14] (Fig. 26.2).

Kinase inhibitors are classified either as type I or type II based on the mode of binding in the adenonsine triphosphate (ATP) pocket. The conformation of the conserved aspartic acid-phenylalanine-glycine (DFG) motif is known to be crucial for selective binding of kinase inhibitors in protein kinases.[15] While type I kinase inhibitors typically bind in and around the adenine binding region of ATP and form 1–3 hydrogen bonds with the hinge region in the active conformation (DFG-in) of a kinase, type II inhibitors use the ATP binding site and a hydrophobic pocket created by the conformational change of the DFG loop ("DGF" out/inactive conformation). BRAF can be targeted by both type I and type II inhibitors.

GW-5074, a type I kinase inhibitor, shows potent inhibition of CRAF (IC$_{50}$ = 9 nM)[8]; its potency was attributed to the 4-OH substitution flanked by two substituents on the benzylidine ring. ZM336372 was identified as a novel inhibitor of CRAF (IC$_{50}$ = 70 nM) with 10-fold selectivity over

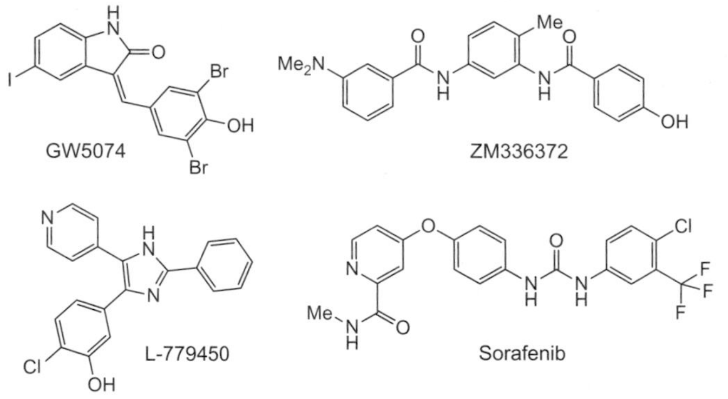

Figure 26.2 Early preclinical and clinical leads.

BRAF.[10] However, incubation of certain tumor cells with ZM336372 induced the activation of CRAF by about 100-fold without an increase in ERK phosphorylation. The ATP-competitive potent RAF kinase inhibitor L-779450, of the type I class ($IC_{50} = 10$ nM), suppressed DNA synthesis and induced apoptosis in cells that proliferate in response to either CRAF or ARAF but not BRAF.[16] Interestingly, three of the early RAF inhibitor leads (GW5074, ZM336372, and L-779450) have a phenol substitution. The first generation RAF inhibitors, despite their potency, lack the relative therapeutic efficacy due to poor bioavailability.

Sorafenib (BAY43-9006, Nexavar), a bis-aryl urea designed as a CRAF inhibitor[10] was also found to inhibit mutant BRAF ($IC_{50} = 38$ nM).[13] It binds to the inactive conformation of wild type and mutant BRAF,[14] typical of a type II inhibitor, and has poor cellular activity in mutant BRAF driven cells.[13] Sorafenib failed to show benefit in a Phase II single agent trial in metastatic melanoma[17] and in a Phase III combination trial as first- or second-line treatment in patients with unresectable stage III or stage IV melanoma.[18] However, in 2005–2007, sorafenib was approved for the treatment of renal and hepatocellular carcinoma. Its efficacy was attributed to potent inhibition of the vascular endothelial growth factor receptor (KDR) kinase activity.

3.3. Screening of scaffold library—discovery of the 7-azaindole scaffold

Traditional drug discovery starts with a specific protein target and the identification of potent lead compounds from a large library of drug-like compounds using high-throughput screening (HTS). In this process, typically, the most potent compounds are selected as leads and optimized for activity towards the specific molecular target. Despite the advances in combinatorial chemistry, robotics technology, and process automation, the pace of new drug discovery has not increased.[19] Fragment-based approaches have utilized libraries with molecules having molecular weights below 150 Da and measured their binding to the target using biophysical methods such as NMR or X-ray for screening.[20] The promise of the fragment-based approaches derives from the idea of linking two separate weakly-binding fragments to yield a larger compound with substantially improved thermodynamic binding properties.

The scaffold-based approach utilizes chemical scaffolds that are larger than traditional fragments and possess functional groups that could form key interactions with the target protein. These weakly active low-molecular-weight inhibitors or modulators bind to the protein target in a region that is conserved within the protein family and serve as a starting point for lead discovery for multiple targets within the family. A strategy has been developed at Plexxikon for identifying such low affinity and low selectivity scaffolds utilizing a combination of biochemical assays and high-throughput co-crystallography as the primary screening method. For this purpose a target naïve library (scaffold library) of more than 20,000 low molecular weight (150–350 Da) compounds has been identified using a novel approach.[21] The process starts with low affinity biochemical screening of the scaffold library compounds at high compound concentrations (100–200 μM) against multiple members of a target protein family, followed immediately by co-crystallography of the screening hits with one or more target proteins. This approach (scaffold-based drug discovery) has been successfully utilized in identifying small molecule inhibitors of phosphodiesterases,[22] and the pan-PPAR agonist indeglitazar.[23] Recently, the expansion of this strategy to discover the 7-azaindole scaffold targeting protein kinases has been described[24] in part using Pim-1 kinase as a robust system for co-crystallography.[25] This novel kinase scaffold, which was one among 70 different compounds from the initial screen that bound in the ATP binding pocket (Fig. 26.3), has been used to rationally design the selective oncogenic BRAF inhibitor vemurafenib, a first-in-class small molecule inhibitor for the treatment of melanoma,[26] and a selective dual FMS–KIT kinase inhibitor PLX647.[27]

3.4. Discovery of the sulfonamide series

Although the 7-azaindole scaffold showed multiple binding orientations in PIM-1, a kinase with an atypical ATP-binding pocket,[25] the co-crystal

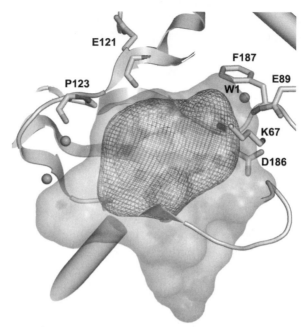

Figure 26.3 Kinase scaffolds bound in the ATP-binding pocket. (See color plate.)

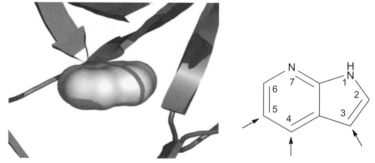

Figure 26.4 7-Azaindole bound in PIM-1 kinase. (See color plate.)

structure of the scaffold with PIM-1 identified the 3, 4, and 5 positions of the 7-azaindole core as productive sites for chemical modification (Fig. 26.4).

Co-crystal structures of selected 3-, 4-, or 5-mono-substituted 7-azaindoles revealed a consistent binding orientation, with the azaindole core making the canonical hydrogen bonding interaction with the kinase hinge region, thereby validating 7-azaindole as a kinase scaffold with potential for modifications to improve potency and selectivity. A proprietary library with diverse chemical modifications built around the 7-azaindole core

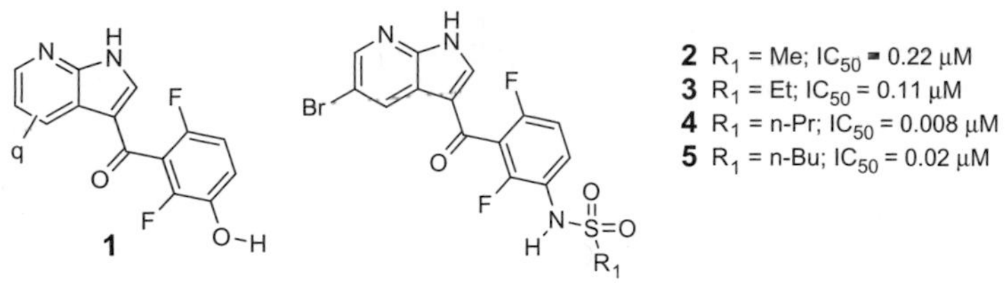

2 R₁ = Me; IC₅₀ = 0.22 μM
3 R₁ = Et; IC₅₀ = 0.11 μM
4 R₁ = n-Pr; IC₅₀ = 0.008 μM
5 R₁ = n-Bu; IC₅₀ = 0.02 μM

Figure 26.5 Early leads—difluorophenols and bio-isosteric replacement.

was screened in the BRAF^{V600E} kinase assay. The initial submicromolar hits contained difluorophenol as the common substituent (Fig. 26.5) connected to the 3-position of the 7-azaindole core through a ketone linker. The co-crystal structure with one of these hits in a surrogate kinase (FGFR) revealed key hydrogen bond interactions of the phenol group with the backbone of the ATP binding pocket. Lead optimization approaches to address potential metabolic liabilities and the observed poor kinase selectivity included replacement of the —OH group with a sulfonamide bioisostere, a widely-used strategy in medicinal chemistry.[28] Methyl sulfonamide replacement of the phenolic group afforded the first sulfonamide lead compound **2** with submicromolar potency (Fig. 26.5). Increasing the sulfonamide chain length to 2 (ethyl), 3 (n-propyl), and 4 (n-butyl) carbon atoms provided compounds **3**, **4**, and **5**, respectively. The n-propyl sulfonamide (compound **4**) was found to be optimal, providing >27X improvement in potency compared to the methyl sulfonamide (compound **2**).

3.5. Optimization of the sulfonamide series

The structure–activity relationship of substitution in the middle phenyl ring was explored (Fig. 26.6). Hydrogen or chlorine replacement of fluorine para (w) to the sulfonamide (compound **7** or **8**) had a moderate impact on the BRAF^{V600E} potency compared to compound **6**. However, replacement of fluorine ortho (q) to the sulfonamide with hydrogen was highly detrimental (compound **9**) thus demonstrating the importance of fluorine at that position.

Replacing the 5-bromo substituent of compound **4** with 5-chloro afforded PLX4720 (compound **10**), which maintained mutant BRAF potency (IC₅₀ = 0.008 μM) and had 10-fold selectivity against wild-type BRAF and >100 fold selectivity against a panel of 70 other kinases. The co-crystal structure of PLX4720 with BRAF (Fig. 26.7) revealed key interactions consistent with its potency and selectivity.[24] PLX4720 preferentially binds to the active conformation ("DFG"-in; Type 1) of the kinase with the

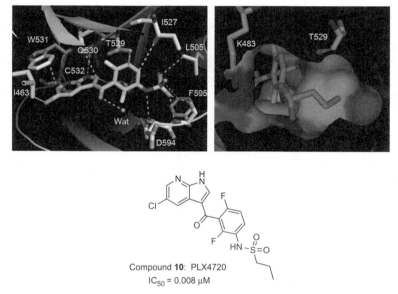

Compound **6**: w=q=F; IC$_{50}$ = 0.11 µM
Compound **7**: w=H q = F; IC$_{50}$ = 0.68 µM
Compound **8**: w=Cl q=F; IC$_{50}$ = 0.24 µM
Compound **9**: w=F q=H; IC$_{50}$ >10 µM

Figure 26.6 SAR of the middle phenyl ring.

Compound **10**: PLX4720
IC$_{50}$ = 0.008 µM

Figure 26.7 Co-crystal structure of compound **10** (PLX4720) in BRAF. (See color plate.)

7-azaindole moiety, anchoring the structure by making the canonical hydrogen bonding interactions with the hinge backbone amide —NH— of Cys-532 and the backbone carbonyl of Gln-530. While the 5-chloro substitution points towards the solvent making van der Waal's contact with Ile-463, the 3-substituent makes extensive interactions with the protein: the oxygen of the ketone linker accepts a hydrogen-bond from a water molecule and the di-fluorophenyl moiety occupies the hydrophobic pocket next to the gate keeper Thr-529 and catalytic Lys-483. The potent inhibition of oncogenic BRAF and the selectivity of PLX4720 versus other kinases can be attributed to the binding interaction between the sulfonamide moiety

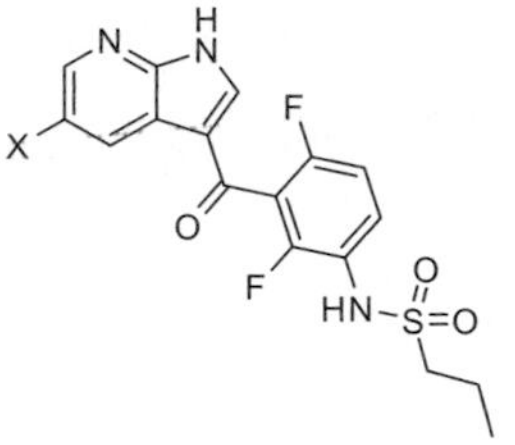

	X	BRAF^{V600E} IC$_{50}$ µM
Compound **11**	Phenyl	0.016
Compound **12**	3-Pyridyl	0.003
Compound **13**	4-NMe2-phenyl	0.055
Compound **14**	4-OMe-phenyl	0.02
Compound **15** Vemurafenib PLX4032	4-Cl-phenyl	0.031

Figure 26.8 5-Position SAR.

with the backbone Asp and Phe of the DFG sequence, which subsequently directs the *n*-propyl chain into a small pocket unique to the RAF family. The nitrogen atom of the sulfonamide tail is within H–bond distance to the main-chain NH-group of Asp-594, suggesting a deprotonated state of the nitrogen atom, while the two oxygen atoms of the sulfonamide form hydrogen-bonds with the backbone NH of Phe-595 and the side chain of Lys-483.

Final optimization was focused on the azaindole-5-position substitution. A number of aromatic and heteroaromatic substituents were synthesized and screened in the BRAF^{V600E} biochemical assay (Fig. 26.8). Although these substituents did not have significant impact on the biochemical potency, effects were observed on the pharmaceutical properties, *in vitro* safety profile, and pharmacokinetic properties.

Vemurafenib (compound **15**, PLX4032) was chosen for further development based on the mutant BRAF potency, selectivity against a panel of >300 kinases, and favorable pharmacokinetic properties both in rats (CL=0.07 L/kg/h; Vss=0.3 L/kg; i.v. $T_{1/2}$=2.5 h; F=>50%) and dogs (CL=0.15 L/kg/h; Vss=0.6 L/kg; i.v. $T_{1/2}$=2.4 h; F=>50%), the preferred species for toxicology studies.

The BRAF^{V600E} kinase domain crystallizes as an asymmetric dimer, often with one protomer in an active (DFG-in) conformation and the other protomer in a DFG-out inactive conformation. The co-crystal structure of vemurafenib with the BRAF^{V600E} kinase domain revealed binding to the active protomer (type 1 inhibitor) of the asymmetric dimer (Fig. 26.9), with the sulfonamide tail forming the same hydrogen bonds with the backbone NHs of the DFG motif[26,29] as PLX4720 in wild type BRAF. This binding causes a shift in the regulatory αC helix, which might explain the destabilizing effect of PLX4720 on CRAF–BRAF kinase domain hetero-dimerization compared to other RAF inhibitors such as AZ-628 and GDC-0879.[30]

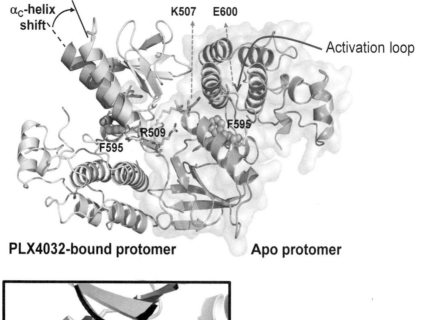

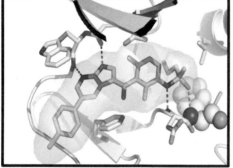

Figure 26.9 Co-crystal structure of vemurafenib. (See color plate.)

4. PRECLINICAL CHARACTERIZATION OF VEMURAFENIB

Vemurafenib displayed similar potency against the three RAF isoforms and selectivity against a majority of kinases (Fig. 26.10 and Table 26.1). The biochemical profile of vemurafenib translated to cellular selectivity, with potent inhibition of cellular proliferation and ERK phosphorylation in BRAF–mutant cell lines (Colo205, Colo829), but minimal effects in cells lacking BRAF mutations (A375, SW620, SKMEL2). Vemurafenib demonstrated dose–dependent inhibition of tumor growth in multiple mutant BRAF bearing colon cancer (Colo205)[29] and melanoma (LOX and Colo829)[31] xenograft models.

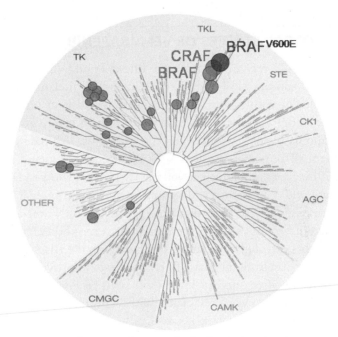

Figure 26.10 Kinome selectivity of vemurafenib. (See color plate.)

Table 26.1 Biochemical and cellular activity of vemurafenib

Biochemical assay	IC$_{50}$ (nM)	Biochemical assay	IC$_{50}$ (nM)	Cellular assay	IC$_{50}$ (nM)
BRAF^{V600E}	31	NEK11	317	Colo205	40
CRAF	48	BLK	547	Colo829	80
BRAF	100	LYNB	599	A375	300
SRMS	18	YES1	604	SW620	7828
ACK1	19	WNK3	877	SKMEL2	7107
MAP4K5 (KHS1)	51	MNK2	1717		
FGR	63	FRK (PTK5)	1884		
LCK	183	CSK	2339		
BRK	213	SRC	2389		

5. CLINICAL STUDIES OF VEMURAFENIB

Vemurafenib was evaluated in a Phase I study in patients with advanced solid tumors (dose-escalation followed by melanoma extension cohort),[32] initially with a crystalline formulation and later with an amorphous formulation.[33] Tumor regression was observed in a majority of metastatic melanoma patients with BRAF^{V600E} mutation. A Phase II study for the treatment of previously treated V600E positive metastatic melanoma patients with vemurafenib given at 960 mg twice daily showed 53% overall response rate and approximately 16 months of median overall survival.[34] The pivotal, randomized, and controlled Phase III study was designed to evaluate the efficacy and safety of vemurafenib as monotherapy in patients with previously untreated metastatic melanoma with the BRAF^{V600E} mutation. Analysis of data from the Phase III trial showed improved rates of overall and progression-free survival in patients who received vemurafenib.[35] The common adverse events associated with vemurafenib treatment were arthralgia, rash, fatigue, alopecia, keratoacanthoma or squamous-cell carcinoma, photosensitivity, nausea, and diarrhea.

A real time PCR reaction was developed to detect the BRAF^{V600E} mutations from formalin-fixed, paraffin-embedded tissue (FFPET).[36] A prototype diagnostic assay was developed concurrent with the Phase I trial. A positive test for the BRAF^{V600E} mutation was an enrollment criterion for the Phase 2 and Phase 3 trials.

6. CONCLUSIONS

Vemurafenib was a culmination of the discovery of oncogenic mutations in BRAF, a novel scaffold-based approach to drug discovery, development of companion diagnostic (cobas® 4800), and a rapid clinical development path. With demonstrated striking tumor shrinkage, a significant survival benefit, and a manageable safety profile, vemurafenib was approved in the United States in 2011 and the European Union in 2012 and is marketed as Zelboraf® for the treatment of patients with unresectable or metastatic melanoma with BRAF^{V600E} mutation as detected by an FDA approved test.

REFERENCES

1. Jerant, A. F.; Johnson, J. T.; Sheridan, C. D.; Caffrey, T. J. *Am. Fam. Physician* **2000**, *62*, 357.
2. Balch, C. M.; Buzaid, A. C.; Soong, S.-J.; Atkins, M. B.; Cascinalli, N.; Coit, D. G.; Fleming, I. D.; Gershenwald, J. E.; Houghton, A.; Kirkwood, J. M.; McMasters, K. M.; Mihm, M. F.; Morton, D. L.; Reintgen, D. S.; Ross, M. I.; Sober, A.; Thompson, J. A.; Thompson, J. F. *J. Clin. Oncol.* **2001**, *19*, 3635.
3. Davies, H.; Bignell, G. R.; Cox, C.; Stephens, P.; Edkins, S.; Clegg, S.; Teague, J.; Woffendin, H.; Garnett, M. J.; Bottomley, W.; Davis, N.; Dicks, E.; Ewing, R.; Floyd, Y.; Gray, K.; Hall, S.; Hawes, R.; Hughes, J.; Kosmidou, V.; Menzies, A.; Mould, C.; Parker, A.; Stevens, C.; Watt, S.; Hooper, S.; Wilson, R.; Jayatilake, H.; Gusterson, B. A.; Cooper, C.; Shipley, J.; Hargrave, D.; Pritchard-Jones, K.; Maitland, N.; Chenevix-Trench, G.; Riggins, G. J.; Bigner, D. D.; Palmieri, G.; Cossu, A.; Flanagan, A.; Nicholson, A.; Ho, J. W. C.; Leung, S. Y.; Yuen, S. T.; Weber, B. L.; Seigler, H. F.; Darrow, T. L.; Paterson, H.; Marais, R.; Marshall, C. J.; Wooster, R.; Stratton, M. R.; Futreal, P. A. *Nature* **2002**, *417*, 949.
4. www.sanger.ac.uk/genetics/CGP/cosmic/.
5. Uribe, P.; Wistuba, I.; Gonzalez, S. *Am. J. Dermatopathol.* **2003**, *25*, 365.
6. Gray-Schopfer, V.; Wellbrock, C.; Marias, R. *Nature* **2007**, *447*, 851.
7. (a) Dankort, D.; Curley, D. P.; Cartlidge, R. A.; Nelson, B.; Karnezis, A. N.; Damsky, W. E.; You, M. J.; DePinho, R. A.; McMahon, M.; Bosenberg, M. *Nat. Genet.* **2009**, *41*, 544. (b) Dhomen, N.; Reis-Filho, J. S.; Dias, S. d.-R.; Hayward, R.; Savage, K.; Delmas, V.; Larue, L.; Pritchard, C.; Marias, R. *Cancer Cell* **2009**, *15*, 294.
8. Lackey, K.; Cory, M.; Davis, R.; Frye, S. V.; Harris, P. A.; Hunter, R. N.; Jung, D. K.; McDonald, O. B.; McNutt, R. W.; Peel, M. R.; Rutkowske, R. D.; Veal, J. M.; Wood, E. R. *Bioorg. Med. Chem. Lett.* **2000**, *10*, 223.
9. Chin, P. C.; Liu, L.; Morrison, B. E.; Siddiq, A.; Ratan, R. R.; Bottiglieri, T.; D'Mello, S. R. *J. Neurochem.* **2004**, *90*, 595.
10. Hall-Jackson, C. A.; Eyers, P. A.; Cohen, P.; Goedert, M.; Boyle, F. T.; Hewitt, N.; Plant, H.; Hedge, P. *Chem. Biol.* **1999**, *6*, 559.
11. Heimbrook, D. C.; Huber, H. E.; Stirdivant, S. M.; Patrick, D. R.; Claremon, D.; Liverton, N.; Selnick, H.; Ahern, J.; Conroy, R.; Drakas, R.; Falconi, N.; Hancock, P.; Robinson, R.; Smith, G.; Olif, A. In: *AACR, New Orleans* 1998, Poster# 3793.
12. Wilhelm, S.; Chein, D.-S. *Curr. Pharm. Des.* **2002**, *8*, 2255.
13. Wilhelm, S.; Carter, C.; Lynch, M.; Lowinger, T.; Dumas, J.; Smith, R. A.; Schwartz, B.; Simantov, R.; Kelly, S. *Nat. Rev. Drug Discov.* **2006**, *5*, 835.
14. Wan, P. T. C.; Garnett, M. J.; Roe, S. M.; Lee, S.; Niculescu-Duvaz, D.; Good, V. M.; Jones, C. M.; Marshall, C. J.; Springer, C. J.; Barford, D.; Marais, R. *Cell* **2004**, *116*, 855.
15. Liu, Y.; Gray, N. S. *Nat. Chem. Biol.* **1996**, *2*, 358.
16. Shelton, J. G.; Moye, P. W.; Steelman, L. S.; Blalock, W. L.; Lee, J. T.; Franklin, R. A.; McMahon, M.; McCubrey, J. A. *Leukemia* **2003**, *17*, 1765.
17. Ott, P. A.; Hamilton, A.; Min, C.; Safarzadeh-Amiri, S.; Goldberg, L.; Yoon, J.; Yee, H.; Buckley, M.; Christos, P. J.; Wright, J. J.; Polsky, D.; Osman, I.; Liebes, L.; Pavlick, A. C. *PLoS One* **2010**, *5*, e15588.
18. Hauschild, A.; Agarwala, S. S.; Trefzer, U.; Hogg, D.; Robert, C.; Hersey, P.; Eggermont, A.; Grabbe, S.; Gonzalez, R.; Gille, J.; Peschel, C.; Schadendrof, D.;

Garbe, C.; O'Day, S.; Daud, A.; White, J. M.; Xia, C.; Patel, K.; Kirkwood, J. M.; Keilholz, U. *J. Clin. Oncol.* **2009**, *27*, 2823.

19. Drews, J. *Science* **2000**, *287*, 1960.
20. Erlanson, D. A.; Braisted, A. C.; Raphael, D. R.; Randal, M.; Stroud, R. M.; Gordon, E. M.; Wells, J. A. *Proc. Natl. Acad. Sci. U.S.A.* **2000**, *97*, 9367.
21. Zhang, K. Y. J.; Milburn, M. V.; Artis, D. R. In: *Scaffold-Based Drug Discovery*; Jhoti, H., Leach, A., Eds.; Springer: Dordrecht/Boston/London, 2007; p 129, Chapter 6.
22. Card, G. L.; Blasdel, L.; England, B. P.; Zhang, C.; Suzuki, Y.; Gillette, S.; Fong, D.; Ibrahim, P. N.; Artis, D. R.; Bollag, G.; Milburn, M. V.; Kim, S.-H.; Schlessinger, J.; Zhang, K. Y. J. *Nat. Biotechnol.* **2005**, *23*, 201.
23. Artis, D. R.; Lin, J. J.; Zhang, C.; Wang, W.; Mehra, U.; Perreault, M.; Erbe, D.; Krupka, H. L.; England, B. P.; Arnold, J.; Plotnikov, A. N.; Marimuthu, A.; Nguyen, H.; Will, S.; Signaevsky, M.; Kral, J.; Cantwell, J.; Settachatgull, C.; Yan, D. S.; Fong, D.; Oh, A.; Shi, S.; Womack, P.; Powell, B.; Habets, G.; West, B. L.; Zhang, K. Y. J.; Milburn, M. V.; Vlasuk, G. P.; Hirth, K. P.; Nolop, K.; Bollag, G.; Ibrahim, P. N.; Tobin, J. F. *Proc. Natl. Acad. Sci. U.S.A.* **2009**, *106*, 262.
24. Tsai, J.; Lee, J. T.; Wang, W.; Zhang, J.; Cho, H.; Mamo, S.; Bremer, R.; Gillette, S.; Kong, J.; Haass, N. K.; Sproesser, K.; Li, L.; Smalley, K. S. M.; Fong, D.; Zhu, Y.-L.; Marimuthu, A.; Nguyen, H.; Lam, B.; Liu, J.; Cheung, I.; Rice, J.; Suzuki, Y.; Luu, C.; Settachatgul, C.; Shellooe, R.; Cantwell, J.; Kim, S.-H.; Schlessinger, J.; Zhang, K. Y. J.; West, B. L.; Powell, B.; Habets, G.; Zhang, C.; Ibrahim, P. N.; Hirth, P.; Artis, D. R.; Herlyn, M.; Bollag, G. *Proc. Natl. Acad. Sci. U.S.A.* **2008**, *105*, 3041.
25. Kumar, A.; Mandiyan, V.; Suzuki, Y.; Zhang, C.; Rice, J.; Tsai, J.; Artis, D. R.; Ibrahim, P.; Bremer, R. *J. Mol. Biol.* **2005**, *348*, 183.
26. Bollag, G.; Tsai, J.; Zhang, J.; Zhang, C.; Ibrahim, P.; Nolop, K.; Hirth, P. *Nat. Rev. Drug Discovery* **2012**, *11*, 873.
27. Zhang, C.; Ibrahim, P.; Zhang, J.; Burton, B.; Habets, G.; Zhang, Y.; Powell, B.; West, B.; Wong, B.; Tsang, G.; Carias, H.; Ngyuyen, H.; Marimuthu, A.; Zhang, K.; Oh, A.; Bremer, R.; Hurt, C.; Wu, G.; Nespi, M.; Spevak, W.; Lin, P.; Nolop, K.; Hirth, P.; Tesch, G. H.; Bollag, G. *Proc. Natl. Acad. Sci. U.S.A.* **2013**, *110*, 5689.
28. Kaiser, C.; Colella, D. F.; Schwartz, M. S.; Garvey, E.; Wardell, J. R. *J. Med. Chem.* **1974**, *17*, 49.
29. Bollag, G.; Hirth, P.; Tsai, J.; Zhang, J.; Ibrahim, P. N.; Cho, H.; Spevak, W.; Zhang, C.; Zhang, Y.; Habets, G.; Burton, E.; Wong, B.; Tsang, G.; West, B. L.; Powell, B.; Shellooe, R.; Marimuthu, A.; Nguyen, H.; Zhang, K. Y. J.; Artis, D. A.; Schlessinger, J.; Su, F.; Higgins, B.; Iyer, R.; D'Andrea, K.; Koehler, A.; Stumm, M.; Lin, P. S.; Lee, R. J.; Grippo, J.; Puzanov, I.; Kim, K. B.; Ribas, A.; McArthur, G. A.; Sosman, J. A.; Chapman, P. B.; Flaherty, K. T.; Xu, X.; Nathanson, K. L.; Nolop, K. *Nature* **2010**, *467*, 596.
30. Hatzivassiliou, G.; Song, K.; Yen, I.; Brandhuber, B. J.; Anderson, D. J.; Alvarado, R.; Ludlam, M. J. C.; Stokoe, D.; Gloor, S. L.; Vigers, G.; Morales, T.; Aliagas, I.; Liu, B.; Sideris, S.; Hoeflich, K. P.; Jaiswal, B. J.; Seshagiri, S.; Koeppen, H.; Belvin, M.; Friedman, L. S.; Malek, S. *Nature* **2010**, *464*, 431.
31. Yang, H.; Higgins, B.; Kolinsky, K.; Packman, K.; Go, Z.; Iyer, R.; Kolis, S.; Zhao, S.; Lee, R.; Grippo, J.; Schostack, K.; Simcox, M. E.; Heimbrook, D.; Bollag, G.; Su, F. *Cancer Res.* **2010**, *17*, 5518.
32. Flaherty, K. T.; Puzanov, I.; Kim, K. B.; Ribas, A.; McArthur, G. A.; Sosman, J. A.; O'Dwyer, P. J.; Lee, R. J.; Grippo, J. F.; Nolop, K.; Chapman, P. B. *N. Engl. J. Med.* **2010**, *363*, 809.

33. Shah, N.; Iyer, R. M.; Mair, H.-J.; Choi, D. S.; Tian, H.; Diodone, R.; Fahnrich, K.; Pabst-Ravot, A.; Tang, K.; Scheubel, E.; Grippo, J. F.; Moreira, S. A.; Go, Z.; Mouskountakis, J.; Louie, T.; Ibrahim, P. N.; Sandhu, H.; Rubia, L.; Chokshi, H.; Singhal, D.; Malick, W. *J. Pharm. Sci.* **2013**, *102*, 967.

34. Sosman, J. A.; Kim, K. B.; Schuchter, L.; Gonzalez, R.; Pavlick, A. C.; Weber, J. S.; McArthur, G. A.; Hutson, T. E.; Moschos, S. J.; Flaherty, K. T.; Hersey, P.; Kefford, R.; Lawrence, D.; Puzanov, I.; Lewis, D. K.; Amaravadi, R. K.; Chmielowski, B.; Lawrence, H. J.; Shyr, Y.; Ye, F.; Li, J.; Nolop, K. B.; Lee, R. J.; Joe, A. K.; Ribas, A. *N. Engl. J. Med.* **2012**, *366*, 707.

35. Chapman, P. B.; Hauschild, A.; Robert, C.; Haanen, J. B.; Ascierto, P.; Larkin, J.; Dummer, R.; Garbe, C.; Testori, A.; Maio, M.; Hogg, D.; Lorigan, P.; Lebbe, C.; Jouary, T.; Schadendorf, D.; Ribas, A.; O'Day, S. J.; Sosman, J. A.; Kirkwood, J. M.; Eggermont, A. M. M.; Dreno, B.; Nolop, K.; Li, J.; Nelson, B.; Hou, J.; Lee, R. J.; Flaherty, K. T.; McArthur, G. A. *N. Engl. J. Med.* **2011**, *364*, 2507.

36. Halait, H.; Demartin, K.; Shah, S.; Soviero, S.; Langland, R.; Cheng, S.; Hillman, G.; Wu, L.; Lawrence, H. J. *Diagn. Mol. Pathol.* **2012**, *21*, 1.

CHAPTER TWENTY-SEVEN

New Chemical Entities Entering Phase III Trials in 2012

Gregory T. Notte
Gilead Sciences Inc., San Mateo, California, USA

Contents

Selection Criteria

- The Phase III clinical trial must have been registered with ClinicalTrials. gov.[a]
- The chemical structure must be available. References describing the medicinal chemistry discovery effort are included if available.
- It must be the first time that this compound has reached Phase III for any indication as a single agent or in combination.
- The compound must be synthetic in origin. The following classes of drugs are not included: biologics, inorganic or organometallic compounds, ssRNA, dendrimers, endogenous substances, radiopharmaceuticals, natural polypeptides, or herbal extracts.
- New formulations or single enantiomers of a previously approved drug are not included; novel pro-drugs are included.
- Compounds meeting the criteria are shown as the free base or free acid except those containing quaternary nitrogens.

---

[a] The selection criteria this year included compounds whose Phase III trial began in 2011, but the trial was not registered with Clinicaltrials.gov until 2012 or the structure was not disclosed until 2012.

Annual Reports in Medicinal Chemistry, Volume 48
ISSN 0065-7743
http://dx.doi.org/10.1016/B978-0-12-417150-3.00027-2

- This list was compiled using publically available information.[b] It is intended to give an overview of small molecule chemical matter entering Phase III and may not be all-inclusive.

Facts and Figures

- In 2012, there were 1426 Phase III trials registered at ClinicalTrials.gov that were classified as having a "drug intervention."
- Of the registered trials, 39 molecules (2.7%) met the selection criteria.
- For the molecules contained herein[c]:
  - Average molecular weight = 505 (range = 243–1681)
  - Average cLogD = 1.97 (range = − 14.1 to 6.76)
- The top four indications from all trials were:
  - Type 1 and 2 diabetes (97 trials)
  - Hepatitis C (35 trials)
  - Chronic obstructive pulmonary disorder (34 trials)
  - Breast cancer (30 trials)
- Of the 42 new chemical entities described in the Phase III chapter last year:
  - 28 (67%) are still in development
  - 7 (17%) have been discontinued
  - 6 (14%) have an NDA filed
  - 1 (2%) has been approved (Bedaquiline)

[b] The following two websites were used extensively in compiling this information: http://www. clinicaltrials.gov/ and http://www.ama-assn.org/ama/pub/physician-resources/medical-science/ united-states-adopted-names-council.page Any additional information could be obtained via a web search or from the Sponsor's website.

[c] Calculated using ACD labs software. Surotomycin and Avibactam were excluded from the cLogD averages due to the large negative values that were obtained.

1. ABT-333[1]

Sponsor: Abbott
MW/cLogD: 493.6/3.64
CAS#: 1132935-63-7
Start/End Date: Oct, 2012–Dec, 2013
Indication: HCV infection
Route of Admin: oral (co-administered with ABT-267, ABT-450, ritonavir, and ribavirin)
MOA: NS5B polymerase inhibitor
ClinicalTrials.gov Identifier: NCT01704755

2. Alisertib (MLN-8237)[2]

Sponsor: Millennium
MW/cLogD: 518.9/2.26
CAS#: 1028486-01-2
Start/End Date: Apr, 2012–Feb, 2017
Indication: Peripheral T-cell lymphoma
Route of Admin: oral, 50 mg, bid
MOA: Aurora-A kinase inhibitors
ClinicalTrials.gov Identifier: NCT01482962

3. Anagliptin (CWP-0403)[3]

Sponsor: JW Pharmaceutical
MW/cLogD: 383.4/−0.44
CAS#: 739366-20-2
Start/End Date: May, 2011–Sept, 2012
Indication: Type II diabetes mellitus
Route of Admin: oral, 100 mg, bid
MOA: Dipeptidyl peptidase IV inhibitor
ClinicalTrials.gov Identifier: NCT01529528

Continued

4. Asunaprevir (BMS-650032)[4]

Sponsor: Bristol-Myers Squibb
MW/cLogD: 748.3/1.11
CAS#: 630420-16-5
Start/End Date: Jan, 2012–May, 2013
Indication: HCV infection
Route of Admin: oral, 60 mg, qd
MOA: NS3 protease inhibitor
ClinicalTrials.gov Identifier: NCT01497834

5 Avibactam (NXL-104)[5]

Sponsor: AstraZeneca
MW/cLogD: 265.2/−13.7
CAS#: 396731-14-9
Start/End Date: Mar, 2012–Jun, 2014
Indication: Complicated intra-abdominal infections
Route of Admin: iv, 500 mg (in combination with ceftazidime and metronidazole)
MOA: Beta-lactamase inhibitor
ClinicalTrials.gov Identifier: NCT01499290

6. Baricitinib (LY-3009104)[6]

Sponsor: Eli Lilly
MW/cLogD: 371.4/−0.07
CAS#: 1187594-09-7
Start/End Date: Oct, 2012–Nov, 2014
Indication: Rheumatoid arthritis
Route of Admin: oral, 4 mg, qd
MOA: Jak1/2 inhibitor
ClinicalTrials.gov Identifier: NCT01710358

7. Betrixaban (PRT-054021)[7]

Sponsor: Portola
MW/cLogD: 451.9/−0.97
CAS#: 330942-05-7
Start/End Date: Mar, 2012–Aug, 2014
Indication: Venous thromboembolism
Route of Admin: oral, 80 mg, qd
MOA: Factor Xa inhibitor
ClinicalTrials.gov Identifier: NCT01583218

8. Bevenopran (CB-5945)[8]

Sponsor: Cubist
MW/cLogD: 386.4/−0.87
CAS#: 676500-67-7
Start/End Date: Oct, 2012–Dec, 2014
Indication: Opioid-induced constipation
Route of Admin: unknown, 0.25 mg, bid
MOA: mu-opioid receptor antagonist
ClinicalTrials.gov Identifier: NCT01696643

9. Buparlisib (NVP-BKM120)[9]

Sponsor: Novartis
MW/cLogD: 410.4/2.07
CAS#: 944396-07-0
Start/End Date: Aug, 2012–Mar, 2017
Indication: Breast cancer
Route of Admin: oral, 100 mg, qd
MOA: PI3Kα inhibitor
ClinicalTrials.gov Identifier: NCT01610284

Continued

10. Cobimetinib (XL-518)[10]

Sponsor: Hoffmann-La Roche
MW/cLogD: 531.3/3.47
CAS#: 934660-93-2
Start/End Date: Nov, 2012–Apr, 2017
Indication: Malignant melanoma
Route of Admin: oral, 60 mg, qd
MOA: MEK inhibitor
ClinicalTrials.gov Identifier: NCT01689519

11. Dapivirine (TMC-120)[11]

Sponsor: International Partnership for
Microbicides
MW/cLogD: 329.4/3.97
CAS#: 244767-67-7
Start/End Date: Mar, 2012–Mar, 2015
Indication: HIV infection
Route of Admin: Vaginal matrix ring
MOA: Non-nucleotide reverse transcriptase
inhibitor
ClinicalTrials.gov Identifier: NCT01337570

12. Delcobuvir (BI-207127)[12]

Sponsor: Boehringer Ingelheim
MW/cLogD: 653.6/3.25
CAS#: 863884-77-9
Start/End Date: Nov, 2012–Feb, 2014
Indication: HCV infection
Route of Admin: oral
MOA: NS5B inhibitor
ClinicalTrials.gov Identifier: NCT01728324

13. Dinaciclib (MK-7965)[13]

Sponsor: Merck
MW/cLogD: 396.5/0.78
CAS#: 779353-01-4
Start/End Date: Aug, 2012–Apr, 2016
Indication: Chronic lymphocytic leukemia
Route of Admin: iv
MOA: CDK 1, 2, 5, and 9 inhibitor
ClinicalTrials.gov Identifier: NCT0158028

14. Ecopipam (SCH-39166)[14]

Sponsor: Psyadon Pharma
MW/cLogD: 313.8/3.54
CAS#: 112108-01-7
Start/End Date: Dec, 2012–Dec, 2013
Indication: Lesch–Nyhan disease
Route of Admin: oral
MOA: Dopamine D1/D5 antagonist
ClinicalTrials.gov Identifier: NCT01751802

15. Elagolix (ABT-620)[15]

Sponsor: Abbott
MW/cLogD: 631.6/4.69
CAS#: 834153-87-6
Start/End Date: May, 2012–Mar, 2014
Indication: Endometriosis
Route of Admin: oral
MOA: Human gonadotropin-releasing
hormone receptor antagonist
ClinicalTrials.gov Identifier: NCT01620528

Continued

*16. Eluxadoline [JNJ-27018966)[16]

Sponsor: Furiex Pharmaceuticals
MW/cLogD: 569.7/1.4
CAS#: 864821-90-9
Start/End Date: May, 2012–Nov, 2013
Indication: Irritable bowel syndrome
Route of Admin: oral, 75–100 mg, bid
MOA: mu-opioid receptor agonist and
delta-opioid receptor antagonist
ClinicalTrials.gov Identifier: NCT01553591

*17. Evacetrapib (LY-2484595)[17]

Sponsor: Eli Lilly
MW/cLogD: 638.6/3.65
CAS#: 1186486-62-3
Start/End Date: Oct, 2012–Sept, 2015
Indication: High-risk vascular disease
Route of Admin: oral, 130 mg, qd
MOA: Cholesteryl ester transfer protein
(CETP) inhibitor
ClinicalTrials.gov Identifier: NCT01687998

*18. EVP-6124 (MT-4666)[18]

Sponsor: EnVivo Pharmaceuticals
MW/cLogD: 320.8/1.86
CAS#: 550999-75-2
Start/End Date: Oct, 2012–Sep, 2014
Indication: Schizophrenia
Route of Admin: oral, qd
MOA: α–7 Nicotinic receptor agonist
ClinicalTrials.gov Identifier: NCT01714661

19. Fexinidazole (HOE-239)[19]
Sponsor: Drugs for Neglected Diseases
MW/cLogD: 279.3/2.28
CAS#: 59729-37-2
Start/End Date: Oct, 2012–Mar, 2016
Indication: African trypanosomiasis
Route of Admin: oral, qd, step dosing
MOA: Antiprotozoal
ClinicalTrials.gov Identifier: NCT01685827

20. FG-4592[20]
Sponsor: FibroGen
MW/cLogD: 352.3/−1.81
CAS#: 808118-40-3
Start/End Date: May, 2012–Oct, 2015
Indication: Anemia
Route of Admin: oral
MOA: HIF prolyl hydroxylase inhibitor
ClinicalTrials.gov Identifier: NCT01630889

21. Ibrutinib (PCI-32765)[21]
Sponsor: Pharmacyclics
MW/cLogD: 440.5/2.92
CAS#: 936563-96-1
Start/End Date: Jun, 2012–July, 2015
Indication: Chronic lymphocytic leukemia
Route of Admin: oral, 320 mg, qd
MOA: Bruton's tyrosine kinase inhibitor
ClinicalTrials.gov Identifier: NCT01578707

Continued

22. Idelalisib (GS-1101)[22]
Sponsor: Gilead Sciences
MW/cLogD: 415.4/2.96
CAS#: 870281-82-6
Start/End Date: Feb, 2012–Feb, 2014
Indication: Chronic lymphocytic leukemia
Route of Admin: oral, 150 mg, bid with rituximab
MOA: PI3Kδ inhibitor
ClinicalTrials.gov Identifier: NCT01539512

23. Ixazomib Citrate (MLN-9708)[23]
Sponsor: Millennium Pharmaceuticals
MW/cLogD: 517.1/–
CAS#: 1201902-80-8
Start/End Date: Jun, 2012–Jun, 2014
Indication: Relapsed and/or refractory multiple myeloma
Route of Admin: oral, 4 mg on days 1, 8, and 15
MOA: Proteasome inhibitor
ClinicalTrials.gov Identifier: NCT01564537

24. Ledipasvir (GS-5885)[24]
Sponsor: Gilead
MW/cLogD: 889.0/6.76
CAS#: 1256388-51-8
Start/End Date: Oct, 2012–Oct, 2014
Indication: HCV infection
Route of Admin: oral, 90 mg, qd
MOA: NS5A inhibitor
ClinicalTrials.gov Identifier: NCT01701401

25. Lifitegrast (SAR-1118)[25]

Sponsor: SARcode Bioscience

MW/cLogD: 615.5/−2.76

CAS#: 1025967-78-5

Start/End Date: Aug, 2011–May, 2012

Indication: Dry eye

Route of Admin: topical, bid

MOA: Lymphocyte function-associated antigen-1 inhibitor

ClinicalTrials.gov Identifier: NCT01421498

26. Nemonoxacin (TG-873870)[26]

Sponsor: TaiGen Biotechnology

MW/cLogD: 371.4/−1.44

CAS#: 378746-64-6

Start/End Date: Apr, 2011–Dec, 2012

Indication: Pneumonia

Route of Admin: oral, 500 mg, qd

MOA: DNA topoisomerase inhibitor

ClinicalTrials.gov Identifier: NCT01529476

27. Obeticholic Acid (INT-747)[27]

Sponsor: Universita degli Studi di Perugia

MW/cLogD: 420.6/3.04

CAS#: 459789-99-2

Start/End Date: Jan, 2012–Jan, 2014

Indication: Primary biliary cirrhosis

Route of Admin: oral, 5–10 mg, qd

MOA: Farnesoid X receptor (FXR) agonist

ClinicalTrials.gov Identifier: NCT01473524

Continued

28. Omarigliptin (MK-3102)[28]

Sponsor: Merck
MW/cLogD: 398.4/−0.92
CAS#: 1226781-44-7
Start/End Date: Sept, 2012–Oct, 2014
Indication: Type 2 diabetes mellitus
Route of Admin: oral, 25 mg, qw
MOA: DPP-IV inhibitor
ClinicalTrials.gov Identifier: NCT01682759

29. Peretinoin (NK-333)[29]

Sponsor: Kowa Company
MW/cLogD: 302.5/4.45
CAS#: 81485-25-8
Start/End Date: Apr, 2012–tbd
Indication: Recurrent liver cancer
Route of Admin: oral, 600 mg, bid
MOA: Retinoic acid receptor agonist
ClinicalTrials.gov Identifier: NCT01640808

30. Ponatinib (AP-24534)[30]

Sponsor: Ariad
MW/cLogD: 532.6/3.39
CAS#: 943319-70-8
Start/End Date: Jun, 2012–Jun, 2021
Indication: Chronic myeloid leukemia
Route of Admin: oral, 45 mg, qd
MOA: Bcr-Abl inhibitor (and other RTKs)
ClinicalTrials.gov Identifier: NCT01650805

31. Pradigastat (LCQ-908)[31]

Sponsor: Novartis
MW/cLogD: 455.5/3.43
CAS#: 956136-95-1
Start/End Date: July, 2012–Nov, 2013
Indication: Familial chylomicronemia syndrome
Route of Admin: oral, qd
MOA: Diacylglycerol acyltransferase type 1 (DGAT-1) inhibitors
ClinicalTrials.gov Identifier: NCT01514461

32. Radotinib (IY-5511)[32]

Sponsor: Il-Yang Pharm. Co.
MW/cLogD: 530.5/4.28
CAS#: 926037-48-1
Start/End Date: Aug, 2011–Nov, 2014
Indication: Chronic myeloid leukemia
Route of Admin: oral, 300–400 mg, bid
MOA: Bcr-Abl kinase inhibitor
ClinicalTrials.gov Identifier: NCT01511289

33. Rolapitant (SCH-619734)[33]

Sponsor: Tesaro, Inc.
MW/cLogD: 500.5/3.28
CAS#: 552292-08-7
Start/End Date: Feb, 2012–Dec, 2012
Indication: Chemotherapy-induced nausea and vomiting
Route of Admin: oral, 200 mg, single dose
MOA: Neurokinin-1 receptor antagonist
ClinicalTrials.gov Identifier: NCT01500226

Continued

34. Siponimod (BAF-312)[34]

Sponsor: Novartis
MW/cLogD: 516.6/4.39
CAS#: 1230487-00-9
Start/End Date: Nov, 2012–Sep, 2016
Indication: Multiple sclerosis
Route of Admin: oral
MOA: Sphingosine 1-phosphate receptor modulator
ClinicalTrials.gov Identifier: NCT01665144

35. Solithromycin (CEM-101)[35]

Sponsor: Cempra
MW/cLogD: 845.0/2.72
CAS#: 760981-83-7
Start/End Date: Dec, 2012–Apr, 2014
Indication: Community-acquired bacterial pneumonia
Route of Admin: oral, qd
MOA: Antibiotic
ClinicalTrials.gov Identifier: NCT01756339

36. Surotomycin (CB-315)[36]

Sponsor: Cubist

MW/cLogD: 1680.7/−15.11

CAS#: 1233389-51-9

Start/End Date: May, 2012–Dec, 2014

Indication: Clostridium difficile associated diarrhea

Route of Admin: oral, 250 mg, bid

MOA: Membrane integrity inhibitor

ClinicalTrials.gov Identifier: NCT01597505

37. Telotristat Ethyl (LX-1606)[37]

Sponsor: Lexicon

MW/cLogD: 575.0/5.41

CAS#: 1033805-22-9

Start/End Date: Oct, 2012–Apr, 2015

Indication: Carcinoid syndrome

Route of Admin: oral, 250–500 mg, tid

MOA: Tryptophan hydroxylase 1 inhibitor

ClinicalTrials.gov Identifier: NCT01677910

Continued

38. Tipiracil (TAS-102)[38]

Sponsor: Taiho Pharma USA

MW/cLogD: 242.7/−3.90

CAS#: 183204-74-2

Start/End Date: Jun, 2012–Jun, 2014

Indication: Refractory metastatic colorectal cancer

Route of Admin: oral (in combination with trifluridine)

MOA: Thymidine phosphorylase inhibitor

ClinicalTrials.gov Identifier: NCT01607957

39. Zabofloxacin (DW-224a)[39]

Sponsor: Dong–Wha

MW/cLogD: 401.4/−0.99

CAS#: 219680-11-2

Start/End Date: Aug, 2012–Mar, 2013

Indication: COPD

Route of Admin: oral

MOA: DNA topoisomerase inhibitor

ClinicalTrials.gov Identifier: NCT01658020

REFERENCES

1. Shekhar, S.; Franczyk, T. S.; Barnes, D. M.; Dunn, D. M.; Haight, A. R.; Chan, V. S. Patent Application WO 2012/009699, 2012.
2. Sloane, D. A.; Trikic, M. Z.; Chu, M. L. H.; Lamers, M. B. A. C.; Mason, C. S.; Mueller, I.; Savory, W. J.; Williams, D. H.; Eyers, P. A. *ACS Chem. Biol.* **2010**, *5*, 563.
3. Kato, N.; Oka, M.; Murase, T.; Yoshida, M.; Sakairi, M.; Yamashita, S.; Yasuda, Y.; Yoshikawa, A.; Hayashi, Y.; Makino, M.; Takeda, Y.; Mirensha, Y.; Kakigami, T. *Bioorg. Med. Chem.* **2011**, *19*, 7221.
4. McPhee, F.; Sheaffer, A. K.; Friborg, J.; Hernandez, D.; Falk, P.; Zhai, G.; Levine, S.; Chaniewski, S.; Yu, F.; Barry, D.; Chen, C.; Lee, M. S.; Mosure, K.; Sun, L.-Q.; Sinz, M.; Meanwell, N. A.; Colonno, R. J.; Knipe, J.; Scola, P. *Antimicrob. Agents Chemother.* **2012**, *56*, 5387.
5. Xu, H.; Hazra, S.; Blanchard, J. S. *Biochemistry* **2012**, *51*, 4551.
6. (a) http://www.ama-assn.org/resources/doc/usan/baricitinib.pdf; (b) Fridman, J. S.; Scherle, P. A.; Collins, R.; Burn, T. C.; Li, Y.; Li, J.; Covington, M. B.; Thomas, B.; Collier, P.; Favata, M. F.; Wen, X.; Shi, J.; McGee, R.; Haley, P. J.; Shepard, S.; Rodgers, J. D.; Yeleswaram, S.; Hollis, G.; Newton, R. C.; Metcalf, B.; Friedman, S. M.; Vaddi, K. *J. Immunol.* **2010**, *184*, 5298.
7. Zhang, P.; Huang, W.; Wang, L.; Bao, L.; Jia, Z. J.; Bauer, S. M.; Goldman, E. A.; Probst, G. D.; Song, Y.; Su, T.; Fan, J.; Wu, Y.; Li, W.; Woolfrey, J.; Sinha, U.; Wong, P. W.; Edwards, S. T.; Arfsten, A. E.; Clizbe, L. A.; Kanter, J.; Pandey, A.; Park, G.; Hutchaleelaha, A.; Lambing, J. L.; Hollenbach, S. J.; Scarborough, R. M.; Zhu, B. Y. *Bioorg. Med. Chem. Lett.* **2009**, *19*, 2179.
8. (a) Woodward, R. M. Patent Application WO 2011/035142, 2011; (b) Singla, N. K.; Techner, L. M.; Gabriel, K.; Mangano, R. *Gastroenterology* **2012**, *143*, e26.
9. Brachmann, S. M.; Kleylein-Sohn, J.; Gaulis, S.; Kauffmann, A.; Blommers, M. J. J.; Kazic-Legueux, M.; Laborde, L.; Hattenberger, M.; Stauffer, F.; Vaxelaire, J.; Romanet, V.; Henry, C.; Murakami, M.; Guthy, D. A.; Sterker, D.; Bergling, S.; Wilson, C.; Brummendorf, T.; Fritsch, C.; Garcia-Echeverria, C.; Sellers, W. R.; Hofmann, F.; Maira, S. M. *Mol. Cancer Ther.* **2012**, *11*, 1747.
10. Rice, K. D.; Aay, N.; Anand, N. K.; Blazey, C. M.; Bowles, O. J.; Bussenius, J.; Costanzo, S.; Curtis, J. K.; Defina, S. C.; Dubenko, L.; Engst, S.; Joshi, A. A.; Kennedy, A. R.; Kim, A. I.; Koltun, E. S.; Lougheed, J. C.; Manalo, J. C. L.; Martini, J. F.; Nuss, J. M.; Peto, C. J.; Tsang, T. H.; Yu, P.; Johnston, S. *ACS Med. Chem. Lett.* **2012**, *3*, 416.
11. Ludovici, D. W.; De Corte, B. L.; Kukla, M. J.; Ye, H.; Ho, C. Y.; Lichtenstein, M. A.; Kavash, R. W.; Andries, K.; de Béthune, M.-P.; Azijn, H.; Pauwels, R.; Lewi, P. J.; Heeres, J.; Koymans, L. M. H.; de Jonge, M. R.; Van Aken, K. J. A.; Daeyaert, F. F. D.; Das, K.; Arnold, E.; Janssen, P. A. J. *Bioorg. Med. Chem. Lett.* **2001**, *11*, 2235.
12. Boecher, W.; Haefner, C.; Kukolj, G. Patent Application WO 2012/041771, 2012.
13. Paruch, K.; Dwyer, M. P.; Alvarez, C.; Brown, C.; Chan, T.-Y.; Doll, R. J.; Keertikar, K.; Knutson, C.; McKittrick, B.; Rivera, J.; Rossman, R.; Tucker, G.; Fischmann, T.; Hruza, A.; Madison, V.; Nomeir, A. A.; Wang, Y.; Kirschmeier, P.; Lees, E.; Parry, D.; Sgambellone, N.; Seghezzi, W.; Schultz, L.; Shanahan, F.; Wiswell, D.; Xu, X.; Zhou, Q.; James, R. A.; Paradkar, V. M.; Park, H.; Rokosz, L. R.; Stauffer, T. M.; Guzi, T. J. *ACS Med. Chem. Lett.* **2010**, *1*, 204.
14. Berger, J. G.; Chang, W. K.; Clader, J. W.; Hou, D.; Chipkin, R. E.; McPhail, A. T. *J. Med. Chem.* **1989**, *32*, 1913.

15. Chen, C.; Wu, D.; Guo, Z.; Xie, Q.; Reinhart, G. J.; Madan, A.; Wen, J.; Chen, T.; Huang, C. Q.; Chen, M.; Chen, Y.; Tucci, F. C.; Rowbottom, M.; Pontillo, J.; Zhu, Y.-F.; Wade, W.; Saunders, J.; Bozigian, H.; Struthers, R. S. *J. Med. Chem.* **2008**, *51*, 7478.

16. Breslin, H. J.; Diamond, C. J.; Kavash, R. W.; Cai, C.; Dyatkin, A. B.; Miskowski, T. A.; Zhang, S.-P.; Wade, P. R.; Hornby, P. J.; He, W. *Bioorg. Med. Chem. Lett.* **2012**, *22*, 4869.

17. Cao, G.; Beyer, T. P.; Zhang, Y.; Schmidt, R. J.; Chen, Y. Q.; Cockerham, S. L.; Zimmerman, K. M.; Karathanasis, S. K.; Cannady, E. A.; Fields, T.; Mantlo, N. B. *J. Lipid Res.* **2011**, *52*, 2169.

18. Prickaerts, J.; van Goethem, N. P.; Chesworth, R.; Shapiro, G.; Boess, F. G.; Methfessel, C.; Reneerkens, O. A. H.; Flood, D. G.; Hilt, D.; Gawryl, M.; Bertrand, S.; Bertrand, D.; König, G. *Neuropharmacology* **2012**, *62*, 1099.

19. Samant, B. S.; Sukhthankar, M. G. *Bioorg. Med. Chem. Lett.* **2011**, *21*, 1015.

20. Kang, X.; Long, W.; Zhang, J.; Hu, Y.; Wang, Y. Patent Application WO 2013/013609, 2013.

21. Pan, Z.; Scheerens, H.; Li, S.-J.; Schultz, B. E.; Sprengeler, P. A.; Burrill, L. C.; Mendonca, R. V.; Sweeney, M. D.; Scott, K. C. K.; Grothaus, P. G.; Jeffery, D. A.; Spoerke, J. M.; Honigberg, L. A.; Young, P. R.; Dalrymple, S. A.; Palmer, J. T. *ChemMedChem* **2007**, *2*, 58.

22. Lannutti, B. J.; Meadows, S. A.; Herman, S. E. M.; Kashishian, A.; Steiner, B.; Johnson, A. J.; Byrd, J. C.; Tyner, J. W.; Loriaux, M. M.; Deininger, M.; Druker, B. J.; Puri, K. D.; Ulrich, R. G.; Giese, N. A. *Blood* **2011**, *117*, 591.

23. (a) http://www.ama-assn.org/resources/doc/usan/ixazomib-citrate.pdf; (b) Bakale, R. P.; Mallamo, J. P.; Roemmele, R. C. Patent Application WO 2012/177835, 2012.

24. (a) http://www.ama-assn.org/resources/doc/usan/ledipasvir.pdf; (b) Lawitz, E. J.; Gruener, D.; Hill, J. M.; Marbury, T.; Moorehead, L.; Mathias, A.; Cheng, G.; Link, J. O.; Wong, K. A.; Mo, H.; McHutchison, J. G.; Brainard, D. M. *J. Hepatol.* **2012**, *57*, 24

25. Zhong, M.; Gadek, T. R.; Bui, M.; Shen, W.; Burnier, J.; Barr, K. J.; Hanan, E. J.; Oslob, J. D.; Yu, C. H.; Zhu, J.; Arkin, M. R.; Evanchik, M. J.; Flanagan, W. M.; Hoch, U.; Hyde, J.; Prabhu, S.; Silverman, J. A.; Wright, J. *ACS Med. Chem. Lett.* **2012**, *3*, 203, Note: this trial began in 2011, but the structure was disclosed in 2012.

26. Arjona, A. *Drugs Fut.* **2009**, *34*, 196, Note: this trial began 2011, but was not registered with ClinicalTrials.gov until Jan, 2012.

27. Pellicciari, R.; Fiorucci, S.; Camaioni, E.; Clerici, C.; Costantino, G.; Maloney, P. R.; Morelli, A.; Parks, D. J.; Willson, T. M. *J. Med. Chem.* **2002**, *45*, 3569.

28. http://www.ama-assn.org/resources/doc/usan/omarigliptin.pdf.

29. Okada, H.; Honda, M.; Campbell, J. S.; Sakai, Y.; Yamashita, T.; Takebuchi, Y.; Hada, K.; Shirasaki, T.; Takabatake, R.; Nakamura, M.; Sunagozaka, H.; Tanaka, T.; Fausto, N.; Kaneko, S. *Cancer Res.* **2012**, *72*, 4459.

30. Huang, W.-S.; Metcalf, C. A.; Sundaramoorthi, R.; Wang, Y.; Zou, D.; Thomas, R. M.; Zhu, X.; Cai, L.; Wen, D.; Liu, S.; Romero, J.; Qi, J.; Chen, I.; Banda, G.; Lentini, S. P.; Das, S.; Xu, Q.; Keats, J.; Wang, F.; Wardwell, S.; Ning, Y.; Snodgrass, J. T.; Broudy, M. I.; Russian, K.; Zhou, T.; Commodore, L.; Narasimhan, N. I.; Mohemmad, Q. K.; Iuliucci, J.; Rivera, V. M.; Dalgarno, D. C.; Sawyer, T. K.; Clackson, T.; Shakespeare, W. C. *J. Med. Chem.* **2010**, *53*, 4701.

31. Wen, H.; Kumaraperumal, N.; Nause, R. Patent Application WO 2012/051488-A1, 2012.

32. Park, J. M.; Lee, J. Y.; Kim, S. Y.; Park, S. B.; Son, J. A.; Kim, H. J.; Cho, D. H.; Bang, S. I.; Jung, M. G.; Ha, S. G. Patent Application KR 2012/130464-A, 2012. Note:

approved in S. Korea in 2012. First US trials conducted in 2011, but not registered with ClinicalTrials.gov until 2012.

33. Gan, T. J.; Gu, J.; Singla, N.; Chung, F.; Pearman, M. H.; Bergese, S. D.; Habib, A. S.; Candiotti, K. A.; Mo, Y.; Huyck, S.; Creed, M. R.; Cantillon, M. *Anesth. Analg.* **2011**, *112*, 804.

34. Pan, S.; Gray, N. S.; Gao, W.; Mi, Y.; Fan, Y.; Wang, X.; Tuntland, T.; Che, J.; Lefebvre, S.; Chen, Y.; Chu, A.; Hinterding, K.; Gardin, A.; End, P.; Heining, P.; Bruns, C.; Cooke, N. G.; Nuesslein-Hildesheim, B. *ACS Med. Chem. Lett.* **2013**, *4*, 333.

35. Fernandes, P.; Pereira, D.; Jamieson, B.; Keedy, K. *Drugs Fut.* **2011**, *36*, 751.

36. Mascio, C. T. M.; Mortin, L. I.; Howland, K. T.; Van Praagh, A. D. G.; Zhang, S.; Arya, A.; Chuong, C. L.; Kang, C.; Li, T.; Silverman, J. A. *Antimicrob. Agents Chemother.* **2012**, *56*, 5023.

37. http://www.ama-assn.org/resources/doc/usan/telotristat-ethyl.pdf.

38. Yano, S.; Kazuno, H.; Sato, T.; Suzuki, N.; Emura, T.; Wierzba, K.; Yamashita, J.; Tada, Y.; Yamada, Y.; Fukushima, M.; Asao, T. *Bioorg. Med. Chem.* **2004**, *12*, 3443.

39. Kwon, A.-R.; Min, Y.-H.; Ryu, J.-M.; Choi, D.-R.; Shim, M.-J.; Choi, E.-C. *J. Antimicrob. Chemother.* **2006**, *58*, 684.

CHAPTER TWENTY-EIGHT

To Market, To Market—2012

Joanne Bronson[*], **Amelia Black**[†], **T. G. Murali Dhar**[‡],
Bruce A. Ellsworth[§], **J. Robert Merritt**[¶]

[*]Bristol-Myers Squibb Company, Wallingford, Connecticut, USA
[†]Bristol-Myers Squibb Company, Redwood City, California, USA
[‡]Bristol-Myers Squibb Company, Princeton, New Jersey, USA
[§]Bristol-Myers Squibb Company, Pennington, New Jersey, USA
[¶]Kean University, Union, New Jersey, USA

Contents

Annual Reports in Medicinal Chemistry, Volume 48
ISSN 0065-7743
http://dx.doi.org/10.1016/B978-0-12-417150-3.00028-4

1. OVERVIEW

This year's To-Market, To-Market chapter provides summaries for 30 compounds (28 small molecules and 2 monoclonal antibodies) that received approval for the first time in any country in 2012.[1] This is a substantial increase relative to new approvals covered in the previous 2 years for this chapter (26 approvals for 2011; 24 for 2010). Twenty-two of the 30 first-time approvals in 2012 came from the United States, followed by 4 from the European Union, 3 from Japan, and 1 from Korea. Note that the United States had a 16-year high with 39 approvals of new products,[2] some of which were previously approved outside of the United States. Cancer treatments again dominated the new entries with a remarkable 14 approvals, including 12 small molecules and 2 monoclonal antibodies. Three new diabetes treatments were introduced, along with three agents for gastrointestinal disorders. The remaining therapeutic areas had 1–2 approvals each. The following overview is organized by therapeutic area, with drugs covered in this year's chapter described first in each area, followed by additional approvals that are of interest but not covered in detail in this chapter (e.g., combination therapies and imaging agents).

This year's class of 14 anticancer drugs is particularly striking for the range of cancers treated, the variety of mechanisms of action, and the focus on treatment of patients who have failed on one or more previous therapies. Six of the anticancer drugs are kinase inhibitors; three of these are Bcr–Abl tyrosine kinase inhibitors (TKIs) for the treatment of chronic myeloid leukemia (CML). Other targeted mechanistic classes within the oncology approvals include a proteosome inhibitor for multiple myeloma (MM), an androgen receptor (AR) antagonist for prostate cancer, a topoisomerase inhibitor for non-Hodgkin's lymphoma (NHL), and a hedgehog pathway inhibitor for basal cell carcinoma (BCC). Two natural products that had previously been used in traditional medicine were approved, one for actinic keratoses and the other for CML. Two of the 2012 anticancer approvals are monoclonal antibodies, with one for T-cell lymphoma and one for metastatic breast cancer.

Kinase inhibitors continue to have a strong showing in the oncology arena. **Axitinib (Inlyta®)** was approved by the US Food and Drug Administration (FDA) for the treatment of advanced renal cell carcinoma (RCC) for patients who have not responded to prior therapy. It is the first kidney cancer drug to be exclusively tested in a second-line setting and the first to compare itself against another targeted therapy (sorafenib). Axitinib binds to the intracellular tyrosine kinase catalytic domain of vascular endothelial growth factor receptors (VEGFRs) leading to blockade of signaling through this angiogenic pathway. The recommended dose of axitinib is 5 mg administered orally twice daily. **Bosutinib (Bosulif®)** has been approved by the US FDA for the treatment of relapsed or refractory CML for patients with resistance or intolerance to prior therapy. Bosutinib is a dual inhibitor of Bcr–Abl and Src family kinases, and inhibits 16 of 18 imatinib-resistant forms of Bcr–Abl expressed in murine myeloid cell lines. Bosutinib was efficacious in patients with imatinib-resistant or -intolerant CML who had chronic, accelerated, or blast phase CML. The recommended dose of bosutinib is 500 mg administered orally once daily, with a meal. **Cabozantinib (Cometriq™)**, a TKI with nanomolar pan-inhibitory activity against VEGFR2, MET, and RET among others, was approved by the US FDA for the treatment of patients with progressive, unresectable, locally advanced, or metastatic medullary thyroid cancer (MTC). Cabozantinib is given as a 140 mg oral capsule once daily. The US FDA approved **ponatinib (Inclusig®)** for the treatment of adult patients with chronic phase, accelerated phase, or blast phase CML. Ponatinib is a pan-Bcr–Abl tyrosine kinase that blocks both the native and mutated kinases, particularly the gatekeeper T315I mutation in CML patients. Ponatinib carries a black-box warning for arterial thrombosis and hepatotoxicity. The recommended dose of ponatinib is 45 mg administered orally once daily, with or without food. **Radotinib (Supect™)**, another inhibitor of Bcr–Abl tyrosine kinase, was approved in Korea as a second-line treatment for CML. Radotinib is a TKI with a similar structure to the second-generation TKI, nilotinib. Radotinib is formulated as 100 and 200 mg oral capsules. The US FDA approved **regorafenib (Stivarga™)** for the treatment of patients with metastatic colorectal cancer (CRC), especially those who have failed on standard therapies. Regorafenib is a multikinase inhibitor with potent inhibitory activity against VEGFRs and platelet-derived growth factor receptors (PDFRs). Both of these classes of receptors are expressed on tumor cells and affect proliferation and angiogenesis. Regorafenib is available in 40 mg tablets and given orally once daily with a low-fat breakfast as a 160 mg dose.

Anticancer agents with nonkinase inhibitor mechanisms of action were also approved in 2012. **Carfilzomib (Kyprolis®)**, an irreversible inhibitor of the proteosome with a favorable safety profile and a low incidence of peripheral neuropathy, was approved by the US FDA for the treatment of patients with MM who have received at least two prior therapies including bortezomib and an immunomodulatory agent and have demonstrated disease progression after completion of the last therapy. Carfilzomib is administered intravenously at 20 mg/m^2 on days 1, 2, 8, 9, 15, and 16 in cycle 1, with escalation to 27 mg/m^2 in subsequent cycles as tolerated every 28 days. The US FDA approved **enzalutamide (Xtandi®)** for the treatment of metastatic castration-resistant prostate cancer (mCRPC) in patients who have previously been treated with docetaxel. Enzalutamide is a potent AR antagonist with improved efficacy relative to prior members of this class. Enzalutamide is approved as a 160 mg total dose (four 40 mg capsules) taken orally once daily, with or without food. **Pixantrone (Pixuvri™)**, an anthracycline analogue that was designed to address the cardiotoxicity seen in earlier anthracyclines, was approved by the European Commission as a single agent for the treatment of relapsed or refractory aggressive B-cell NHL in adult patients who have failed on at least two previous therapies. Pixantrone inhibits topoisomerase II by intercalation with DNA and is also believed to form covalent adducts with guanine. The recommended dose is 85 mg/m^2 of pixantrone dimaleate on days 1, 8, and 15 of each 28-day cycle for up to six cycles. The US FDA approved **vismodegib (Erivedge™)** for the treatment of adults with metastatic BCC, with locally advanced BCC that has recurred following surgery, or who are not candidates for surgery or radiation. Vismodegib inhibits the Hedgehog signaling pathway by functioning as an antagonist of the seven transmembrane receptor smoothened (SMO), thereby inhibiting the activation of Hedgehog target genes and decreasing downstream production of proliferation factors. Vismodegib carries a black-box warning for embryo-fetal death and severe birth defects. The recommended dose of vismodegib is 150 mg administered orally once daily.

The two natural products approved as anticancer agents in 2012 are **ingenol mebutate** and **omacetaxine mepesuccinate**. **Ingenol mebutate (Picato®)** was approved for the topical treatment of actinic keratosis (AK) of the face/scalp and trunk in the United States, EU, Australia, and Brazil. Ingenol mebutate is a terpenoid natural product that has a novel mechanism of action involving initial plasma membrane disruption, rapid loss of mitochondrial membrane potential, and cell death by primary necrosis within 1 h; a subsequent tumor-specific immune response results in antibody-dependent cellular toxicity that eliminates residual cells. Among the advantages to ingenol mebutate are short duration of treatment (2–3 days), high efficacy, relatively rapid resolution of local reactions, and high adherence to therapy. Ingenol mebutate gel was approved for topical application at two concentrations: 0.15% for the treatment of AK lesion of the face and scalp; 0.05% for AK lesions of the trunk and extremities. **Omacetaxine mepesuccinate (Synribo™)** was approved by the US FDA for the treatment of patients with chronic or accelerated phase CML with resistance or intolerance to at least two TKIs. Omacetaxine is a naturally occurring alkaloid that acts on the initial step of protein translation and results in the rapid loss of a number of short-lived proteins that regulate proliferation and cell survival. Omacetaxine is supplied in a single-use vial containing 3.5 mg of omacetaxine mepesuccinate as a lyophilized powder and is administered via subcutaneous injection.

Two monoclonal antibodies were approved as anticancer agents in 2012. **Mogamulizumab (Poteligeo®)** is a CC chemokine receptor 4 (CCR4) targeting antibody that was approved in Japan for the treatment of relapsed or refractory CCR4+ adult T-cell leukemia–lymphoma. Mogamulizumab is a humanized monoclonal antibody that has been glycoengineered to remove fucose residues from the Fc region oligosaccharides, resulting in an enhanced ability to mediate antibody-dependent cellular cytotoxicity (ADCC). It is the first glycoengineered antibody to reach the market. Mogamulizumab is administered intravenously once weekly (eight administrations) at a dose of 1.0 mg/kg. **Pertuzumab (Perjeta®)** is an anti-human epidermal growth factor receptor 2 (HER2) antibody that was approved by the US FDA for use in combination with trastuzumab (Herceptin®) and docetaxel chemotherapy for the treatment of people with HER2-positive (HER2+) metastatic breast cancer who have not received prior anti-HER2 therapy or chemotherapy for metastatic disease. Pertuzumab is the first HER2 dimerization inhibitor to reach the market. Pertuzumab is given at an initial dose of 840 mg administered as a 60-min intravenous infusion, followed every 3 weeks by 420 mg administered as a 30–60-min intravenous infusion.

The US FDA approved **Choline C11** as a positron emission tomographic (PET) imaging agent to detect recurrent prostate cancer in patients with suspected prostate cancer recurrence based on elevated blood prostate-specific antigen levels following initial therapy for prostate cancer.[3] Choline C11 is administered intravenously, with immediate imaging for 0–15 min after injection. In clinical studies, at least half of the patients who had abnormalities detected on PET scans also had recurrent prostate cancer that was confirmed by histological tissue sampling. PET scan errors also were reported, highlighting the need for histological verification of findings.

There were two approvals in the immunology area: one for multiple sclerosis (MS) and one for arthritis. **Teriflunomide (Aubagio®)** is the second approved oral therapy for the treatment of relapsing forms of MS, the first being Gilenya®, which was approved in 2010. Teriflunomide's mechanism of action is via inhibition of *de novo* pyrimidine synthesis, primarily targeting activated lymphocytes in the periphery. The US FDA approved teriflunomide for the treatment of relapsing forms of MS at 7 or 14 mg administered orally once daily. **Tofacitinib (Xeljanz®)** has been approved by the US FDA for the treatment of adult patients with moderate to severe rheumatoid arthritis (RA) who have had an inadequate response or intolerance to methotrexate. It is the first kinase inhibitor approved for the treatment of RA and works by inhibiting the Janus kinases (JAK-1, 2, 3, and Tyk-2), indirectly affecting the production of inflammatory mediators and suppressing STAT1-dependent genes in joint tissue. Efficacy with tofacitinib as monotherapy and in combination with methotrexate is comparable to that seen with antibody treatment. The recommended dose of tofacitinib is 5 mg administered orally twice daily either as monotherapy or in combination with methotrexate or other nonbiologic disease-modifying antirheumatic drugs.

In the infectious diseases area, **bedaquiline (Sirturo™)** was approved by the US FDA as part of combination therapy for the treatment of multidrug-resistant tuberculosis (MDR-TB), becoming the first drug approved for MDR-TB. It is also the first approval from a new class of anti-tuberculosis agents in the past 40 years. The mechanism of action of bedaquiline is unique amongst anti-TB drugs and involves inhibition of mycobacterial ATP synthase. Bedaquiline has a black-box warning for increased rate of death and QTc prolongation. Bedaquiline was approved at 400 mg daily for 2 weeks followed by 200 mg three times weekly for 22 weeks, as part of a combination therapy for MDR-TB. **Stribild™**, a once-a-day, four-drug combination pill, was approved by the US FDA for the treatment of HIV-1 infection in adults who are antiretroviral treatment-naïve.[4] Stribild combines the previously approved drugs emtricitabine (a nucleoside analogue reverse transcriptase inhibitor) and tenofovir disoproxil fumarate (a nucleotide analogue reverse transcriptase inhibitor) with two new drugs, elvitegravir (an HIV integrase inhibitor) and cobicistat (a Cyp 3A4 inhibitor to reduce metabolism of elvitegravir). The dual combination of emtricitabine and tenofovir disoproxil fumarate (Truvada®) was approved in 2004. Clinical results with Stribild™ showed that 88–90% of patients had an undetectable amount of HIV in their blood after 48 weeks of treatment, compared with 84% treated with Atripla® and 87% treated with Truvada® plus atazanavir and ritonavir. One tablet of Stribild™ is composed of 150 mg of elvitegravir, 150 mg of cobicistat, 200 mg of emtricitabine, and 300 mg of tenofovir disoproxil fumarate and is taken once daily with food.

In the endocrine diseases area, first-time approvals were obtained for three drugs for the treatment of type 2 diabetes mellitus (T2DM) and one for the treatment of Cushing's disease (CD). **Dapagliflozin (Forxiga®)** is a first-in-class selective sodium-dependent glucose transporter 2 (SGLT2) inhibitor that was approved in Australia and Europe as an adjunct to diet and exercise for the treatment of type 2 diabetes. Dapagliflozin has a glucose-sensitive and insulin-independent mechanism of action in which SGLT2 inhibition lowers the renal threshold for reabsorption of glucose and allows excess glucose to be eliminated via the kidneys. The aryl C-glucoside linkage found in dapagliflozin confers resistance to glucosidase-mediated metabolism. Dapagliflozin was approved at a 10 mg usual daily dose (5 mg starting dose for severe hepatic impairment) as oral monotherapy for the treatment of diabetes, and as add-on therapy with other glucose-lowering agents. **Anagliptin (Suiny)** and **teneligliptin (Tenelia®)** are two of a growing number of dipeptidyl peptidase-4 (DPP-4) inhibitors to be approved worldwide for the treatment of patients with T2DM. DPP-4 inhibitors block degradation of glucagon-like peptide 1 (GLP-1), a 30-amino acid peptide that is secreted in response to food intake. GLP-1 stimulates insulin secretion and inhibits glucagon secretion, which leads to lower levels of plasma glucose. Anagliptin is a potent and selective DPP-4 inhibitor that was approved in Japan for the treatment of T2DM at 100 mg given once daily. Teneligliptin is also a potent and selective DPP-4 inhibitor approved in Japan as 20 mg tablets taken orally once daily for the treatment of patients with T2DM that is not controlled by diet or exercise. The US FDA approved two therapies for T2DM that combine previously approved DPP-4 inhibitors with metformin, an older antidiabetic agent that lowers blood glucose by reducing glucose production by the liver. **Jentadueto®** is a combination of the DPP-4 inhibitor linagliptin with metformin that is taken twice daily with meals.[5] The starting dose is based on the individual's current therapy, with maximum recommended dose of 2.5 mg linagliptin/1000 mg metformin. **Janumet® XR** is a new formulation of the sitagliptin/metformin combination that is taken once daily with the evening meal.[6] The starting dose depends on current treatment, with a maximum recommended daily dose of 100 mg sitagliptin/2000 mg metformin. There are two previously approved combinations of DPP-4 inhibitors with metformin: Kombiglyze (saxagliptin/metformin, approved in 2010) and Janumet (sitagliptin/metformin, approved in 2007). **Pasireotide (Signifor®)** was approved in Europe and the US for the treatment of CD in adult patients who have not responded to surgical intervention or for whom surgery is not an option. CD is a relatively rare disease that is caused by pituitary tumors that secrete adrenocorticotropic hormone (ACTH), leading to production of excess cortisol and a number of adverse health consequences. Pasireotide is a cyclohexapeptide that acts as a somatostatin analogue to inhibit the release of ACTH. The approved initial dosage is either 600 or 900 µg given by subcutaneous injection twice daily, with titration of dosing based on treatment response and tolerability.

In the metabolic diseases area, there was one first-time approval for an antiobesity agent and one for an antihypercholesteremic agent. The US FDA approved **lorcaserin (Belviq®)** for chronic weight management in adult patients who are characterized as overweight or obese and have at least one comorbid, weight-related condition. Lorcaserin is the first $5\text{-}HT_{2C}$ agonist approved for chronic weight management since the withdrawal of the $5\text{-}HT_{2C}$ agonist fenfluramine in 1997 due to rare cases of cardiac valvulopathy. Lorcaserin has improved selectivity for $5\text{-}HT_{2C}$, which is found primarily in the hypothalamus and regulates appetite and feeding behavior, over $5HT_{2A}$ and $5HT_{2B}$, which are believed to be related to safety issues. Incidences of cardiac valvulopathy occurred in 2% of patients who received placebo or lorcaserin at 10 mg twice daily. Lorcaserin is given orally twice daily as a 10 mg tablet. A combination of phentermine and topiramate, **Qsymia™**, was approved in an extended release formulation for chronic weight management as an adjunct to a reduced calorie diet and exercise in obese adults or overweight adults with at least one weight-related condition such as high blood pressure (hypertension), type 2 diabetes, or high cholesterol.[7] Phentermine was already approved as a single agent for short-term weight loss in overweight/obese adults, while topiramate is an antiepileptic and antimigraine agent. In two randomized, placebo-controlled trials, Qsymia™ given at the recommended daily dose of 7.5 mg phentermine and 46 mg of topiramate resulted in loss of at least 5% of body weight for 62% of subjects compared with 20% on placebo. The FDA approved Qsymia™ with a risk evaluation and mitigation strategy (REMS) due to potential for fetal toxicity. **Lomitapide (Juxtapid™)** was approved by the US FDA for the treatment of patients with familial hypercholesteremia in conjunction with a low-fat diet and other lipid-lowering treatments. Lomitapide is an inhibitor of microsomal triglyceride transfer protein (MTP), which is a viable target for reducing cholesterol based on genetic evidence. The US FDA approved lomitapide (5, 10, and 20 mg capsules), a first-in-class MTP inhibitor, contingent on the implementation of a REMS.

There were three approvals in the gastrointestinal area: one for irritable bowel syndrome, one for short bowel syndrome (SBS), and one for diarrhea in patients on HIV/AIDS therapy. The US FDA approved **linaclotide (Linzess™)**, a first-in-class, orally administered 14-amino acid peptide, as a therapy for patients suffering from chronic idiopathic constipation (CIC) and irritable bowel syndrome with constipation (IBS-C). Linaclotide mimics the actions of the endogenous intestinal peptides guanylin and uroguanylin, leading to the activation of cystic fibrosis transmembrane regulator (CFTR) ion channel, which increases levels of HCO_3^-, Cl^-, and water in the intestinal lumen and accelerates gastrointestinal transit. Linaclotide carries a black-box warning for pediatric risk. The recommended dose of linaclotide is 290 µg administered orally once daily for IBS-C and 145 µg administered orally once daily for CIC. **Teduglutide (Gattex®)**, a glucagon peptide 2 (GLP-2) agonist, was approved in Europe and the United States for the treatment of adults with SBS who are dependent on parenteral nutrition (PN). GLP-2 is a 33-amino acid endogenous peptide that is released upon eating a meal and acts to slow gastric emptying, reduce gastric secretions, and stimulate growth and repair of intestinal epithelium. Teduglutide is a GLP-2 analogue in which glycine has been substituted for alanine at position 2 from the N-terminus, thereby conferring resistance to degradation. Due to potential risk of developing cancer and abnormal growths in the intestine, teduglutide was approved with a REMS. Teduglutide is administered at a dose of 0.05 mg/kg, which is prepared by reconstitution of lyophilized powder with distilled water and given as a subcutaneous injection once daily. **Crofelemer (Fulyzaq™)**, an oligomeric proanthocyanidin natural product that is isolated from the red latex of *Croton lechleri*, was approved by the US FDA for the treatment of diarrhea in HIV/AIDS patients on antiretroviral therapy. Crofelemer is an inhibitor of CFTR and calcium-stimulated chloride intestinal channels. Crofelemer is not absorbed to appreciable amounts, with systemic exposures below the limit of detection in plasma. The approved use of crofelemer for the treatment of patients with HIV/AIDS-related diarrhea is one 125 mg delayed release tablet, taken twice daily.

There were two drugs approved for the first time in the respiratory diseases area. **Aclidinium bromide (Tudorza Pressair™** in the US; **Eklira®/Bretaris® Genuair®** in the EU) was approved in the US and the EU for long-term maintenance treatment of bronchospasm associated with chronic obstructive pulmonary disease (COPD). Aclidinium bromide is a selective muscarinic M3 receptor antagonist that is given by inhalation to limit systemic exposure. In addition, aclidinium bromide undergoes rapid hydrolysis in human plasma, providing inactive by-products and reducing the potential for systemic side effects. Aclidinium bromide was approved at a 400 µg dose given twice daily as an inhaled powder. The US FDA approved **ivacaftor (Kalydeco®)** for the treatment of cystic fibrosis (CF) in patients who have the G551D mutation of the CFTR and are at least 6 years old. Ivacaftor is a CFTR potentiator that increases the open probability of CFTR, thus increasing chloride secretion, particularly in the 5% of CF patients with the G551D gating mutation. In Phase III trials, ivacaftor improved several parameters of lung function. Ivacaftor is given orally twice daily with food as a 150 mg tablet.

In the central nervous system area, **perampanel (Fycompa™)** was approved in the United States for the treatment of partial onset seizures in epileptic patients who are at least 12 years old. Perampanel is the first α-amino-3-hydroxyl-5-methyl-4-isoxazole-propionic acid (AMPA) receptor antagonist to receive FDA approval as an antiepileptic drug (AED). Perampanel prevents ion channel opening and reduces propagation of action potentials. In a double-blind, placebo-controlled Phase III study, the 50% responder rate after 1 and 2 years was 47.6% and 63.2%, respectively, with 7% of subjects reporting a year-long seizure-free period after the first year. An increase in anger, confusion, and depression were observed at the 12 mg dose, which resulted in a boxed warning for serious psychiatric events. Perampanel is given orally once daily as a 2, 4, 6, 8, 10, or 12 mg tablet. A radiopharmaceutical agent for PET imaging of β-amyloid neuritic plaque in the brain was approved by the US FDA. **Florbetapir F18 (Amyvid™)** is indicated for imaging of cognitively impaired patients undergoing evaluation for Alzheimer's disease (AD).[8,9] Florbetapir F18 is given as a single intravenous bolus injection, followed by 10-min PET scanning 30–50 min after administration. A negative scan indicates that cognitive impairment may not be due to AD and that other causes for cognitive decline should be investigated. A positive scan is consistent with AD, but is not conclusive for an AD diagnosis since β-amyloid neuritic plaques are found in other neurological diseases as well as in AD. Furthermore, β-amyloid plaques may be found in normal elderly individuals. In the main clinical studies supporting FDA approval, the accuracy of florbetapir scans was assessed in the brains of terminally ill patients who participated in a brain-donation program. In 59 patients who underwent florbetapir scans and autopsy, scan sensitivity for the detection of moderate to frequent β-amyloid neuritic plaques was 92% (range 69–95), and scan specificity was 95% (range 90–100), on the basis of the median assessment among five highly trained readers.

An additional entry in 2012 was approval in the United States of **peginesatide (Omontys®)** for the treatment of anemia in patients with chronic kidney disease (CKD) who require dialysis. Peginesatide is a 54 kDa pegylated peptide containing unnatural amino acids that activates the erythropoietin receptor, but it is not structurally related to EPO. In clinical trials, peginesatide was shown to be noninferior to epoietin when given once monthly at doses tritrated to maintain hemoglobin levels of 10–12 dg/mL. In February 2013, all doses of peginesatide were recalled in a voluntary move by the marketing companies due to postmarketing reports of serious hypersensitivity reactions, including anaphylaxis, in 0.02% of the patient population.

2. ACLIDINIUM BROMIDE (CHRONIC OBSTRUCTIVE PULMONARY DISEASE)[10–19]

Class: Muscarinic M3 receptor antagonist
Country of origin: Spain
Originator: Almirall
First introduction: United States
Introduced by: Forest (US); Almirall (EU)
Trade name: Tudorza Pressair™ (US)
Eklira®/Bretaris® Genuair® (EU)
CAS registry no: 320345-99-1

In July 2012, aclidinium bromide was approved in the US and the EU for long-term maintenance treatment of bronchospasm associated with chronic obstructive pulmonary disorder (COPD). COPD is a progressive inflammatory disease that is characterized by chronic obstructed airflow that makes breathing difficult. COPD encompasses two main forms of disease that often coexist: chronic bronchitis, which involves a long-term cough with mucus, and emphysema, which involves progressive destruction of the lungs. COPD occurs in 3–8% of adults in the United States and varies from 4% to 10% of adults in European countries.[10,11] The primary risk factor for COPD is cigarette smoking (85–90% of cases), but COPD is also associated with exposure to environmental pollutants, recurrent infection, diet, and genetic factors. There is no cure for COPD, but treatment can control symptoms and slow disease progression. The main pharmacological treatment for managing COPD symptoms is use of bronchodilators, such as long-acting β2-selective adrenoceptor agonists and long-acting muscarinic antagonists.[12] Aclidinium bromide is in the latter category, acting as a selective antagonist for the muscarinic M3 receptor.[13,14] M3 receptors are localized in airway smooth muscle, and are the primary subtype responsible for bronchial and tracheal smooth muscle contraction. Muscarinic antagonists are well-established bronchodilators that are effective for treating COPD, but these agents have unwanted side effects if systemically absorbed. Systemic exposure can be limited by inhaled administration, which is the route of delivery for aclidinium bromide. In addition, aclidinium bromide was designed to undergo rapid hydrolysis in human plasma, providing inactive acid and alcohol products, and reducing the potential for systemic side effects.[15] Aclidinium bromide was identified amongst a series of quaternary ammonium (3R)-quinuclidinol esters as having the best combination of high potency (M3 $K_i = 0.14$ nM), long duration of action (29 h for 50% reduction of therapeutic effect in a guinea pig bronchoconstriction model), low oral absorption, and rapid plasma degradation.[15,16] The synthesis of aclidinium bromide was achieved by reaction of dimethyl oxalate with 2-thienylmagnesium bromide followed by treatment of the resulting methyl ester with (3R)-quinuclidinol in the presence of sodium hydride. Quaternization of the amine was achieved by treatment with 3-phenoxypropyl bromide to give aclidinium bromide.[15]

The safety and pharmacokinetics of aclidinium bromide were assessed in multiple Phase I studies in healthy volunteers and patients at doses ranging from 200 to 800 µg given once or over multiple days of dosing.[14] Following inhalation of a 200 µg dose of radiolabeled aclidinium bromide, ~30% of the dose was deposited in the lung and 55% in the oropharynx. There was low systemic bioavailability following inhalation (<5%) due to rapid and extensive hydrolysis of aclidinium bromide via nonenzymatic and enzymatic pathways. Of the absorbed dose, maximum plasma concentrations were achieved 5–15 min after dosing. Adverse event (AE) frequency was similar between treatment and placebo groups. The efficacy of aclidinium bromide was evaluated in two initial double-blind, placebo-controlled Phase III trials in which the drug was given to patients with moderate to severe COPD at a dose of 200 µg once daily (ACCLAIM I and II).[17] Efficacy was assessed at 12 and 28 weeks, as measured by improvement in forced expiratory volume in one second (FEV$_1$). Aclidinium bromide-treated patients showed increases in FEV$_1$ that were statistically significant, although the clinical significance of these improvements was not established. In subsequent Phase III trials, 200 and 400 µg doses of aclidinium bromide were given twice daily for 12–24 weeks.[14] In the 12-week ACCORD I trial, significant improvements in bronchodilation, health status, and COPD symptoms were seen at both doses, which were well tolerated and had safety profiles similar to placebo.[18] In the 24-week ATTAIN trial, the 200 and 400 µg doses resulted in statistically significant improvement in trough FEV$_1$, with the 400 µg dose giving clinically significant improvements in lung function, health status, and symptoms that were not observed with the lower dose.[19] Both doses were well tolerated, with no differences in safety profile between the two doses. The incidence of anticholinergic AEs in both groups was low and similar to placebo. Aclidinium bromide was approved at a 400 µg dose given twice daily as an inhaled powder.

3. ANAGLIPTIN (ANTIDIABETIC)[20–29]

Class: Dipeptidyl peptidase-4 inhibitor
Country of origin: Japan
Originator: Sanwa Kagaku Kenkyusho
First introduction: Japan
Introduced by: Kowa, Sanwa
Trade name: Suiny
CAS registry no: 739366-20-2
Molecular weight: 383.45

Anagliptin is a dipeptidyl peptidase-4 (DPP-4) inhibitor that was approved in Japan in November 2012 for the treatment of patients with Type 2 diabetes mellitus (T2DM). Diabetes is a chronic disease in which blood sugar levels are raised due to insufficient levels of insulin, the hormone that regulates blood sugar levels, or due to insulin resistance. As a result of damage to nerves and blood vessels, diabetes leads to increased risk of serious health problems including heart disease and stroke, neuropathy, blindness, and kidney failure. It is estimated that nearly 350 million people worldwide have diabetes, with type 2 being the most common form.[20] In Japan, the incidence of T2DM grew from 10.1% in 1993 to an estimated 13.5% in 2002; the WHO projects that there will be nearly 9 million people with diabetes in Japan in 2030.[20,21] Anagliptin (also known as SK-0403) is a treatment for diabetes based on inhibition of DPP-4, an enzyme that is responsible for degradation of glucagon-like peptide 1 (GLP-1), a 30-amino acid peptide that is secreted in response to food intake. GLP-1 stimulates insulin secretion and inhibits glucagon secretion, which leads to lower levels of plasma glucose. Following the introduction of the first DPP-4 inhibitor, sitagliptin, in 2006, several members of the gliptin class have been approved worldwide.[22] Anagliptin was discovered from an effort to replace a metabolically labile isoindoline group from an earlier DPP-4 inhibitor series with a stable bioisostere.[23] Anagliptin is a potent DPP-4 inhibitor, with an $IC_{50} = 3.8$ nM and >10,000-fold selectivity over inhibition of DPP-8 and DPP-9. Anagliptin was efficacious in the oral glucose tolerance test model in rats at oral doses as low as 0.1 mg/kg.[24] Anagliptin was also effective in a preclinical model of diabetes with glucokinase knockout mice on a high-fat diet.[25] Anagliptin is not an inhibitor or inducer of Cyp isoforms, does not inhibit hERG channels, was negative in the Ames assay, and showed high selectivity against a panel of receptors and ion channels.[23] Anagliptin showed high oral bioavailability (100%) and low clearance in dogs, but lower bioavailability (23%) and higher clearance in rats. Protein binding is ~77% in rat serum, 52–65% in dog serum, and 26–29% in human serum. The major metabolite was the carboxylic acid resulting from hydrolysis of the cyano group.[24] Anagliptin was prepared from t-butyloxycarbamate-protected 2-amino-2-methyl-propylamine by coupling with (S)-1-(2-chloroacetyl)-pyrrolidine-2-carbonitrile, deprotection, and coupling of the amine with pyrazolo[1,5-]pyrimidine carboxylic acid.[23]

A Phase I study in six healthy male subjects given [^{14}C]-anagliptin showed rapid absorption ($T_{max} = 1.8$ h).[26] About 50% of the dose was eliminated unchanged (46.6% in urine and 4.1% in feces); metabolism of the cyano group to the carboxylic acid accounted for 29% of the dose. The terminal half-life was 4.4 h for anagliptin and 9.9 h for acid metabolite. The safety, tolerability, pharmacokinetics, and pharmacodynamics of anagliptin were studied in a Phase I double-blind, placebo-controlled trial in 48 healthy Japanese men at single ascending doses ranging from 10 to 400 mg.[27] Anagliptin was rapidly absorbed ($T_{max} = 1$–2 h); C_{max} and AUC showed dose proportional increases over the dosing range. Dose-dependent increases in plasma DPP-4 inhibition were also observed. The AUC of GLP-1 increased two- to threefold relative to placebo when anagliptin was given at 50, 100, or 200 mg doses just before a meal. Anagliptin was generally well tolerated up to 400 mg. In a second Phase I double-blind, placebo-controlled study, nine subjects were given repeated doses of anagliptin ($n=6$) or placebo ($n=3$) over a 9-day period (day 0, 200 mg qd or placebo; days 3–8, 200 mg bid or placebo; day 9, 200 mg qd or placebo just before a meal).[27] In this study, the time course of DPP-4 inhibition and GLP-1 level increases was shown to be similar on day 0 and day 9. In both studies in healthy patients, there were no effects on serum insulin, plasma C-peptide, plasma glucagon, and blood glucose. Anagliptin has been evaluated in several open-label studies in Japanese patients with T2DM. For example, in 20 subjects given 100 mg anagliptin twice daily or 50 mg sitagliptin once daily for 3 days, blood glucose levels were significantly lowered in both treatment arms. Plasma levels of active GLP-1 increased with anagliptin treatment and were significantly higher than in the sitagliptin treatment arm.[28] In a 52-week randomized, open-label trial in T2DM patients, treatment with 100 mg of anagliptin twice daily before or after meals significantly improved glycemic control.[29] Anagliptin was generally well tolerated, with <1% occurrence of hypoglycemia. Anagliptin was approved for the treatment of T2DM at 100 mg given once daily.

4. AXITINIB (ANTICANCER)[30–35]

Class: Tyrosine kinase inhibitor
Country of origin: United States
Originator: Pfizer
First introduction: United States
Introduced by: Pfizer
Trade name: Inlyta®
CAS registry no: 319460-85-0
Molecular weight: 386.47

In January 2012, the US FDA approved axitinib (also referred to as AG-013736) for the treatment of advanced renal cell carcinoma (RCC) for patients who have not responded to prior therapy. RCC is a form of kidney cancer that affects the proximal convoluted tubules of the kidney. Based on 2010 statistics, an estimated 58,240 Americans were diagnosed with renal cancer, of which 92% of cases were accounted for by RCC.[30] Chemotherapeutic options include treatment with sorafenib, sunitinib, temsirolimus, everolimus, bevacizumab, and pazopanib. Expression levels of vascular endothelial growth factor (VEGF) are high in RCC as well as in many other malignancies. Axitinib is a pan VEGF inhibitor and functions by binding to the intracellular tyrosine kinase catalytic domain of VEGF leading to blockade of signaling through this angiogenic pathway. Axitinib is ~50–400 times more potent for VEGF (enzyme K_i and cellular IC_{50s} for VEGF 1, 2, and 3 are ~0.1 nM) than first-generation inhibitors like sorafenib and sunitinib. Axitinib also inhibits c-Kit and PDGFR(α/β) with enzyme K_i's of ~2 nM and was selective when tested against a broad panel of other protein kinases.[31] Axitinib was discovered by a structure-based drug design approach and binds to the kinase domain of VEGF in a DFG-out conformation.[32] Axitinib blocks VEGF-2 phosphorylation up to 7 h postdose *in vivo* and inhibits endothelial cell proliferation in xenograft tumors implanted in mice.[33] Synthetic routes to axitinib employing a Migita coupling to form the diaryl sulfide and a Heck reaction to install the 2-styrylpyridine moiety have been reported.[34a,b]

The pharmacokinetics of axitinib was established from pooled data of 17 trials in healthy volunteers and cancer patients. Following a single oral dose of 5 mg, the median T_{max} ranged from 2.5 to 4.1 h, the mean absolute bioavailability was 58%, and the plasma half-life ranged from 2.5 to 6.1 h. Axitinib is highly protein bound (>99%). Following a 5 mg dose in the fed state in patients with RCC, the mean C_{max} and AUC_{0-24} were 27.8 ng/mL and 265 ng h/mL, respectively. The clearance and apparent volume of distribution were 38 L/h and 160 L, respectively. The efficacy of axitinib in RCC was assessed in a randomized, open-label, multicenter Phase III study in patients whose disease had progressed after treatment with one prior systemic therapy.[35] Patients were randomized to either receive sorafenib ($N=362$) or axitinib ($N=361$) with progression-free survival (PFS) as the primary end point and objective response rate (ORR) and overall survival (OS) as secondary end points. Statistically significant improvements in PFS were observed in patients randomly assigned to axitinib (median PFS of 6.7 months) compared to with sorafenib (median PFS of 4.7 months). There was no statistically significant difference between the arms in OS. Diarrhea, hypertension, decreased appetite, fatigue, nausea, dysphonia, palmar–plantar erythrodysesthesia (hand–foot) syndrome, weight loss, vomiting, weakness, and constipation were the most common AEs (≥20%) observed during axitinib clinical trials. The recommended dose of axitinib is 5 mg administered orally twice daily, ~12 h apart with or without a meal.

5. BEDAQUILINE (ANTIBACTERIAL)[36–45]

Class: ATP synthase inhibitor
Country of origin: Belgium
Originator: Janssen
First introduction: United States
Introduced by: Janssen
Trade name: Sirturo™
CAS registry no: 843663-66-1
Molecular weight: 555.51

In December 2012, the US FDA approved bedaquiline as part of combination therapy for the treatment of multi-drug resistant tuberculosis (MDR-TB). Bedaquiline is the first drug approved for MDR-TB and is the first approval from a new class of antituberculosis agents in the past 40 years. Due to the high unmet medical need for treating MDR-TB, the FDA granted bedaquiline accelerated approval based on Phase II results, providing patients access to the drug while additional clinical studies are carried out. Tuberculosis (TB) is caused by infection with *Mycobacterium tuberculosis*, an airborne pathogen that primarily affects the lung and is spread by individuals with active TB. TB is the second leading cause of death from infectious disease worldwide, with 1.4 million people dying from TB in 2011.[36] Latent TB occurs in up to one-third of the world's population, with an estimate that 10% of people with latent infection will develop active infection in their lifetime. The current standard of care for TB is a combination of isoniazid, rifampin, ethambutol, and pyrazinamide given for 6 months. While the combination drug treatment is effective when taken properly, emergence of MDR-TB has occurred due to poor compliance and premature discontinuation of therapy. In 2011, the WHO estimated that there were 310,000 cases of MDR-TB. Options for the treatment of MDR-TB are limited, and require prolonged treatment with second-line agents that are often less effective, expensive, and associated with adverse effects. Significant efforts have been undertaken to identify new anti-TB agents, particularly for use in treating MDR-TB.[37,38] Bedaquiline (also known as TMC207 and R207910) is a diarylquinoline that was discovered from a high-throughput, whole-cell screening strategy with *Mycobacterium smegmatis* used as a surrogate for *M. tuberculosis*.[39,40] Bedaquiline is a single enantiomer of an initial screening hit. Bedaquiline has potent and selective activity against mycobacteria, and is active against both drug-sensitive and drug-resistant *M. tuberculosis*. The mechanism of action of bedaquiline is unique amongst anti-TB drugs and involves inhibition of mycobacterial ATP synthase; it is not active against human ATP synthase.[39,41] Bedaquiline has *in vivo* activity in numerous preclinical models of TB infection, alone and in combination with other anti-TB agents, and has bactericidal activity in established TB infection models.[39,42] Bedaquiline is synthesized in five steps from 3-phenylpropionic acid and *para*-bromoaniline. Following amide formation, treatment with POCl$_3$ and DMF under Vilsmeier–Hack conditions gave a 2-chloroquinoline product. Treatment with sodium methoxide, followed by condensation with 3-(dimethylamino)-1-(naphthalen-1-yl)propan-1-one, and separation of isomers gave bedaquiline.[40]

In a Phase I trial in healthy human subjects, bedaquiline was found to be well absorbed after a single oral dose, with C_{max} reached at 5 h and with linear pharmacokinetics up to the highest dose of 700 mg.[39] AEs were mild to moderate and were seen in subjects receiving bedaquiline and placebo. Bedaquiline is metabolized by Cyp3A4. The plasma protein binding is >99.9% and the volume of distribution is 164 L. Approval of bedaquiline was based on the efficacy and safety results from three Phase II trials; Phase III trials were anticipated to begin in 2013.[43] In the first trial (C208 Stage 1), 47 newly diagnosed MDR-TB patients were randomly assigned to receive bedaquiline (400 mg daily for 2 weeks, followed by 200 mg three times a week for 6 weeks) or placebo in combination with a five-drug background regimen (BR).[44,45] Addition of bedaquiline to the combination therapy significantly reduced the time to conversion to a negative sputum culture (primary endpoint) and decreased the proportion of patients with conversion of sputum culture (48% vs. 9% for placebo). Most AEs were mild to moderate, with nausea being the most frequently observed in bedaquiline-treated (26%) compared with placebo-treated (4%) patients. The second trial (C208 Stage 2), which included 160 subjects, was similar to the first trial with the bedaquiline arm receiving drug at 400 mg daily for 2 weeks and then 200 mg three times weekly for 22 additional weeks.[43] The median time to negative culture conversion was 73 days for bedaquiline-treated patients versus 125 days for placebo. At week 24, 79% of bedaquiline-treated patients converted to negative culture versus 58% in the placebo group. Follow-up at week 72 showed that the response was maintained. The third trial (C209) was a single-armed, open-label trial in 233 subjects who were newly diagnosed or treatment-experienced MDR-TB patients.[43] Patients received bedaquiline for up to 24 weeks in combination with an individualized

BR. Median time to negative culture conversion was 57 days in this trial. Although the overall safety profile of bedaquiline and placebo were similar, elevation of transaminases and QTc prolongation were observed with bedaquiline treatment. In addition, in Trial C208 Stage 2, there were deaths in 9 of 79 bedaquiline-treated patients (11.5%) compared with 2 of 81 patients in the placebo-treated group (2.5%) although no clear relationship between mortality and bedaquiline treatment was identified given that the deaths were due to different causes. These findings lead to a black-box warning for increased rate of death in the drug-treatment group compared with placebo and for QTc prolongation. Bedaquiline was approved at 400 mg daily for 2 weeks followed by 200 mg three times weekly for 22 weeks, as part of a combination therapy for MDR-TB.

6. BOSUTINIB (ANTICANCER)[46–51]

Class: Tyrosine kinase inhibitor
Country of origin: United States
Originator: Wyeth
First introduction: United States
Introduced by: Pfizer
Trade name: Bosulif®
CAS registry no: 380843-75-4
Molecular weight: 548.46 (hydrate) and
530.46 (anhydrous)

In September 2012, the US FDA approved bosutinib (also referred to as SKI-607) for the treatment of relapsed or refractory chronic myeloid leukemia (CML) for patients with resistance or intolerance to prior therapy. The National Cancer Institute estimates that in 2012, 5430 men and women were diagnosed with CML and 610 died of CML.[46] First and second-line therapies for the treatment of CML include imatinib, dasatinib, and nilotinib. CML is a form of leukemia that affects myeloid cells in the bone marrow. In CML patients, the bone marrow makes an uncontrolled number of immature white blood cells (blast cells), which are incapable of fighting infections as effectively as normal mature white blood cells. The production of blast cells is a result of the expression of the Bcr–Abl fusion oncoprotein, formed by reciprocal translocation between chromosomes 9 and 22 and also referred to as the Philadelphia chromosome. This oncoprotein is a deregulated tyrosine kinase that activates several proliferative and antiapoptotic signaling pathways. Imatinib was the first targeted therapy for CML. However, acquired resistance to imatinib necessitated the development of second-generation inhibitors. Bosutinib, which was initially identified as a Src inhibitor,[47] was later found to be a dual inhibitor of Bcr–Abl ($IC_{50} = 1.4$ nM) and Src family kinases ($IC_{50} = 3.5$ nM).[48] Bosutinib inhibits 16 of 18 imatinib-resistant forms of Bcr–Abl expressed in murine myeloid cell lines. In preclinical *in vivo* studies, bosutinib at 15 mg/kg administered orally for 5 days caused regression of K562 CML tumors in nude mice and in BaF3 tumors expressing wild type or different imatinib-resistant Bcr–Abl mutants at varying doses.[48] A manufacturing process for the synthesis of bosutinib monohydrate has been reported that employs a key three-component cyclization reaction involving an aniline, a cyanoacetamide intermediate, and triethyl orthoformate followed by cyclization using phosphorous oxytrichloride and an optimized hydration procedure.[49]

The pharmacokinetics of bosutinib was established in a randomized, double-blind, placebo-controlled, single-ascending dose, sequential-group study at doses ranging from 200 to 800 mg in 55 healthy adult volunteers. Bosutinib exposures (C_{max} and AUC) were linear and dose proportional from 200 to 800 mg with food. Median time to C_{max} was 6 h. Apparent volume of distribution (Vz/F) was 131–214 L/kg, mean apparent clearance (Cl/F) was 2.25–3.81 L/h/kg, and mean terminal elimination half-life ($t_{1/2}$) was 32–39 h.[50] The safety and efficacy of bosutinib was evaluated in a Phase I/II single arm, open-label, multicenter trial in 546 patients with imatinib-resistant or -intolerant CML who had chronic, accelerated, or blast phase CML.[51] In patients with chronic phase CML, 34% of patients who had been previously treated with imatinib achieved major cytogenetic response (MCyR, efficacy end point) after 24 weeks. Among patients previously treated with imatinib followed by dasatinib and/or nilotinib, about 27% achieved MCyR within the first 24 weeks of treatment. In patients with accelerated CML previously treated with at least imatinib, 33% had a complete hematologic response (CHR) and 55% achieved normal blood counts with no evidence of leukemia (overall hematologic response) within the first 48 weeks of treatment. In patients with blast phase CML, 15% and 28% achieved CHR and overall hematologic response, respectively. Diarrhea, nausea, thrombocytopenia, vomiting, abdominal pain, rash, anemia, pyrexia, and fatigue (>20%) were the most common AEs observed during bosutinib clinical trials. The recommended dose of bosutinib is 500 mg administered orally once daily with a meal.

7. CABOZANTINIB (ANTICANCER)[52–61]

Class: Tyrosine kinase inhibitor
Country of origin: United States
Originator: Exelixis
First introduction: United States
Introduced by: Exelixis
Trade name: Cometriq™
CAS registry no: 849217-68-1 (free
base) and 1140909-48-3 (salt)
Molecular weight: 501.51 (free base)
and 635.59 (salt)

Cabozantinib was approved in November 2012 for the treatment of patients with progressive, unresectable, locally advanced, or metastatic medullary thyroid cancer (MTC). The US National Cancer Institute estimates that there will be ∼60,000 new cases of thyroid cancer in 2013, with nearly 2000 deaths due to the disease.[52] MTC comprises 3–4% of thyroid cancers, and 25% of the incidence has familial linkage. Current treatments include thyroidectomy, external radiation therapy, and palliative chemotherapy with vandetanib. Vandetanib has been found to increase QT interval and to cause increased incidence of Torsades de pointes and sudden death in patients.[53] Cabozantinib was granted orphan drug status by the FDA to facilitate development of new treatment options for patients with MTC. It is a member of a class of tyrosine kinase inhibitors (TKIs) with nanomolar pan-inhibitory activity against VEGFR2, MET, and RET among others.[54] Inhibition of the VEGF pathway has been shown preclinically to initially slow tumor growth, but rapid revascularization is followed by aggressive tumor growth. The MET pathway has been implicated in the development of VEGF resistance, so dual VEGF/MET activity is viewed as desirable.[54,55] In addition, mutations in RET play a particular role in MTC, with 25% of the tumors inheriting a germline mutation in the proto-oncogene,[53] so multiple tyrosine kinase inhibition may be viewed as particularly beneficial for the treatment of MTC. The medicinal chemistry program to discover cabozantinib has not been described in the literature; however, its synthesis is reported in a patent application.[56] In this disclosure, ∼300 examples were synthesized around the quinoline core with substantial variation at the O-alkyl and N-acylated portions of the molecule. Cabozantinib was synthesized via 5-step convergent sequence involving S$_N$Ar displacement of 4-chloro-6,7-dimethylquinoline with 4-aminophenol to form the aminobiaryl ether. Temperature controlled amide bond formation between 4-fluoroaninline with cyclopropane-1,1-dicarboxylic acid gave a mono-carboxylic acid that was coupled to the aminobiaryl ether intermediate to form cabozantinib.[57] *In vivo* characterization of cabozantinib at 30 and 60 mg/kg qd (plasma concentrations of 9–16 µM) was performed over 12 days in nu/nu mice implanted with tumor cells, with cabozantinib demonstrating disease stabilization or tumor regression in certain cases. The antitumor efficacy is suggested to be the result of decreased angiogenesis, rather than a blockade of cellular proliferation.[54]

In a Phase I clinical trial, administration of cabozantinib at 0.08–0.32 mg/kg gave plasma C_{max} values between 34.2 and 189.3 ng/mL and an elimination half-life of 91.3 ± 33.3 h.[58] Cabozantinib is highly protein bound (>99.7% in human plasma) and is primarily metabolized via oxidation of the quinoline to its N–oxide via CYP3A4.[59] In a Phase III clinical trial, 330 patients were registered and genotyped, including 48% of patients with the RET positive mutation, 12% RET negative, and 39% unknown. Cabozantinib treatment resulted in 28% partial regression and a median progression-free survival (PFS) of 11.2 months versus 4.0 months for placebo.[53] The most commonly reported AEs (>25%) were diarrhea, stomatitis, palmar–plantar erythrodysesthesia syndrome, decreased weight and appetite, nausea, fatigue, oral pain, hair color changes, dysgeusia, hypertension, abdominal pain, and constipation.[60] Overall rates of survival were not statistically significant different for cabozantinib treatment compared with placebo,[59] but objective response rates were improved (28% vs. 0% for placebo). Cabozantinib as a 140 mg daily oral capsule was approved by the FDA for the treatment of patients with MTC; further clinical trials are registered to evaluate cabozantinib in other oncology settings.[61]

8. CARFILZOMIB (ANTICANCER)[62–68]

Class: Proteosome inhibitor
Country of origin: United States
Originator: Proteolix Inc.
First introduction: United States
Introduced by: Onyx
Trade name: Kyprolis®
CAS registry no: 868540-17-4
Molecular weight: 719.91

In July 2012, the US FDA approved carfilzomib (also referred to as PR-171) for the treatment of patients with multiple myeloma (MM) who have received at least two prior therapies including bortezomib and an immunomodulatory agent and have demonstrated disease progression within 60 days of completion of the last therapy. MM is the second most prevalent blood cancer after non-Hodgkin lymphoma (NHL). In 2012, the American Cancer Society estimated that 21,700 new cases of MM would be diagnosed in the United States, including 12,190 cases in men and 9500 cases in women, and that MM would cause an estimated 10,710 deaths.[62] Chemotherapeutic options include treatment with thalidomide, lenalidomide, bortezomib, and doxil. Of these therapies, bortezomib was the first-generation proteosome inhibitor that provided proof-of-principle for proteosome inhibition as a therapeutic approach in MM. Carfilzomib binds to and irreversibly inhibits the chymotrypsin-like protease activity of the constitutive proteosome ($\beta5$) and immunoproteosome ($\beta5i$) via its epoxyketone pharmacophore.[63] Proteosome inhibition results in the accumulation of polyubiquitinated proteins and induction of apoptosis through activation of both the intrinsic and extrinsic caspase pathways. Carfilzomib inhibits chymotrypsin activity with an IC_{50} of 6 nM and is less potent an inhibitor of trypsin and caspase (IC_{50s} of 3600 and 2400 nM, respectively).[64] Cell cycle arrest and apoptosis are seen in a variety of hematologic and solid tumor cell lines (e.g., MM, acute myeloid leukemia (AML), pancreatic cancer, lung cancer) treated with carfilzomib. The synthesis of carfilzomib is reported in the patent literature [65a] and a route to the intermediate epoxy ketone from a Boc-leucine-derived α,β-unsaturated ketone is also described.[65b]

The pharmacokinetics of carfilzomib was established in an open-label, multicenter Phase I trial in 29 patients with refractory or relapsed hematologic malignancies, including MM.[66] Following an intravenous dose that ranged from 1.2 to 20 mg/m^2 every day for 5 consecutive days in 14-day cycles, the mean maximum concentration was 325.9 ng/mL for the maximum tolerated dose in this study, 15 mg/m^2. T_{max} was 5.8 min, the elimination half-life was 28.9 min, and the volume of distribution at steady state was 942 L. The efficacy of carfilzomib in patients with relapsed or refractory MM was supported by data from 526 patients enrolled in four Phase II studies. In the pivotal Phase IIb, open-label, single arm study with 266 patients, carfilzomib, administered intravenously at 20 mg/m^2 on days 1, 2, 8, 9, 15, and 16 of a 28-day cycle for 12 cycles showed significant, durable responses in patients with relapsed or refractory MM who were previously treated with at least two therapies (bortezomib and either thalidomide or lenalidomide; and an alkylator and/or anthracycline).[67] The overall response rate was 24% with median duration of response of 7.4 months. The median

OS rate was 15.5 months (95% CI, 12.7–19.0). Fatigue, anemia, nausea, thrombocytopenia, dyspnea, diarrhea, and pyrexia ($\geq 30\%$) were the most common AEs observed during carfilzomib clinical trials. Carfilzomib is administered intravenously at 20 mg/m^2 over 2–10 min on days 1, 2, 8, 9, 15, and 16 in cycle 1, with escalation to 27 mg/m^2 in subsequent cycles as tolerated every 28 days.[68]

9. CROFELEMER (ANTIDIARRHEAL)[69–77]

Class: Chloride ion channel inhibitor (CTFR and calcium-stimulated chloride channel)
Country of origin: South America (natural product)/United States (Shaman Pharmaceuticals)
Originator: Shaman Pharmaceuticals
First introduction: United States
Introduced by: Salix Pharmaceuticals
Trade name: Fulyzaq™
CAS registry no: 148465-45-6
Molecular weight: 860–9100 (polymer range); 2200 Da (average)

R = H, OH
average degree of oligomerization (n) = 5–7.5.

Crofelemer was approved by the US FDA in December 2012 for the treatment of diarrhea in HIV/AIDS patients on antiretroviral therapy. An estimated 34 million people worldwide live with HIV/AIDS.[69] Patients experience a high incidence of diarrhea;[70] 50–90% of patients with low CD4 lymphocyte count are affected. Crofelemer is indicated for a subset of the population (30–40% of HIV/AIDS cases) whose diarrhea is determined to be noninfective in origin.[71] Crofelemer is an oligomeric proanthocyanidin natural product that is isolated from the red latex of *Croton lechleri* and has been historically used by traditional practitioners of medicine in South America to treat diarrhea.[72] Previously known as SP-303, crofelemer has been shown to inhibit chloride secretion across epithelial cell monolayers, which causes concomitant movement of water and sodium. It achieves this action via inhibition of cystic fibrosis transmembrane regulator (CFTR) ($IC_{50} = 7$ μM, 90% maximal inhibition) and calcium-stimulated chloride intestinal channels, such as TMEM16A, a Ca^{2+}-gated chloride channel ($IC_{50} = 6.5$ μM, $>90\%$ maximal inhibition).[73] In contrast to other antidiarrheal agents, drugs such as crofelemer with antisecretory mechanisms do not affect gut peristalsis.[74] In preclinical studies, mice were treated with cholera toxin (CT) to produce significant fluid accumulation (FA); administration of 100 mg/kg of crofelemer restored FA ratios to normal levels (FA ratios: normal control $= 0.63 \pm 0.07$; CT-treated $= 1.86 \pm 0.2$; CT + crofelemer 100 mg/kg $= 0.7 \pm 0.19$).[73]

Crofelemer (125 mg, bid) is not absorbed to appreciable amounts, with systemic exposures below the 50 ng/mL limit of detection in plasma.[72,75] It inhibits CYP3A4 and transporters MRP2 and OATP1A2 at expected concentrations in the gut, but is not expected to have systemic inhibitory effects on CYPs or transporters. Efficacy was assessed in Phase II double-blind, placebo-controlled clinical trials of 136 HIV-positive patients receiving 125 mg of crofelemer twice daily. These patients experienced reduced incidence of watery bowel movements by 17.6% (vs. 8.0% in placebo). The most prevalent adverse reaction in three aggregated clinical trials with 125 mg crofelemer bid was upper respiratory tract infection (5.7% with crofelemer vs. 1.5% for placebo). In addition, bronchitis, cough, flatulence, and increased bilirubin occurred more frequently with crofelemer (3.1–3.9%) than with placebo (0–1.1%).[75,76] Trials of crofelemer in other clinical settings of diarrhea have been reviewed.[77] The currently approved use for the treatment of patients with HIV/AIDS-related diarrhea is one 125 mg delayed release tablet of crofelemer, taken twice daily.

10. DAPAGLIFLOZIN (ANTIDIABETIC)[78–86]

Class: Sodium-dependent glucose transporter 2 (SGLT2) inhibitor
Country of origin: United States

Continued

—cont'd

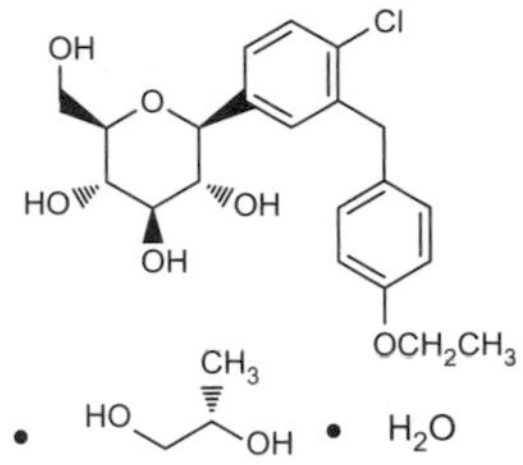

Originator: Bristol-Myers Squibb
First introduction: European Union
Introduced by: Bristol-Myers Squibb and Astra Zeneca
Trade name: Forxiga®
CAS registry no: 461432-26-8 (neat form) and 960404-48-2 (co-crystal)
Molecular weight: 408.87 (neat form) and 502.98 (co-crystal)

The Australian Therapeutic Goods Administration (TGA) and the European Commission approved dapagliflozin in October and November 2012, respectively, as an adjunct to diet and exercise for the treatment of type 2 diabetes. The incidence of Type 2 diabetes mellitus (T2DM) was estimated at 371 million worldwide in 2012 and is expected to rise to 552 million by 2030.[78] In addition to lifestyle modifications, there are a number of pharmaceutical treatments to manage diabetes, including insulin, metformin, sulfonylureas, GLP-1 mimetics, DPP-4 inhibitors, thiazolidinediones, and alpha glucosidase inhibitors.[79,80] However, ~40% of patients do not achieve desired glycated hemoglobin (HbA_{1c}) levels. Dapagliflozin is a potentially attractive therapy due to its glucose-sensitive and insulin-independent mechanism of action. It is a first-in-class selective SGLT2 inhibitor ($IC_{50}=1.1$ nM; selectivity vs. SGLT1 >1000) that lowers the renal threshold for reabsorption of glucose, allowing excess glucose to be eliminated via the kidneys. In normal rats, administration of dapagliflozin promotes dose-dependent excretion of up to 1900 mg of glucose over a 24 h period, with a maximal effect at 3 mg/kg. In a rat model of diabetes, pretreatment with the pancreatic toxin streptozotocin results in hyperglycemia that is reduced 55% by administration of a single 0.1 mg/kg dose of dapagliflozin compared with vehicle.[81] Aryl O-glucoside SGLT2 inhibitors were early entrants into the clinic, but the aryl C-glucoside linkage found in dapagliflozin confers resistance to glucosidase-mediated metabolism leading to improved clinical utility relative to aryl O-glucosides.[81] The modified carbohydrate–aglycone linkage required concomitant adjustment from an *ortho*- to a *meta*-substituted arylglucoside to achieve potent SGLT2 inhibition.[82] Dapagliflozin was synthesized in several steps via reaction of an aryllithium with per-silylated gluconolactone to form the key C-glucoside linkage. An alpha-selective reduction of the resultant anomeric glycoside gave the desired beta-C-arylglucoside.[81] The main circulating (inactive) metabolite is the result of 3-O-glucuronidation of the glucosyl moiety.[83] Of the minority metabolites, the main oxidative species result from O-dealkylation of the ethoxy-group and hydroxylation of the biarylmethane moiety.[84] The stability of the C-glucosyl linkage leads to favorable PK in normal healthy humans ($\%F = 78\%$,[84] $C_{max} = 136$ ng/mL, $T_{max} = 1$ h, $T_{1/2} = 12.9$ h, at 10 mg dose).[83]

Clinical efficacy and safety of dapagliflozin were assessed in multiple Phase III placebo-controlled trials, and the results were recently reviewed.[85] Patients receiving dapagliflozin experienced improved glucose control as demonstrated by placebo-subtracted HbA_{1c} reductions of $>0.5\%$. Patients experienced additional benefits of weight loss (2–3%) and reductions in systolic blood pressure (-4.4 mm Hg)[86] that are attributed to caloric loss (glucosuria) and osmotic diuresis, respectively. The incidence of urinary tract and genital infections were modestly increased over background. Several studies have demonstrated the additional glucose-lowering effect of dapagliflozin on a background of insulin or oral glucose-lowering agents, indicating that the mechanism of action is complimentary to other antidiabetic therapies.[85] Dapagliflozin was approved by the Australian TGA and the European Commission in a 10 mg usual daily dose (5 mg starting dose for severe hepatic impairment) as oral monotherapy for the treatment of diabetes, and as add-on therapy with other glucose-lowering agents.

11. ENZALUTAMIDE (ANTICANCER)[87–94]

Class: Androgen receptor antagonist
Country of origin: United States
Originator: University of California
First introduction: United States
Introduced by: Medivation and Astellas Pharmaceuticals
Trade name: Xtandi®
CAS registry no: 915087-33-1
Molecular weight: 464.44

In August 2012, the US FDA approved enzalutamide for the treatment of metastatic castration-resistant prostate cancer (mCRPC) in patients who have previously been treated with docetaxel. Prostate cancer is the most prevalent cancer in North American and European men. An estimated 238,590 new cases of prostate cancer will occur in the United States alone in 2013.[87] First-line therapy for prostate cancer includes suppression of gonadal androgens, which are implicated in promoting tumor growth. Progression to mCRPC often results in poor outcomes, including high rates of mortality.[88] Two approved products, cabazitaxel and abiraterone acetate are currently marketed for the treatment of mCRPC that continues to progress while still on or following after docetaxel treatment. Several androgen receptor (AR) antagonists/partial agonists are in clinical use, including cyproterone acetate, flutamide, nilutamide, and bicalutamide; however, resistance to these treatments arises after 2–4 years of treatment due to upregulation of the AR. In cellular systems with AR upregulation, bicalutamide and flutamide can demonstrate agonist and proliferative activity, rather than antagonist activity.[89] To surmount the development of resistance, increased AR antagonist potency was sought through SAR explorations around RU59063, a potent AR agonist.[90] Synthesis of enzalutamide was achieved by a triply convergent route that employed a Strecker condensation, followed by isothiocyanate condensation and hydrolysis to form the thiohydantoin moiety. In LNCaP/AR cells with high expression of AR, enzalutamide demonstrated potent inhibition of 16β-[18F]-5α-dihydrotestosterone binding ($IC_{50}=21$ nM compared with bicalutamide $IC_{50}=160$ nM), and inhibited AR translocation to the nucleus more potently than bicalutamide.[89] These cellular effects translated to significantly improved cytostatic activity *in vivo*. At 10 mg/kg, enzalutamide demonstrated cytostatic activity in xenograft tumors over 31 days, whereas bicalutamide (10 mg/kg) did not demonstrate statistically significant reduction in tumor volume in this study.

In early clinical trials, enzalutamide elimination half-life was shown to be approximately 1-week following a single oral dose.[91] With a daily dosing regimen, steady state concentrations are achieved by day 28. The primary metabolite is the result of CYP2C8-mediated N-demethylation; enzalutamide is primarily eliminated by hepatic metabolism.[92] In an interim analysis of the AFFIRM trial (NCT00974311), in which 800 men with mCRPC received enzalutamide 160 mg and 399 patients received placebo, the enzalutamide group demonstrated improved survival (18.4 months vs. 13.6 months). There were also improvements in secondary endpoints, including reduction of prostate-specific antigen (PSA, a gene product that is regulated by AR) by 54% versus 2% for placebo treatment. Rates of fatigue, diarrhea, and hot flashes were higher in the drug-treated group, and five patients (0.6%) had reported seizures; however, overall rates of grade ≥ 3 AEs were lower than placebo (45.3% vs. 53.1%, respectively).[93] The US FDA approved enzalutamide without an advisory committee meeting because, as the letter states, "...there were no controversial issues that would benefit from advisory committee discussion."[94] Enzalutamide is approved as a 160 mg total dose (four 40 mg capsules) taken orally once daily, with or without food.

12. INGENOL MEBUTATE (ANTICANCER)[95–103]

Class: Cytotoxic
Country of origin: United States
Originator: Peplin
First introduction: United States
Introduced by: Leo Pharma Inc.
Trade name: Picato®
CAS registry no: 75567-37-2
Molecular weight: 430.53

Ingenol mebutate (also known as PEP005) was approved in January 2012 by the US FDA for the topical treatment of actinic keratoses (AK). Ingenol mebutate also received approval in 2012 in the European Union, Australia, and Brazil for the same indication. Actinic keratoses are premalignant skin lesions that are common in sun-exposed skin in older, fair-skinned people.[95] It is estimated that in northern hemisphere populations, 11–25% of adults have at least one AK lesion. The estimate is 40–60% in Australia, which is reported to have the highest incidence in the world.[96] AKs are the result of abnormal keratinocyte proliferation, and may progress to squamous cell carcinoma (SCC). In the United States, SCC is the second leading cause of skin cancer death after melanoma and it is estimated that 65% of SCC arise from preexisting AKs. AKs can be prevented by avoiding exposure of the skin to sun. Once AKs develop, it is unknown which lesions may progress to SCC, so the goal of treatment is to completely eliminate AKs, while obtaining cosmetically acceptable outcomes. Treatment options range from cryotherapy, surgical removal, and laser resurfacing to topical treatments, such as chemical peels, topical immunomodulators, and topical chemotherapeutic agents. Drawbacks to current treatments include scarring, lengthy treatment, painful skin irritation, and recurrence of lesions.[95] Ingenol mebutate, a diterpene ester natural product, is the active agent in the sap of the plant *Euphorbia peplus*, which has long been used as a traditional remedy for skin lesions. Ingenol mebutate has a novel mechanism of action involving initial plasma membrane disruption, rapid loss of mitochondrial membrane potential, and cell death by primary necrosis within 1 h;

a subsequent tumor-specific immune response results in antibody-dependent cellular toxicity that eliminates residual cells.[97] Ingenol mebutate is obtained by extraction from the dried, milled aerial parts of *Euphorbia peplus* followed by a series of purification steps.[98] An efficient semi-synthesis from ingenol via protection, ester formation under conditions that minimize double bond isomerization, and deprotection has recently been reported.[99]

The efficacy of *Euphorbia peplus* sap was demonstrated in a Phase I/II study in which the sap was applied once daily for 3 days to superficial non-melanoma skin cancer lesions; more than 40% of the patients in this study had previously failed to respond with other therapies.[100] Complete clearance was observed in more than 50% of lesions over a mean follow-up period of 15 months, supporting further studies with the active agent ingenol mebutate. An open-label Phase I clinical trial with a single application of 0.01% ingenol mebutate gel in 16 patients with AK lesions showed a favorable safety profile and clinically relevant effect on lesions within 21 days.[101] Subsequent Phase II trials showed that concentrations of 0.0025%, 0.01%, and 0.05% ingenol mebutate gel were safe and well tolerated, with maximal efficacy achieved using the 0.05% formulation.[97] There was minimal systemic absorption of ingenol mebutate following topical application. Four multicenter, randomized, double-blinded Phase III trials were conducted, with patients assigned to treatment regimens based on the site of their lesion (face/scalp and trunk/extremities).[102] Ingenol mebutate gel (0.015%) was self-applied once daily for 3 days in two trials for the treatment of face/scalp AKs with assessment at 57 days posttreatment. More than 98% of patients completed the 3-day regimen. Pooled analysis of the two trials showed complete clearance of lesions in 42% of patients treated with ingenol mebutate (277 patients) versus 3.7% in the placebo group (270 patients). Partial lesion reduction (at least 75%) was seen in 63% of drug-treated patients compared with 7.4% for placebo-treated patients. The overall median lesion reduction was 83% in drug-treated patients. Of the 42% of patients showing complete response (CR), 87% of the lesions at the treatment area were still clear after 12 months. The primary AEs were application site conditions, including pain (13.9%), pruritis (8%), and irritation (1.8%). There was minimal change in pigmentation and minimal scarring. In two trials in patients with AKs of the trunk/extremities, 0.05% ingenol mebutate was self-applied once daily for 2 days. The CR rate was 34% with ingenol mebutate treatment (226 patients) versus 4.7% for placebo (232 patients); 85% of the lesion sites remained clear after 12 months in the CR group. The partial response (PR) rate was 49% for ingenol mebutate treatment and 6.9% for placebo.

The clinical response rates following 2–3 days of treatment with ingenol mebutate are similar to other chemotherapeutic agents that require 1–4 months of treatment. Among the advantages to the short treatment regimen with ingenol mebutate are relatively rapid resolution of local reactions and high adherence to therapy.[103] The most common AEs were pruritis (8.4%), irritation (3.6%), and pain (2.2%). No serious AEs related to study treatment occurred. Ingenol mebutate gel was approved at two concentrations: 0.15% for treatment of AK lesion of the face and scalp; 0.05% for AK lesions of the trunk and extremities.

13. IVACAFTOR (CYSTIC FIBROSIS)[104–113]

Class: Cystic fibrosis transmembrane
conductance regulator (CFTR) potentiator
Country of origin: United States
Originator: Vertex Pharmaceuticals
First introduction: United States
Introduced by: Vertex Pharmaceuticals
Trade name: Kalydeco®
CAS registry no: 873054-44-5
Molecular weight: 392.49

In January 2012, the US FDA approved ivacaftor for the treatment of cystic fibrosis (CF) in patients who have the G551D mutation of the CF transmembrane regulator (CFTR) and are at least 6 years old.[104] CF is an inherited disorder in which both copies of the CFTR gene are mutated. Class III mutations of CFTR decrease transport of chloride ions across the epithelium, which leads to decreased fluid transport and thickening of secretions.[105] CF patients experience recurrent respiratory tract infections, pancreatic malabsorption, and infertility.[105] In the US, there are ~30,000 individuals with CF, and an estimated 1000 new cases are diagnosed each year.[106] Standard treatments for CF include mucolytics for thinning of secretions and antibiotics to treat and prevent lung infections.[107] Ivacaftor (also known as VX-770) is a CFTR potentiator that increases the open probability of CFTR, thus increasing chloride secretion particularly in the 5% of CF patients with the G551D/F508 gating/processing mutation.[108] Ivacaftor was discovered by medicinal chemistry optimization of a lead scaffold identified through high-throughput screening of a 228,000 compound collection.[109] In cultured bronchial epithelial cells from a CF patient with F508del, ivacaftor increased chloride secretion

(EC$_{50}$ = 81 nM).[110] Preparation of ivacaftor is accomplished via a multistep synthesis of two intermediates, 4-oxo-1,4-dihydroquinoline-3-carboxylic acid and 5-amino-di-tert-butylphenyl methyl carbonate, which are coupled using propane phosphonic acid anhydride (T3P) to afford the amide; deprotection of the phenol then provides ivacaftor.[111]

When ivacaftor was given to healthy volunteers at a single daily dose of 150 mg, T_{max} was 4 h, C_{max} was 768 ng/mL, half-life was 12 h, clearance was 17.3 L/h, and AUC was 10,600 ng h/mL.[111] The package insert recommends that ivacaftor be taken with fat-containing foods as this can increase drug exposure by two- to fourfold. Ivacaftor is highly protein bound in human plasma (99%) and extensively metabolized by CYP3A yielding two major, less potent metabolites. Metabolites and parent are primarily eliminated in the feces.[112] The efficacy and safety of ivacaftor was assessed in a 48-week STRIVE Phase III study in patients at least 12 years of age with CF and either G551D or F508-CFTR mutation.[113] Subjects received a 150 mg dose of ivacaftor or placebo every 12 h. The primary endpoint, change in the percent of predicted FEV$_1$, improved by 10.6 percentage points in the ivacaftor group over the control group through week 24 of the study. Additionally, patients in the ivacaftor group were 55% less likely to have a pulmonary exacerbation and scored 8.6 points higher on the respiratory-symptoms domain of the Cystic Fibrosis Questionnaire (100-point scale). The most common AEs were headache, upper respiratory tract infection, nasal congestion, rash, and dizziness.[113] Ivacaftor is given orally twice daily with food as a 150 mg tablet.[112]

14. LINACLOTIDE (IRRITABLE BOWEL SYNDROME)[114–122]

Class: Guanylate cyclase-C agonist
Country of origin: United States
Originator: Ironwood Pharmaceuticals and Forest Pharmaceuticals
First introduction: United States
Introduced by: Ironwood Pharmaceuticals and Forest Pharmaceuticals
Trade name: Linzess™
CAS registry no: 851199-59-2
Molecular weight: 1526.74

H–Cys–Cys–Glu–Tyr–Cys–Cys–Asn–Pro—

Ala–Cys–Thr–Gly–Cys–Tyr–OH

In August 2012, the US FDA approved linaclotide (also referred to as MD-1100), a first-in-class, orally administered 14-amino acid peptide as a therapy for patients suffering from chronic idiopathic constipation (CIC) and irritable bowel syndrome with constipation (IBS-C). It is the second approved therapy that is on the market for the treatment of both IBS-C and CIC (the first being lubiprostone). Both CIC and IBS-C are characterized by hard stools, infrequent bowel movements, rectal pressure, abdominal pain, incomplete evacuation, and occasional manual intervention to evacuate stool. It is estimated that ~15% of the US population suffers from chronic constipation.[114] Linaclotide and its active metabolite MM-419447, which results from the cleavage of the C-terminal tyrosine residue by carboxypeptidase A,[115] mimic the actions of the endogenous intestinal peptides guanylin (15 amino acids) and uroguanylin (16 amino acids) by activating guanylyl cyclase C (GC-C) on the intestinal epithelium. Activation of GC-C leads to increased intra- and extracellular levels of cGMP and activation of the CFTR ion channel, resulting in increased levels of HCO_3^-, Cl^-, and water in the intestinal lumen and accelerated gastrointestinal transit.[116] Based on an *in vitro* assay measuring the accumulation of cGMP in T84 cell exposed to an agonist, the EC_{50} of linaclotide at pH 7.0 was 99 ± 17.5 nM.[117] In preclinical studies in mice using the transit of activated charcoal as a measure of efficacy, linaclotide at 100 µg/kg significantly accelerated transit compared to wild-type mice treated with charcoal only or GC-C null mice treated with and without linaclotide.[118] Efficacy was also seen in rats treated with linaclotide at doses of 5, 10, and 20 µg/kg.[117] Linaclotide has been synthesized using conventional solid-phase peptide technology.[119]

The pharmacokinetics of linaclotide was established in Phase I trials in healthy volunteers at doses ranging from 30 to 1000 µg. Linaclotide was safe and well tolerated; plasma exposures were below the limits of quantitation for both linaclotide and its active metabolite MM-419447 following oral administration suggesting that there is no systemic exposure of parent or metabolite.[120] The efficacy and safety of linaclotide in IBS-C was assessed in two double-blind, placebo-controlled, randomized, multicenter trials in adult patients (Trials 1 and 2, $n = 1604$).[121a,b] Patients were randomized to either receive placebo ($n = 798$) or 290 µg of drug ($n = 806$). Both trials demonstrated superior efficacy of the linaclotide-treated arm compared to placebo in terms of improvement from baseline in abdominal pain and CSBMs (complete spontaneous bowel movements) frequency. The efficacy of linaclotide in CIC was assessed in two double-blind, placebo-controlled, randomized, multicenter trials in adult patients (Trials 3 and 4, $n = 1276$).[122] Patients were randomized to either receive placebo ($n = 424$) or 290 µg of drug ($n = 422$) or 145 µg ($n = 430$) of drug. In both trials, the primary efficacy end point (three CSBMs per week and an increase of at least one CSBM per week from baseline for 9 of 12 weeks in the study period) was met for the linaclotide-treated group compared with placebo. For example, in Trial 3, the primary endpoint was met for 21.2% of patients in the linaclotide 145 μg group and 19.4% of patients in the linaclotide 290 μg group, compared with 3.3% in placebo group ($P<0.01$). Diarrhea, abdominal pain, flatulence, and abdominal distension were the most common AEs observed during linaclotide clinical trials. Linaclotide carries a black-box warning for pediatric use and is contraindicated in pediatric patients up to 6 years of age. The recommended dose of linaclotide is 290 μg administered orally once daily for IBS-C and 145 μg administered orally once daily for CIC.

15. LOMITAPIDE MESYLATE (ANTIHYPERCHOLESTEREMIC)[123–129]

Class: Microsomal triglyceride transfer protein inhibitor
Country of origin: United States
Originator: Bristol-Myers Squibb
First introduction: United States
Introduced by: Aegerion
Trade name: Juxtapid™
CAS registry no: 182431-12-5 (neutral form) and 202914-84-9 (methanesulfonate salt)
Molecular weight: 693.72 (neutral form) and 789.83 (methanesulfonate salt)

Lomitapide was approved by the US FDA in December 2012 for the treatment of patients with familial hypercholesteremia (referred to as HoFH) in conjunction with a low-fat diet and other lipid-lowering treatments. The incidence of HoFH is estimated to be 1 in 1,000,000 globally, and patients with HoFH are characterized with untreated low-density lipoprotein (LDL) cholesterol concentrations of ≥ 500 mg/dL or treated levels of ≥ 300 mg/dL.[123] Due to the congenital nature of the disease,[124] hypercholesteremia often presents itself early in life for HoFH patients and, left untreated, HoFH patients have a high incidence of early death in their 20s and 30s. Typically, the HoFH population has a high incidence of coronary atherosclerosis and obstructive coronary artery disease as a result of hypercholesteremia. Current treatments for HoFH involve statin therapy (e.g., simvastatin) in combination with other lipid-lowering agents such as ezetimibe, niacin, bile acid sequestrants, fibrates, and omega-3 fatty acids. In patients who are improperly controlled or are intolerant to statin treatment, LDL apheresis may be initiated to reduce acute hypercholesteremia;

however, this treatment is not widely available, it is invasive, and high LDL levels tend to be reestablished rapidly (1–2 weeks), requiring frequent treatment.[123]

Genetic evidence in patients with abetalipoproteinemia,[125] who have characteristically low LDL levels as the result of mutations in microsomal triglyceride transfer protein (MTP), provides support for the inhibition of MTP as a target to reduce hypercholesteremia. Lomitapide was discovered from a high-throughput screen that identified several structurally distinct MTP inhibitors. Combination of key structural features from two structurally distinct HTS hits provided potent MTP inhibitors. Parallel analog synthesis led to lomitapide as an optimized structure. Lomitapide was synthesized via alkylation of 9-fluorenylcarboxylic acid with 1,4-dibromobutane which, after trifluoroethylamide formation, provided a bromide intermediate that was displaced by Boc-4-aminopiperidine. Introduction of the 4'-trifluoromethylbiphenylcarboxamide gave lomitapide,[126] which was found to inhibit MTP with an IC_{50} of 0.5 nM and to exhibit good cholesterol-lowering efficacy in Sprague–Dawley rats (intravenous and oral $ED_{50} \sim 0.2$ mg/kg).[125] In a model with Watanabe-heritable hyperlipidemic rabbits, which are thought to serve as a good preclinical model of HoFH, treatment for 14 days with 10 mg/kg of lomitapide resulted in 60–90% reduction in LDL levels versus pretreatment levels. In this same model, lomitapide had an ED_{50} for cholesterol lowering of 1.9 mg/kg.[125]

Early clinical studies of lomitapide focused on its use at high doses (25, 50, and 100 mg) as monotherapy for cholesterol reduction. Upon oral administration, 33% of the dose is absorbed; however, due to high hepatic extraction, lomitapide has relatively low (7%) absolute bioavailability.[127] Lomitapide has an elimination half-life in humans of 34.4 h with excretion in feces (52.9%) and urine (33%, predominantly as metabolites). The site of action is believed to be in the liver and intestine, so circulating plasma levels may not be reflective of levels required for efficacy. Initial clinical trials had high dropout rates due to increases ($\geq$ 3-fold) in alanine aminotransferase (ALT), and significant gastrointestinal AEs. In Phase III clinical trials in HoFH patients, 29 patients were treated with lomitapide with 23 completing the initial 26-week study.[124] Subsets of patients continued lomitapide treatment through 56 and 78 weeks. Doses started at 5 mg and were escalated in some patients to 60 mg, based on tolerability. At the end of 26 weeks, 19/23 patients had >25% cholesterol lowering, and 12/23 had >50% lowering. A majority of patients experienced gastrointestinal adverse events that were generally characterized as mild in severity. Three of the six patient withdrawals in the first 26-week period were due to severe gastrointestinal disorders, and ten patients experienced increases in serum ALT and/or AST of >3-fold above the upper limit of normal. These AEs were managed by dose reduction or temporary interruption in lomitapide treatment. The US FDA approved lomitapide (5, 10, and 20 mg capsules), a first-in-class MTP inhibitor, contingent on the implementation of a Risk Evaluation and Mitigation Strategy (REMS).[128] Phase I clinical trials were recently initiated in healthy Caucasian and Japanese subjects to study the pharmacokinetics, safety, tolerability, and pharmacodynamic effects in non-HoFH patients with elevated LDL cholesterol.[129]

16. LORCASERIN HYDROCHLORIDE (ANTIOBESITY)[130–139]

Class: 5-HT$_{2C}$ receptor agonist
Country of origin: United States
Originator: Arena Pharmaceuticals
First introduction: United States
Introduced by: Eisai, Inc.
Trade name: Belviq®
CAS registry no: 846589-98-8 (salt)
Molecular weight: 232.15 (salt)

In June 2012, the US FDA approved lorcaserin for chronic weight management in adult patients who are characterized as overweight or obese and have at least one comorbid, weight-related condition.[130] Such comorbid conditions include type 2 diabetes, cardiovascular disease, osteoarthritis, depression, sleep apnea, and malignancies. The World Health Organization estimations from 2008 indicate that 1.4 billion adults are overweight and over 10% of the world population are obese.[131] A 5% weight reduction in obese patients can result in improvement of most comorbid conditions. Existing FDA-approved medications for long-term weight loss therapy include sibutrimine, a serotonin and norepinephrine reuptake inhibitor with use-limiting side effects, and orlistat, a lipase inhibitor with gastrointestinal side effects. Lorcaserin (also known as APD356) is the first 5-HT$_{2C}$ agonist approved for chronic weight management since the withdrawal of the 5-HT$_{2C}$ agonist fenfluramine in 1997 due to rare cases of cardiac valvulopathy. The serotonin receptor 5-HT$_{2C}$ is found primarily in the hypothalamus and regulates appetite and feeding behavior. Safety issues with fenfluramine were associated with poor selectivity for 5-HT$_{2C}$ versus 5HT$_{2A}$ and 5HT$_{2B}$. Ring constraint afforded by the benzazepine in lorcaserin improved selectivity for 5HT$_{2C}$ by over 2 log units.[132] At therapeutically relevant doses, no evidence of QT prolongation was observed in Purkinje fiber and hERG assays.[132] In a rat acute food intake study at doses of 25–100 μmol/kg, lorcaserin inhibited food intake by 50% over 6 h ($ED_{50} = 18$ mg/kg).[132] The synthesis of lorcaserin is accomplished efficiently via a 4-step route utilizing an intramolecular Friedel–Crafts alkylation to form the 7-membered ring from [2-(4-chlorophenyl)-ethyl]-(2-chloropropyl)-ammonium chloride followed by chiral resolution of a tartrate salt to yield the desired enantiomer.[133]

When lorcaserin was given to healthy volunteers at a 10 mg dose, steady state was achieved by day 4.[134] In otherwise healthy obese patients receiving 10 mg bid, C_{max} at steady state was 131 nmol/L.[134] The half-life of lorcaserin is 10–11 h with peak plasma concentrations achieved at ~2 h.[135] Lorcaserin is readily metabolized yielding a sulfamate ester and an N-carbamoyl glucuronide as the major metabolites in plasma and urine, respectively.[135] Lorcaserin is a CYP2D6 inhibitor and may increase exposure of drugs that are CYP2D6 substrates.[136] The safety and efficacy of lorcaserin for weight reduction was assessed in a Phase IIa trial in 469 obese men and women for 12 weeks. Progressive weight loss of 1.8, 2.6, and 3.6 kg was observed at doses of 10 mg qd, 15 mg qd, and 10 mg bid, respectively.[134] In a 1-year Phase III BLOSSOM trial, 4008 obese patients with a comorbid condition received lorcaserin or placebo along with diet and exercise counseling. At doses of 10 mg bid and qd, 40–47% of patients lost at least 5% of baseline body weight.[137] The most common AEs were headache, nausea, and dizziness.[137] Incidences of cardiac valvulopathy occurred in 2% of patients who received placebo or lorcaserin at 10 mg bid.[137] Sixty-eight percent of patients who lost at least 5% of their baseline weight at year 1 and continued to receive lorcaserin in year 2 maintained their weight loss.[138] Approximately 55% of patients receiving lorcaserin continued through year 2 of the trial versus only 45% of those receiving placebo.[138] Lorcaserin is given orally twice daily as a 10 mg tablet.[139]

17. MOGAMULIZUMAB (ANTICANCER)[140–146]

Class: Recombinant monoclonal antibody
Type: Humanized, defucosylated IgG1κ, anti-CCR4
Country of origin: Japan
Originator: Kyowa Hakko
First introduction: Japan
Introduced by: Kyowa Hakko Kirin

Continued

—cont'd
Trade name: Poteligeo®
CAS registry no: 1159266-37-1
Expression System: Recombinant Chinese Hamster Ovary CHO-cell line
Molecular weight: ~146 kDa

Mogamulizumab is a CC chemokine receptor 4 (CCR4) targeting antibody that was approved in March 2012 by the Japanese Ministry of Health, Labour and Welfare for the treatment of relapsed or refractory CCR4+ adult T-cell leukemia–lymphoma (ATL). ATL is an aggressive peripheral T-cell malignancy caused by human T-cell lymphotropic virus type 1 (HTLV-1). The virus is endemic in certain regions, particularly in southwestern Japan, West Africa, the Caribbean islands, and Brazil, with seroprevalence rates of 2–10%.[140,141] There are an estimated 1.2 million HTLV-1 carriers in Japan, with a small percentage developing ATL and ~700–1000 deaths per year. The 5-year overall survival (OS) rate for ATL is 14%. The disease is largely resistant to standard chemotherapy agents and current treatment options are limited.[142] A recent Phase III clinical trial using dose-intensified, multidrug chemotherapy regimen (vincristine, cyclophosphamide, doxorubicin, prednisone, ranimustine, vindesine, etoposide, and carboplatin) resulted in median PFS and OS time of 7 and 12.7 months, respectively.[143] In Western countries, combination chemotherapy with interferon-α and zidovudine has been widely used. CCR4 is expressed on normal Th2 and T regulatory cells and on tumor cells from various types of peripheral T-cell lymphoma, including the majority of ATL. Mogamulizumab is a humanized monoclonal antibody that recognizes the N-terminal region of human CCR4. It has been glycoengineered to remove fucose residues from the Fc region oligosaccharides, resulting in an enhanced ability to mediate ADCC. It is the first glycoengineered antibody to reach the market.[144] In preclinical studies, mogamulizumab showed potent antitumor activity mediated by enhanced ADCC against ATL cell lines and primary ATL cells in vitro and in vivo.[145]

A Phase I clinical study was conducted in 19 patients with CCR4+ peripheral T-cell lymphoma, including ATL.[146] Mogamulizumab was intravenously administered once a week for 4 weeks at four dose levels (0.01, 0.1, 0.5, and 1 mg/kg) according to a conventional 3+3 trial design. Pharmacokinetic analysis revealed plasma mogamulizumab trough concentrations of 7.5–19.6 µg/mL after the first to the fourth administration at the 1 mg/kg dose. The main Phase II trial with mogamulizumab was conducted in 26 evaluable patients with CCR4+ ATL that had recurred or relapsed after previous chemotherapy and was the basis of Kyowa Hakko Kirin's new drug application. Mogamulizumab was intravenously administered once a week for 8 weeks at a dose of 1 mg/kg. The mean maximum and trough plasma concentrations of the eight infusions were 42.9 ± 14.2 and 33.6 ± 10.6 µg/mL, respectively. The plasma half-life was ~18 days, comparable to the half-life of endogenous human IgG (14–21 days). Blood T-cell subset distribution was compared with values from healthy volunteers. Prior to treatment, the numbers of circulating CD4+/CCR4+ cells were significantly higher in patients with ATL than in healthy donors, but were significantly reduced from baseline following the first infusion with mogamulizumab, with the reduction lasting at least 4 months after the eighth infusion.[145] The primary efficacy endpoint of the trial was the best overall response rate (ORR). Secondary endpoints included the best response at each disease site, PFS, and OS. The best ORR was 50% (13/26) including eight with CR. With the lower limit of the 95% confidence interval exceeding the threshold response rate of 5%, the clinical efficacy of mogamulizumab was confirmed. CRs and PRs according to disease sites were 100% for peripheral blood (13 patients, all CR), 63% for skin (8 patients, 3 CR and 2 PR), and 25% for nodal and extranodal lesions (12 patients, 3 CR/CR unverified). Median PFS and OS were 5.2 and 13.7 months, respectively. The best ORR was calculated also for each disease subtype, yielding 43% in patients with acute type (14 patients, 5 CR and 1 PR), 33% in patients with lymphoma type (6 patients, 1 CR and 1 PR), and 83% in patients with unfavorable chronic type (6 patients, 2 CR and 3 PR). In addition, the best ORR was 39% for patients younger than 65 years and 62% for patients 65 years or older. The most common nonhematologic AE was an infusion reaction (89%). At least 80% of the AEs occurred along with the infusion reaction including fever, chills, tachycardia, hypertension, nausea, and hypoxemia. These events occurred primarily at the first infusion and all patients recovered, although some needed systemic steroids. The mechanism of the infusion reaction is unclear, but may be caused by cytokines and related cytotoxic molecules released from highly activated NK cells. Skin rashes were another frequent nonhematologic AE (63%), mostly occurring at the fourth or subsequent infusions. Other noninfusion–related AEs ≥grade 2 were increases in liver enzymes (aspartate aminotransferase, 26%; alanine aminotransferase, 22%), hypoxemia (19%), pruritus (15%), increased γ-glutamyltransferase levels (15%), hypophosphatemia, lymphopenia (96%), leukopenia (56%), neutropenia (33%), thrombocytopenia (26%), and anemia (15%). There have been no reported deaths. In addition, Kyowa Medex has developed two companion diagnostic tests for identifying relevant subpopulations of patients with ATL who are most likely to respond to mogamulizumab. These tests assay the presence or absence of CCR4-expressing ATL and have received regulatory approval in Japan. Mogamulizumab is administered intravenously once weekly (eight administrations) at a dose of 1.0 mg/kg.

18. OMACETAXINE MEPESUCCINATE (ANTICANCER)[147-157]

Class: Protein synthesis inhibitor
Country of origin: China/United States
Originator: Chinese Academy of Medical Sciences/ChemGenex
First introduction: United States
Introduced by: Teva
Trade name: Synribo™
CAS registry no: 26833-87-4
Molecular weight: 545.62

Omacetaxine mepesuccinate (also known as homoharringtonine) was approved by the US FDA in October 2012 for the treatment of patients with chronic or accelerated phase chronic myeloid leukemia (CML) with resistance or intolerance to at least two tyrosine kinase inhibitors (TKIs). CML accounts for ~10% of all new cases of leukemia. CML mainly affects adults (average age at diagnosis is 65 years old) and is more common in men than women. For 2013, the American Cancer Society projects the diagnosis of about 5920 new cases of CML and about 610 deaths due to the disease.[147] The current standard of treatment for CML is imatinib, an inhibitor of BCR–ABL1 tyrosine kinase, a fusion oncoprotein that results from a translocation between chromosomes 9 and 22 and is present in 95% of patients with CML.[148] BCR–ABL1 has deregulated tyrosine kinase activity, leading to increased protein synthesis and cellular proliferation. The second-generation BCR–ABL1 TKIs dasatinib and nilotinib are used for patients who develop resistance to imatinib or are intolerant to the drug, although resistance often develops to these agents as well. Mutations in the kinase domain of BCR–ABL1 are a primary cause of resistance, with the T315I mutation conferring resistance to all current TKIs except ponatinib. Omacetaxine is a protein synthesis inhibitor that was studied in the 1970s for the treatment of acute myeloid leukemia (AML) and in the 1990s for

CML.[149] Emergence of resistance to first- and second-generation TKIs has lead to renewed interest in omacetaxine due to its differentiated mode of action. Omacetaxine acts on the initial step of protein translation and results in the rapid loss of a number of short-lived proteins that regulate proliferation and cell survival.[149,150] Omacetaxine induces apoptosis and shows *in vitro* activity in a number of leukemia cell lines and in murine leukemia models.[151,152] Omacetaxine is a naturally occurring alkaloid isolated from *Cephalotaxus* coniferous shrubs that are indigenous to Asia. Extracts of the bark have been used by practitioners of traditional Chinese medicine for the treatment of cancer. Although omacetaxine could be isolated directly from bark and roots, a more efficient approach is semi-synthesis by esterification of the abundant biosynthetic precursor cephalotaxine, which can be extracted from leaves rather than non-renewable sources.[153] Esterification is carried out with an activated ester in which the diol side-chain is protected as a tetrahydropyran; after ester formation, the diol is released in two steps under mild conditions.

Early clinical studies utilized short intravenous (iv) infusions of omacetaxine, however subcutaneous dosing was subsequently found to reduce the dose-limiting cardiotoxicity seen with intravenous administration.[152] A Phase I study in patients with advanced AML showed a maximum tolerated subcutaneous dose of 5 mg/m^2 for 9 days.[154] Maximum drug concentrations were reached 0.6–1 h after administration, the mean half-life was 11 h, and the volume of distribution at steady state was 2 L/kg. A small Phase I study in CML patients who failed on prior therapy showed the maximum tolerated dose of omacetaxine to be 1.25 mg/m^2 given subcutaneously twice daily.[155] In patients given a loading intravenous dose of 2.5 mg/m^2 over 24 h, followed by 1.25 mg/m^2 subcutaneously for 14 days every 28 days until remission, complete hematologic remission was obtained in all five evaluable patients. All patients developed myelosuppression and three had their dose reduced due to prolonged neutropenia. A Phase II open-label trial (CML-202) assessed the efficacy of omacetaxine in chronic phase CML patients with the T315I mutation who had failed to respond to TKI treatment.[156] Sixty-two patients received omacetaxine subcutaneously at 1.25 mg/m^2 twice daily for 14 days every 28 days until hematologic response or a maximum of six cycles, and then for 7 days every 28 days as maintenance. Complete hematologic response (CHR) was achieved in 48 patients (77%) with a median response duration of 9.1 months. Fourteen patients (23%) achieved major cytogenetic response (MCyR), including complete cytogenetic response in ten patients. Median progression-free survival (PFS) was 7.7 months. AEs that were managed by dose reduction included thrombocytopenia (76%), neutropenia (44%), and anemia (39%).

Nonhematologic AEs included infection (42%), diarrhea (40%), and nausea (34%). Additional open-label trials have also demonstrated meaningful response to omacetaxine treatment in patients with chronic and acute phase CML that have failed at least two TKIs.[157] Omacetaxine is supplied in a single-use vial containing 3.5 mg of omacetaxine mepesuccinate as a lyophilized powder.

19. PASIREOTIDE (CUSHING'S DISEASE)[158–166]

Class: Somatostatin receptor agonist
Country of origin: Switzerland
Originator: Novartis
First introduction: European Union
Introduced by: Novartis
Trade name: Signifor®
CAS registry no: 396091-73-9
Molecular weight: 1047.21

In April 2012, the European Commission approved pasireotide for the treatment of Cushing's Disease (CD) in adult patients who have not responded to surgical intervention or for whom surgery is not an option.[158] Pasireotide was approved for the same indication by the US FDA in December of 2012. CD is a relatively rare disease that is caused by typically benign pituitary tumors that secrete adrenocorticotropic hormone (ACTH), leading to the production of excess cortisol by the adrenal glands.[159] The consequences of chronic exposure to high cortisol levels include weight gain, bruising, muscle weakness, hypertension, diabetes, cognitive dysfunction, and mood alterations. CD typically occurs in adults aged 20–50, with an estimated 10–15 per million people being affected each year.[160] Surgical removal of the pituitary adenoma is the first-line treatment of CD, with remission rates varying between 60% and 90%. Up to 25% of patients develop recurrent adenoma. Pasireotide (also known as SOM230) is a cyclohexapeptide that acts as a somatostatin analogue to inhibit the release of ACTH. Somatostatins are cyclic peptides of 14 and 28 amino acids that play an important role in regulating endocrine and exocrine release in many tissues through an inhibitory mechanism. There are five known subtypes of somatostatin receptors (SSTRs). Natural somatostatins bind with high affinity to all five subtypes, however, their therapeutic use is limited by rapid degradation in plasma. Pasireotide arose from efforts to identify a somatostatin mimetic with long-lasting inhibitory effects.[161,162] Starting with a 14-amino acid somatostatin peptide, a systematic alanine scan revealed residues that were essential for receptor sub-type binding, including key β-turn regions and adjacent residues. Placing the key structural elements as unnatural amino acids in a cyclohexapeptide backbone gave pasireotide. Pasireotide binds with high affinity to four of the five SSTRs, including SSTR5 ($IC_{50} = 0.16$ nM), which is highly expressed in tumor cells from patients with CD. Pasireotide demonstrated a prolonged duration of action in preclinical models as shown by inhibition of the release of hormones regulated by somatostatin 6 h postadministration.[161,162] The terminal half-life of pasireotide in rats was 23 h.

In healthy male subjects given pasireotide subcutaneously at doses ranging from 50 to 600 μg once a day, peak levels were reached within 1 h, with a half-life of 10 h.[163] In another Phase I study, healthy volunteers were given 900, 1200, or 1500 μg of pasireotide subcutaneously as a single dose or two divided doses 12 h apart.[164] Pasireotide showed rapid absorption ($T_{max} = 0.56–0.69$ h), low clearance (8–9 L/h), a long terminal half-life (54–97 h), and a favorable tolerability profile. Pasireotide is 88% protein bound in human plasma. In a Phase II proof-of-concept study, CD patients with moderate to severe disease were given 600 μg pasireotide subcutaneously twice daily for 15 days.[165] Steady state was achieved within 5 days, with a C_{min} of ~5 μg/mL and a C_{max} of ~20 μg/mL. In the primary efficacy analysis, 22 of 29 patients showed a reduction in urinary free cortisol (UFC), with 5 of these patients (responders) showing normal UFC levels. Serum cortisol and plasma ACTH levels were also reduced, with greater reductions seen in responders than nonresponders. Pasireotide was generally well tolerated with gastrointestinal events being the most common AEs. In a double-blind, uncontrolled Phase III study, 162 patients with CD (persistent, recurrent, or newly diagnosed) were randomized to either 600 or 900 μg of pasireotide given subcutaneously twice daily for 6 months followed by open-label treatment through 12 months.[166] At baseline, 78% of patients had moderate-to-very severe hypercortisolism. The primary endpoint (UFC level ≤ the upper limit of the normal range at 6 months) was achieved in 12 of the 82 patients at the 600 μg pasireotide dose and in 21 of the 80 patients at the 900 μg dose. Normalization of UFC was more likely to be achieved in patients with lower baseline levels. Serum and salivary cortisol and plasma corticotropin levels were decreased, and clinical signs and symptoms of CD diminished. The most common AEs were transient gastrointestinal disturbances, and the most frequently reported grade 3 and 4 AEs were hyperglycemia (13%) and diabetes mellitus (7%). Pasireotide was associated with hyperglycemic events of all grades in 118 of 162 patients (73%); 6% of patients discontinued the treatment because of a hyperglycemic AE. Treatment with a glucose-lowering medication was initiated in 74 of 162 patients. Overall, the efficacy of pasireotide supported its approval for the treatment of CD. The approved initial dosage is either 600 or 900 μg given by subcutaneous injection twice daily, with titration of dosing based on treatment response and tolerability.

20. PEGINESATIDE (HEMATOPOIETIC)[167–177]

Class: Erythropoiesis stimulating agent
Country of origin: United States
Originator: Affymax
First introduction: United States
Introduced by: Affymax and Takeda
Trade name: Omontys®
CAS registry no: 1350810-60-4
Molecular weight: 54 kDa

Peginesatide was approved by the US FDA in March 2012 to treat anemia in patients with chronic kidney disease (CKD) requiring dialysis. Anemia is a common complication of CKD: by the time of dialysis and end-stage kidney disease, anemia affects 60–80% of patients.[167] Current treatments for anemia in CKD include blood transfusion, recombinant erythropoietin (EPO) α or β, and darbepoetin alfa. However, EPO has been known to cause pure red–cell aplasia in a small number of patients with chronic renal failure.[168] There are also concerns about all-cause mortality and thrombotic events with EPO.[169] Peginesatide activates the erythropoietin receptor, but it is not structurally related to EPO, so it was proposed to avoid the liabilities associated with EPO. Peginesatide was discovered by screening of a recombinant $>10^{10}$ membered–peptide library against the human erythropoietin receptor. The hits were optimized by introducing nonnatural amino acids (1–Nap-Ala), resulting in a dimeric peptide with little sequence identity to natural erythropoietin.[170] To improve the circulating half-life of peginesatide, the modified peptide dimer was coupled to the NHS-activated ester of *bis*-N-PEG-carbamoylated lysine (PEG$_2$–Lys–NHS). The peptide coupling partner was synthesized via solid-phase peptide synthesis to construct 21-mer peptides that were coupled in 2:1 ratio to NHS-activated esters of Boc-β-alanine-N-diacetic diacid. Boc-deprotection produced the penultimate free amine for coupling to PEG$_2$–Lys–NHS to give peginesatide.[171] The incorporation of two PEG units gave peginesatide an elimination half-life of between 17.5 and 30.7 h in rats. Once-weekly administration of 1.35 mg/kg peginesatide to rats produced an erythropoietic response with no detected antibodies to the pegylated peptide.[172]

When administered intravenously to healthy individuals, peginesatide had an elimination half-life of 18.9 h.[173] Complete pharmacokinetic studies were not performed in humans, but in monkeys, peginesatide was not appreciably metabolized, and urinary excretion was the primary route of elimination. In CKD patients on dialysis, decreased renal extraction leads to a significantly lengthened elimination half-life of 34.9 h relative to healthy individuals.[174] The efficacy of peginesatide was compared to epoietin in the Phase III EMERALD 1 and 2 clinical trials in 1608 patients undergoing hemodialysis, wherein noninferiority was demonstrated.[175] In EMERALD 1, patients were titrated with treatment to maintain hemoglobin levels between 10 and 12 g/dL, which required 5.7 mg of peginesatide per injection once monthly and 9900 U epoietin per week to give similar hemoglobin levels. In EMERALD 2, comparable hemoglobin levels were achieved with 4.8 mg peginesatide once monthly and 6805 U epoietin per week. The incidence of withdrawal (22–30% peginesatide, 22–25% epoietin) and AEs (94.6% peginesatide, 93.0% epoietin) were similar across EMERALD studies. Drug-specific antibodies developed in 12 patients (1.1%); antibodies in 8 patients were determined to be neutralizing. No cases of pure red-cell aplasia or antierythropoietin antibodies were reported, and allergic reactions including anaphylaxis were not reported in patients developing antibodies. AEs observed in $\geq 10\%$ of dialysis patients treated with peginesatide were diarrhea, vomiting, high blood pressure (hypertension), and joint, back, leg, or arm pain (arthralgia).[176] In February 2013, all doses of peginesatide were recalled in a voluntary move by the marketing companies due to postmarketing reports of serious hypersensitivity reactions, including anaphylaxis, in 0.02% of the patient population.[177]

21. PERAMPANEL (ANTICONVULSANT)[178-186]

Class: AMPA glutamate receptor antagonist
Country of origin: Japan
Originator: Eisai
First introduction: United States
Introduced by: Eisai
Trade name: Fycompa™
CAS registry no: 380917-97-5
Molecular weight: 349.384

In October 2012, the US FDA approved perampanel for the treatment of partial onset seizures in epileptic patients who are at least 12 years old. Partial onset or focal seizures, unlike generalized seizures, do not cause total loss of consciousness but may be associated with repetitive movements, unusual behavior and memory loss during the episode.[178] The worldwide incidence of partial seizures is ~30/100,000 per year.[179] There are many FDA approved antiepileptic drugs (AEDs) for the treatment of partial seizures, however one-third of patients do not respond to monotherapy.[180] Furthermore, many AEDs for partial seizures are sodium channel blockers. Perampanel is the first AMPA receptor antagonist to receive FDA approval as an AED. AMPA glutamate receptors are found primarily on postsynaptic neurons in the brain. As a selective, noncompetitive antagonist of AMPA, parampanel prevents ion channel opening and reduces propagation of action potential.[181] Parampanel was discovered through lead optimization of a commercially available compound, 2,4-diphenyl-4H-[1,3,4]oxadiazin-5-one, which was identified by high-throughput screening of a compound collection employing a rat cortical neuron AMPA-induced cell-death assay.[182] Modifications of aromatic rings at positions 1, 3, and 5 while changing the core to pyridone led to parampanel which inhibited AMPA-induced calcium influx (IC$_{50}$ = 60 nM). Parampanel had a minimum effective oral dose of 2 mg/kg in an AMPA-induced mouse seizure model. The synthesis of parampanel was accomplished via a 6-step route utilizing Suzuki–Miyaura couplings and modified Ullmann reactions for incorporation of aryl groups.[182]

In a Phase I study, when parampanel was given to healthy male volunteers at repeated doses from 0.2 to 8 mg once daily for 14 days, the plasma half-life was 66–90 h and the maximum plasma concentrations were achieved at 1 h with sedative effects observed at doses of 2 mg or greater.[183] Parampanel is a CYP3A4 substrate, and strong CYP3A4 inducers such as the AEDs phenytoin and oxcarbazepine cause a 50% reduction in parampanel concentrations.[184] Parampanel was evaluated in subjects with uncontrolled, refractory partial–onset seizures who were receiving concomitant therapy with 1–3 other AEDs. In this double-blind, placebo-controlled Phase III study, the 50% responder rate after 1 and 2 years was 47.6% and 63.2%, respectively; 7% of subjects reported a year-long seizure-free period after the first year with continued treatment.[185] Parampanel was well tolerated at doses of 4–12 mg with only mild to moderate AEs, such as headache, somnolence, and fatigue.[185] An increase in anger, confusion, and depression were observed at the 12 mg dose, which resulted in a boxed warning for serious psychiatric events.[184] Parampanel is given orally once daily as a 2, 4, 6, 8, 10, or 12 mg tablet.[186]

22. PERTUZUMAB (ANTICANCER)[187–192]

Class: Recombinant monoclonal antibody
Type: Humanized, IgG1κ, anti–HER2
Country of origin: United States
Originator: Genentech, a member of the Roche Group
First introduction: United States
Introduced by: Genentech
Trade name: Perjeta®
CAS registry no: 380610-27-5
Expression System: Recombinant Chinese Hamster Ovary CHO-cell line
Molecular weight: ∼148 kDa

Pertuzumab is an anti-human epidermal growth factor receptor 2 (HER2) antibody that was approved by the US FDA in June 2012 for use in combination with trastuzumab (Herceptin®) and docetaxel chemotherapy for the treatment of people with HER2-positive (HER2$^+$) metastatic breast cancer (mBC) who have not received prior anti-HER2 therapy or chemotherapy for metastatic disease. According to the American Cancer Society, ∼229,000 people in the United States would be diagnosed with breast cancer with 40,000 deaths in 2012. Approximately 25% of breast cancers are HER2$^+$ which is associated with a more aggressive phenotype and a poor prognosis.[187] Despite treatment with the standard of care (trastuzumab in combination with chemotherapy), almost all patients with HER2[+] mBC eventually progress.[188] Trastuzumab and pertuzumab both target HER2, but they bind to distinct epitopes on HER2. Trastuzumab inhibits HER2-mediated proliferation by activating antibody-dependent cellular cytotoxicity (ADCC), preventing ligand-independent HER2 signaling, and inhibiting HER2-mediated angiogenesis. Pertuzumab inhibits HER2 dimer formation with other HER members, such as HER1 and HER3, thus inhibiting the downstream signaling processes associated with tumor growth and progression. Pertuzumab is the first HER2 dimerization inhibitor to reach the market. The mechanism of actions of trastuzumab and pertuzumab are believed to complement each other and provide a more comprehensive inhibition of tumor growth than either agent alone.[189] Preclinical data showed that the combination of trastuzumab and pertuzumab has a strongly enhanced antitumor effect and induces tumor regression in xenograft models, something that cannot be achieved by either monotherapy alone.[190]

The Phase I trial was a dose-escalation study of pertuzumab (0.5–15 mg/kg) given intravenously every 3 weeks to 21 patients with incurable solid tumors that had progressed.[191] The mean elimination half-life was 18.9 ± 8 days. At doses ≥ 5 mg/kg of pertuzumab, serum concentrations were maintained above 20 µg/mL for the first two treatment cycles. A randomized, double-blind, placebo-controlled Phase III trial, CLEOPATRA (Clinical Evaluation of Pertuzumab and TRAstuzumab), was conducted in 808 patients with HER2+ mBC who had not received chemotherapy or biologic therapy for their metastatic disease.[187] Patients were randomly assigned in a 1:1 ratio to receive placebo + trastuzumab + docetaxel or pertuzumab + trastuzumab + docetaxel. Patients received a loading dose of 8 mg/kg trastuzumab, followed by maintenance doses of 6 mg/kg every 3 weeks until disease progression or development of toxic effects that could not be effectively managed. Docetaxel was administered every 3 weeks at a starting dose of 75 mg/m^2. The investigator could change the dose to 55–100 mg/m^2 depending on how well docetaxel was tolerated. If chemotherapy was discontinued due to side effects, antibody therapy was continued. Pertuzumab or placebo was given at a loading dose of 840 mg, followed by 420 mg every 3 weeks until disease progression or development of toxic effects that could not be effectively managed. All drugs were administered intravenously. The study was designed to have 80% power to detect a 33% improvement in median PFS in the pertuzumab group. The median number of study treatment cycles was 15 and 18 cycles and the median study duration was 11.8 and 18.1 months in the control and pertuzumab groups, respectively. The primary endpoint was independently assessed PFS. Secondary endpoints included OS, PFS assessed by the investigator, objective response rate, and safety. Treatment with pertuzumab + trastuzumab + docetaxel as compared with placebo + trastuzumab + docetaxel significantly increased PFS. The median independently assessed PFS was prolonged from 12.4 months in the control group to 18.5 months in the pertuzumab group. The median investigator-assessed PFS was 12.4 months in the control group, as compared with 18.5 months in the control group. The benefit of pertuzumab + trastuzumab + docetaxel in PFS was observed across all predefined subgroups of patients. The interim analysis of OS showed a trend toward survival benefit for the pertuzumab group: 96 deaths occurred in the control group as compared with 69 deaths in the pertuzumab group. The AE profile was generally balanced between the two groups. The incidence of any grade AE of diarrhea, rash, mucosal inflammation, febrile neutropenia, and dry skin were at least 5% higher in the pertuzumab group and incidence of grade 3 or higher of febrile neutropenia and diarrhea were at least 2% higher in the pertuzumab group than in the control group. Among toxicities of HER2 targeted therapy, cardiac dysfunction is of particular concern as HER family members play a crucial role in cardiac development.[192] Left ventricular systolic dysfunction was reported more frequently in the control versus the pertuzumab group (8.3% vs.4.4% for any grade and 2.8% vs. 1.2% for $\geq$ grade 3.) More deaths were attributed to disease progression in the control group (81 deaths, 20.4% of patients) than in the pertuzumab group (57 deaths, 14% of patients). Deaths from causes other than disease progression were similar in the two groups. Deaths related to AEs were 2.5% in the control group and 2% in the pertuzumab group. Infections were the most common cause of death due to an AE. Pertuzumab is recommended to be given at an initial dose of 840 mg administered as a 60-min intravenous infusion, followed every 3 weeks by 420 mg administered as a 30–60-min intravenous infusion.

23. PIXANTRONE DIMALEATE (ANTICANCER)[193–202]

Class: Topoisomerase II inhibitor
Country of origin: United States
Originator: University of Vermont
First introduction: European Union
Introduced by: Cell Therapeutics Inc.
Trade name: Pixuvri™
CAS registry no: 144675-97-8 (dimaleate salt) and
144510-96-3 (free base)
Molecular weight: 557.51 (dimaleate salt) and
325.37 (free base)

In May 2012, pixantrone was approved by the European Commission as a single agent for the treatment of relapsed or refractory aggressive B-cell non–Hodgkin lymphoma (NHL) in adult patients who have failed on at least two previous therapies. NHLs encompass a diverse group of white blood cell cancers that are classified by the immune cell of origin, with B-cell lymphomas accounting for 85% of all NHLs and T-cell lymphomas as the remaining 15%. NHL can occur at any age and is the tenth most common cancer worldwide.[193] NHL accounts for 4% of all cancers in the United States, with estimates that ~70,000 people will be diagnosed with NHL and ~19,000 will die from NHL in the United States in 2013.[194] For patients with aggressive NHL, multidrug regimens are the first-line treatment although nearly half of patients will relapse or are refractory to therapy.[195] Anthracyclines, which are part of the combination therapy, are among the most active single agents against NHL, but are associated with cardiotoxicity that limits their use, particularly in second-line therapy. Pixantrone (also known as BBR–2778) is an anthracycline analogue that was specifically designed to address the cardiotoxicity seen in earlier agents by replacement of a 1,4-dihydroxyanthracene-9,10-dione core with a benzoisoquinoline-5,10-dione ring system.[196] Pixantrone inhibits topoisomerase II by intercalation with DNA[197] and is also believed to form covalent adducts with the N–2 amino group of guanine via a formaldehyde aminal formed with the primary amino groups.[198] Pixantrone is less cytotoxic than other anthracycline derivatives, but shows good antitumor activity *in vivo* in a variety of preclinical tumor models, including leukemia and lymphoma models.[195,199] Pixantrone also demonstrated significantly reduced cardiotoxicity in preclinical models compared with the anthracyclines doxorubicin and mitroxantrone. Pixantrone was synthesized by Friedel–Crafts reaction of pyridine-3,4-dicarboxylic acid anhydride with 1,4-difluorobenzene to give a ketoacid that was cyclized to the tricyclic core by treatment with fuming sulfuric acid at $140\,^\circ$C. Reaction of the resulting 6,9-difluorobenzoisoquinoline-5-10-dione with ethylenediamine followed by careful pH adjustment and treatment with maleic acid gave pixantrone in good overall yield and purity.[196,200]

In a Phase I study in 24 patients with advanced or refractory NHL, 84 mg/m^2 of pixantrone dimaleate administered once weekly for a cycle of 3 weeks was shown to give a balance of efficacy and manageable side effects.[201] The C_{max} at this dose was 2.24 µM, with a half-life of 17.5 h, an AUC of 3.68 µM h, and a volume of distribution of 13.9 L/kg. The dose limiting toxicity was myelosuppression. In this limited trial of patients with poor prognosis based on multiple prior treatment failures, there were three CRs and two PRs. A key Phase III clinical study, PIX301 (EXTEND), was an open-label, multicenter, randomized trial in which pixantrone dimaleate was given as a single agent to patients with aggressive NHL who had relapsed after two or more previous chemotherapy regimens.[202] Sixty-eight patients were treated with pixantrone dimaleate at 85 mg/m^2 intravenously on days 1, 8, and 15 of a 28-day cycle, for up to six cycles; sixty-seven patients received a comparator drug administered at standard doses and schedules. The primary endpoint was the proportion of patients who achieved a CR or unconfirmed complete response (CRu). In the pixantrone dimaleate arm, 14 patients (20%) achieved CR/CRu compared with 4 patients (5.7%) in the comparator arm. The most common grade 3/4 AEs (>10%) in the pixantrone dimaleate arm were neutropenia, leucopenia, and thrombocytopenia. Neutropenia and leucopenia was seen more frequently in the pixantrone dimaleate arm than in the comparator group. More cardiac AEs occurred with pixantrone dimaleate than with comparator agents, although these were mostly grade 1 and 2, and were predominantly asymptomatic decreases in left ventricular ejection fraction, which was not associated with clinical evidence of cardiac impairment. Pixantrone was approved by the EC on the basis of the PIX-301 trial. The recommended dose is 85 mg/m^2 of pixantrone dimaleate (equivalent to 50 mg/m^2 of free base) on days 1, 8, and 15 of each 28-day cycle for up to six cycles.

24. PONATINIB (ANTICANCER)[46,203–207]

Class: Tyrosine kinase inhibitor
Country of origin: United States
Originator: Ariad
First introduction: United States
Introduced by: Ariad
Trade name: Iclusig®
CAS registry no: 1114544–31–8 (HCl salt) and 943319–70–8 (free base)
Molecular weight: 569.02 (HCl salt) and 532.56 (free base)

In December 2012, the US FDA approved ponatinib (also referred to as AP 24534) for the treatment of adult patients with chronic phase, accelerated phase, or blast phase chronic myeloid leukemia (CML). In 2012, the National Cancer Institute estimated that 5430 men and women would be diagnosed with CML and 610 of them would have died of the disease.[46] First and second-line therapies for the treatment of CML include imatinib, dasatinib, and nilotinib. Fusion of the Bcr region of chromosome 22 with the Abl region of chromosome 9 results in the Bcr–Abl fusion oncoprotein, a deregulated tyrosine kinase that activates several proliferative and antiapoptotic signaling pathways. Resistance to tyrosine kinase inhibitors (TKI's), imatinib, dasatinib, and nilotinib, is a major issue that needs to be addressed for treatment of all forms of CML. One of the clinically important mutations, T315I (threonine to isoleucine) found in ~20% of patients, confers resistance to all TKIs (including bosutinib, approved in 2012).[203] Approved TKI inhibitors make a key hydrogen bond with the side chain of the gatekeeper residue Thr315 that contributes to the potency and selectivity of these inhibitors. Mutation of the Thr315 residue to isoleucine not only negates the formation of H-bond, but also blocks entry to an adjacent hydrophobic pocket. Ponatinib is a pan-Bcr–Abl TKI that blocks both the native ($IC_{50} = 0.4$ nM) and Bcr–Abl^{T315I} mutated kinases ($IC_{50} = 2.0$ nM) in addition to other mutated kinases in CML patients. In the Ba/F3 cell proliferation assay, ponatinib inhibits ABL and the T315I Abl mutant with IC_{50s} of 1.2 and 8.8 nM, respectively.[204] Ponatinib was identified by a structure-based drug design approach. Ponatinib binds to the kinase domain in a DFG-out conformation; the ethynyl moiety helps the inhibitor evade the mutant gatekeeper isoleucine residue at position 315.[204] In addition to Abl and the T315I mutant of Abl, ponatinib inhibits VEGFR, PDGFR, FGFR, SRC, KIT, RET, TIE2, FLT3, and EPH receptors at concentrations ranging from 0.1 to 20 nM. In preclinical *in vivo* studies, ponatinib at 30 mg/kg administered orally once daily increased the median survival rate significantly in an aggressive mouse model of CML driven by the T315I mutation.[3] Synthetic routes to ponatinib employing a Sonagashira coupling of 3-ethynylimidazo[1,2-b]pyridazine and 3-iodo-4-methylbenzoic acid as a key step have been reported.[204,205]

The C_{max} and AUC of ponatinib in patients with advanced hematologic malignancies treated with 45 mg qd were 73 ng/mL and 1253 ng h/mL, respectively. Median time to C_{max} was ~6 h. The steady state volume of distribution was 1223 L and the terminal half-life was ~24 h.[206] The safety and efficacy of ponatinib was evaluated in a Phase II (ongoing), single arm, open-label, international multicenter trial in 449 patients. Patients were assigned to one of six cohorts: chronic phase CML (CP-CML), accelerated phase CML (AP-CML), or blast phase CML (BP-CML)/(Ph + ALL), resistance or intolerance (R/I) to prior TKI therapy, and the presence of the T315I mutation. The primary efficacy endpoint in CP-CML was major cytogenetic response (MCyR) and the primary efficacy endpoint in AP-CML, BP-CML, and Ph + ALL was major hematologic response (MaHR). In patients with CP-CML who achieved MCyR, the median time to MCyR was 84 days (range 49–334 days). The median time to MaHR in patients with AP-CML, BP-CML, and Ph + ALL was 21 days (range 12–176 days), 29 days (range 12–113 days), and 20 days (range 11–168 days), respectively.[207] Hypertension, rash, abdominal pain, fatigue, headache, dry skin, constipation, arthralgia, nausea, pyrexia, thrombocytopenia, anemia, neutropenia, lymphopenia, and leukopenia (> 20%) were the most common AEs observed during ponatinib clinical trials. Ponatinib carries a black–box warning for arterial thrombosis and hepatotoxicity. The recommended dose of ponatinib is 45 mg administered orally once daily, with or without food.

25. RADOTINIB (ANTICANCER)[46,208–214]

Class: Tyrosine kinase inhibitor	
Country of origin: Korea	
Originator: Il-Yang	
First introduction: Korea	
Introduced by: Il-Yang	
Trade name: Supect™	
CAS registry no: 926037-48-1	
Molecular weight: 530.50	

Radotinib, an inhibitor of Bcr–Abl tyrosine kinase, was approved in January 2012 in Korea as a second-line treatment for chronic myeloid leukemia (CML). CML is a malignant myeloproliferative disorder of hematopoietic stem cells that affects 1–2 people per 100,000 annually and is reported to account for 15–20% of adult leukemia cases.[46] The genetic hallmark of CML is a fusion between the breakpoint cluster region (BCR) on chromosome 22 and the Abelson tyrosine kinase (ABL) on chromosome 9, resulting in a translocation known as the Philadelphia chromosome. This translocation leads to constitutive activation of tyrosine kinases that promote cell proliferation, differentiation, and growth. Several tyrosine kinase inhibitors (TKIs) have been approved for the treatment of CML, the first being imatinib in 2001. Approval of a number of second-generation and novel TKIs has followed to address primary or acquired resistance to imatinib.[208] Radotinib is a TKI with a similar structure to the second-generation TKI, nilotinib,[208] in which a pyridyl group has been replaced with a pyrazine moiety.[209] The *in vitro* activity of radotinib against a variety of tumor cell lines is disclosed in an issued patent.[210] Radotinib was significantly more potent than imatinib in all of the cell lines tested. The synthesis of radotinib via amide coupling is described in the patent literature.[210,211]

Interim results were reported from a Phase II clinical trial with radotinib in 77 adult patients with chronic phase, Philadelphia chromosome positive CML who failed or were unable to tolerate TK inhibitors.[212] Patients were treated with radotinib 400 mg twice daily for 12 cycles (1 cycle = 4 weeks). The primary end point was an achievement of major cytogenetic response (MCyR) by 12 months. At the time of the interim analysis, 46 patients were continuing treatment, 29 patients had discontinued treatment, and 2 patients had died. The MCyR rate was 63.6%, with 35 patients having a complete response (CR) and 14 patients having a partial response (PR). The median time to MCyR was 2.8 months. The most common grade 3/4 AEs were thrombocytopenia (27.3%) and hyperbilirubinemia (31.2%). Common nonhematologic AEs (>10%) included rash, fatigue, nausea/vomiting, headache, and pruritus. Most of the AEs occurred in the early treatment period were tolerable, and were controlled by dose interruption or dose reduction. Clinical trials with radotinib are ongoing.[213,214] Radotinib, which is formulated as 100 and 200 mg oral capsules, was approved in Korea as a second-line treatment for patients with Philadelphia chromosome-positive chronic myelogenous leukemia.

26. REGORAFENIB (ANTICANCER)[215-225]

Class: Kinase inhibitor
Country of origin: Germany
Originator: Bayer
First introduction: United States
Introduced by: Bayer and Onyx Pharmaceuticals
Trade name: Stivarga™
CAS registry no: 755037-03-7
Molecular weight: 482.82

In September 2012, the US FDA approved regorafenib for the treatment of patients with metastatic colorectal cancer (CRC), especially those for whom standard therapies have failed, including fluoropyrimidine-, oxaliplatin-, and irinotecan-based chemotherapy, an anti-VEGF therapy, and, if KRAS wild type, an anti-EGFR therapy. CRC is the second leading cause of cancer-related deaths in the United States.[215] In 2007, 142,672 Americans were diagnosed with CRC.[215] Approximately half of patients develop metastases and have unresectable tumors.[216] Chemotherapy with fluoropyrimidines such as 5-fluorouracil (5-FU) is the typical first-line treatment for these patients[216] and, in combination with panitumumab, a monoclonal antibody for epidermal growth factor receptor, can increase survival rate from 1 to 2 years.[217] Regorafenib is a multikinase inhibitor with potent inhibitory activity versus VEGFRs and PDFRs.[218] Both of these classes of receptors are expressed on tumor cells and affect proliferation and angiogenesis.[219] Regorafenib inhibited growth in murine xenograft models for colon, breast, renal, lung, melanoma, pancreatic, and ovarian tumors when dosed at 10–30 mg/kg.[220] Regorafenib is a fluorinated analog of sorafenib, a multikinase inhibitor co-marketed by Bayer and Onyx for the treatment of kidney and liver cancer.[221] The synthesis of regorafenib is accomplished in two steps from commercially available starting materials. 4-Aminophenol is coupled to 4-chloro-N-methyl-2-pyridinecarboxamide to give 4-(2-(N-methylcarbamoyl)-4-pyridyloxy) aniline. Subsequent treatment with 4-chloro-3-(trifluoromethyl)phenyl isocycanate affords the urea, regorafenib.[222]

In a Phase I study in 28 patients with CRC, regorafenib, at an oral dose of 160 mg daily for a median duration of 53 days, provided 107 days of progression-free survival (PFS) along with AEs including hand–foot–skin reaction, voice change, fatigue, and rash.[223] Plasma exposure of regorafenib and its two major metabolites, M2 (N-oxide) and M5 (des-methyl, N-oxide), was evaluated after 21 days of treatment. After the first cycle of treatment, $AUCs_{0-24h}$ (mg h/L) were 50, 48, and 65 for regorafenib, M2 and M5 respectively, with a C_{max} (mg/L) of 3.4, 3.2, and 4.0.[9] There was significant variability in AUCs and C_{max} for regorafenib and its metabolites between patients. The two metabolites are suspected to contribute to clinical activity since both were found to be pharmacologically active in vitro and in xenograft studies.[223] A subsequent Phase III study (CORRECT) evaluated patients receiving regorafenib (160 mg daily) or placebo (best supportive care) after failure of standard therapies for metastatic CRC. The primary endpoint, overall survival (OS), was achieved with a 23% reduction in risk of deaths for treated patients. While little reduction in tumor growth was observed, 41% of patients achieved disease stabilization.[224] Regorafenib is available in 40 mg tablets and given orally once daily with a low-fat breakfast as a 160 mg dose.[225]

27. TEDUGLUTIDE (SHORT BOWEL SYNDROME)[226-235]

Class: GLP-2 receptor agonist
Country of origin: Canada
Originator: NPS Allelix
First introduction: European Union
Introduced by: NPS Pharmaceuticals, Takeda
Trade name: Gattex®
CAS registry no: 197922-42-2
Molecular weight: 3752.1
H₂N-His-Gly-Asp-Gly-Ser-Phe-Ser-Asp-Glu-Met-Asn-Thr-Ile-Leu-
Asp-Asn-Leu-Ala-Ala-Arg-Asp-Phe-Ile-Asn-Trp-Leu-Ile-Gln-Thr-Lys-
Ile-Thr-Asp-OH

Teduglutide was approved in August 2012 by the European Commission and in December 2012 by the US FDA for the treatment of adults with short bowel syndrome (SBS) who are dependent on parenteral nutrition (PN). The bowel is comprised of the small and large intestine, with the majority of digestion and nutrient uptake occurring in the small intestine. SBS is a condition in which nutrients are not properly absorbed because a large part of the small intestine is missing or has been surgically removed.[226] Treatment of patients with SBS typically involves specialized diets, antidiarrheal and antisecretory agents, and intravenous PN given 5–7 days a week for 10 or more hours a day. While some patients may be gradually weaned from PN as intestinal adaptation occurs, many patients will require PN throughout their lifetime. Glucagon-like peptide 2 (GLP-2) is a 33-amino acid endogenous peptide that is stimulated upon eating a meal and acts to slow gastric emptying, reduce gastric secretions, and stimulate growth and repair of intestinal epithelium.[227] Administration of GLP-2 to SBS patients was shown to improve parameters associated with nutrient and fluid absorption, however GLP-2 is rapidly degraded *in vivo* by the action of dipeptidyl peptidase 4 (DPP-4), limiting its utility.[228] Teduglutide (also known as ALX-0600) is a GLP-2 analogue in which glycine has been substituted for alanine at position 2 from the N-terminus, thereby conferring resistance to degradation by DPP-4.[228] Teduglutide has similar binding affinity, agonist potency, and agonist efficacy as GLP-2 in assays with the rat GLP-2 receptor.[229] Administration of teduglutide to rats gave a significant increase in small intestine weight and an increase in the height of intestinal villi.[228] Teduglutide was prepared by solid-phase peptide synthesis.[230]

The pharmacokinetics, safety, and tolerability of teduglutide were determined in a double-blinded, randomized, placebo-controlled Phase I trial in healthy subjects ($n=64$) following subcutaneous administration for 8 days at doses of 10–80 mg per day.[231] Teduglutide was rapidly absorbed ($T_{max}=2.8$–6.0 h) with a half-life of 3.0–4.6 h. Pharmacokinetic parameters were linear over the dose range and no accumulation was observed. Teduglutide was safe and well tolerated, with all AEs being mild or moderate, except for one vasovagal-related apparent seizure. A Phase II open-label dose-ranging study in SBS patients ($n=18$) showed doses of 0.03, 0.10, and 0.15 mg/kg/day of teduglutide for 21 days to be efficacious as determined by increased intestinal wet weight absorption, increased urine weight and urine sodium excretion, and decreased fecal weight and fecal energy secretion.[232] Colonic biopsies showed relevant histological changes, including increased villus height and crypt depth. A placebo-controlled, randomized Phase III study was carried out in SBS patients treated once daily for 24 weeks with doses of 0.05 mg/kg/day ($n=35$), 0.10 mg/kg/day ($n=32$), or placebo ($n=16$).[233] Responders were subjects who demonstrated reductions of $\geq 20\%$ in parenteral fluid volumes from baseline at weeks 20 and 24. Both teduglutide doses resulted in a reduction in volumes of PN compared with placebo, although the 0.10 mg/kg/day group did not reach statistical significance. Two patients in the 0.05 mg/kg/day group were weaned from PN versus none in placebo. A follow-up 24-week, Phase III study confirmed the efficacy of the 0.05 mg/kg/day dose of teduglutide, with the responder rate in the 0.05 mg/kg/day treatment (27/43; 63%) being more than twofold greater than for placebo (13/43; 30%).[234] The percentage of patients with a 1 day or more reduction in the weekly need for parental support was greater in the teduglutide group (54%) than in the placebo group (23%). Patients on the 0.05 mg/kg/day dose from both Phase II trials were able to enroll in extension studies to determine longer term benefits of teduglutide. A reduction in days on PN support of at least 2 days was achieved in 13/34 patients after treatment with teduglutide for 1 year.[235] Teduglutide was safe and well tolerated across studies, with the most frequently reported AEs ($>20\%$) being injection site pain, abdominal distension, constipation, and headache. Intestinal biopsies showed no evidence of dysplasia or malignancy. Due to potential risk of developing cancer and abnormal growths in the intestine, teduglutide was approved with a Risk Evaluation and Mitigation Strategy (REMS), including a recommendation for a colonoscopy no less frequently than every 5 years. Teduglutide is administered at a dose of 0.05 mg/kg, which is prepared by reconstitution of lyophilized powder with distilled water and given as a subcutaneous injection once daily.

28. TENELIGLIPTIN (ANTIDIABETIC)[22,236–240]

Class: Dipeptidyl peptidase-4 inhibitor
Country of origin: Japan
Originator: Mitsubishi Tanabe Pharma
First introduction: Japan
Introduced by: Mitsubishi Tanabe Pharma
Trade name: Tenelia®
CAS registry no: 760937-92-6
Molecular weight: 426.58

Teneligliptin was approved in September 2012 in Japan for the treatment of patients with Type 2 diabetes mellitus (T2DM). T2DM is a chronic disease characterized by hyperglycemia due to decreased secretion of insulin, the hormone that controls blood sugar levels, and resistance to the effects of insulin. Diabetes can cause serious health issues, including heart disease and stroke, blindness, kidney failure, and neuropathy leading to lower limb amputation. Risk factors for T2DM include obesity, lack of physical activity, family history of diabetes, and older age. The International Diabetes Federation has reported the global prevalence of diabetes to be 8.3%, with an additional 6.4% having impaired glucose tolerance, which is a risk factor for developing diabetes and cardiovascular disease.[236] Japan ranks sixth worldwide in number of people with diabetes (10.7 million in 2011). For comparison, in 2011, China ranked first with 90 million people with diabetes, while the United States ranked third with 23.7 million people with diabetes. The incidence of diabetes is increasing in all countries. Teneligliptin is a member of the dipeptidyl peptidase 4 (DPP-4) inhibitor class of antidiabetes agents. DPP-4 is an enzyme that degrades GLP-1, a 30-amino acid peptide that is secreted in response to food intake. GLP-1 stimulates insulin secretion and inhibits glucagon secretion, which leads to lower levels of plasma glucose. Teneligliptin is one of a growing numbers of DPP-4 inhibitors to be approved worldwide.[22] The discovery of teneligliptin was guided by structure-based design, with a key element being binding of the phenyl group in the S2 pocket, which not only increases potency for DPP-4, but also improves selectivity versus DPP-8 and DPP-9.[237] Teneligliptin is a potent inhibitor of DPP-4 in the enzyme inhibition assay ($IC_{50} = 0.37$ nM). Teneligliptin showed good oral bioavailability in rats

(63–86%) with a T_{max} of 0.75–0.88 h and a terminal half-life of 8–16 h. In monkeys, the oral bioavailability of teneligliptin was 44–83%, with a T_{max} of 0.5–1.38 h and a terminal half-life of 15–19 h. In preclinical *in vivo* models,[237,238] teneligliptin was effective in inhibiting DPP-4 activity in a dose-dependent manner at doses ranging from 0.03 to 1 mg/kg. At the 1 mg/kg dose, >50% inhibition of DPP-4 activity was sustained for 24 h. Teneligliptin was also efficacious in improving glucose tolerance in Zucker fatty rats following a 1 g/kg glucose challenge, with normalization of glucose levels achieved at doses as low as 0.03 mg/kg. At 1 mg/kg, the glucose-lowering effects lasted for 12 h postadministration. Repeated administration of teneligliptin for 2 weeks in Zucker rats reduced glucose excursions, and furthermore, decreased plasma levels of triglycerides and free fatty acids under nonfasting conditions.[238] Key steps in the synthesis of teneligliptin include formation of 1-phenyl-3-methyl-5-piperazinylpyrazole from a keto-amide precursor and reductive amination of the resulting piperazine with a keto-proline intemediate.[237]

The pharmacokinetic profile of 10 and 20 mg doses of teneligliptin was reported as part of a 4-week randomized, double-blind, placebo-controlled trial in patients with T2DM.[239] At the 10 mg dose ($n = 33$ patients), a C_{max} of 125 ng/mL was achieved at 1 h and the terminal half-life was 20.8 h. For the 20 mg dose ($n = 33$ patients), C_{max} was 275 ng/mL at 1 h and the terminal half-life was 18.9 h. The AUC_{0-24h} increased in a dose-proportional manner from 830.9 ng h/mL at the 10 mg dose to 1625.1 ng h/mL at the 20 mg dose. Teneligliptin is metabolized in the liver and is primarily eliminated in feces. The maximum inhibition of plasma DPP-4 activity occurred within 2 h of administration (10 mg, 81.3% inhibition; 20 mg, 89.7% inhibition). In addition, plasma levels of active GLP-1 were higher for both doses through 24 h after drug administration than in the placebo group. The primary efficacy endpoints in this study included changes in levels of glucose (2 h postmeal, 24-h mean, fasting) and changes in AUC_{0-2h} for glucose, insulin, and glucagon. The differences in glucose levels 2 h postmeal for 10 mg of teneligliptin versus placebo were −50.7, −34.8, and −37.5 mg/dL for breakfast, lunch, and dinner, respectively (compared with a baseline fasting plasma glucose level of 169.2 mg/dL). For the 20 mg dose group, the differences versus placebo were −38.1, −28.6, and −36.1 mg/dL for breakfast, lunch, and dinner, respectively. The 24-h mean glucose levels and the fasting glucose levels also decreased significantly for the teneligliptin treated groups compared with placebo. Insulin levels increased significantly in the teneligliptin 10 mg group after dinner but not after other meals, while glucagon levels decreased significantly after breakfast and lunch in both groups, and after dinner in the 20 mg group compared with placebo. The incidence of AEs was not significantly different between placebo and either drug–treatment group. None of the patients in any group experienced hypo-glycemic symptoms or serious AEs. Teneligliptin was also evaluated in a 12-week, randomized, double-blind, placebo–controlled study with 324 Japanese patients with T2DM.[240] Patients were randomized to receive 10, 20, or 40 mg of teneligliptin or placebo. All doses of teneligliptin pro-duced significant and clinically meaningful reductions in hemoglobin (Hb) A1c, a marker of plasma glucose levels, and fasting plasma glucose levels. There were no significant between–group differences for the three teneligliptin doses in terms of efficacy, but there were slightly more adverse drug related events in the 40 mg group. Overall, the results of both trials showed that once–daily teneligliptin provides sustained lowering of glucose throughout the day without increasing the risk of hypoglycemia. Teneligliptin was well tolerated in both studies. Teneligliptin was approved in Japan as 20 mg tablets taken orally once daily for the treatment of patients with T2DM that is not controlled by diet or exercise.

29. TERIFLUNOMIDE (MULTIPLE SCLEROSIS)[241–247]

Class: Dihydroorotate dehydrogenase inhibitor
Country of origin: United States
Originator: Genzyme
First introduction: United States
Introduced by: Genzyme/Sanofi Aventis
Trade name: Aubagio®
CAS registry no: 163451–81–8
Molecular weight: 270.21

In September 2012, the US FDA approved teriflunomide (also referred to as HMR–1726) as a therapy for patients with relapsing forms of multiple sclerosis (MS). Teriflunomide is the second approved oral treatment option for MS, after Gilenya® which was approved in 2010. MS affects more than 400,000 people in the United States and ~2.5 million worldwide.[241] Teriflunomide, which is the active metabolite of leflunomide (a marketed drug for the treatment of rheumatoid and psoriatic arthritis), is a non-competitive and selective inhibitor of dihydroorotate dehydrogenase (DHODH), the rate–limiting enzyme in the *de novo* synthesis of pyrimidines.

Although the net effect of inhibition of DHODH by teriflunomide and its therapeutic effect in MS are not clear, it is hypothesized that by inhibiting *de novo* pyrimidine synthesis, the effector functions of activated lymphocytes are suppressed, thus dampening the effect of an overactive immune system.[242] In preclinical *in vivo* studies in rats, teriflunomide at oral doses of 1, 3, and 10 mg/kg once daily showed a dose-dependent reduction in clinical signs of disease both when given pre- and post-onset of experimental autoimmune encephalomyelitis (EAE), a preclinical model of MS. Histology of the EAE animals indicated reduced demyelination and axonal loss in the spinal cord.[243] A few different approaches have been reported for the synthesis of teriflunomide.[244a,b] In one approach, reaction of leflunomide (prodrug of teriflunomide) with sodium hydroxide in refluxing methanol–water results in the opening of the isoxazole ring leading to the formation of teriflunomide.[244a]

The pharmacokinetic profile of teriflunomide was established based on eleven trials in healthy volunteers and one trial in MS patients.[245] The doses tested ranged from 7 to 100 mg; the median time to reach peak plasma concentrations ranged from 1 to 4 h. The median terminal half-life was 18 and 19 days after repeated doses of 7 and 14 mg, respectively. The clearance and volume of distribution after a single intravenous administration of teriflunomide were 30.5 mL/h and 11 L, respectively. Enterohepatic recirculation is the mechanism responsible for the long half-life of teriflunomide. Teriflunomide is highly protein bound (>99%) and is mainly eliminated via biliary excretion. The safety and efficacy of teriflunomide was assessed in placebo-controlled, double-blind, parallel–group Phase III trials involving a total of 1088 patients suffering from relapsing forms of MS who were randomized to receive once-daily dosing of placebo ($n=363$), teriflunomide 7 mg ($n=366$), or teriflunomide 14 mg ($n=359$) for 108 weeks.[246] The annualized relapse rate (primary end point) was reduced by 31% in patients treated with 7 or 14 mg of teriflunomide compared with the placebo-treated patients. Teriflunomide-treated patients also experienced a delay in the progression of disability (secondary end point) relative to placebo, with progression occurring in 27.3% of patients with placebo, 21.7% with teriflunomide 7 mg ($P=0.08$), and 20.2% with teriflunomide 14 mg ($P=0.03$). Magnetic resonance imaging end points were also evaluated. A significantly lower change in total lesion volume from baseline in the 7- and 14-mg groups compared with the placebo group was noted. Headache, ALT increase, alopecia, diarrhea, influenza, nausea, and paresthesia were the most common AEs observed during teriflunomide clinical trials.

Because of hepatotoxicity and teratogenicity risks, teriflunomide carries a black-box warning. The recommended dose of teriflunomide is 7 or 14 mg administered orally once daily.[247]

30. TOFACITINIB (ANTIARTHRITIC)[248–255]

Class: Janus kinase inhibitor
Country of origin: United States
Originator: Pfizer
First introduction: United States
Introduced by: Pfizer
Trade name: Xeljanz®
CAS registry no: 540737-29-9 (citrate salt) and 477600-75-2 (free base)
Molecular weight: 504.5 (citrate salt) and 312.4 (free base)

In November 2012, the US FDA approved tofacitinib (also referred to as CP-690550) for the treatment of adult patients with moderate to severely active rheumatoid arthritis (RA) who have had an inadequate response or intolerance to methotrexate. The prevalence of RA is estimated to be ∼0.3–1% around the world, with ∼1% of adults in northern Europe and North America affected.[248] Genetic, environmental, and hormonal factors are thought to contribute to RA, while the disease itself is characterized by synovial inflammation and destruction of joint cartilage and bone. Therapeutic paradigms for RA involve treatment with disease-modifying antirheumatic drugs (DMARDs) such as methotrexate followed by injectable biologics such as etanercept, anakinra, tocilizumab, abatacept, and rituximab. Tofacitinib is the first small molecule kinase inhibitor approved for the treatment of RA. Tofacitinib is an inhibitor of the four subtypes of Janus kinase (JAK): JAK1, JAK2, JAK3, and Tyk2. The JAKs are intracellular, nonreceptor tyrosine kinases that play important roles in the signal transduction pathway of many cytokines (e.g., interleukins 2, 4, 7, 9, 15, and 21) and are involved in the propagation of inflammation in RA. Tofacitinib acts by inhibiting the phosphorylation and activation of signal transducers and activators of transcription (STATs), thereby suppressing the production of inflammatory mediators in joint tissue. At the enzyme level, tofacitinib inhibits JAKs 1, 2, 3, and Tyk2 with IC_{50s} of 3.2, 4.1,

1.6, and 34 nM, respectively.[249] At the cellular level, tofacitinib inhibits the *in vitro* activities of JAK1/JAK2, JAK1/JAK3, and JAK2/JAK2 combinations with IC_{50s} of 406, 56, and 1377 nM, respectively. Preclinical *in vivo* studies in an established adjuvant-induced arthritis model indicated that tofacitinib administered orally to rats twice daily starting on day 14 resulted in dose-dependent inhibition of arthritis ($ED_{50} = 0.55$ mg/kg).[250] A kilogram -scale synthesis of the key *cis*-3-methylamino-4-methylpiperidine intermediate via hydrogenation of a substituted pyridine derivative followed by resolution using di-*p*-toluoyl-L-tartaric acid been reported.[251] The cyanoacetamide moiety is installed employing a DBU catalyzed amidation of ethyl cyanoacetate.[252]

The pharmacokinetic profile of tofacitinib was established from pooled data of trials in healthy volunteers, RA patients, patients with Crohn's disease, patients with ulcerative colitis, and patients with stable renal transplants at different dose ranges and frequency of administration (BID and QD). Tofacitinib is rapidly absorbed with a T_{max} of ~0.5 h and a half-life of ~3 h. The absolute oral bioavailability of tofacitinib is 74% and the volume of distribution after intravenous administration is 87 L. Protein binding for tofacitinib is ~40% and hepatic metabolism contributes to nearly 70% of the clearance of the drug. The safety and efficacy of tofacitinib was evaluated in seven clinical trials in adult patients with moderate to severely active RA. Study 1032 (Step study) enrolled 399 participants who had a previous inadequate response to Tumor Necrosis Factor (TNF) inhibitor therapy.[253] Patients were randomly assigned to tofacitinib 5 or 10 mg orally twice daily or placebo and received background methotrexate therapy. At 3 months, ACR20 (an American College of Rheumatology scale) was achieved in 41.7% of tofacitinib 5-mg patients ($P < 0.05$) and 48.1% of tofacitinib 10-mg patients ($P < 0.0001$), compared with 24.4% of placebo patients. Study 1045 (Solo study) enrolled 611 participants who had previous inadequate response to DMARD therapy.[254] Patients were randomly assigned to either placebo or tofacitinib 5 or 10 mg orally twice daily as monotherapy, with no background treatment. At 3 months, ACR20 was achieved in 59.8% and 65.7% of tofacitinib 5- and 10-mg patients, respectively ($P < 0.001$, both groups), compared with 26.7% of placebo patients. Upper respiratory tract infections, headache, diarrhea, and nasopharyngitis were the most common AEs observed during tofacitinib clinical trials. The recommended dose of tofacitinib is 5 mg administered orally twice daily either as monotherapy or in combination with methotrexate or other nonbiologic DMARDs.[255]

31. VISMODEGIB (ANTICANCER)[256–263]

Class: Hedgehog pathway inhibitor
Country of origin: United States
Originator: Curis/Genentech
First introduction: United States
Introduced by: Genentech (member of the Roche group)
Trade name: Erivedge™
CAS registry no: 879085-55-9
Molecular weight: 421.30

In January 2012, the US FDA approved vismodegib (also referred to as GDC-0449) for the treatment of adults with metastatic basal cell carcinoma (BCC), with locally advanced BCC that has recurred following surgery or who are not candidates for surgery or radiation. Skin cancer is the most common form of cancer with ~3.5 million cases diagnosed annually.[256] Melanoma is a lethal form of skin cancer accounting for <5% of all skin cancers. BCC accounts for ~80% of nonmelanoma skin cancers and is associated with a risk of developing metastatic disease of <0.55%.[257] First-line therapies for the treatment of BCC include surgery, radiation therapy, photodynamic therapy or treatment with topical imiquimod or 5-fluorouracil. However, for metastatic disease, where the median survival rate is 6 months to 3.6 years, there was no FDA-approved therapy until vismodegib. Aberrant signaling of the hedgehog (Hh) pathway is the key molecular event responsible for BCC. In patients with Gorlin's syndrome (who suffer from BCC) a loss-of-function mutation of Patched 1 gene (PTCH1) on chromosome 9 predisposes them to BCC. PTCH1 is the receptor for Hh ligands and is a negative regulator of the seven transmembrane receptor SMO. Mutations in the Hh pathway that inactivate PTCH1 can cause uncontrolled SMO signaling resulting in abnormal proliferation of cells, leading to BCC and several other cancers.[258] Vismodegib inhibits the Hh signaling pathway by functioning as an antagonist of SMO thereby inhibiting the activation of Hedgehog target genes, resulting in decreased downstream production of proliferation factors. The IC_{50} of vismodegib in a Hedgehog-responsive cell line derived from human embryonic palatal mesenchyme cells was 2.8 nM.[4] In preclinical *in vivo* studies, vismodegib at 12.5 mg/kg (bid) caused complete regression of tumors in a Hh pathway dependent medulloblastoma allograft model generated from Ptch+/− mice.[259] A synthesis of vismodegib starting from 2-chloro-5-nitro aniline and employing a Negishi coupling with 2-pyridyl zinc iodide as a key step has been reported.[262]

The pharmacokinetics of vismodegib was evaluated in patients with locally advanced or metastatic solid tumors. At a single oral dose of 150 mg, the C_{max} of 3.58 μmol/L was reached in 2.43 days and the mean AUC was 322 μmol h/L.[260] The mean volume of distribution was 16.4–26.8 L[261] and the estimated elimination half-life was 4 days after continuous once-daily dosing and 12 days after a single dose.[262] The safety and efficacy of vismodegib was evaluated in a single arm, open-label, multicenter, nonrandomized, two-cohort Phase II trial in 104 patients with either metastatic BCC ($n=33$) or locally advanced BCC ($n=71$) at an oral dose of 150 mg.[263] In the 33 patients with metastatic BCC, the objective response rate (primary endpoint) was 30% ($P=0.001$) and in the 63 patients with locally advanced BCC, the response rate was 43% ($P<0.001$) with complete responses in 13 patients. The median duration of response in both cohorts was 7.6 months. Muscle spasms, alopecia, dysgeusia, weight loss, fatigue, nausea, diarrhea, decreased appetite, constipation, arthralgias, vomiting, and ageusia ($\geq 10\%$) were the most common AEs observed during vismodegib clinical trials. Vismodegib carries a black-box warning for embryo-fetal death and severe birth defects. The recommended dose of vismodegib is 150 mg administered orally once daily.

REFERENCES

1. The collection of new therapeutic entities first launched in 2012 originated from the following sources: Prous Integrity Database; Thomson-Reuters Pipeline Database; The Pink Sheet; Drugs@FDA Website; FDA News Releases.
2. http://cen.acs.org/articles/91/i5/New-Drug-Approvals-Hit-16.html.
3. http://www.accessdata.fda.gov/drugsatfda_docs/label/2012/203155s000lbl.pdf.
4. http://www.accessdata.fda.gov/drugsatfda_docs/label/2012/203100s000lbl.pdf.
5. http://www.accessdata.fda.gov/drugsatfda_docs/label/2012/201281s000lbl.pdf.
6. http://www.accessdata.fda.gov/drugsatfda_docs/label/2012/202270s000lbl.pdf.
7. http://www.accessdata.fda.gov/drugsatfda_docs/label/2013/022580s004lbl.pdf.
8. Yang, L.; Rieves, D.; Ganley, C. N. Engl. J. Med. 2012, 367, 885.
9. http://www.accessdata.fda.gov/drugsatfda_docs/label/2012/202008s000lbl.pdf.
10. Raherison, C.; Girodet, P.-O. Eur. Respir. Rev. 2009, 18, 213.
11. American Lung Association, State of Lung Disease in Diverse Communities, http://www.lung.org/assets/documents/publications/solddc-chapters/copd.pdf, 2010; Centers for Disease Control, http://www.cdc.gov/nchs/data/databriefs/db63.htm.
12. Cazzola, M.; Page, C. P.; Calzetta, L.; Matera, M. G. Pharmacol. Rev. 2012, 64, 450.
13. Cazzola, M.; Page, C.; Matera, M. G. Pulm. Pharmacol. Ther. 2013, 26, 307.
14. Frampton, J. E. Drugs 2012, 72, 1999.
15. Prat, M.; Fernández, D.; Buil, M. A.; Crespo, M. I.; Casals, G.; Ferrer, M.; Tort, L.; Castro, J.; Monleón, J. M.; Gavaldà, A.; Miralpeix, M.; Ramos, I.; Doménech, T.; Vilella, D.; Antón, F.; Huerta, J. M.; Espinosa, S.; López, M.; Sentellas, S.;

González, M.; Albertí, J.; Segarra, V.; Cárdenas, A.; Beleta, J.; Ryder, H. *J. Med. Chem.* **2009**, *52*, 5076.

16. Gavaldà, A.; Miralpeix, M.; Ramos, I.; Otal, R.; Carreño, C.; Viñals, M.; Doménech, T.; Carcasona, C.; Reyes, B.; Vilella, D.; Gras, J.; Cortijo, J.; Morcillo, E.; Llenas, J.; Ryder, H.; Beleta, J. *J. Pharm. Exp. Ther.* **2009**, *331*, 740.

17. Jones, P. W.; Rennard, S. I.; Agusti, A.; Chanez, P.; Magnussen, H.; Fabbri, L.; Donohue, J. F.; Bateman, E. D.; Gross, N. J.; Lamarca, R.; Caracta, C.; Gil, E. G. *Respir. Res.* **2011**, *12*, 55.

18. Kerwin, E. M.; D'Urzo, A. D.; Gelb, A. F.; Lakkis, H.; Gil, E. G.; Caracta, C. F. ACCORD I Study Investigators. *COPD* **2012**, *9*, 90.

19. Jones, P. W.; Singh, D.; Bateman, E. D.; Agusti, A.; Lamarca, R.; de Miguel, G.; Segarra, R.; Caracta, C.; Gil, E. G. *Eur. Respir. J.* **2012**, *40*, 830.

20. http://www.who.int/diabetes/en/.

21. Neville, S. E.; Boye, K. S.; Montgomery, W. S.; Iwamoto, K.; Okamura, M.; Hayes, R. P. *Diabetes Metab. Res. Rev.* **2009**, *25*, 705.

22. Scheen, A. J. *Expert Opin. Pharmacother.* **2012**, *13*, 81.

23. Kato, N.; Oka, M.; Murase, T.; Yoshida, M.; Sakairi, M.; Yamashita, S.; Yasuda, Y.; Yoshikawa, A.; Hayashi, Y.; Makino, M.; Takeda, M.; Mirensha, Y.; Kakigami, T. *Bioorg. Med. Chem.* **2011**, *19*, 7221.

24. Furuta, S.; Tamura, M.; Hirooka, H.; Mizuno, Y.; Miyoshi, M.; Furata, Y. *Eur. J. Drug Metab. Pharmacokinet.* **2013**, *38*, 87.

25. Nakaya, K.; Kubota, N.; Takamoto, I.; Kubota, T.; Katsuyama, H.; Sato, H.; Tokuyama, K.; Hashimoto, S.; Goto, M.; Jomori, T.; Ueki, K.; Kadowaki, T. *Metabolism* **2013**, *62*, 939.

26. Furuta, S.; Smart, C.; Hackett, A.; Benning, R.; Warrington, S. *Xenobiotica* **2013**, *43*, 432.

27. Sunami, Y.; Yoshioka, N.; Hayashi, I.; Ishida, T. American Diabetes Association 67th Scientific Session, 2007; Poster 0482-P, http://professional.diabetes.org/Content/Post ers/2007/p0482-P.pdf.

28. Uchino, H.; Kaku, K. *Jpn. Pharmacol. Ther.* **2012**, *40*, 859.

29. Kaku, K. *Jpn. Pharmacol. Ther.* **2012**, *40*, 733.

30. American Cancer Society, Cancer Facts & Figures 2010. www.cancer.org/acs/groups/content/@epidemiologysurveilance/documents/document/acspc-026238.pdf.

31. Hu-Lowe, D.; Zou, H. Y.; Grazzini, M. L.; Hallin, M. E.; Wickman, G. R.; Amundson, K.; Chen, J. H.; Rewolinski, D. A.; Yamazaki, S.; Wu, E. Y.; McTigue, A.; Murray, B. W.; Kania, R. S.; O'Connor, P.; Shalinsky, D. R.; Bender, S. L. *Clin. Cancer Res.* **2008**, *14*, 7272.

32. Kania, R. S. In *Kinase Inhibitor Drugs*; Li, R., Stafford, J.A., Eds.; Wiley: Hoboken, NJ, 2009; p 167.

33. Hu-Lowe, D.; Heller, D.; Brekken, J.; Feeley, R.; Amundson, K.; Haines, Wi; Troche, G.; Kim, Y.; Gonzalez, D.; Herrman, M.; Batugo, W.; Vekich, S.; Kania, R.; McTigue, M.; Gregory, S.; Bender, S.; Shalinsky, D. In: *93rd Annual Meeting of the American Association for Cancer Research, San Francisco, CA,* 2002. Abstract 5357.

34. (a) Kalia, R. S.; Bender, S. L.; Borchardt, A. J.; Cripps, S. J.; Hua, Y.; Johnson, T. O.; Johnson, M. D.; Luu, H. T.; Palmer, C. L.; Reich, S. H.; Tempczyk-Russell, A. M.; Teng, M.; Thomas, C.; Varney, M. D.; Wallace, M. B.; Collins, M. R. U.S. Patent 6,534,524, 2003; (b) Singer, R. In: *244th ACS National Meeting Philadelphia, PA* 2012. Abstract ORGN-273

35. Rini, B. I.; Escudier, B.; Tomczak, P.; Kaprin, A.; Szczylik, C.; Hutson, T. E.; Michaelson, M. D.; Gorbunova, V. A.; Gore, M. E.; Rusakov, I. G.; Negrier, S.; Ou, Y. C.; Castellano, D.; Lim, H. Y.; Uemura, H.; Tarazi, J.; Cella, D.; Chen, C.; Rosbrook, B.; Kim, S.; Motzer, R. J. *Lancet* **2011**, *378*, 1931.

36. http://www.who.int/tb/publications/factsheet_global.pdf—WHO global tuberculosis report, 2012.
37. Villemagne, B.; Crauste, C.; Flipo, M.; Baulard, A. R.; Déprez, B.; Willand, N. *Eur. J. Med. Chem.* **2012**, *51*, 1.
38. Almeida Da Silva, P. E.; Palomino, J. C. *J. Antimicrob. Chemother.* **2011**, *66*, 1417.
39. Andries, K.; Verhasselt, P.; Guillemont, J.; Göhlmann, H. W. H.; Neefs, J.-M.; Winkler, H.; Van Gestel, J.; Timmerman, P.; Zhu, M.; Lee, E.; Williams, P.; de Chaffoy, D.; Huitric, E.; Hoffner, S.; Cambau, E.; Truffot-Pernot, C.; Lounis, N.; Jarlier, V. *Science* **2005**, *307*, 223.
40. Guillemont, J.; Meyer, C.; Poncelet, A.; Bourdrez, X.; Andries, K. *Future Med. Chem.* **2011**, *3*, 1345.
41. Haagsma, A. C.; Podasca, I.; Koul, A.; Andries, K.; Guillemont, J.; Lill, H.; Bald, D. *PLoS One* **2011**, *6*, e23575.
42. Rouan, M.-C.; Lounis, N.; Gevers, T.; Dillen, L.; Gillssen, R.; Raoof, A.; Andries, K. *Antimicrob. Agents Chemother.* **2012**, *56*, 1444.
43. http://www.fda.gov/downloads/AdvisoryCommittees/CommitteesMeetingMaterials/Drugs/Anti-InfectiveDrugsAdvisoryCommittee/UCM329260.pdf.
44. Diacon, A. H.; Pym, A.; Grobusch, M.; Patientia, R.; Rustomjee, R.; Page-Shipp, L.; Pistorius, C.; Krause, R.; Bogoshi, M.; Churchyard, G.; Venter, A.; Allen, J.; Palomino, J. C.; De Marez, T.; van Heeswijk, R. P. G.; Lounis, N.; Meyvisch, P.; Verbeeck, J.; Parys, W.; de Beule, K.; Andries, K.; Mc Neeley, D. F. *N. Engl. J. Med.* **2009**, *360*, 2397.
45. Diacon, A. H.; Donald, P. R.; Pym, A.; Grobusch, M.; Patienta, R. F.; Mahanyele, R.; Bantubani, N.; Narasimooloo, R.; De Marez, T.; van Heeswijk, R.; Lounis, N.; Meyvisch, P.; Andries, K.; McNeeley, D. F. *Antimicrob. Agents Chemother.* **2012**, *56*, 3271.
46. http://seer.cancer.gov/statfacts/html/cmyl.html.
47. Boschelli, D. H.; Ye, F.; Wang, Y. D.; Dutia, M.; Johnson, S. L.; Wu, B.; Miller, K.; Powell, D. W.; Yaczko, D.; Young, M.; Tischler, M.; Arndt, K.; Discafani, C.; Etienne, C.; Gibbons, J.; Grod, J.; Lucas, J.; Weber, J. M.; Boschelli, F. *J. Med. Chem.* **2001**, *44*, 3965.
48. Boschelli, F.; Arndt, K.; Gambacorti-Passerini, C. *Eur. J. Cancer* **2010**, *46*, 1781.
49. Withbroe, G. J.; Seadeek, C.; Girard, K. P.; Guinness, S. M.; Vanderplas, B. C.; Vaidyanathan, R. *Org. Process Res. Dev.* **2013**, *17*, 500.
50. Abbas, R.; Hug, B. A.; Leister, C.; Gaaloul, M. E.; Chalon, S.; Sonnichsen, D. *Cancer Chemother. Pharmacol.* **2012**, *69*, 221.
51. Khoury, H. J.; Cortes, J. E.; Kantarjian, H. M.; Gambacorti-Passerini, C.; Baccarani, M.; Kim, D.-W.; Zaritskey, A.; Countouriotis, A.; Besson, N.; Leip, E.; Kelly, V.; Brummendorf, T. H. *Blood* **2012**, *119*, 3403.
52. US National Cancer Institute, NIH website: http://www.cancer.gov/cancertopics/pdq/treatment/thyroid/HealthProfessional.
53. Nagilla, M.; Brown, R. L.; Cohen, E. E. W. *Adv. Ther.* **2012**, *29*, 925.
54. Yakes, F. M.; Chen, J.; Tan, J.; Yamaguchi, K.; Shi, Y.; Yu, P.; Qian, F.; Chu, F.; Bentzien, F.; Cancilla, B.; Orf, J.; You, A.; Laird, A. D.; Engst, S.; Lee, L.; Lesch, J.; Chou, Y.-C.; Joly, A. H. *Mol. Cancer Ther.* **2011**, *10*, 2298.
55. Peters, S.; Adjei, A. A. *Nat. Rev. Clin. Oncol.* **2012**, *9*, 314.
56. Bannen, L. C.; Chan, D. S.-M.; Forsyth, T. P.; Khoury, R. G.; Leahy, J. W.; Mac, M. B.; Mann, L. W.; Nuss, J. M.; Parks, J. J.; Wang, Y.; Xu, W. U.S. Patent 7,579,473 B2, 2009.
57. Wilson, J. A. WO Patent Application 2012/109510A1, 2012.
58. Kurzrock, R.; Camacho, L.; Hong, D.; Ng, C.; Janisch, L.; Ratain, M. J.; Salgia, R. *Eur. J. Cancer Suppl.* **2006**, *4*, 124–125, Abst. 405.
59. Giusti, R. M. US FDA Center for Drug Evaluation and Research, NDA 203756 Medical Review, Nov 4, 2012.

60. Cometriq prescribing information, www.accessdata.fda.gov/drugsatfda_docs/label/2012/203756lbl.pdf.
61. Bowles, D. W.; Kessler, E. R.; Jimeno, A. *Drugs Today* **2011**, *47*, 857.
62. American Cancer Society. Multiple Myeloma. Available at http://www.cancer.org/Cancer/MultipleMyeloma/DetailedGuide/multiple-myeloma-key-statistics.
63. Parlati, F.; Lee, S. J.; Aujay, M.; Suzuki, E.; Levitsky, K.; Lorens, J. B.; Micklem, D. R.; Ruurs, P.; Sylvain, C.; Lu, Y.; Shenk, K. D.; Bennett, M. K. *Blood* **2009**, *114*, 3439.
64. Moreau, P.; Richardson, P. G.; Cavo, M.; Orlowski, R. Z.; San Miguel, J. F.; Palumbo, A.; Harousseau, J.-L. *Blood* **2012**, *120*, 947.
65. (a) Smyth, M. S.; Laidig, G. J. Patent Application WO2006/017842, 2006; (b) Sin, N.; Kim, K. B.; Elofsson, M.; Meng, L.; Auth, H.; Kwok, B. H.; Crews, C. M. *Bioorg. Med. Chem. Lett.* **1999**, *9*, 2283.
66. O'Connor, O. A.; Stewart, A. K.; Vallone, M.; Molineaux, C. J.; Kunkel, L. A.; Gerecitano, J. F.; Orlowski, R. Z. *Clin. Cancer Res.* **2009**, *15*, 7085.
67. Vij, R.; Wang, M.; Kaufman, J. L.; Lonial, S.; Jakubowiak, A. J.; Stewart, A. K.; Kukreti, V.; Jagannath, S.; McDonagh, K. T.; Alsina, M.; Bahlis, N. J.; Reu, F. J.; Gabrail, N. Y.; Belch, A.; Matous, J. V.; Lee, P.; Rosen, P.; Sebag, M.; Vesole, D. H.; Kunkel, L. A.; Wear, S. M.; Wong, A. F.; Orlowski, R. Z.; Siegel, D. S. *Blood* **2012**, *119*, 5661.
68. Abdelghany, S.; Adib, R.; Patel, K. *Formulary* **2012**, *47*, 282.
69. US FDA, Centers for Disease Control and Prevention, www.cdc.gov/globalaids/Global-HIV-AIDS-at-CDC/default.html. U.S. President's Emergency Plan for AIDS Relief fact sheet (accessed March 2013).
70. Siddiqui, U.; Bini, E. J.; Chandarana, K.; Leong, J.; Ramsetty, S.; Schiliro, D.; Poles, M. *J. Clin. Gastroenterol.* **2007**, *41*, 484.
71. Klein, R.; Struble, K. US FDA press release, December 31, 2012, Silver Spring, MD. http://www.fda.gov/NewsEvents/Newsroom/PressAnnouncements/ucm333701.htm.
72. Gabriel, S. E.; Davenport, S. E.; Steagall, R. J.; Vimal, V.; Carlson, T.; Rozhon, E. J. *Am. J. Physiol.* **1999**, *276*, G58.
73. Tradtrantip, L.; Namkung, W.; Verkman, A. S. *Mol. Pharmacol.* **2010**, *77*, 69.
74. DiCesare, D.; DuPont, H. L.; Mathewson, J. J.; Ashley, D.; Martinez-Sandoval, F.; Pennington, J. E.; Porter, S. B. *Am. J. Gastroenterol.* **2002**, *97*, 2585.
75. Beitz, J.G. US FDA Fulyzaq label, reference ID 3238051, 2012, Silver Spring, MD, www.accessdata.fda.gov/drugsatfda_docs/label/2012/202292s000lbl.pdf.
76. Liscinsky, M. US FDA News Release ucm333701, Dec 31, 2012, Silver Spring, MD.
77. Crutchley, R. D.; Miller, J.; Garey, K. W. *Ann. Pharmacother.* **2010**, *44*, 878.
78. International Diabetes Federation, *IDF Diabetes Atlas*, 5th ed.; International Diabetes Federation: Brussels, Belgium, 2012.
79. American Diabetes Association website: http://www.diabetes.org/living-with-diabetes/treatment-and-care/medication/oral-medications/what-are-my-options.html.
80. Rotenstein, L. S.; Kozak, B. M.; Shivers, J. P.; Yarchoan, M.; Close, J.; Close, K. L. *Clin. Diabetes* **2012**, *30*, 44.
81. Meng, W.; Ellsworth, B. A.; Nirschl, A. A.; McCann, P. J.; Patel, M.; Girotra, R. N.; Wu, G.; Sher, P. M.; Morrison, E. P.; Biller, S. A.; Zahler, R.; Deshpande, P. P.; Pullockaran, A.; Hagan, D. L.; Morgan, N.; Taylor, J. R.; Obermeier, M. T.; Humphreys, W. G.; Khanna, A.; Discenza, L.; Robertson, J. G.; Wang, A.; Han, S.; Wetterau, J. R.; Janovitz, E. B.; Flint, O. P.; Whaley, J. M.; Washburn, W. N. *J. Med. Chem.* **2008**, *51*, 1145.
82. Washburn, W. N. *J. Med. Chem.* **2009**, *52*, 1785.
83. Kasichayanula, S.; Liu, X.; Zhang, W.; Pfister, M.; LeCreta, F. P.; Boulton, D. W. *Clin. Ther.* **2011**, *33*, 1798.

84. Obermeier, M.; Yao, M.; Khanna, A.; Koplowitz, B.; Zhu, M.; Li, W.; Komoroski, B.; Kasichayanula, S.; Discenza, L.; Washburn, W.; Meng, W.; Ellsworth, B. A.; Whaley, J. M.; Humphreys, W. G. *Drug Metab. Dispos.* **2010**, *38*, 405.
85. Washburn, W. N.; Poucher, S. M. *Expert Opin. Investig. Drugs* **2013**, *22*, 463.
86. Forxiga prescribing label, European Medicines Agency, http://www.ema.europa. eu/ema/index.jsp?curl=pages/medicines/human/medicines/002322/human_med_ 001546.jsp&mid=WC0b01ac058001d124 (last accessed April 2013).
87. American Cancer Society. Cancer Facts & Figures 2013. American Cancer Society: Atlanta, 2013. http://www.cancer.org/acs/groups/content/@epidemiologysurveilance/ documents/document/acspc-036845.pdf.
88. Mukherji, D.; Eichholz, A.; De Bono, J. S. *Drugs* **2012**, *72*, 1011.
89. Tran, C.; Ouk, S.; Clegg, N. J.; Chen, Y.; Watson, P. A.; Arora, V.; Wongvipat, J.; Smith-Jones, P. M.; Yoo, D.; Kwon, A.; Wasielewska, T.; Welsbie, D.; Chen, C. D.; Higano, C. S.; Beer, T. M.; Hung, D. T.; Scher, H. I.; Jung, M. E.; Sawyers, C. L. *Science* **2009**, *324*, 787.
90. Jung, M. E.; Ouk, S.; Yoo, D.; Sawyers, C. L.; Chen, C.; Tran, C.; Wongvipat, J. *J. Med. Chem.* **2010**, *53*, 2779.
91. Scher, H. I.; Beer, T.; Higano, C.; Efstathiou, E.; Anand, A.; Hirmand, M.; Hung, D.; Steve, L.; Fleisher, M.; Sawyers, C. *J. Urol.* **2009**, *181*, 230, Abst. 641.
92. US FDA Center for Drug Evaluation and Research, application number 203415, Clinical Pharmacology and Biopharmaceutics Review.
93. Scher, H. I.; Fizazi, K.; Saad, F.; Taplin, M.-E.; Sternberg, C. N.; Miller, K.; de Wit, R.; Mulders, P.; Chi, K. N.; Shore, N. D.; Armstrong, A. J.; Flaig, T. W.; Flechon, A.; Mainwaring, P.; Fleming, M.; Hainsworth, J. D.; Hirmand, M.; Selby, B.; Seely, L.; de Bono, J. S. *N. Engl. J. Med.* **2012**, *367*, 1187.
94. Pazdur, R. US FDA, NDA 203415 approval letter, August 31, 2012, Silver Spring, MD.
95. Berman, B.; Amini, S. *Expert Opin. Pharmacother.* **2012**, *13*, 1847.
96. Alam, M. *Adv. Stud. Med.* **2006**, *6*(8A), S785.
97. Rosen, R. H.; Gupta, A. K.; Tyring, S. K. *J. Am. Acad. Dermatol.* **2012**, *66*, 486.
98. Keating, G. M. *Drugs* **2012**, *72*, 2397.
99. Liang, X.; Grue-Sørensen, G.; Petersen, A. K.; Högberg, T. *Synlett.* **2012**, *23*, 2647.
100. Ramsey, J. R.; Suhrbier, A.; Aylward, J. H.; Ogbourne, S.; Cozzi, S. J.; Poulsen, M. G.; Baumann, K. C.; Welburn, P.; Redlich, G. L.; Parsons, P. G. *Br. J. Dermatol.* **2011**, *164*, 633.
101. Hampson, P.; Wang, K.; Lord, J. M. *Drugs Future* **2005**, *30*, 1003.
102. Lebwohl, M.; Swanson, N.; Anderson, L. L.; Melgaard, A.; Xu, Z.; Berman, B. *N. Engl. J. Med.* **2012**, *366*, 1010.
103. Martin, G.; Swanson, N. *J. Am. Acad. Dermatol.* **2013**, *68*, S39.
104. http://www.accessdata.fda.gov/drugsatfda_docs/appletter/2012/203188Orig1s001ltr. pdf (last accessed 19 Nov 2012).
105. Jones, A. M.; Helm, J. M. *Drugs* **2009**, *69*, 1903.
106. http://www.lung.org/assets/documents/publications/solddc-chapters/cf.pdf.
107. http://www.cff.org/treatments/Therapies/ (last accessed 19 Nov 2012).
108. Song, J. C.; Chiu, H.; Yoon, J. *Formulary* **2012**, *47*, 132.
109. Van Goor, F.; Straley, K. S.; Cao, D.; Gonzalez, J.; Hadida, S.; Hazelwood, A.; Joubran, J.; Knapp, T.; Makings, L. R.; Miller, M.; Nueberger, T.; Olson, E.; Panchenko, V.; Rader, J.; Singh, A.; Stack, J. H.; Tung, R.; Grootenhuis, P. D. J.; Negulescu, P. *Am. J. Physiol. Lung Cell. Mol. Physiol.* **2006**, *290*, L1117.
110. Van Goor, F.; Hadida, S.; Grootenhuis, P. D. J.; Burton, B.; Cao, D.; Neuberger, T.; Turnbull, A.; Singh, A.; Joubran, J.; Hazlewood, A.; Zhou, J.; McCartney, J.;

Arumugam, V.; Decker, C.; Yang, J.; Young, C.; Olson, E. R.; Wine, J. J.; Frizzell, R. A.; Ashlock, M.; Negulescu, P. *Proc. Natl. Acad. Sci.* **2009**, *106*, 18825.

111. Arekar, S. G.; Johnston, S. C.; Krawiec, M.; Madek, A.; Mudunuri, P.; Sullivan, M. J. U.S. Patent Application 2011/0230519, 2011.

112. http://pi.vrtx.com/files/uspi_ivacaftor.pdf (last accessed 28 Nov 2012).

113. Ramsey, B. W.; Davies, J.; McElvaney, N. G.; Tullis, E.; Bell, S. C.; Drevinek, P.; Griese, M.; McKone, E. F.; Wainwright, C. E.; Konstan, M. W.; Moss, R.; Ratjen, F.; Sermet-Gaudelus, I.; Rowe, S. M.; Dong, Q.; Rodriguez, S.; Yen, K.; Ordonez, C.; Elborn, J. S. *N. Engl. J. Med.* **2011**, *365*, 1663.

114. McCallum, I. J. D.; Ong, S.; Mercer-Jones, M. *Br. Med. J.* **2009**, *338*, b831.

115. Busby, R. W.; Bryant, A. P.; Bartolini, W. P.; Cordero, E. A.; Kessler, M. M.; Pierce, C. M.; Tobin, J. V.; Wakefield, J. D.; Fretzen, A.; Kurtz, C. B.; Currie, M. G. *Drug Metab. Rev.* **2006**, *38*(Suppl 2), 96.

116. Busby, R. W.; Bryant, A. P.; Cordero, E. A.; Kessler, M. M.; Pierce, C. M.; Tobin, J. V.; Cohen, M. B.; Currie, M. G. *Gastroenterology* **2005**, *128*, T1136.

117. Busby, R. W.; Bryant, A. P.; Bartolini, W. P.; Cordero, E. A.; Hannig, G.; Kessler, M. M.; Mahajan-Miklos, S.; Pierce, C. M.; Solinga, R. M.; Sun, L. J.; Tobin, J. V.; Kurtz, C. B.; Currie, M. G. *Eur. J. Pharmacol.* **2010**, *649*, 328.

118. Bryant, A. P.; Busby, R. W.; Bartolini, W. P.; Cordero, E. A.; Hannig, G.; Kessler, M. M.; Pierce, C. M.; Solinga, R. M.; Tobin, J. V.; Mahajan-Miklos, S.; Cohen, M. B.; Kurtz, C. B.; Currie, M. G. *Life Sci.* **2010**, *86*, 760.

119. (a) Currie, M. G.; Mahajan-Miklos, S.; Fretzen, A.; Sun, L. J.; Milne, G. T.; Norman, T.; Kurtz, C. Patent Application WO2005/087797, 2005; (b) Currie, M. G.; Mahajan-Miklos, S.; Fretzen, A.; Sun, L. J.; Kurtz, C.; Milne, T. G.; Norman, T.; Roberts, S.; Sullivan, K. E. Patent Application WO 2007/022531, 2007.

120. Kurtz, C. B.; Fitch, D.; Busby, R. W.; Fretzen, A.; Geis, S.; Currie, M. G. *Gastroenterology* **2006**, *130*(4, Suppl 2), Abst. 132.

121. (a) Chey, W. D.; Lembo, A.; MacDougall, J. E.; Lavins, B. J.; Schneier, H.; Johnston, J. M. *Gastroenterology* **2011**, *140*, S135; (b) Rao, S.; Lembo, A.; Shiff, S. J.; Shi, K.; Johnston, J. M.; Schneier, H. *Gastroenterology* **2011**, *140*, S138.

122. Lembo, A. J.; Schneier, H. A.; Shiff, S. J.; Kurtz, C. B.; MacDougall, J. E.; Jia, X. D.; Shao, J. Z.; Lavins, B. J.; Currie, M. G.; Fitch, D. A.; Jeglinski, B. I.; Eng, P.; Fox, S. M.; Johnston, J. M. *N. Engl. J. Med.* **2011**, *365*, 527.

123. Raal, F. J.; Santos, R. D. *Atherosclerosis* **2012**, *223*, 262.

124. Cuchel, M.; Meagher, E. A.; Theron, H. d. T.; Blom, D. J.; Marais, A. D.; Hegele, R. A.; Averna, M. R.; Sirtori, C. R.; Shah, P. K.; Gaudet, D.; Stefanutti, C.; Vigna, G. B.; Du Plessis, A. M. E.; Propert, K. J.; Sasiela, W. J.; Bloedon, L. T.; Rader, D. J. *Lancet* **2013**, *381*, 40.

125. Wetterau, J. R.; Gregg, R. E.; Harrity, T. W.; Arbeeny, C.; Cap, M.; Connolly, F.; Chu, C.-H.; George, R. J.; Gordon, D. A.; Jamil, H.; Jolibois, K. G.; Kunselman, L. K.; Lan, S.-J.; Maccagnan, T. J.; Ricci, B.; Yan, M.; Young, D.; Chen, Y.; Fryszman, O. M.; Logan, J. V. H.; Musial, C. L.; Poss, M. A.; Robl, J. A.; Simpkins, L. M.; Slusarchyk, W. A.; Sulsky, R.; Taunk, P.; Magnin, D. R.; Tino, J. A.; Lawrence, R. M.; Dickson, J. K.; Biller, S. A. *Science* **1998**, *282*, 751.

126. Biller, S. A.; Dickson, J. K.; Lawrence, R. M.; Magnin, D. R.; Poss, M. A.; Robl, J. A.; Sulsky, R. B.; Tino, J. A. U.S. Patent 5,712,279, 1998.

127. Smith, J. P. US FDA Centers for Drug Evaluation and Research, Endocrinologic and Metabolic Drugs Advisory Committee Meeting briefing document, NDA203858. October 17, 2012.

128. US FDA approval letter, December 21, 2012, NDA 203858.

129. www.clinicaltrials.gov, ClinicalTrials.gov Identifier: NCT01760187.

130. http://www.accessdata.fda.gov/drugsatfda_docs/appletter/2012/022529Orig1s000ltr. pdf (last accessed 9 Nov 2012).
131. http://www.who.int/mediacentre/factsheets/fs311/en/ (last accessed 9 Nov 2012).
132. Smith, B. M.; Smith, J. M.; Tsai, J. H.; Schultz, J. A.; Gilson, C. A.; Estrada, S. A.; Chen, R. R.; Park, D. M.; Prieto, E. B.; Gallardo, C. S.; Sengupta, D.; Dosa, P. I.; Covel, J. A.; Ren, A.; Webb, R. R.; Beeley, N. R.; Martin, M.; Morgan, M.; Espitia, S.; Saldana, H. R.; Bjenning, C.; Whelan, K. T.; Grottick, A. J.; Menzaghi, F.; Thomsen, W. J. *J. Med. Chem.* **2008**, *51*, 305.
133. Weigl, U.; Porstmann, F.; Straessler, C.; Ulmer, L.; Koetz, U. U.S. Patent 8,168,782, 2012.
134. Smith, S. R.; Prosser, W. A.; Donahue, D. J.; Morgan, M. E.; Anderson, C. M.; Shanahan, W. R.; APD356-004 Study Group, *Obesity* **2008**, *17*, 494.
135. Bays, H. E. *Expert Rev. Cardiovasc. Ther.* **2009**, *7*, 1429.
136. Bai, B.; Wang, Y. *Drug Des. Devel. Ther.* **2011**, *5*, 1.
137. Fidler, M. C.; Sanchez, M.; Raether, B.; Weissman, N. J.; Smith, S. R.; Shanahan, W. R.; Anderson, C. M.; BLOSSOM Clinical Trial Group. *J. Clin. Endocrinol. Metab.* **2011**, *96*, 3067.
138. Smith, S. R.; Weissman, N. J.; Anderson, C. M.; Sanchez, M.; Chuang, E.; Stubbe, S.; Bays, H.; Shanahan, W. R. *N. Eng. J. Med.* **2010**, *363*, 245.
139. http://us.eisai.com/package_inserts/BelviqPI.pdf (last accessed 9 Nov 2012).
140. Ascani, S.; Zinzani, P. L.; Gherlinzoni, F.; Sabattini, E.; Briskomatis, A.; de Vivo, A.; Piccioli, M.; Fraternali Orcioni, G.; Pieri, F.; Goldoni, A.; Piccaluga, P. P.; Zallocco, D.; Burnelli, R.; Leoncini, L.; Falini, B.; Tura, S.; Pileri, S. A. *Ann. Oncol.* **1998**, *8*, 583.
141. Proietti, F. A.; Carnieiro-Proietti, A. B. F.; Catalan-Soares, B. C.; Murphy, E. L. *Oncogene* **2005**, *24*, 6058.
142. Tobinai, K.; Takahashi, T.; Akinaga, S. *Curr. Hematol. Malig. Rep.* **2012**, *7*, 235.
143. Ishida, T.; Joh, T.; Uike, N.; Yamamoto, K.; Utsunomiya, A.; Yoshida, S.; Saburi, Y.; Miyamoto, T.; Takemoto, S.; Suzushima, H.; Tsukasaki, K.; Nosaka, K.; Fujiwara, H.; Ishitsuka, K.; Inagaki, H.; Ogura, M.; Akinaga, S.; Tomonaga, M.; Tobinai, K.; Ueda, R. *J. Clin. Oncol.* **2012**, *8*, 837.
144. Beck, A.; Reichert, J. M. *MAbs* **2012**, *4*, 419.
145. Subramaniam, J. M.; Whiteside, G.; McKeage, K.; Croxtall, J. C. *Drugs* **2012**, *72*, 1293.
146. Yamamoto, K.; Utsunomiya, A.; Tobinai, K.; Tsukasaki, K.; Uike, N.; Uozumi, K.; Yamaguchi, K.; Yamada, Y.; Hanada, S.; Tamura, K.; Nakamura, S.; Inagaki, H.; Ohshima, K.; Kiyoi, H.; Ishida, T.; Matsushima, K.; Akinaga, S.; Ogura, M.; Tomonaga, M.; Ueda, R. *J. Clin. Oncol.* **2010**, *28*, 1591.
147. http://www.cancer.org/cancer/leukemia-chronicmyeloidcml/detailedguide/leukemia-chronic-myeloid-myelogenous-key-statistic.
148. Cortes, J.; Goldman, J. M.; Hughes, T. *J. Natl. Compr. Canc. Netw.* **2012**, *10*(Suppl 3), S-1.
149. Wetzler, M.; Segal, D. *Curr. Pharm. Design* **2011**, *17*, 59.
150. Allen, E. K.; Holyoake, T. L.; Craig, A. R.; Jørgensen, H. G. *Leukemia* **2011**, *25*, 985.
151. Kim, T. D.; Frick, M.; le Coutre, P. *Expert Opin. Pharmacother.* **2011**, *12*, 2381.
152. Quintas-Cardama, A.; Cortes, J. *Expert Opin. Pharmacother.* **2008**, *9*, 1029.
153. Robin, J.-P.; Dhal, R.; Dujardin, G.; Girodier, L.; Mevellec, L.; Poutot, S. *Tetrahedron Lett.* **1999**, *40*, 2931.
154. Levy, V.; Zohar, S.; Bardin, C.; Vekhoff, A.; Chaoui, D.; Rio, B.; Legrand, O.; Sentenac, S.; Rousselot, P.; Raffoux, E.; Chast, F.; Chevret, S.; Marie, J. P. *Br. J. Cancer* **2006**, *95*, 253.
155. Quintas-Cardama, A.; Kantarjian, H.; Garcia-Manero, G.; O'Brien, S.; Faderl, S.; Estrov, Z.; Giles, F.; Murgo, A.; Ladie, N.; Verstovsek, S.; Cortes, J. *Cancer* **2007**, *109*, 248.

156. Cortes, J.; Lipton, J. H.; Rea, D.; Digumarti, R.; Chuah, C.; Nanda, N.; Benichou, A.-C.; Craig, A. R.; Michallet, M.; Nicolini, F. E.; Kantarjian, H.; On behalf of the Omacetaxine 202 Study Group, *Blood* **2012**, *120*, 2573.

157. Cortes, J. E.; Nicolini, F. E.; Wetzler, M.; Lipton, J. H.; Akard, L. P.; Craig, A.; Nanda, N.; Dial, C.; Benichou, A.-C.; Cairati, K.; Baccarani, M.; Kennealey, G. T.; Kantarjian, H. M. In: *53rd Annual Meeting of the American Society of Hematology*, 2011. Abstract 3761.

158. Feelders, R. A.; Yasothan, U.; Kirkpatrick, P. *Nat. Rev. Drug Discov.* **2012**, *11*, 597.

159. Feelders, R. A.; Hofland, L. J. *J. Clin. Endocrinol. Metab.* **2013**, *98*, 425.

160. http://pituitary.mgh.harvard.edu/cushings.htm.

161. Bruns, C.; Lewis, I.; Briner, U.; Meno-Tetang, G.; Weckbecker, G. *Eur. J. Endocrinol.* **2002**, *146*, 707.

162. Lewis, I.; Rainer, A.; Bauer, W.; Chandramouli, N.; Pless, J.; Oberer, L.; Bovermann, G.; van der Hoek, J.; Boerlin, V.; Lamberts, S. W. J.; Schmid, H. A.; Weckbecker, G.; Bruns, C. *Chimia* **2004**, *58*, 222.

163. Arnaldi, G.; Boscaro, M. *Expert Opin. Investig. Drugs* **2010**, *19*, 889.

164. Petersenn, S.; Hu, K.; Maldonado, M.; Zhang, Y.; Lasher, J.; Bouillaud, E.; Wang, Y.; Mann, K.; Unger, N. *Clin. Ther.* **2012**, *34*, 677.

165. Boscaro, M.; Ludlam, W. H.; Atkinson, B.; Glusman, J. E.; Petersenn, S.; Reincke, M.; Snyder, P.; Tabarin, A.; Biller, B. M.; Findling, J.; Melmed, S.; Darby, C. H.; Hu, K.; Wang, Y.; Freda, P. U.; Grossman, A. B.; Frohman, L. A.; Bertherat, J. *J. Clin. Endocrinol. Metab.* **2009**, *94*, 115.

166. Colao, A.; Petersenn, S.; Newell-Price, J.; Findling, J. W.; Gu, F.; Maldonado, M.; Schoenherr, U.; Mills, D.; Salgado, L. R.; Biller, B. M. K.; For the Pasireotide B2305 Study Group, *N. Engl. J. Med.* **2012**, *366*, 914.

167. Strippoli, G. F.; Naveneethan, S. D.; Craig, J. C. *Cochrane Database Syst. Rev.* **2006**, CD003967. http://dx.doi.org/10.1002/14651858.CD003967.pub2.

168. Schellekens, H. *Nephrol. Dial. Transplant.* **2003**, *18*, 1257.

169. US FDA Centers for Drug Evaluation and Research, Oncolytic Drugs Advisory Committee Briefing Document UCM301424, 2012, Silver Spring, MD. www.fda.gov/downloads/AdvisoryCommittees/CommitteesMeetingMaterials/Drugs/OncolyticDrugsAdvisoryCommittee/UCM301424.pdf.

170. Woodburn, K. W.; Fan, Q.; Leuther, K. K.; Holmes, C. P.; Zhang, J. J.; Velkovska, S.; Chen, M.-J.; Schatz, P. J. *Blood* **2004**, *104*, 2904 (ASH Annual Meeting Abstracts).

171. Holmes, C. P.; Yin, Q.; Lalonde, G.; Schatz, P. J.; Tumelty, D.; Palani, B.; Zemede, G. U.S. Patent 7,084,245B2, 2006.

172. Woodburn, K. W.; Holmes, C. P.; Wilson, S. D.; Fong, K.-L.; Press, R. J.; Moriya, Y.; Tagawa, Y. *Xenobiotica* **2012**, *42*, 660.

173. Moon, Y. J.; Bullock, J.; Earp, J.; Garnett, C. US FDA Centers for Drug Evaluation and Research, Clinical Pharmacology Review, NDA 202.

174. www.fda.gov/downloads/Drugs/DrugSafety/UCM297877.pdf.

175. Fishbane, S.; Schiller, B.; Locatelli, F.; Covic, A. C.; Provenzano, R.; Wiecek, A.; Levin, N. W.; Kaplan, M.; Macdougall, I. C.; Francisco, C.; Mayo, M. R.; Polu, K. R.; Duliege, A.-M.; Besarab, A.; EMERALD Study Groups, *N. Engl. J. Med.* **2013**, *368*, 307.

176. http://www.fda.gov/NewsEvents/Newsroom/PressAnnouncements/ucm297464.htm.

177. US FDA press release, February 23, 2013, Affymax and Takeda Announce a Nationwide Voluntary Recall of All Lots of OMONTYS® (peginesatide) Injection, Silver Spring, MD.

178. Berg, A. T.; Scheffer, I. E. *Epilepsia* **2011**, *52*, 1058.

179. Kotsopoulos, I. A.; van Merode, T.; Kessels, F. G.; de Krom, M. C.; Knottnerus, J. A. *Epilepsia* **2002**, *43*, 1402.
180. Guimaraes, J.; Mendes, J. A. *Neurologist* **2010**, *16*, 353.
181. Rogawski, M. A. *Epilepsy Curr.* **2011**, *11*, 56.
182. Hibi, S.; Ueno, K.; Nagato, S.; Kawano, K.; Ito, K.; Norimine, Y.; Takenaka, O.; Hanada, T.; Yonaga, M. *J. Med. Chem.* **2012**, *55*, 10584.
183. Templeton, D. *Epilepsia* **2009**, *50,* (Suppl. 11), 1.
184. Cheng, C.; Gonyeau, M. J.; Kirwin, J. L. *Formulary* **2013**, *48*, 19.
185. Krauss, G. L.; Bar, B.; Klapper, J. A.; Rektor, I.; Vaiciene-Magistris, N.; Squillacote, D.; Kumar, D. *Acta Neurol. Scand.* **2012**, *125*, 8.
186. http://www.fycompa.eu/docs/PIL/English-PIL.pdf (last accessed 19 March 2013).
187. Baselga, J.; Cortes, J.; Kim, S. B.; Im, S. A.; Hegg, R.; Im, Y. H.; Roman, L.; Pedrini, J. L.; Pienkowski, T.; Knott, A.; Clark, E.; Benyunes, M. C.; Ross, G.; Swain, S. M.; CLEOPATRA Study Group, *N. Engl. J. Med.* **2012**, *366*, 109.
188. Baselga, J.; Swain, S. M. *Clin. Breast Cancer* **2010**, *10*, 489.
189. Baselga, J.; Gelmon, K. A.; Verma, S.; Wardley, A.; Conte, P.; Miles, D.; Bianchi, G.; Cortes, J.; McNally, V. A.; Ross, G. A.; Fumoleau, P.; Gianni, L. *J. Clin. Oncol.* **2010**, *28*, 1138.
190. Scheuer, W.; Friess, T.; Burtscher, H.; Bossenmaier, B.; Endl, J.; Hasmann, M. *Cancer Res.* **2009**, *69*, 9330.
191. Agus, D. B.; Gordon, M. S.; Taylor, C.; Natale, R. B.; Karlan, B.; Mendelson, D. S.; Press, M. F.; Allison, D. E.; Sliwkowski, M. X.; Lieberman, G.; Kelsey, S. M.; Fyfe, G. *J. Clin. Oncol.* **2005**, *23*, 2534.
192. Portera, C. C.; Walshe, J. M.; Rosing, D. R.; Denduluri, N.; Berman, A. W.; Vatas, U.; Velarde, M.; Chow, C. K.; Steinberg, S. M.; Nguyen, D.; Yang, S. X.; Swain, S. M. *Clin. Cancer Res.* **2008**, *14*, 2710.
193. http://www.cancerresearchuk.org/cancer-info/cancerstats/types/nhl/incidence/uk-nonhodgkin-lymphoma-incidence-statistics#world.
194. http://www.cancer.gov/cancertopics/types/non-hodgkin.
195. Mukherji, D.; Pettengell, R. *Expert Opin. Pharmacother.* **2010**, *11*, 1915.
196. Krapcho, A. P.; Petry, M. E.; Getahun, A.; Landi, J. J.; Stallman, J.; Polsenberg, J. F.; Gallagher, C. E.; Maresch, M. J.; Hacker, M. P. *J. Med. Chem.* **1994**, *37*, 828.
197. Hazelhurst, L. A.; Krapcho, A. P.; Hacker, M. P. *Cancer Lett.* **1995**, *91*, 115.
198. Evison, B. J.; Chiu, F.; Pezzoni, G.; Phillips, D. R.; Cutts, S. M. *Mol. Pharmacol.* **2008**, *74*, 184.
199. El-Helw, L. M.; Hancock, B. W. *Expert Opin. Investig. Drugs* **2007**, *16*, 1683.
200. Spinelli, S.; DiDomenico, R. U.S. Patent 5,717,099, 1998.
201. Borchmann, P.; Schnell, R.; Knippertz, R.; Staak, J. O.; Camboni, G. M.; Bernareffi, A.; Hübel, K.; Staib, P.; Schulz, A.; Diehl, V.; Engert, A. *Ann. Oncol.* **2001**, *12*, 661.
202. Pettengell, R.; Coiffier, B.; Narayanan, G.; de Mendoza, F. H.; Digumarti, R.; Gomez, H.; Zinzani, P. L.; Schiller, G.; Rizzieri, D.; Boland, G.; Cernohous, P.; Wang, L.; Kuefer, C.; Gorbatchevsky, I.; Singer, J. *Lancet Oncol.* **2012**, *13*, 696.
203. O'Hare, T.; Eide, C. A.; Deininger, M. W. *Blood* **2007**, *110*, 2242.
204. Huang, W.-S.; Metcalf, C. A.; Sundaramoorthi, R.; Wang, Y.; Zou, D.; Thomas, R. M.; Zhu, X.; Cai, L.; Wen, D.; Liu, S.; Romero, J.; Qi, J.; Chen, I.; Banda, G.; Lentini, S. P.; Das, S.; Xu, Q.; Keats, J.; Wang, F.; Wardwell, S.; Ning, Y.; Snodgrass, J. T.; Broudy, M. I.; Russian, K.; Zhou, T.; Commodore, L.; Narsimhan, N. I.; Mohemmad, Q. K.; Iuliucci, J.; Rivera, V.; Dalgarno, D. C.; Sawyer, T. K.; Clackson, T.; Shakespeare, W. C. *J. Med. Chem.* **2010**, *53*, 4701.

205. Zou, D.; Huang, W.-S.; Thomas, R. M.; Romero, J. A. C.; Qi, J.; Wang, Y.; Zhu, X.; Shakespeare, W. C.; Sundaromoorthi, R.; Metcalf, C. A.; Dalgarno, D. C.; Sawyer, T. K. Patent Application WO2007/075869, 2007.

206. Cortes, J. E.; Kantarjian, H.; Shah, N. P.; Bixby, D.; Mauro, M. J.; Flinn, I.; O'Hare, T.; Hu, S.; Narasimhan, N. I.; Rivera, V. M.; Clackson, T.; Turner, C. D.; Haluska, F. G.; Druker, B. J.; Deininger, M. W. N.; Talpaz, M. N. *Engl. J. Med.* **2012**, *367*, 2075.

207. Iclusig® Package Insert; http://www.accessdata.fda.gov/drugsatfda_docs/label/2012/203469lbl.pdf.

208. Hedge, S.; Schmidt, M. *Annu. Rep. Med. Chem.* **2008**, *43*, 480.

209. Thienelt, C. D.; Green, K.; Bowles, D. W. *Drugs Today* **2012**, *48*, 601.

210. Kim, D. Y.; Cho, D. J.; Lee, G. Y.; Kim, H. Y.; Woo, S. H.; Kim, Y. S.; Lee, S. A.; Han, B. C. U.S. Patent 7,501,424 B2, 2009.

211. Kim, D. Y.; Cho, D. J.; Lee, G. Y.; Kim, H. Y.; Woo, S. H. Patent Application WO2010/018895, 2010.

212. Kim, S.-H.; Menon, H.; Jootar, S.; Saikia, T.; Kwak, J.-Y.; Sohn, S. K.; Park, J. S.; Kim, H. J.; Oh, S. J.; Kim, H.; Zang, D. Y.; Park, S.; Lee, H. A.; Park, H. L.; Woo, S. H.; Kim, H. Y.; Lee, G. Y.; Cho, D. J.; Shin, J. S.; Kim, D. Y.; Kim, D.-W. In: *54th American Society of Hematology Annual Meeting and Exposition*, 2012. Abstract 695, https//ash.confex.com/ash/2012/webprogram/Paper51024.html.

213. http://clinicaltrials.gov/show/NCT01511289.

214. http://clinicaltrials.gov/show/NCT01602952.

215. http://www.cdc.gov/cancer/colorectal/ (last accessed 22 March 2013).

216. Van Cutsem, E.; Nordlinger, B.; Cervantes, A. *Ann. Oncol.* **2010**, *21*, 93.

217. Douillard, J.-Y.; Salvatore, S.; Cassidy, J.; Tabernero, J.; Burkes, R.; Barugel, M.; Humblet, Y.; Bodoky, G.; Cunningham, D.; Jassem, J.; Rivera, F.; Kocakova, I.; Ruff, P.; Blasinka-Morawiec, M.; Smakal, M.; Canon, J.-L.; Rother, M.; Oliner, K. S.; Wolf, M.; Gansert, J. *J. Clin. Oncol.* **2010**, *28*, 4697.

218. Waddell, T.; Cunningham, D. *Lancet* **2013**, *381*, 273.

219. Cherrington, J. M.; Strawn, L. M.; Shawver, L. K. *Adv. Cancer Res.* **2000**, *79*, 1.

220. Wilhelm, S. M.; Dumas, J.; Adnane, L.; Lynch, M.; Carter, C. A.; Schutz, G.; Thierauch, K. H.; Zopf, D. *Int. J. Cancer* **2011**, *129*, 245.

221. Hedge, S.; Schmidt, M. *Annu. Rep. Med. Chem.* **2005**, *41*, 466.

222. Riedl, B.; Dumas, J.; Khire, U.; Lowinger, T. B.; Scott, W. J.; Smith, R. A.; Wood, J. E.; Monahan, M.-K.; Natero, R.; Renick, J.; Sibley, R. N. U.S. Patent 7,235,576, 2007.

223. Strumberg, D.; Scheulen, M. E.; Schultheis, B.; Richly, H.; Frost, A.; Buchert, M.; Christensen, O.; Jeffers, M.; Heinig, R.; Boix, O.; Mross, K. *Br. J. Cancer* **2012**, *106*, 1722.

224. Grothey, A.; Van Cutsem, E.; Sobrero, A.; Siena, S.; Falcone, A.; Ychou, M.; Humblet, Y.; Bouche, O.; Mineur, L.; Barone, C.; Adenis, A.; Tabernero, J.; Yoshino, T.; Lenz, H.-J.; Goldberg, R. M.; Sargent, D. J.; Cihon, F.; Cupit, L.; Wagner, A.; Laurent, D. *Lancet* **2013**, *381*, 303.

225. http://labeling.bayerhealthcare.com/html/products/pi/Stivarga_PI.pdf (last accessed 1 April 2013).

226. Buchman, A. L. *MedGenMed* **2004**, *6*, 12 (http://www.ncbi.nlm.nih.gov/pmc/articles/PMC1395790/).

227. Drucker, D. J.; Ehrlich, P.; Asa, S. L.; Brubaker, P. L. *Proc. Natl. Acad. Sci. U.S.A.* **1996**, *93*, 7911.

228. Drucker, D. J.; Shi, Q.; Crivici, A.; Sumner-Smith, M.; Tavares, W.; Hill, M.; DeForest, L.; Cooper, S.; Brubaker, P. L. *Nat. Biotechnol.* **1997**, *15*, 673.

229. DaCambra, M. P.; Yusta, B.; Sumner-Smith, M.; Crivici, A.; Drucker, D. J.; Brubaker, P. L. *Biochemistry* **2000**, *39*, 8888.

230. Drucker, D. J.; Crivici, A. E.; Sumner-Smith, M. U.S. Patent 6,184,201, 2001.

231. Marier, J.-F.; Beliveau, M.; Mouksassi, M.-S.; Shaw, P.; Cyran, J.; Kesavan, J.; Wallens, J.; Zahir, H.; Wells, D.; Caminis, J. *J. Clin. Pharmacol.* **2008**, *48*, 1289.

232. Jeppesen, P. B.; Sanguinetti, E. L.; Buchman, A.; Howard, L.; Scolapio, J. S.; Ziegler, T. R.; Gregory, J.; Tappenden, K. A.; Holst, J.; Mortensen, P. B. *Gut* **2005**, *54*, 1224.

233. Jeppesen, P. B.; Gilroy, R.; Pertkiewicz, M.; Allard, J. P.; Messing, B.; O'Keefe, S. J. *Gut* **2011**, *60*, 902.

234. Jeppesen, P. B.; Pertkiewicz, M.; Messing, B.; Iyer, K.; Seidner, D. L.; O'Keefe, S. J.; Forbes, A.; Heinze, H.; Joelsson, B. *Gastroenterology* **2012**, *143*, 1473.

235. GATTEX® Briefing Document, Advisory Committee Meeting, 16 October 2012; http://www.fda.gov/downloads/AdvisoryCommittees/CommitteesMeetingMaterials/Drugs/GastrointestinalDrugsAdvisoryCommittee/UCM323506.pdf.

236. http://www.idf.org/diabetesatlas/5e/the-global-burden.

237. Yoshida, T.; Akahoshi, F.; Sakashita, H.; Kitajima, H.; Nakamura, M.; Sonda, S.; Takeuchi, M.; Tanaka, Y.; Ueda, N.; Sekiguchi, S.; Ishige, T.; Shima, K.; Nabeno, M.; Abe, Y.; Anabuki, J.; Soejima, A.; Yoshida, K.; Takashina, Y.; Ishii, S.; Kiuchi, S.; Fukuda, S.; Tsutsumiuchi, R.; Kosaka, K.; Murozono, T.; Nakamaru, Y.; Utsumi, H.; Masutomi, N.; Kishida, H.; Miyaguchi, I.; Hayashi, Y. *Bioorg. Med. Chem.* **2012**, *20*, 5705.

238. Fukuda-Tsuru, S.; Anabuki, J.; Abe, Y.; Yoshida, K.; Ishii, S. *Eur. J. Pharmacol.* **2012**, *696*, 194.

239. Eto, T.; Inoue, S.; Kadowaki, T. *Diabetes Obes. Metab.* **2012**, *14*, 1040.

240. Kadowaki, T.; Kondo, K. *Diabetes Obes. Metab.* **2013**, *15*, 810.

241. Compston, A. *McAlpine's Multiple Sclerosis*; Elsevier/Churchill Livingstone: Philadelphia, 2006.

242. Claussen, M. C.; Korn, T. *Clin. Immunol.* **2012**, *142*, 49.

243. Merrill, J. E.; Hanak, S.; Pu, S. F.; Liang, J.; Dang, C.; Iglesias-Bregna, D.; Harvey, B.; Zhu, B.; McMonagle-Strucko, K. *J. Neurol.* **2009**, *256*, 89.

244. (a) Bartlett, R. R.; Kämmerer, F. J. Patent Application WO1991/017748, 1991; (b) Hachtel, J.; Neises, B.; Schwab, W.; Utz, R.; Zahn, M. Patent Application WO2004/083165, 2004.

245. Limsakun, T.; Menguy-Vacheron, F. *Neurology* **2010**, *74*, A415.

246. O'Connor, P.; Wolinsky, J. S.; Confavreux, C.; Comi, G.; Kappos, L.; Olsson, T. P.; Benzerdjeb, H.; Truffinet, P.; Wang, L.; Miller, A.; Freedman, M. S. *N. Engl. J. Med.* **2011**, *365*, 1293.

247. Wojtusik, A.; Feret, B. *Formulary* **2012**, *47*, 97.

248. Scott, D. L.; Wolfe, F.; Huizinga, T. W. J. *Lancet* **2010**, *376*, 1094.

249. Flanagan, M. E.; Blumenkopk, T. A.; Brissette, W. H.; Brown, M. F.; Casavant, J. M.; Shang-Poa, C.; Doty, J. L.; Elliott, E. A.; Fisher, M. B.; Hines, M.; Kent, C.; Kudlacz, E. M.; Lillie, B. M.; Magnuson, K. S.; McCurdy, S. P.; Munchhof, M. J.; Perry, B. D.; Sawyer, P. S.; Strelevitz, T. J.; Subramanyam, C.; Sun, J.; Whipple, D. A.; Changelian, P. S. *J. Med. Chem.* **2010**, *53*, 8468.

250. Meyer, D. M.; Jesson, M. I.; Li, X.; Elrick, M. M.; Funckes-Shippy, C. L.; Warner, J. D.; Gross, C. J.; Dowty, M. E.; Ramaiah, S. K.; Hirsch, J. L.; Saabye, M. J.; Barks, J. L.; Kishore, N.; Morris, D. L. *J. Inflamm.* **2010**, *7*, 41.

251. Cai, W.; Colony, J. L.; Frost, H.; Hudspeth, J. P.; Kendall, P. M.; Krishnan, A. M.; Makowski, T.; Mazur, D. J.; Phillips, J.; Ripin, D. H. B.; Ruggeri, S. G.; Stearns, J. F.; White, T. D. *Org. Process Res. Dev.* **2005**, *9*, 51.

252. Price, K. E.; Larrivee-Aboussafy, C.; Lillie, B. M.; McLaughlin, R. W.; Mustakis, J.; Hettenbach, K. W.; Hawkins, J. M.; Vaidyanathan, R. *Org. Lett.* **2009**, *11*, 2003.

253. Burmester, G. R.; Blanco, R.; Charles-Schoeman, C.; Wollenhaupt, J.; Zerbini, C.; Benda, B.; Gruben, D.; Wallenstein, G.; Krishnaswami, S.; Zwillich, S. H.; Koncz, T.; Soma, K.; Bradley, J.; Mebus, C. *Lancet* **2013**, *381*, 451.

254. Fleischmann, R.; Kremer, J.; Cush, J.; Schulze-Koops, H.; Connell, C. A.; Bradley, J. D.; Gruben, D.; Wallenstein, G. V.; Zwillich, S. H.; Kanik, K. S. *N. Engl. J. Med.* **2012**, *367*, 495.

255. O'Dell, K. M. *Formulary* **2012**, *47*, 350.

256. American Cancer Society. Skin Cancer: Basal and Squamous Cell. www.cancer.org/Cancer/SkinCancer-BasalandSquamousCell/DetailedGuide/index.

257. Madan, V.; Lear, J. T.; Szeimies, R. M. *Lancet* **2010**, *375*, 673.

258. Cirrone, F.; Harris, C. M. *Clin. Ther.* **2012**, *34*, 2039.

259. Robarge, K. D.; Brunton, S. A.; Castanedo, G. M.; Cui, Y.; Dina, M. S.; Goldsmith, R.; Gould, S. E.; Guichert, O.; Gunzner, J. L.; Halladay, J.; Jia, W.; Khojasteh, C.; Koehler, M. F. T.; Kotkow, K.; La, H.; LaLonde, R. L.; Lau, K.; Lee, L.; Marshall, D.; Marsters, J. C.; Murray, L. J.; Qian, C.; Rubin, L. L.; Salphati, L.; Stanley, M. S.; Stibbard, J. H. A.; Sutherlin, D. P.; Ubhayaker, S.; Wang, S.; Wong, S.; Xie, M. *Bioorg. Med. Chem. Lett.* **2009**, *19*, 5576.

260. Graham, R. A.; Lum, B. L.; Cheeti, S.; Jin, J. Y.; Jorga, K.; Von Hoff, D. D.; Rudin, C. M.; Reddy, J. C.; Low, J. A.; LoRusso, P. M. *Clin. Cancer Res.* **2011**, *17*, 2512.

261. Graham, R. A.; Hop, C. E. C. A.; Borin, M. T.; Lum, B. L.; Colburn, D.; Chang, I.; Shin, Y. G.; Malhi, V.; Low, J. A.; Dresser, M. J. *Br. J. Clin. Pharmacol.* **2012**, *74*, 788.

262. Erivedge™ Package Insert.

263. Sekulic, A.; Migden, M. R.; Oro, A. E.; Dirix, L.; Lewis, K. D.; Hainsworth, J. D.; Solomon, J. A.; Yoo, S.; Arron, S. T.; Friedlander, P. A.; Marmur, E.; Rudin, C. M.; Chang, A. L. S.; Low, J. A.; Mackey, H. M.; Yauch, R. L.; Graham, R. A.; Reddy, J. C.; Hauschild, A. *N. Engl. J. Med.* **2012**, *366*, 2171.

KEYWORD INDEX, VOLUME 48

Note: Page numbers followed by "*f*" indicate figures, "*t*" indicate tables, and "*np*" indicate footnotes.

CUMULATIVE CHAPTER TITLES KEYWORD INDEX, VOLUME 1 – 48

CUMULATIVE NCE INTRODUCTION INDEX, 1983–2012

GENERIC NAME	INDICATION	YEAR INTRODUCED	ARMC VOL., (PAGE)
abacavir sulfate	antiviral	1999	35 (333)
abarelix	anticancer	2004	40 (446)
abatacept	antiarthritic	2006	42 (509)
abiraterone acetate	anticancer	2011	47 (505)
acarbose	antidiabetic	1990	26 (297)
aceclofenac	antiinflammatory	1992	28 (325)
acemannan	wound healing agent	2001	37 (259)
acetohydroxamic acid	urinary tract/bladder disorders	1983	19 (313)
acetorphan	antidiarrheal	1993	29 (332)
acipimox	antihypercholesterolemic	1985	21 (323)
acitretin	antipsoriasis	1989	25 (309)
aclidinium bromide	chronic obstructive pulmonary disorder	2012	48 (481)
acrivastine	antiallergy	1988	24 (295)
actarit	antiinflammatory	1994	30 (296)
adalimumab	antiarthritic	2003	39 (267)
adamantanium bromide	antibacterial	1984	20 (315)
adefovir dipivoxil	antiviral	2002	38 (348)
adrafinil	sleep disorders	1986	22 (315)
AF-2259	antiinflammatory	1987	23 (325)
aflibercept	ophthalmologic, macular degeneration	2011	47 (507)
afloqualone	muscle relaxant	1983	19 (313)
agalsidase alfa	Fabry's disease	2001	37 (259)
alacepril	antihypertensive	1988	24 (296)
alcaftadine	ophthalmologic (allergic conjunctivitis)	2010	46 (444)
alclometasone dipropionate	antiinflammatory	1985	21 (323)
alefacept	antipsoriasis	2003	39 (267)
alemtuzumab	anticancer	2001	37 (260)
alendronate sodium	osteoporosis	1993	29 (332)
alfentanil hydrochloride	analgesic	1983	19 (314)
alfuzosin hydrochloride	antihypertensive	1988	24 (296)
alglucerase	Gaucher's disease	1991	27 (321)
alglucosidase alfa	Pompe disease	2006	42 (511)
aliskiren	antihypertensive	2007	43 (461)
alitretinoin	anticancer	1999	35 (333)
alminoprofen	analgesic	1983	19 (314)
almotriptan	antimigraine	2000	36 (295)

GENERIC NAME	INDICATION	YEAR INTRODUCED	ARMC VOL., (PAGE)
alogliptin	antidiabetic	2010	46 (446)
alosetron hydrochloride	irritable bowel syndrome	2000	36 (295)
alpha-1 antitrypsin	emphysema	1988	24 (297)
alpidem	anxiolytic	1991	27 (322)
alpiropride	antimigraine	1988	24 (296)
alteplase	antithrombotic	1987	23 (326)
alvimopan	post-operative ileus	2008	44 (584)
ambrisentan	pulmonary hypertension	2007	43 (463)
amfenac sodium	antiinflammatory	1986	22 (315)
amifostine	cytoprotective	1995	31 (338)
aminoprofen	antiinflammatory	1990	26 (298)
amisulpride	antipsychotic	1986	22 (316)
amlexanox	antiasthma	1987	23 (327)
amlodipine besylate	antihypertensive	1990	26 (298)
amorolfine hydrochloride	antifungal	1991	27 (322)
amosulalol	antihypertensive	1988	24 (297)
ampiroxicam	antiinflammatory	1994	30 (296)
amprenavir	antiviral	1999	35 (334)
amrinone	congestive heart failure	1983	19 (314)
amrubicin hydrochloride	anticancer	2002	38 (349)
amsacrine	anticancer	1987	23 (327)
amtolmetin guacil	antiinflammatory	1993	29 (332)
anagliptin	antidiabetic	2012	48 (483)
anagrelide hydrochloride	antithrombotic	1997	33 (328)
anakinra	antiarthritic	2001	37 (261)
anastrozole	anticancer	1995	31 (338)
angiotensin II	anticancer adjuvant	1994	30 (296)
anidulafungin	antifungal	2006	42 (512)
aniracetam	cognition enhancer	1993	29 (333)
anti-digoxin polyclonal antibody	antidote, digoxin poisoning	2002	38 (350)
APD	osteoporosis	1987	23 (326)
apixaban	antithrombotic	2011	47 (509)
apraclonidine hydrochloride	antiglaucoma	1988	24 (297)
aprepitant	antiemetic	2003	39 (268)
APSAC	antithrombotic	1987	23 (326)
aranidipine	antihypertensive	1996	32 (306)
arbekacin	antibacterial	1990	26 (298)
arformoterol	antiasthma	2007	43 (465)
argatroban	antithrombotic	1990	26 (299)
arglabin	anticancer	1999	35 (335)

GENERIC NAME	INDICATION	YEAR INTRODUCED	ARMC VOL., (PAGE)
aripiprazole	antipsychotic	2002	38 (350)
armodafinil	sleep disorders	2009	45 (478)
arotinolol hydrochloride	antihypertensive	1986	22 (316)
arteether	antimalarial	2000	36 (296)
artemisinin	antimalarial	1987	23 (327)
asenapine	antipsychotic	2009	45 (479)
aspoxicillin	antibacterial	1987	23 (328)
astemizole	antiallergy	1983	19 (314)
astromycin sulfate	antibacterial	1985	21 (324)
atazanavir	antiviral	2003	39 (269)
atomoxetine	attention deficit hyperactivity disorder	2003	39 (270)
atorvastatin calcium	antihypercholesterolemic	1997	33 (328)
atosiban	premature labor	2000	36 (297)
atovaquone	antiparasitic	1992	28 (326)
auranofin	antiarthritic	1983	19 (314)
avanafil	male sexual dysfunction	2011	47 (512)
axitinib	anticancer	2012	48 (485)
azacitidine	anticancer	2004	40 (447)
azelaic acid	acne	1989	25 (310)
azelastine hydrochloride	antiallergy	1986	22 (316)
azelnidipine	antihypertensive	2003	39 (270)
azilsartan	antihypertensive	2011	47 (514)
azithromycin	antibacterial	1988	24 (298)
azosemide	diuretic	1986	22 (316)
aztreonam	antibacterial	1984	20 (315)
balofloxacin	antibacterial	2002	38 (351)
balsalazide disodium	ulcerative colitis	1997	33 (329)
bambuterol	antiasthma	1990	26 (299)
barnidipine hydrochloride	antihypertensive	1992	28 (326)
beclobrate	antihypercholesterolemic	1986	22 (317)
bedaquiline	antibacterial	2012	48 (487)
befunolol hydrochloride	antiglaucoma	1983	19 (315)
belatacept	immunosuppressant	2011	47 (516)
belimumab	lupus	2011	47 (519)
belotecan	anticancer	2004	40 (449)
benazepril hydrochloride	antihypertensive	1990	26 (299)
benexate hydrochloride	antiulcer	1987	23 (328)
benidipine hydrochloride	antihypertensive	1991	27 (322)

GENERIC NAME	INDICATION	YEAR INTRODUCED	ARMC VOL., (PAGE)
beraprost sodium	antiplatelet	1992	28 (326)
besifloxacin	antibacterial	2009	45 (482)
betamethasone butyrate propionate	antiinflammatory	1994	30 (297)
betaxolol hydrochloride	antihypertensive	1983	19 (315)
betotastine besilate	antiallergy	2000	36 (297)
bevacizumab	anticancer	2004	40 (450)
bevantolol hydrochloride	antihypertensive	1987	23 (328)
bexarotene	anticancer	2000	36 (298)
biapenem	antibacterial	2002	38 (351)
bicalutamide	anticancer	1995	31 (338)
bifemelane hydrochloride	nootropic	1987	23 (329)
bilastine	antiallergy	2010	46 (449)
bimatoprost	antiglaucoma	2001	37 (261)
binfonazole	sleep disorders	1983	19 (315)
binifibrate	antihypercholesterolemic	1986	22 (317)
biolimus drug-eluting stent	coronary artery disease, antirestenotic	2008	44 (586)
bisantrene hydrochloride	anticancer	1990	26 (300)
bisoprolol fumarate	antihypertensive	1986	22 (317)
bivalirudin	antithrombotic	2000	36 (298)
blonanserin	antipsychotic	2008	44 (587)
boceprevir	antiviral	2011	47 (521)
bopindolol	antihypertensive	1985	21 (324)
bortezomib	anticancer	2003	39 (271)
bosentan	antihypertensive	2001	37 (262)
bosutinib	anticancer	2012	48 (489)
brentuximab	anticancer	2011	47 (523)
brimonidine	antiglaucoma	1996	32 (306)
brinzolamide	antiglaucoma	1998	34 (318)
brodimoprin	antibacterial	1993	29 (333)
bromfenac sodium	antiinflammatory	1997	33 (329)
brotizolam	sleep disorders	1983	19 (315)
brovincamine fumarate	cerebral vasodilator	1986	22 (317)
bucillamine	immunomodulator	1987	23 (329)
bucladesine sodium	congestive heart failure	1984	20 (316)
budipine	Parkinson's disease	1997	33 (330)
budralazine	antihypertensive	1983	19 (315)
bulaquine	antimalarial	2000	36 (299)
bunazosin hydrochloride	antihypertensive	1985	21 (324)

GENERIC NAME	INDICATION	YEAR INTRODUCED	ARMC VOL., (PAGE)
bupropion hydrochloride	antidepressant	1989	25 (310)
buserelin acetate	hormone therapy	1984	20 (316)
buspirone hydrochloride	anxiolytic	1985	21 (324)
butenafine hydrochloride	antifungal	1992	28 (327)
butibufen	antiinflammatory	1992	28 (327)
butoconazole	antifungal	1986	22 (318)
butoctamide	sleep disorders	1984	20 (316)
butyl flufenamate	antiinflammatory	1983	19 (316)
cabazitaxel	anticancer	2010	46 (451)
cabergoline	antiprolactin	1993	29 (334)
cabozantinib	anticancer	2012	48 (491)
cadexomer iodine	wound healing agent	1983	19 (316)
cadralazine	antihypertensive	1988	24 (298)
calcipotriol	antipsoriasis	1991	27 (323)
camostat mesylate	anticancer	1985	21 (325)
canakinumab	antiinflammatory	2009	45 (484)
candesartan cilexetil	antihypertensive	1997	33 (330)
capecitabine	anticancer	1998	34 (319)
captopril	antihypertensive	1982	13 (086)
carboplatin	antibacterial	1986	22 (318)
carfilzomib	anticancer	2012	48 (492)
carperitide	congestive heart failure	1995	31 (339)
carumonam	antibacterial	1988	24 (298)
carvedilol	antihypertensive	1991	27 (323)
caspofungin acetate	antifungal	2001	37 (263)
catumaxomab	anticancer	2009	45 (486)
cefbuperazone sodium	antibacterial	1985	21 (325)
cefcapene pivoxil	antibacterial	1997	33 (330)
cefdinir	antibacterial	1991	27 (323)
cefditoren pivoxil	antibacterial	1994	30 (297)
cefepime	antibacterial	1993	29 (334)
cefetamet pivoxil hydrochloride	antibacterial	1992	28 (327)
cefixime	antibacterial	1987	23 (329)
cefmenoxime hydrochloride	antibacterial	1983	19 (316)
cefminox sodium	antibacterial	1987	23 (330)
cefodizime sodium	antibacterial	1990	26 (300)
cefonicid sodium	antibacterial	1984	20 (316)
ceforanide	antibacterial	1984	20 (317)
cefoselis	antibacterial	1998	34 (319)
cefotetan disodium	antibacterial	1984	20 (317)

GENERIC NAME	INDICATION	YEAR INTRODUCED	ARMC VOL., (PAGE)
cefotiam hexetil hydrochloride	antibacterial	1991	27 (324)
cefozopran hydrochloride	antibacterial	1995	31 (339)
cefpimizole	antibacterial	1987	23 (330)
cefpiramide sodium	antibacterial	1985	21 (325)
cefpirome sulfate	antibacterial	1992	28 (328)
cefpodoxime proxetil	antibacterial	1989	25 (310)
cefprozil	antibacterial	1992	28 (328)
ceftaroline fosamil	antibacterial	2010	46 (453)
ceftazidime	antibacterial	1983	19 (316)
cefteram pivoxil	antibacterial	1987	23 (330)
ceftibuten	antibacterial	1992	28 (329)
ceftobiprole medocaril	antibacterial	2008	44 (589)
cefuroxime axetil	antibacterial	1987	23 (331)
cefuzonam sodium	antibacterial	1987	23 (331)
celecoxib	antiarthritic	1999	35 (335)
celiprolol hydrochloride	antihypertensive	1983	19 (317)
centchroman	contraception	1991	27 (324)
centoxin	immunomodulator	1991	27 (325)
cerivastatin	antihypercholesterolemic	1997	33 (331)
certolizumab pegol	irritable bowel syndrome	2008	44 (592)
cetirizine hydrochloride	antiallergy	1987	23 (331)
cetrorelix	infertility	1999	35 (336)
cetuximab	anticancer	2003	39 (272)
cevimeline hydrochloride	antixerostomia	2000	36 (299)
chenodiol	gallstones	1983	19 (317)
CHF-1301	Parkinson's disease	1999	35 (336)
choline alfoscerate	cognition enhancer	1990	26 (300)
choline fenofibrate	antihypercholesterolemic	2008	44 (594)
cibenzoline	antiarrhythmic	1985	21 (325)
ciclesonide	antiasthma	2005	41 (443)
cicletanine	antihypertensive	1988	24 (299)
cidofovir	antiviral	1996	32 (306)
cilazapril	antihypertensive	1990	26 (301)
cilostazol	antithrombotic	1988	24 (299)
cimetropium bromide	antispasmodic	1985	21 (326)
cinacalcet	hyperparathyroidism	2004	40 (451)
cinildipine	antihypertensive	1995	31 (339)
cinitapride	gastroprokinetic	1990	26 (301)
cinolazepam	anxiolytic	1993	29 (334)
ciprofibrate	antihypercholesterolemic	1985	21 (326)

GENERIC NAME	INDICATION	YEAR INTRODUCED	ARMC VOL., (PAGE)
ciprofloxacin	antibacterial	1986	22 (318)
cisapride	gastroprokinetic	1988	24 (299)
cisatracurium besilate	muscle relaxant	1995	31 (340)
citalopram	antidepressant	1989	25 (311)
cladribine	anticancer	1993	29 (335)
clarithromycin	antibacterial	1990	26 (302)
clevidipine	antihypertensive	2008	44 (596)
clevudine	antiviral	2007	43 (466)
clobenoside	antiinflammatory	1988	24 (300)
cloconazole hydrochloride	antifungal	1986	22 (318)
clodronate disodium	calcium regulation	1986	22 (319)
clofarabine	anticancer	2005	41 (444)
clopidogrel hydrogensulfate	antithrombotic	1998	34 (320)
cloricromen	antithrombotic	1991	27 (325)
clospipramine hydrochloride	antipsychotic	1991	27 (325)
colesevelam hydrochloride	antihypercholesterolemic	2000	36 (300)
colestimide	antihypercholesterolemic	1999	35 (337)
colforsin daropate hydrochloride	congestive heart failure	1999	35 (337)
conivaptan	hyponatremia	2006	42 (514)
corifollitropin alfa	infertility	2010	46 (455)
crizotinib	anticancer	2011	47 (525)
crofelemer	antidiarrheal	2012	48 (494)
crotelidae polyvalent immune fab	antidote, snake venom poisoning	2001	37 (263)
cyclosporine	immunosuppressant	1983	19 (317)
cytarabine ocfosfate	anticancer	1993	29 (335)
dabigatran etexilate	anticoagulant	2008	44 (598)
dalfampridine	multiple sclerosis	2010	46 (458)
dalfopristin	antibacterial	1999	35 (338)
dapagliflozin	antidiabetic	2012	48 (495)
dapiprazole hydrochloride	antiglaucoma	1987	23 (332)
dapoxetine	premature ejaculation	2009	45 (488)
daptomycin	antibacterial	2003	39 (272)
darifenacin	urinary tract/bladder disorders	2005	41 (445)
darunavir	antiviral	2006	42 (515)
dasatinib	anticancer	2006	42 (517)
decitabine	myelodysplastic syndromes	2006	42 (519)
defeiprone	iron chelation therapy	1995	31 (340)

GENERIC NAME	INDICATION	YEAR INTRODUCED	ARMC VOL., (PAGE)
deferasirox	iron chelation therapy	2005	41 (446)
defibrotide	antithrombotic	1986	22 (319)
deflazacort	antiinflammatory	1986	22 (319)
degarelix acetate	anticancer	2009	45 (490)
delapril	antihypertensive	1989	25 (311)
delavirdine mesylate	antiviral	1997	33 (331)
denileukin diftitox	anticancer	1999	35 (338)
denopamine	congestive heart failure	1988	24 (300)
denosumab	osteoporosis	2010	46 (459)
deprodone propionate	antiinflammatory	1992	28 (329)
desflurane	anesthetic	1992	28 (329)
desloratadine	antiallergy	2001	37 (264)
desvenlafaxine	antidepressant	2008	44 (600)
dexfenfluramine	antiobesity	1997	33 (332)
dexibuprofen	antiinflammatory	1994	30 (298)
dexlansoprazole	antiulcer	2009	45 (492)
dexmedetomidine hydrochloride	sleep disorders	2000	36 (301)
dexmethylphenidate hydrochloride	attention deficit hyperactivity disorder	2002	38 (352)
dexrazoxane	cardioprotective	1992	28 (330)
dezocine	analgesic	1991	27 (326)
diacerein	antiinflammatory	1985	21 (326)
didanosine	antiviral	1991	27 (326)
dilevalol	antihypertensive	1989	25 (311)
diquafosol tetrasodium	ophthalmologic (dry eye)	2010	46 (462)
dirithromycin	antibacterial	1993	29 (336)
disodium pamidronate	osteoporosis	1989	25 (312)
divistyramine	antihypercholesterolemic	1984	20 (317)
docarpamine	congestive heart failure	1994	30 (298)
docetaxel	anticancer	1995	31 (341)
dofetilide	antiarrhythmic	2000	36 (301)
dolasetron mesylate	antiemetic	1998	34 (321)
donepezil hydrochloride	Alzheimer's disease	1997	33 (332)
dopexamine	congestive heart failure	1989	25 (312)
doripenem	antibacterial	2005	41 (448)
dornase alfa	cystic fibrosis	1994	30 (298)
dorzolamide hydrochloride	antiglaucoma	1995	31 (341)
dosmalfate	antiulcer	2000	36 (302)
doxacurium chloride	muscle relaxant	1991	27 (326)
doxazosin mesylate	antihypertensive	1988	24 (300)
doxefazepam	anxiolytic	1985	21 (326)
doxercalciferol	hyperparathyroidism	1999	35 (339)

GENERIC NAME	INDICATION	YEAR INTRODUCED	ARMC VOL., (PAGE)
doxifluridine	anticancer	1987	23 (332)
doxofylline	antiasthma	1985	21 (327)
dronabinol	antiemetic	1986	22 (319)
dronedarone	antiarrhythmic	2009	45 (495)
drospirenone	contraception	2000	36 (302)
drotrecogin alfa	antisepsis	2001	37 (265)
droxicam	antiinflammatory	1990	26 (302)
droxidopa	Parkinson's disease	1989	25 (312)
duloxetine	antidepressant	2004	40 (452)
dutasteride	benign prostatic hyperplasia	2002	38 (353)
duteplase	anticoagulant	1995	31 (342)
ebastine	antiallergy	1990	26 (302)
eberconazole	antifungal	2005	41 (449)
ebrotidine	antiulcer	1997	33 (333)
ecabet sodium	antiulcer	1993	29 (336)
ecallantide	angioedema, hereditary	2009	46 (464)
eculizumab	hemoglobinuria	2007	43 (468)
edaravone	neuroprotective	2001	37 (265)
edoxaban	antithrombotic	2011	47 (527)
efalizumab	antipsoriasis	2003	39 (274)
efavirenz	antiviral	1998	34 (321)
efonidipine	antihypertensive	1994	30 (299)
egualen sodium	antiulcer	2000	36 (303)
eldecalcitol	osteoporosis	2011	47 (529)
eletriptan	antimigraine	2001	37 (266)
eltrombopag	antithrombocytopenic	2009	45 (497)
emedastine difumarate	antiallergy	1993	29 (336)
emorfazone	analgesic	1984	20 (317)
emtricitabine	antiviral	2003	39 (274)
enalapril maleate	antihypertensive	1984	20 (317)
enalaprilat	antihypertensive	1987	23 (332)
encainide hydrochloride	antiarrhythmic	1987	23 (333)
enfuvirtide	antiviral	2003	39 (275)
enocitabine	anticancer	1983	19 (318)
enoxacin	antibacterial	1986	22 (320)
enoxaparin	anticoagulant	1987	23 (333)
enoximone	congestive heart failure	1988	24 (301)
enprostil	antiulcer	1985	21 (327)
entacapone	Parkinson's disease	1998	34 (322)
entecavir	antiviral	2005	41 (450)
enzalutamide	anticancer	2012	48 (497)
epalrestat	antidiabetic	1992	28 (330)
eperisone hydrochloride	muscle relaxant	1983	19 (318)

GENERIC NAME	INDICATION	YEAR INTRODUCED	ARMC VOL., (PAGE)
epidermal growth factor	wound healing agent	1987	23 (333)
epinastine	antiallergy	1994	30 (299)
epirubicin hydrochloride	anticancer	1984	20 (318)
eplerenone	antihypertensive	2003	39 (276)
epoprostenol sodium	antiplatelet	1983	19 (318)
eprosartan	antihypertensive	1997	33 (333)
eptazocine hydrobromide	analgesic	1987	23 (334)
eptilfibatide	antithrombotic	1999	35 (340)
erdosteine	expectorant	1995	31 (342)
eribulin mesylate	anticancer	2010	46 (465)
erlotinib	anticancer	2004	40 (454)
ertapenem sodium	antibacterial	2002	38 (353)
erythromycin acistrate	antibacterial	1988	24 (301)
erythropoietin	hematopoietic	1988	24 (301)
escitalopram oxalate	antidepressant	2002	38 (354)
eslicarbazepine acetate	anticonvulsant	2009	45 (498)
esmolol hydrochloride	antiarrhythmic	1987	23 (334)
esomeprazole magnesium	antiulcer	2000	36 (303)
eszopiclone	sleep disorders	2005	41 (451)
ethyl icosapentate	antithrombotic	1990	26 (303)
etizolam	anxiolytic	1984	20 (318)
etodolac	antiinflammatory	1985	21 (327)
etoricoxibe	antiarthritic	2002	38 (355)
etravirine	antiviral	2008	44 (602)
everolimus	immunosuppressant	2004	40 (455)
exemestane	anticancer	2000	36 (304)
exenatide	antidiabetic	2005	41 (452)
exifone	cognition enhancer	1988	24 (302)
ezetimibe	antihypercholesterolemic	2002	38 (355)
factor VIIa	haemophilia	1996	32 (307)
factor VIII	hemostatic	1992	28 (330)
fadrozole hydrochloride	anticancer	1995	31 (342)
falecalcitriol	hyperparathyroidism	2001	37 (266)
famciclovir	antiviral	1994	30 (300)
famotidine	antiulcer	1985	21 (327)
fasudil hydrochloride	amyotrophic lateral sclerosis	1995	31 (343)
febuxostat	gout	2009	45 (501)
felbamate	anticonvulsant	1993	29 (337)
felbinac	antiinflammatory	1986	22 (320)
felodipine	antihypertensive	1988	24 (302)
fenbuprol	biliary tract dysfunction	1983	19 (318)
fenoldopam mesylate	antihypertensive	1998	34 (322)

GENERIC NAME	INDICATION	YEAR INTRODUCED	ARMC VOL., (PAGE)
fenticonazole nitrate	antifungal	1987	23 (334)
fesoterodine	urinary tract/bladder disorders	2008	44 (604)
fexofenadine	antiallergy	1996	32 (307)
fidaxomicin	antibacterial	2011	47 (531)
filgrastim	immunostimulant	1991	27 (327)
finasteride	benign prostatic hyperplasia	1992	28 (331)
fingolimod	multiple sclerosis	2010	46 (468)
fisalamine	antiinflammatory	1984	20 (318)
fleroxacin	antibacterial	1992	28 (331)
flomoxef sodium	antibacterial	1988	24 (302)
flosequinan	congestive heart failure	1992	28 (331)
fluconazole	antifungal	1988	24 (303)
fludarabine phosphate	anticancer	1991	27 (327)
flumazenil	antidote, benzodiazepine overdose	1987	23 (335)
flunoxaprofen	antiinflammatory	1987	23 (335)
fluoxetine hydrochloride	antidepressant	1986	22 (320)
flupirtine maleate	analgesic	1985	21 (328)
flurithromycin ethylsuccinate	antibacterial	1997	33 (333)
flutamide	anticancer	1983	19 (318)
flutazolam	anxiolytic	1984	20 (318)
fluticasone furoate	antiallergy	2007	43 (469)
fluticasone propionate	antiinflammatory	1990	26 (303)
flutoprazepam	anxiolytic	1986	22 (320)
flutrimazole	antifungal	1995	31 (343)
flutropium bromide	antiasthma	1988	24 (303)
fluvastatin	antihypercholesterolemic	1994	30 (300)
fluvoxamine maleate	antidepressant	1983	19 (319)
follitropin alfa	infertility	1996	32 (307)
follitropin beta	infertility	1996	32 (308)
fomepizole	antidote, ethylene glycol poisoning	1998	34 (323)
fomivirsen sodium	antiviral	1998	34 (323)
fondaparinux sodium	antithrombotic	2002	38 (356)
formestane	anticancer	1993	29 (337)
formoterol fumarate	chronic obstructive pulmonary disorder	1986	22 (321)
fosamprenavir	antiviral	2003	39 (277)
fosaprepitant dimeglumine	antiemetic	2008	44 (606)
foscarnet sodium	antiviral	1989	25 (313)
fosfluconazole	antifungal	2004	40 (457)

GENERIC NAME	INDICATION	YEAR INTRODUCED	ARMC VOL., (PAGE)
fosfosal	analgesic	1984	20 (319)
fosinopril sodium	antihypertensive	1991	27 (328)
fosphenytoin sodium	anticonvulsant	1996	32 (308)
fotemustine	anticancer	1989	25 (313)
fropenam	antibacterial	1997	33 (334)
frovatriptan	antimigraine	2002	38 (357)
fudosteine	expectorant	2001	37 (267)
fulveristrant	anticancer	2002	38 (357)
gabapentin	anticonvulsant	1993	29 (338)
gabapentin Enacarbil	restless leg syndrome	2011	47 (533)
gadoversetamide	diagnostic	2000	36 (304)
gallium nitrate	calcium regulation	1991	27 (328)
gallopamil hydrochloride	antianginal	1983	19 (3190)
galsulfase	mucopolysaccharidosis VI	2005	41 (453)
ganciclovir	antiviral	1988	24 (303)
ganirelix acetate	infertility	2000	36 (305)
garenoxacin	antibacterial	2007	43 (471)
gatilfloxacin	antibacterial	1999	35 (340)
gefitinib	anticancer	2002	38 (358)
gemcitabine hydrochloride	anticancer	1995	31 (344)
gemeprost	abortifacient	1983	19 (319)
gemifloxacin	antibacterial	2004	40 (458)
gemtuzumab ozogamicin	anticancer	2000	36 (306)
gestodene	contraception	1987	23 (335)
gestrinone	contraception	1986	22 (321)
glatiramer acetate	multiple sclerosis	1997	33 (334)
glimepiride	antidiabetic	1995	31 (344)
glucagon, rDNA	antidiabetic	1993	29 (338)
GMDP	immunostimulant	1996	32 (308)
golimumab	antiinflammatory	2009	45 (503)
goserelin	hormone therapy	1987	23 (336)
granisetron hydrochloride	antiemetic	1991	27 (329)
guanadrel sulfate	antihypertensive	1983	19 (319)
gusperimus	immunosuppressant	1994	30 (300)
halobetasol propionate	antiinflammatory	1991	27 (329)
halofantrine	antimalarial	1988	24 (304)
halometasone	antiinflammatory	1983	19 (320)
histrelin	precocious puberty	1993	29 (338)
hydrocortisone aceponate	antiinflammatory	1988	24 (304)
hydrocortisone butyrate	antiinflammatory	1983	19 (320)

GENERIC NAME	INDICATION	YEAR INTRODUCED	ARMC VOL., (PAGE)
irsogladine	antiulcer	1989	25 (315)
isepamicin	antibacterial	1988	24 (305)
isofezolac	antiinflammatory	1984	20 (319)
isoxicam	antiinflammatory	1983	19 (320)
isradipine	antihypertensive	1989	25 (315)
itopride hydrochloride	gastroprokinetic	1995	31 (344)
itraconazole	antifungal	1988	24 (305)
ivabradine	antianginal	2006	42 (522)
ivacaftor	cystic fibrosis	2012	48 (501)
ivermectin	antiparasitic	1987	23 (336)
ixabepilone	anticancer	2007	43 (473)
ketanserin	antihypertensive	1985	21 (328)
ketorolac tromethamine	analgesic	1990	26 (304)
kinetin	dermatologic, skin photodamage	1999	35 (341)
lacidipine	antihypertensive	1991	27 (330)
lacosamide	anticonvulsant	2008	44 (610)
lafutidine	antiulcer	2000	36 (307)
lamivudine	antiviral	1995	31 (345)
lamotrigine	anticonvulsant	1990	26 (304)
landiolol	antiarrhythmic	2002	38 (360)
laninamivir octanoate	antiviral	2010	46 (470)
lanoconazole	antifungal	1994	30 (302)
lanreotide acetate	growth disorders	1995	31 (345)
lansoprazole	antiulcer	1992	28 (332)
lapatinib	anticancer	2007	43 (475)
laronidase	mucopolysaccharidosis I	2003	39 (278)
latanoprost	antiglaucoma	1996	32 (311)
lefunomide	antiarthritic	1998	34 (324)
lenalidomide	myelodysplastic syndromes, multiple myeloma	2006	42 (523)
lenampicillin hydrochloride	antibacterial	1987	23 (336)
lentinan	immunostimulant	1986	22 (322)
lepirudin	anticoagulant	1997	33 (336)
lercanidipine	antihypertensive	1997	33 (337)
letrazole	anticancer	1996	32 (311)
leuprolide acetate	hormone therapy	1984	20 (319)
levacecarnine hydrochloride	cognition enhancer	1986	22 (322)
levalbuterol hydrochloride	antiasthma	1999	35 (341)
levetiracetam	anticonvulsant	2000	36 (307)
levobunolol hydrochloride	antiglaucoma	1985	21 (328)

GENERIC NAME	INDICATION	YEAR INTRODUCED	ARMC VOL., (PAGE)
levobupivacaine hydrochloride	anesthetic	2000	36 (308)
levocabastine hydrochloride	antiallergy	1991	27 (330)
levocetirizine	antiallergy	2001	37 (268)
levodropropizine	antitussive	1988	24 (305)
levofloxacin	antibacterial	1993	29 (340)
levosimendan	congestive heart failure	2000	36 (308)
lidamidine hydrochloride	antidiarrheal	1984	20 (320)
limaprost	antithrombotic	1988	24 (306)
linaclotide	irritable bowel syndrome	2012	48 (502)
linagliptin	antidiabetic	2011	47 (540)
linezolid	antibacterial	2000	36 (309)
liraglutide	antidiabetic	2009	45 (507)
liranaftate	antifungal	2000	36 (309)
lisdexamfetamine	attention deficit hyperactivity disorder	2007	43 (477)
lisinopril	antihypertensive	1987	23 (337)
lobenzarit sodium	antiinflammatory	1986	22 (322)
lodoxamide tromethamine	antiallergy	1992	28 (333)
lomefloxacin	antibacterial	1989	25 (315)
lomerizine hydrochloride	antimigraine	1999	35 (342)
lomitapide mesylate	antihypercholersteremic	2012	48 (504)
lonidamine	anticancer	1987	23 (337)
lopinavir	antiviral	2000	36 (310)
loprazolam mesylate	sleep disorders	1983	19 (321)
loprinone hydrochloride	congestive heart failure	1996	32 (312)
loracarbef	antibacterial	1992	28 (333)
loratadine	antiallergy	1988	24 (306)
lorcaserin hydrochloride	antiobesity	2012	48 (506)
lornoxicam	antiinflammatory	1997	33 (337)
losartan	antihypertensive	1994	30 (302)
loteprednol etabonate	antiallergy	1998	34 (324)
lovastatin	antihypercholesterolemic	1987	23 (337)
loxoprofen sodium	antiinflammatory	1986	22 (322)
lulbiprostone	constipation	2006	42 (525)
luliconazole	antifungal	2005	41 (454)
lumiracoxib	antiinflammatory	2005	41 (455)
lurasidone hydrochloride	antipsychotic	2010	46 (473)
Lyme disease vaccine	Lyme disease	1999	35 (342)

GENERIC NAME	INDICATION	YEAR INTRODUCED	ARMC VOL., (PAGE)
mabuterol hydrochloride	antiasthma	1986	22 (323)
malotilate	hepatoprotective	1985	21 (329)
manidipine hydrochloride	antihypertensive	1990	26 (304)
maraviroc	antiviral	2007	43 (478)
masoprocol	anticancer	1992	28 (333)
maxacalcitol	hyperparathyroidism	2000	36 (310)
mebefradil hydrochloride	antihypertensive	1997	33 (338)
medifoxamine fumarate	antidepressant	1986	22 (323)
mefloquine hydrochloride	antimalarial	1985	21 (329)
meglutol	antihypercholesterolemic	1983	19 (321)
melinamide	antihypercholesterolemic	1984	20 (320)
meloxicam	antiarthritic	1996	32 (312)
mepixanox	respiratory stimulant	1984	20 (320)
meptazinol hydrochloride	analgesic	1983	19 (321)
meropenem	antibacterial	1994	30 (303)
metaclazepam	anxiolytic	1987	23 (338)
metapramine	antidepressant	1984	20 (320)
methylnaltrexone bromide	constipation	2008	44 (612)
mexazolam	anxiolytic	1984	20 (321)
micafungin	antifungal	2002	38 (360)
mifamurtide	anticancer	2009	46 (476)
mifepristone	abortifacient	1988	24 (306)
miglitol	antidiabetic	1998	34 (325)
miglustat	Gaucher's disease	2003	39 (279)
milnacipran	antidepressant	1997	33 (338)
milrinone	congestive heart failure	1989	25 (316)
miltefosine	anticancer	1993	29 (340)
minodronic acid	osteoporosis	2009	45 (509)
miokamycin	antibacterial	1985	21 (329)
mirabegron	urinary tract/bladder disorders	2011	47 (542)
mirtazapine	antidepressant	1994	30 (303)
misoprostol	antiulcer	1985	21 (329)
mitiglinide	antidiabetic	2004	40 (460)
mitoxantrone hydrochloride	anticancer	1984	20 (321)
mivacurium chloride	muscle relaxant	1992	28 (334)
mivotilate	hepatoprotective	1999	35 (343)
mizolastine	antiallergy	1998	34 (325)

GENERIC NAME	INDICATION	YEAR INTRODUCED	ARMC VOL., (PAGE)
mizoribine	immunosuppressant	1984	20 (321)
moclobemide	antidepressant	1990	26 (305)
modafinil	sleep disorders	1994	30 (303)
moexipril hydrochloride	antihypertensive	1995	31 (346)
mofezolac	analgesic	1994	30 (304)
mogamulizumab	anticancer	2012	48 (507)
mometasone furoate	antiinflammatory	1987	23 (338)
montelukast sodium	antiasthma	1998	34 (326)
moricizine hydrochloride	antiarrhythmic	1990	26 (305)
mosapride citrate	gastroprokinetic	1998	34 (326)
moxifloxacin hydrochloride	antibacterial	1999	35 (343)
moxonidine	antihypertensive	1991	27 (330)
mozavaptan	hyponatremia	2006	42 (527)
mupirocin	antibacterial	1985	21 (330)
muromonab–CD3	immunosuppressant	1986	22 (323)
muzolimine	diuretic	1983	19 (321)
mycophenolate mofetil	immunosuppressant	1995	31 (346)
mycophenolate sodium	immunosuppressant	2003	39 (279)
nabumetone	antiinflammatory	1985	21 (330)
nadifloxacin	antibacterial	1993	29 (340)
nafamostat mesylate	pancreatitis	1986	22 (323)
nafarelin acetate	hormone therapy	1990	26 (306)
naftifine hydrochloride	antifungal	1984	20 (321)
naftopidil	urinary tract/bladder disorders	1999	35 (344)
nalfurafine hydrochloride	pruritus	2009	45 (510)
nalmefene hydrochloride	addiction, opioids	1995	31 (347)
naltrexone hydrochloride	addiction, opioids	1984	20 (322)
naratriptan hydrochloride	antimigraine	1997	33 (339)
nartograstim	leukopenia	1994	30 (304)
natalizumab	multiple sclerosis	2004	40 (462)
nateglinide	antidiabetic	1999	35 (344)
nazasetron	antiemetic	1994	30 (305)
nebivolol	antihypertensive	1997	33 (339)
nedaplatin	anticancer	1995	31 (347)
nedocromil sodium	antiallergy	1986	22 (324)
nefazodone	antidepressant	1994	30 (305)
nelarabine	anticancer	2006	42 (528)

GENERIC NAME	INDICATION	YEAR INTRODUCED	ARMC VOL., (PAGE)
nelfinavir mesylate	antiviral	1997	33 (340)
neltenexine	cystic fibrosis	1993	29 (341)
nemonapride	antipsychotic	1991	27 (331)
nepafenac	antiinflammatory	2005	41 (456)
neridronic acide	calcium regulation	2002	38 (361)
nesiritide	congestive heart failure	2001	37 (269)
neticonazole hydrochloride	antifungal	1993	29 (341)
nevirapine	antiviral	1996	32 (313)
nicorandil	antianginal	1984	20 (322)
nif ekalant hydrochloride	antiarrhythmic	1999	35 (344)
nilotinib	anticancer	2007	43 (480)
nilutamide	anticancer	1987	23 (338)
nilvadipine	antihypertensive	1989	25 (316)
nimesulide	antiinflammatory	1985	21 (330)
nimodipine	cerebral vasodilator	1985	21 (330)
nimotuzumab	anticancer	2006	42 (529)
nipradilol	antihypertensive	1988	24 (307)
nisoldipine	antihypertensive	1990	26 (306)
nitisinone	antityrosinaemia	2002	38 (361)
nitrefazole	addiction, alcohol	1983	19 (322)
nitrendipine	antihypertensive	1985	21 (331)
nizatidine	antiulcer	1987	23 (339)
nizofenzone	nootropic	1988	24 (307)
nomegestrol acetate	contraception	1986	22 (324)
norelgestromin	contraception	2002	38 (362)
norfloxacin	antibacterial	1983	19 (322)
norgestimate	contraception	1986	22 (324)
OCT-43	anticancer	1999	35 (345)
octreotide	growth disorders	1988	24 (307)
ofatumumab	anticancer	2009	45 (512)
ofloxacin	antibacterial	1985	21 (331)
olanzapine	antipsychotic	1996	32 (313)
olimesartan Medoxomil	antihypertensive	2002	38 (363)
olopatadine hydrochloride	antiallergy	1997	33 (340)
omacetaxine mepesuccinate	anticancer	2012	48 (510)
omalizumab allergic	antiasthma	2003	39 (280)
omeprazole	antiulcer	1988	24 (308)
ondansetron hydrochloride	antiemetic	1990	26 (306)
OP-1	osteoinductor	2001	37 (269)
orlistat	antiobesity	1998	34 (327)

GENERIC NAME	INDICATION	YEAR INTRODUCED	ARMC VOL., (PAGE)
ornoprostil	antiulcer	1987	23 (339)
osalazine sodium	antiinflammatory	1986	22 (324)
oseltamivir phosphate	antiviral	1999	35 (346)
oxaliplatin	anticancer	1996	32 (313)
oxaprozin	antiinflammatory	1983	19 (322)
oxcarbazepine	anticonvulsant	1990	26 (307)
oxiconazole nitrate	antifungal	1983	19 (322)
oxiracetam	cognition enhancer	1987	23 (339)
oxitropium bromide	antiasthma	1983	19 (323)
ozagrel sodium	antithrombotic	1988	24 (308)
paclitaxal	anticancer	1993	29 (342)
palifermin	mucositis	2005	41 (461)
paliperidone	antipsychotic	2007	43 (482)
palonosetron	antiemetic	2003	39 (281)
panipenem/ betamipron carbapenem	antibacterial	1994	30 (305)
panitumumab	anticancer	2006	42 (531)
pantoprazole sodium	antiulcer	1995	30 (306)
parecoxib sodium	analgesic	2002	38 (364)
paricalcitol	hyperparathyroidism	1998	34 (327)
parnaparin sodium	anticoagulant	1993	29 (342)
paroxetine	antidepressant	1991	27 (331)
pasireotide	Cushing's Disease	2012	48 (512)
pazopanib	anticancer	2009	45 (514)
pazufloxacin	antibacterial	2002	38 (364)
pefloxacin mesylate	antibacterial	1985	21 (331)
pegademase bovine	immunostimulant	1990	26 (307)
pegaptanib	ophthalmologic (macular degeneration)	2005	41 (458)
pegaspargase	anticancer	1994	30 (306)
peginesatide	hematopoietic	2012	48 (514)
pegvisomant	growth disorders	2003	39 (281)
pemetrexed	anticancer	2004	40 (463)
pemirolast potassium	antiasthma	1991	27 (331)
penciclovir	antiviral	1996	32 (314)
pentostatin	anticancer	1992	28 (334)
peramivir	antiviral	2010	46 (477)
perampanel	anticonvulsant	2012	48 (516)
pergolide mesylate	Parkinson's disease	1988	24 (308)
perindopril	antihypertensive	1988	24 (309)
perospirone hydrochloride	antipsychotic	2001	37 (270)
pertuzumab	anticancer	2012	48 (517)
picotamide	antithrombotic	1987	23 (340)

GENERIC NAME	INDICATION	YEAR INTRODUCED	ARMC VOL., (PAGE)
pidotimod	immunostimulant	1993	29 (343)
piketoprofen	antiinflammatory	1984	20 (322)
pilsicainide hydrochloride	antiarrhythmic	1991	27 (332)
pimaprofen	antiinflammatory	1984	20 (322)
pimecrolimus	immunosuppressant	2002	38 (365)
pimobendan	congestive heart failure	1994	30 (307)
pinacidil	antihypertensive	1987	23 (340)
pioglitazone hydrochloride	antidiabetic	1999	35 (346)
pirarubicin	anticancer	1988	24 (309)
pirfenidone	pulmonary fibrosis, idiopathic	2008	44 (614)
pirmenol	antiarrhythmic	1994	30 (307)
piroxicam cinnamate	antiinflammatory	1988	24 (309)
pitavastatin	antihypercholesterolemic	2003	39 (282)
pivagabine	antidepressant	1997	33 (341)
pixantrone dimaleate	anticancer	2012	48 (519)
plaunotol	antiulcer	1987	23 (340)
plerixafor hydrochloride	stem cell mobilizer	2009	45 (515)
polaprezinc	antiulcer	1994	30 (307)
ponatinib	anticancer	2012	48 (521)
porfimer sodium	anticancer	1993	29 (343)
posaconazole	antifungal	2006	42 (532)
pralatrexate	anticancer	2009	45 (517)
pramipexole hydrochloride	Parkinson's disease	1997	33 (341)
pramiracetam sulfate	cognition enhancer	1993	29 (343)
pramlintide	antidiabetic	2005	41 (460)
pranlukast	antiasthma	1995	31 (347)
prasugrel	antiplatelet	2009	45 (519)
pravastatin	antihypercholesterolemic	1989	25 (316)
prednicarbate	antiinflammatory	1986	22 (325)
pregabalin	anticonvulsant	2004	40 (464)
prezatide copper acetate	wound healing agent	1996	32 (314)
progabide	anticonvulsant	1985	21 (331)
promegestrone	contraception	1983	19 (323)
propacetamol hydrochloride	analgesic	1986	22 (325)
propagermanium	antiviral	1994	30 (308)
propentofylline propionate	cerebral vasodilator	1988	24 (310)
propiverine hydrochloride	urinary tract/bladder disorders	1992	28 (335)
propofol	anesthetic	1986	22 (325)

GENERIC NAME	INDICATION	YEAR INTRODUCED	ARMC VOL., (PAGE)
sarpogrelate hydrochloride	antithrombotic	1993	29 (344)
saxagliptin	antidiabetic	2009	45 (521)
schizophyllan	immunostimulant	1985	22 (326)
seratrodast	antiasthma	1995	31 (349)
sertaconazole nitrate	antifungal	1992	28 (336)
sertindole	antipsychotic	1996	32 (318)
setastine hydrochloride	antiallergy	1987	23 (342)
setiptiline	antidepressant	1989	25 (318)
setraline hydrochloride	antidepressant	1990	26 (309)
sevoflurane	anesthetic	1990	26 (309)
sibutramine	antiobesity	1998	34 (331)
sildenafil citrate	male sexual dysfunction	1998	34 (331)
silodosin	urinary tract/bladder disorders	2006	42 (540)
simvastatin	antihypercholesterolemic	1988	24 (311)
sipuleucel-t	anticancer	2010	46 (484)
sitafloxacin hydrate	antibacterial	2008	44 (621)
sitagliptin	antidiabetic	2006	42 (541)
sitaxsentan	pulmonary hypertension	2006	42 (543)
sivelestat	antiinflammatory	2002	38 (366)
SKI-2053R	anticancer	1999	35 (348)
sobuzoxane	anticancer	1994	30 (310)
sodium cellulose phosphate	urinary tract/bladder disorders	1983	19 (323)
sofalcone	antiulcer	1984	20 (323)
solifenacin	urinary tract/bladder disorders	2004	40 (466)
somatomedin-1	growth disorders	1994	30 (310)
somatotropin	growth disorders	1994	30 (310)
somatropin	growth disorders	1987	23 (343)
sorafenib	anticancer	2005	41 (466)
sorivudine	antiviral	1993	29 (345)
sparfloxacin	antibacterial	1993	29 (345)
spirapril hydrochloride	antihypertensive	1995	31 (349)
spizofurone	antiulcer	1987	23 (343)
stavudine	antiviral	1994	30 (311)
strontium ranelate	osteoporosis	2004	40 (466)
succimer	antidote, lead poisoning	1991	27 (333)
sufentanil	analgesic	1983	19 (323)
sugammadex	neuromuscular blockade, reversal	2008	44 (623)
sulbactam sodium	antibacterial	1986	22 (326)
sulconizole nitrate	antifungal	1985	21 (332)
sultamycillin tosylate	antibacterial	1987	23 (343)

GENERIC NAME	INDICATION	YEAR INTRODUCED	ARMC VOL., (PAGE)
sumatriptan succinate	antimigraine	1991	27 (333)
sunitinib	anticancer	2006	42 (544)
suplatast tosilate	antiallergy	1995	31 (350)
suprofen	analgesic	1983	19 (324)
surfactant TA	respiratory surfactant	1987	23 (344)
tacalcitol	antipsoriasis	1993	29 (346)
tacrine hydrochloride	Alzheimer's disease	1993	29 (346)
tacrolimus	immunosuppressant	1993	29 (347)
tadalafil	male sexual dysfunction	2003	39 (284)
tafamidis	neurodegeneration	2011	47 (550)
tafluprost	antiglaucoma	2008	44 (625)
talaporfin sodium	anticancer	2004	40 (469)
talipexole	Parkinson's disease	1996	32 (318)
taltirelin	neurodegeneration	2000	36 (311)
tamibarotene	anticancer	2005	41 (467)
tamsulosin hydrochloride	benign prostatic hyperplasia	1993	29 (347)
tandospirone	anxiolytic	1996	32 (319)
tapentadol hydrochloride	analgesic	2009	45 (523)
tasonermin	anticancer	1999	35 (349)
tazanolast	antiallergy	1990	26 (309)
tazarotene	antipsoriasis	1997	33 (343)
tazobactam sodium	antibacterial	1992	28 (336)
teduglutide	short bowel syndrome	2012	48 (526)
tegaserod maleate	irritable bowel syndrome	2001	37 (270)
teicoplanin	antibacterial	1988	24 (311)
telaprevir	antiviral	2011	47 (552)
telavancin	antibacterial	2009	45 (525)
telbivudine	antiviral	2006	42 (546)
telithromycin	antibacterial	2001	37 (271)
telmesteine	expectorant	1992	28 (337)
telmisartan	antihypertensive	1999	35 (349)
temafloxacin hydrochloride	antibacterial	1991	27 (334)
temocapril	antihypertensive	1994	30 (311)
temocillin disodium	antibacterial	1984	20 (323)
temoporphin	anticancer	2002	38 (367)
temozolomide	anticancer	1999	35 (349)
temsirolimus	anticancer	2007	43 (490)
teneligliptin	antidiabetic	2012	48 (528)
tenofovir disoproxil fumarate	antiviral	2001	37 (271)
tenoxicam	antiinflammatory	1987	23 (344)
teprenone	antiulcer	1984	20 (323)

GENERIC NAME	INDICATION	YEAR INTRODUCED	ARMC VOL., (PAGE)
terazosin hydrochloride	antihypertensive	1984	20 (323)
terbinafine hydrochloride	antifungal	1991	27 (334)
terconazole	antifungal	1983	19 (324)
teriflunomide	multiple sclerosis	2012	48 (530)
tertatolol hydrochloride	antihypertensive	1987	23 (344)
tesamorelin acetate	lipodystrophy	2010	46 (486)
thrombin alfa	hemostatic	2008	44 (627)
thrombomodulin (recombinant)	anticoagulant	2008	44 (628)
thymopentin	immunomodulator	1985	21 (333)
tiagabine	anticonvulsant	1996	32 (319)
tiamenidine hydrochloride	antihypertensive	1988	24 (311)
tianeptine sodium	antidepressant	1983	19 (324)
tibolone	hormone therapy	1988	24 (312)
ticagrelor	antithrombotic	2010	46 (488)
tigecycline	antibacterial	2005	41 (468)
tilisolol hydrochloride	antihypertensive	1992	28 (337)
tiludronate disodium	Paget's disease	1995	31 (350)
timiperone	antipsychotic	1984	20 (323)
tinazoline	nasal decongestant	1988	24 (312)
tioconazole	antifungal	1983	19 (324)
tiopronin	urolithiasis	1989	25 (318)
tiotropium bromide	chronic obstructive pulmonary disorder	2002	38 (368)
tipranavir	antiviral	2005	41 (470)
tiquizium bromide	antispasmodic	1984	20 (324)
tiracizine hydrochloride	antiarrhythmic	1990	26 (310)
tirilazad mesylate	subarachnoid hemorrhage	1995	31 (351)
tirofiban hydrochloride	antithrombotic	1998	34 (332)
tiropramide hydrochloride	muscle relaxant	1983	19 (324)
tizanidine	muscle relaxant	1984	20 (324)
tofacitinib	antiarthritic	2012	48 (532)
tolcapone	Parkinson's disease	1997	33 (343)
toloxatone	antidepressant	1984	20 (324)
tolrestat	antidiabetic	1989	25 (319)
tolvaptan	hyponatremia	2009	45 (528)
topiramate	anticonvulsant	1995	31 (351)
topotecan hydrochloride	anticancer	1996	32 (320)
torasemide	diuretic	1993	29 (348)
toremifene	anticancer	1989	25 (319)
tositumomab	anticancer	2003	39 (285)

GENERIC NAME	INDICATION	YEAR INTRODUCED	ARMC VOL., (PAGE)
tosufloxacin tosylate	antibacterial	1990	26 (310)
trabectedin	anticancer	2007	43 (492)
trandolapril	antihypertensive	1993	29 (348)
travoprost	antiglaucoma	2001	37 (272)
treprostinil sodium	antihypertensive	2002	38 (368)
tretinoin tocoferil	antiulcer	1993	29 (348)
trientine hydrochloride	antidote, copper poisoning	1986	22 (327)
trimazosin hydrochloride	antihypertensive	1985	21 (333)
trimegestone	contraception	2001	37 (273)
trimetrexate glucuronate	antifungal	1994	30 (312)
troglitazone	antidiabetic	1997	33 (344)
tropisetron	antiemetic	1992	28 (337)
trovafloxacin mesylate	antibacterial	1998	34 (332)
troxipide	antiulcer	1986	22 (327)
ubenimex	immunostimulant	1987	23 (345)
udenafil	male sexual dysfunction	2005	41 (472)
ulipristal acetate	contraception	2009	45 (530)
unoprostone isopropyl ester	antiglaucoma	1994	30 (312)
ustekinumab	antipsoriasis	2009	45 (532)
vadecoxib	antiarthritic	2002	38 (369)
vaglancirclovir hydrochloride	antiviral	2001	37 (273)
valaciclovir hydrochloride	antiviral	1995	31 (352)
valrubicin	anticancer	1999	35 (350)
valsartan	antihypertensive	1996	32 (320)
vandetanib	anticancer	2011	47 (555)
vardenafil	male sexual dysfunction	2003	39 (286)
varenicline	addiction, nicotine	2006	42 (547)
vemurafenib	anticancer	2011	47 (556)
venlafaxine	antidepressant	1994	30 (312)
vernakalant	antiarrhythmic	2010	46 (491)
verteporfin	ophthalmologic (macular degeneration)	2000	36 (312)
vesnarinone	congestive heart failure	1990	26 (310)
vigabatrin	anticonvulsant	1989	25 (319)
vilazodone	antidepressant	2011	47 (558)
vildagliptin	antidiabetic	2007	43 (494)
vinflunine	anticancer	2009	46 (493)
vinorelbine	anticancer	1989	25 (320)
vismodegib	anticancer	2012	48 (534)
voglibose	antidiabetic	1994	30 (313)

GENERIC NAME	INDICATION	YEAR INTRODUCED	ARMC VOL., (PAGE)
voriconazole	antifungal	2002	38 (370)
vorinostat	anticancer	2006	42 (549)
xamoterol fumarate	congestive heart failure	1988	24 (312)
ximelagatran	anticoagulant	2004	40 (470)
zafirlukast	antiasthma	1996	32 (321)
zalcitabine	antiviral	1992	28 (338)
zaleplon	sleep disorders	1999	35 (351)
zaltoprofen	antiinflammatory	1993	29 (349)
zanamivir	antiviral	1999	35 (352)
ziconotide	analgesic	2005	41 (473)
zidovudine	antiviral	1987	23 (345)
zileuton	antiasthma	1997	33 (344)
zinostatin stimalamer	anticancer	1994	30 (313)
ziprasidone hydrochloride	antipsychotic	2000	36 (312)
zofenopril calcium	antihypertensive	2000	36 (313)
zoledronate disodium	osteoporosis	2000	36 (314)
zolpidem hemitartrate	sleep disorders	1988	24 (313)
zomitriptan	antimigraine	1997	33 (345)
zonisamide	anticonvulsant	1989	25 (320)
zopiclone	sleep disorders	1986	22 (327)
zucapsaicin	analgesic	2010	46 (495)
zuclopenthixol acetate	antipsychotic	1987	23 (345)

CUMULATIVE NCE INTRODUCTION INDEX, 1983–2012 (BY INDICATION)

GENERIC NAME	INDICATION	YEAR INTRODUCED	ARMC VOL., (PAGE)
gemeprost	abortifacient	1983	19 (319)
mifepristone	abortifacient	1988	24 (306)
azelaic acid	acne	1989	25 (310)
nitrefazole	addiction, alcohol	1983	19 (322)
varenicline	addiction, nicotine	2006	42 (547)
naltrexone hydrochloride	addiction, opioids	1984	20 (322)
nalmefene hydrochloride	addiction, opioids	1995	31 (347)
tacrine hydrochloride	Alzheimer's disease	1993	29 (346)
donepezil hydrochloride	Alzheimer's disease	1997	33 (332)
rivastigmin	Alzheimer's disease	1997	33 (342)
fasudil hydrochloride	amyotrophic lateral sclerosis	1995	31 (343)
riluzole	amyotrophic lateral sclerosis	1996	32 (316)
alfentanil hydrochloride	analgesic	1983	19 (314)
alminoprofen	analgesic	1983	19 (314)
meptazinol hydrochloride	analgesic	1983	19 (321)
sufentanil	analgesic	1983	19 (323)
suprofen	analgesic	1983	19 (324)
emorfazone	analgesic	1984	20 (317)
fosfosal	analgesic	1984	20 (319)
flupirtine maleate	analgesic	1985	21 (328)
propacetamol hydrochloride	analgesic	1986	22 (325)
eptazocine hydrobromide	analgesic	1987	23 (334)
ketorolac tromethamine	analgesic	1990	26 (304)
dezocine	analgesic	1991	27 (326)
mofezolac	analgesic	1994	30 (304)
remifentanil hydrochloride	analgesic	1996	32 (316)
parecoxib sodium	analgesic	2002	38 (364)
ziconotide	analgesic	2005	41 (473)
tapentadol hydrochloride	analgesic	2009	45 (523)
zucapsaicin	analgesic	2010	46 (495)
propofol	anesthetic	1986	22 (325)
sevoflurane	anesthetic	1990	26 (309)
desflurane	anesthetic	1992	28 (329)

GENERIC NAME	INDICATION	YEAR INTRODUCED	ARMC VOL., (PAGE)
ropivacaine	anesthetic	1996	32 (318)
levobupivacaine hydrochloride	anesthetic	2000	36 (308)
icatibant	angioedema, hereditary	2008	44 (608)
ecallantide	angioedema, hereditary	2009	46 (464)
astemizole	antiallergy	1983	19 (314)
azelastine hydrochloride	antiallergy	1986	22 (316)
nedocromil sodium	antiallergy	1986	22 (324)
cetirizine hydrochloride	antiallergy	1987	23 (331)
repirinast	antiallergy	1987	23 (341)
setastine hydrochloride	antiallergy	1987	23 (342)
acrivastine	antiallergy	1988	24 (295)
loratadine	antiallergy	1988	24 (306)
ebastine	antiallergy	1990	26 (302)
tazanolast	antiallergy	1990	26 (309)
levocabastine hydrochloride	antiallergy	1991	27 (330)
lodoxamide tromethamine	antiallergy	1992	28 (333)
emedastine difumarate	antiallergy	1993	29 (336)
epinastine	antiallergy	1994	30 (299)
suplatast tosilate	antiallergy	1995	31 (350)
fexofenadine	antiallergy	1996	32 (307)
olopatadine hydrochloride	antiallergy	1997	33 (340)
loteprednol etabonate	antiallergy	1998	34 (324)
mizolastine	antiallergy	1998	34 (325)
betotastine besilate	antiallergy	2000	36 (297)
ramatroban	antiallergy	2000	36 (311)
desloratadine	antiallergy	2001	37 (264)
levocetirizine	antiallergy	2001	37 (268)
rupatadine fumarate	antiallergy	2003	39 (284)
fluticasone furoate	antiallergy	2007	43 (469)
bilastine	antiallergy	2010	46 (449)
gallopamil hydrochloride	antianginal	1983	19 (3190
nicorandil	antianginal	1984	20 (322)
ivabradine	antianginal	2006	42 (522)
ranolazine	antianginal	2006	42 (535)
cibenzoline	antiarrhythmic	1985	21 (325)
encainide hydrochloride	antiarrhythmic	1987	23 (333)
esmolol hydrochloride	antiarrhythmic	1987	23 (334)

GENERIC NAME	INDICATION	YEAR INTRODUCED	ARMC VOL., (PAGE)
moricizine hydrochloride	antiarrhythmic	1990	26 (305)
tiracizine hydrochloride	antiarrhythmic	1990	26 (310)
pilsicainide hydrochloride	antiarrhythmic	1991	27 (332)
pirmenol	antiarrhythmic	1994	30 (307)
ibutilide fumarate	antiarrhythmic	1996	32 (309)
nif ckalant hydrochloride	antiarrhythmic	1999	35 (344)
dofetilide	antiarrhythmic	2000	36 (301)
landiolol	antiarrhythmic	2002	38 (360)
dronedarone	antiarrhythmic	2009	45 (495)
vernakalant	antiarrhythmic	2010	46 (491)
auranofin	antiarthritic	1983	19 (314)
meloxicam	antiarthritic	1996	32 (312)
lefunomide	antiarthritic	1998	34 (324)
celecoxib	antiarthritic	1999	35 (335)
rofecoxib	antiarthritic	1999	35 (347)
anakinra	antiarthritic	2001	37 (261)
etoricoxibe	antiarthritic	2002	38 (355)
vadecoxib	antiarthritic	2002	38 (369)
adalimumab	antiarthritic	2003	39 (267)
abatacept	antiarthritic	2006	42 (509)
iguratimod	antiarthritic	2011	47 (535)
tofacitinib	antiarthritic	2012	48 (532)
oxitropium bromide	antiasthma	1983	19 (323)
doxofylline	antiasthma	1985	21 (327)
mabuterol hydrochloride	antiasthma	1986	22 (323)
amlexanox	antiasthma	1987	23 (327)
flutropium bromide	antiasthma	1988	24 (303)
ibudilast	antiasthma	1989	25 (313)
bambuterol	antiasthma	1990	26 (299)
salmeterol hydroxynaphthoate	antiasthma	1990	26 (308)
pemirolast potassium	antiasthma	1991	27 (331)
pranlukast	antiasthma	1995	31 (347)
seratrodast	antiasthma	1995	31 (349)
zafirlukast	antiasthma	1996	32 (321)
zileuton	antiasthma	1997	33 (344)
montelukast sodium	antiasthma	1998	34 (326)
levalbuterol hydrochloride	antiasthma	1999	35 (341)
omalizumab allergic	antiasthma	2003	39 (280)
ciclesonide	antiasthma	2005	41 (443)

GENERIC NAME	INDICATION	YEAR INTRODUCED	ARMC VOL., (PAGE)
arformoterol	antiasthma	2007	43 (465)
cefmenoxime hydrochloride	antibacterial	1983	19 (316)
ceftazidime	antibacterial	1983	19 (316)
norfloxacin	antibacterial	1983	19 (322)
adamantanium bromide	antibacterial	1984	20 (315)
aztreonam	antibacterial	1984	20 (315)
cefonicid sodium	antibacterial	1984	20 (316)
ceforanide	antibacterial	1984	20 (317)
cefotetan disodium	antibacterial	1984	20 (317)
temocillin disodium	antibacterial	1984	20 (323)
astromycin sulfate	antibacterial	1985	21 (324)
cefbuperazone sodium	antibacterial	1985	21 (325)
cefpiramide sodium	antibacterial	1985	21 (325)
imipenem/cilastatin	antibacterial	1985	21 (328)
miokamycin	antibacterial	1985	21 (329)
mupirocin	antibacterial	1985	21 (330)
ofloxacin	antibacterial	1985	21 (331)
pefloxacin mesylate	antibacterial	1985	21 (331)
rifaximin	antibacterial	1985	21 (332)
carboplatin	antibacterial	1986	22 (318)
ciprofloxacin	antibacterial	1986	22 (318)
enoxacin	antibacterial	1986	22 (320)
rokitamycin	antibacterial	1986	22 (325)
sulbactam sodium	antibacterial	1986	22 (326)
aspoxicillin	antibacterial	1987	23 (328)
cefixime	antibacterial	1987	23 (329)
cefminox sodium	antibacterial	1987	23 (330)
cefpimizole	antibacterial	1987	23 (330)
cefteram pivoxil	antibacterial	1987	23 (330)
cefuroxime axetil	antibacterial	1987	23 (331)
cefuzonam sodium	antibacterial	1987	23 (331)
lenampicillin hydrochloride	antibacterial	1987	23 (336)
rifaximin	antibacterial	1987	23 (341)
sultamycillin tosylate	antibacterial	1987	23 (343)
azithromycin	antibacterial	1988	24 (298)
carumonam	antibacterial	1988	24 (298)
erythromycin acistrate	antibacterial	1988	24 (301)
flomoxef sodium	antibacterial	1988	24 (302)
isepamicin	antibacterial	1988	24 (305)
rifapentine	antibacterial	1988	24 (310)
teicoplanin	antibacterial	1988	24 (311)
cefpodoxime proxetil	antibacterial	1989	25 (310)
lomefloxacin	antibacterial	1989	25 (315)

GENERIC NAME	INDICATION	YEAR INTRODUCED	ARMC VOL., (PAGE)
RV-11	antibacterial	1989	25 (318)
arbekacin	antibacterial	1990	26 (298)
cefodizime sodium	antibacterial	1990	26 (300)
clarithromycin	antibacterial	1990	26 (302)
tosufloxacin tosylate	antibacterial	1990	26 (310)
cefdinir	antibacterial	1991	27 (323)
cefotiam hexetil hydrochloride	antibacterial	1991	27 (324)
temafloxacin hydrochloride	antibacterial	1991	27 (334)
cefetamet pivoxil hydrochloride	antibacterial	1992	28 (327)
cefpirome sulfate	antibacterial	1992	28 (328)
cefprozil	antibacterial	1992	28 (328)
ceftibuten	antibacterial	1992	28 (329)
fleroxacin	antibacterial	1992	28 (331)
loracarbef	antibacterial	1992	28 (333)
rifabutin	antibacterial	1992	28 (335)
rufloxacin hydrochloride	antibacterial	1992	28 (335)
tazobactam sodium	antibacterial	1992	28 (336)
brodimoprin	antibacterial	1993	29 (333)
cefepime	antibacterial	1993	29 (334)
dirithromycin	antibacterial	1993	29 (336)
levofloxacin	antibacterial	1993	29 (340)
nadifloxacin	antibacterial	1993	29 (340)
sparfloxacin	antibacterial	1993	29 (345)
cefditoren pivoxil	antibacterial	1994	30 (297)
meropenem	antibacterial	1994	30 (303)
panipenem/ betamipron carbapenem	antibacterial	1994	30 (305)
cefozopran hydrochloride	antibacterial	1995	31 (339)
cefcapene pivoxil	antibacterial	1997	33 (330)
flurithromycin ethylsuccinate	antibacterial	1997	33 (333)
fropenam	antibacterial	1997	33 (334)
cefoselis	antibacterial	1998	34 (319)
trovafloxacin mesylate	antibacterial	1998	34 (332)
dalfopristin	antibacterial	1999	35 (338)
gatilfloxacin	antibacterial	1999	35 (340)
moxifloxacin hydrochloride	antibacterial	1999	35 (343)
quinupristin	antibacterial	1999	35 (338)

GENERIC NAME	INDICATION	YEAR INTRODUCED	ARMC VOL., (PAGE)
linezolid	antibacterial	2000	36 (309)
telithromycin	antibacterial	2001	37 (271)
balofloxacin	antibacterial	2002	38 (351)
biapenem	antibacterial	2002	38 (351)
ertapenem sodium	antibacterial	2002	38 (353)
pazufloxacin	antibacterial	2002	38 (364)
prulifloxacin	antibacterial	2002	38 (366)
daptomycin	antibacterial	2003	39 (272)
gemifloxacin	antibacterial	2004	40 (458)
doripenem	antibacterial	2005	41 (448)
tigecycline	antibacterial	2005	41 (468)
garenoxacin	antibacterial	2007	43 (471)
retapamulin	antibacterial	2007	43 (486)
ceftobiprole medocaril	antibacterial	2008	44 (589)
sitafloxacin hydrate	antibacterial	2008	44 (621)
besifloxacin	antibacterial	2009	45 (482)
telavancin	antibacterial	2009	45 (525)
ceftaroline fosamil	antibacterial	2010	46 (453)
fidaxomicin	antibacterial	2011	47 (531)
bedaquiline	antibacterial	2012	48 (487)
enocitabine	anticancer	1983	19 (318)
flutamide	anticancer	1983	19 (318)
epirubicin hydrochloride	anticancer	1984	20 (318)
mitoxantrone hydrochloride	anticancer	1984	20 (321)
camostat mesylate	anticancer	1985	21 (325)
amsacrine	anticancer	1987	23 (327)
doxifluridine	anticancer	1987	23 (332)
lonidamine	anticancer	1987	23 (337)
nilutamide	anticancer	1987	23 (338)
ranimustine	anticancer	1987	23 (341)
pirarubicin	anticancer	1988	24 (309)
fotemustine	anticancer	1989	25 (313)
interleukin-2	anticancer	1989	25 (314)
toremifene	anticancer	1989	25 (319)
vinorelbine	anticancer	1989	25 (320)
bisantrene hydrochloride	anticancer	1990	26 (300)
idarubicin hydrochloride	anticancer	1990	26 (303)
fludarabine phosphate	anticancer	1991	27 (327)
interferon, gamma-1	anticancer	1992	28 (332)
masoprocol	anticancer	1992	28 (333)
pentostatin	anticancer	1992	28 (334)

GENERIC NAME	INDICATION	YEAR INTRODUCED	ARMC VOL., (PAGE)
cladribine	anticancer	1993	29 (335)
cytarabine ocfosfate	anticancer	1993	29 (335)
formestane	anticancer	1993	29 (337)
miltefosine	anticancer	1993	29 (340)
paclitaxal	anticancer	1993	29 (342)
porfimer sodium	anticancer	1993	29 (343)
irinotecan	anticancer	1994	30 (301)
pegaspargase	anticancer	1994	30 (306)
sobuzoxane	anticancer	1994	30 (310)
zinostatin stimalamer	anticancer	1994	30 (313)
anastrozole	anticancer	1995	31 (338)
bicalutamide	anticancer	1995	31 (338)
docetaxel	anticancer	1995	31 (341)
fadrozole hydrochloride	anticancer	1995	31 (342)
gemcitabine hydrochloride	anticancer	1995	31 (344)
nedaplatin	anticancer	1995	31 (347)
letrazole	anticancer	1996	32 (311)
oxaliplatin	anticancer	1996	32 (313)
raltitrexed	anticancer	1996	32 (315)
topotecan hydrochloride	anticancer	1996	32 (320)
capecitabine	anticancer	1998	34 (319)
alitretinoin	anticancer	1999	35 (333)
arglabin	anticancer	1999	35 (335)
denileukin diftitox	anticancer	1999	35 (338)
OCT-43	anticancer	1999	35 (345)
SKI-2053R	anticancer	1999	35 (348)
tasonermin	anticancer	1999	35 (349)
temozolomide	anticancer	1999	35 (349)
valrubicin	anticancer	1999	35 (350)
bexarotene	anticancer	2000	36 (298)
exemestane	anticancer	2000	36 (304)
gemtuzumab ozogamicin	anticancer	2000	36 (306)
alemtuzumab	anticancer	2001	37 (260)
imatinib mesylate	anticancer	2001	37 (267)
amrubicin hydrochloride	anticancer	2002	38 (349)
fulveristrant	anticancer	2002	38 (357)
gefitinib	anticancer	2002	38 (358)
ibritunomab tiuxetan	anticancer	2002	38 (359)
temoporphin	anticancer	2002	38 (367)
bortezomib	anticancer	2003	39 (271)
cetuximab	anticancer	2003	39 (272)

GENERIC NAME	INDICATION	YEAR INTRODUCED	ARMC VOL., (PAGE)
tositumomab	anticancer	2003	39 (285)
abarelix	anticancer	2004	40 (446)
azacitidine	anticancer	2004	40 (447)
belotecan	anticancer	2004	40 (449)
bevacizumab	anticancer	2004	40 (450)
erlotinib	anticancer	2004	40 (454)
pemetrexed	anticancer	2004	40 (463)
talaporfin sodium	anticancer	2004	40 (469)
clofarabine	anticancer	2005	41 (444)
sorafenib	anticancer	2005	41 (466)
tamibarotene	anticancer	2005	41 (467)
dasatinib	anticancer	2006	42 (517)
nelarabine	anticancer	2006	42 (528)
nimotuzumab	anticancer	2006	42 (529)
panitumumab	anticancer	2006	42 (531)
sunitinib	anticancer	2006	42 (544)
vorinostat	anticancer	2006	42 (549)
ixabepilone	anticancer	2007	43 (473)
lapatinib	anticancer	2007	43 (475)
temsirolimus	anticancer	2007	43 (490)
trabectedin	anticancer	2007	43 (492)
catumaxomab	anticancer	2009	45 (486)
degarelix acetate	anticancer	2009	45 (490)
ofatumumab	anticancer	2009	45 (512)
pazopanib	anticancer	2009	45 (514)
pralatrexate	anticancer	2009	45 (517)
abiraterone acetate	anticancer	2011	47 (505)
brentuximab	anticancer	2011	47 (523)
crizotinib	anticancer	2011	47 (525)
ipilimumab	anticancer	2011	47 (537)
ruxolitinib	anticancer	2011	47 (548)
vandetanib	anticancer	2011	47 (555)
vemurafenib	anticancer	2011	47 (556)
axitinib	anticancer	2012	48 (485)
bosutinib	anticancer	2012	48 (489)
cabozantinib	anticancer	2012	48 (491)
carfilzomib	anticancer	2012	48 (492)
enzalutamide	anticancer	2012	48 (497)
ingenol mebutate	anticancer	2012	48 (499)
mogamulizumab	anticancer	2012	48 (507)
omacetaxine mepesuccinate	anticancer	2012	48 (510)
pertuzumab	anticancer	2012	48 (517)
pixantrone dimaleate	anticancer	2012	48 (519)
ponatinib	anticancer	2012	48 (521)

GENERIC NAME	INDICATION	YEAR INTRODUCED	ARMC VOL., (PAGE)
radotinib	anticancer	2012	48 (523)
regorafenib	anticancer	2012	48 (524)
vismodegib	anticancer	2012	48 (534)
nilotinib	anticancer	2007	43 (480)
mifamurtide	anticancer	2009	46 (476)
romidepsin	anticancer	2009	46 (482)
vinflunine	anticancer	2009	46 (493)
cabazitaxel	anticancer	2010	46 (451)
eribulin mesylate	anticancer	2010	46 (465)
sipuleucel–t	anticancer	2010	46 (484)
angiotensin II	anticancer adjuvant	1994	30 (296)
enoxaparin	anticoagulant	1987	23 (333)
parnaparin sodium	anticoagulant	1993	29 (342)
reviparin sodium	anticoagulant	1993	29 (344)
duteplase	anticoagulant	1995	31 (342)
lepirudin	anticoagulant	1997	33 (336)
ximelagatran	anticoagulant	2004	40 (470)
dabigatran etexilate	anticoagulant	2008	44 (598)
rivaroxaban	anticoagulant	2008	44 (617)
thrombomodulin (recombinant)	anticoagulant	2008	44 (628)
progabide	anticonvulsant	1985	21 (331)
vigabatrin	anticonvulsant	1989	25 (319)
zonisamide	anticonvulsant	1989	25 (320)
lamotrigine	anticonvulsant	1990	26 (304)
oxcarbazepine	anticonvulsant	1990	26 (307)
felbamate	anticonvulsant	1993	29 (337)
gabapentin	anticonvulsant	1993	29 (338)
topiramate	anticonvulsant	1995	31 (351)
fosphenytoin sodium	anticonvulsant	1996	32 (308)
tiagabine	anticonvulsant	1996	32 (319)
levetiracetam	anticonvulsant	2000	36 (307)
pregabalin	anticonvulsant	2004	40 (464)
rufinamide	anticonvulsant	2007	43 (488)
lacosamide	anticonvulsant	2008	44 (610)
eslicarbazepine acetate	anticonvulsant	2009	45 (498)
retigabine	anticonvulsant	2011	47 (544)
perampanel	anticonvulsant	2012	48 (516)
fluvoxamine maleate	antidepressant	1983	19 (319)
indalpine	antidepressant	1983	19 (320)
tianeptine sodium	antidepressant	1983	19 (324)
metapramine	antidepressant	1984	20 (320)
toloxatone	antidepressant	1984	20 (324)
fluoxetine hydrochloride	antidepressant	1986	22 (320)

GENERIC NAME	INDICATION	YEAR INTRODUCED	ARMC VOL., (PAGE)
medifoxamine fumarate	antidepressant	1986	22 (323)
bupropion hydrochloride	antidepressant	1989	25 (310)
citalopram	antidepressant	1989	25 (311)
setiptiline	antidepressant	1989	25 (318)
moclobemide	antidepressant	1990	26 (305)
setraline hydrochloride	antidepressant	1990	26 (309)
paroxetine	antidepressant	1991	27 (331)
mirtazapine	antidepressant	1994	30 (303)
nefazodone	antidepressant	1994	30 (305)
venlafaxine	antidepressant	1994	30 (312)
milnacipran	antidepressant	1997	33 (338)
pivagabine	antidepressant	1997	33 (341)
reboxetine	antidepressant	1997	33 (342)
escitalopram oxolate	antidepressant	2002	38 (354)
duloxetine	antidepressant	2004	40 (452)
desvenlafaxine	antidepressant	2008	44 (600)
vilazodone	antidepressant	2011	47 (558)
tolrestat	antidiabetic	1989	25 (319)
acarbose	antidiabetic	1990	26 (297)
epalrestat	antidiabetic	1992	28 (330)
glucagon, rDNA	antidiabetic	1993	29 (338)
voglibose	antidiabetic	1994	30 (313)
glimepiride	antidiabetic	1995	31 (344)
insulin lispro	antidiabetic	1996	32 (310)
troglitazone	antidiabetic	1997	33 (344)
miglitol	antidiabetic	1998	34 (325)
repaglinide	antidiabetic	1998	34 (329)
nateglinide	antidiabetic	1999	35 (344)
pioglitazone hydrochloride	antidiabetic	1999	35 (346)
rosiglitazone maleate	antidiabetic	1999	35 (348)
mitiglinide	antidiabetic	2004	40 (460)
exenatide	antidiabetic	2005	41 (452)
pramlintide	antidiabetic	2005	41 (460)
sitagliptin	antidiabetic	2006	42 (541)
vildagliptin	antidiabetic	2007	43 (494)
liraglutide	antidiabetic	2009	45 (507)
saxagliptin	antidiabetic	2009	45 (521)
alogliptin	antidiabetic	2010	46 (446)
linagliptin	antidiabetic	2011	47 (540)
anagliptin	antidiabetic	2012	48 (483)
dapagliflozin	antidiabetic	2012	48 (495)
teneligliptin	antidiabetic	2012	48 (528)
lidamidine hydrochloride	antidiarrheal	1984	20 (320)

GENERIC NAME	INDICATION	YEAR INTRODUCED	ARMC VOL., (PAGE)
acetorphan	antidiarrheal	1993	29 (332)
crofelemer	antidiarrheal	2012	48 (494)
flumazenil	antidote, benzodiazepine overdose	1987	23 (335)
trientine hydrochloride	antidote, copper poisoning	1986	22 (327)
anti-digoxin polyclonal antibody	antidote, digoxin poisoning	2002	38 (350)
fomepizole	antidote, ethylene glycol poisoning	1998	34 (323)
succimer	antidote, lead poisoning	1991	27 (333)
crotelidae polyvalent immune fab	antidote, snake venom poisoning	2001	37 (263)
dronabinol	antiemetic	1986	22 (319)
ondansetron hydrochloride	antiemetic	1990	26 (306)
granisetron hydrochloride	antiemetic	1991	27 (329)
tropisetron	antiemetic	1992	28 (337)
nazasetron	antiemetic	1994	30 (305)
ramosetron	antiemetic	1996	32 (315)
dolasetron mesylate	antiemetic	1998	34 (321)
aprepitant	antiemetic	2003	39 (268)
palonosetron	antiemetic	2003	39 (281)
indisetron	antiemetic	2004	40 (459)
fosaprepitant dimeglumine	antiemetic	2008	44 (606)
oxiconazole nitrate	antifungal	1983	19 (322)
terconazole	antifungal	1983	19 (324)
tioconazole	antifungal	1983	19 (324)
naftifine hydrochloride	antifungal	1984	20 (321)
sulconizole nitrate	antifungal	1985	21 (332)
butoconazole	antifungal	1986	22 (318)
cloconazole hydrochloride	antifungal	1986	22 (318)
fenticonazole nitrate	antifungal	1987	23 (334)
fluconazole	antifungal	1988	24 (303)
itraconazole	antifungal	1988	24 (305)
amorolfine hydrochloride	antifungal	1991	27 (322)
terbinafine hydrochloride	antifungal	1991	27 (334)
butenafine hydrochloride	antifungal	1992	28 (327)
sertaconazole nitrate	antifungal	1992	28 (336)
neticonazole hydrochloride	antifungal	1993	29 (341)

GENERIC NAME	INDICATION	YEAR INTRODUCED	ARMC VOL., (PAGE)
lanoconazole	antifungal	1994	30 (302)
trimetrexate glucuronate	antifungal	1994	30 (312)
flutrimazole	antifungal	1995	31 (343)
liranaftate	antifungal	2000	36 (309)
caspofungin acetate	antifungal	2001	37 (263)
micafungin	antifungal	2002	38 (360)
voriconazole	antifungal	2002	38 (370)
fosfluconazole	antifungal	2004	40 (457)
eberconazole	antifungal	2005	41 (449)
luliconazole	antifungal	2005	41 (454)
anidulafungin	antifungal	2006	42 (512)
posaconazole	antifungal	2006	42 (532)
befunolol hydrochloride	antiglaucoma	1983	19 (315)
levobunolol hydrochloride	antiglaucoma	1985	21 (328)
dapiprazole hydrochloride	antiglaucoma	1987	23 (332)
apraclonidine hydrochloride	antiglaucoma	1988	24 (297)
unoprostone isopropyl ester	antiglaucoma	1994	30 (312)
dorzolamide hydrochloride	antiglaucoma	1995	31 (341)
brimonidine	antiglaucoma	1996	32 (306)
latanoprost	antiglaucoma	1996	32 (311)
brinzolamide	antiglaucoma	1998	34 (318)
bimatoprost	antiglaucoma	2001	37 (261)
travoprost	antiglaucoma	2001	37 (272)
tafluprost	antiglaucoma	2008	44 (625)
lomitapide mesylate	antihypercholersteremic	2012	48 (504)
meglutol	antihypercholesterolemic	1983	19 (321)
divistyramine	antihypercholesterolemic	1984	20 (317)
melinamide	antihypercholesterolemic	1984	20 (320)
acipimox	antihypercholesterolemic	1985	21 (323)
ciprofibrate	antihypercholesterolemic	1985	21 (326)
beclobrate	antihypercholesterolemic	1986	22 (317)
binifibrate	antihypercholesterolemic	1986	22 (317)
ronafibrate	antihypercholesterolemic	1986	22 (326)
lovastatin	antihypercholesterolemic	1987	23 (337)
simvastatin	antihypercholesterolemic	1988	24 (311)
pravastatin	antihypercholesterolemic	1989	25 (316)
fluvastatin	antihypercholesterolemic	1994	30 (300)
atorvastatin calcium	antihypercholesterolemic	1997	33 (328)

GENERIC NAME	INDICATION	YEAR INTRODUCED	ARMC VOL., (PAGE)
cerivastatin	antihypercholesterolemic	1997	33 (331)
colestimide	antihypercholesterolemic	1999	35 (337)
colesevelam hydrochloride	antihypercholesterolemic	2000	36 (300)
ezetimibe	antihypercholesterolemic	2002	38 (355)
pitavastatin	antihypercholesterolemic	2003	39 (282)
rosuvastatin	antihypercholesterolemic	2003	39 (283)
choline fenofibrate	antihypercholesterolemic	2008	44 (594)
captopril	antihypertensive	1982	13 (086)
betaxolol hydrochloride	antihypertensive	1983	19 (315)
budralazine	antihypertensive	1983	19 (315)
celiprolol hydrochloride	antihypertensive	1983	19 (317)
guanadrel sulfate	antihypertensive	1983	19 (319)
enalapril maleate	antihypertensive	1984	20 (317)
terazosin hydrochloride	antihypertensive	1984	20 (323)
bopindolol	antihypertensive	1985	21 (324)
bunazosin hydrochloride	antihypertensive	1985	21 (324)
ketanserin	antihypertensive	1985	21 (328)
nitrendipine	antihypertensive	1985	21 (331)
trimazosin hydrochloride	antihypertensive	1985	21 (333)
arotinolol hydrochloride	antihypertensive	1986	22 (316)
bisoprolol fumarate	antihypertensive	1986	22 (317)
bevantolol hydrochloride	antihypertensive	1987	23 (328)
enalaprilat	antihypertensive	1987	23 (332)
lisinopril	antihypertensive	1987	23 (337)
pinacidil	antihypertensive	1987	23 (340)
tertatolol hydrochloride	antihypertensive	1987	23 (344)
alacepril	antihypertensive	1988	24 (296)
alfuzosin hydrochloride	antihypertensive	1988	24 (296)
amosulalol	antihypertensive	1988	24 (297)
cadralazine	antihypertensive	1988	24 (298)
cicletanine	antihypertensive	1988	24 (299)
doxazosin mesylate	antihypertensive	1988	24 (300)
felodipine	antihypertensive	1988	24 (302)
nipradilol	antihypertensive	1988	24 (307)
perindopril	antihypertensive	1988	24 (309)
rilmenidine	antihypertensive	1988	24 (310)
tiamenidine hydrochloride	antihypertensive	1988	24 (311)

GENERIC NAME	INDICATION	YEAR INTRODUCED	ARMC VOL., (PAGE)
delapril	antihypertensive	1989	25 (311)
dilevalol	antihypertensive	1989	25 (311)
isradipine	antihypertensive	1989	25 (315)
nilvadipine	antihypertensive	1989	25 (316)
quinapril	antihypertensive	1989	25 (317)
ramipril	antihypertensive	1989	25 (317)
amlodipine besylate	antihypertensive	1990	26 (298)
benazepril hydrochloride	antihypertensive	1990	26 (299)
cilazapril	antihypertensive	1990	26 (301)
manidipine hydrochloride	antihypertensive	1990	26 (304)
nisoldipine	antihypertensive	1990	26 (306)
benidipine hydrochloride	antihypertensive	1991	27 (322)
carvedilol	antihypertensive	1991	27 (323)
fosinopril sodium	antihypertensive	1991	27 (328)
lacidipine	antihypertensive	1991	27 (330)
moxonidine	antihypertensive	1991	27 (330)
barnidipine hydrochloride	antihypertensive	1992	28 (326)
tilisolol hydrochloride	antihypertensive	1992	28 (337)
imidapril hydrochloride	antihypertensive	1993	29 (339)
trandolapril	antihypertensive	1993	29 (348)
efonidipine	antihypertensive	1994	30 (299)
losartan	antihypertensive	1994	30 (302)
temocapril	antihypertensive	1994	30 (311)
cinildipine	antihypertensive	1995	31 (339)
moexipril hydrochloride	antihypertensive	1995	31 (346)
spirapril hydrochloride	antihypertensive	1995	31 (349)
aranidipine	antihypertensive	1996	32 (306)
valsartan	antihypertensive	1996	32 (320)
candesartan cilexetil	antihypertensive	1997	33 (330)
eprosartan	antihypertensive	1997	33 (333)
irbesartan	antihypertensive	1997	33 (336)
lercanidipine	antihypertensive	1997	33 (337)
mebefradil hydrochloride	antihypertensive	1997	33 (338)
nebivolol	antihypertensive	1997	33 (339)
fenoldopam mesylate	antihypertensive	1998	34 (322)
telmisartan	antihypertensive	1999	35 (349)
zofenopril calcium	antihypertensive	2000	36 (313)
bosentan	antihypertensive	2001	37 (262)
olimesartan Medoxomil	antihypertensive	2002	38 (363)

GENERIC NAME	INDICATION	YEAR INTRODUCED	ARMC VOL., (PAGE)
treprostinil sodium	antihypertensive	2002	38 (368)
azelnidipine	antihypertensive	2003	39 (270)
eplerenone	antihypertensive	2003	39 (276)
aliskiren	antihypertensive	2007	43 (461)
clevidipine	antihypertensive	2008	44 (596)
azilsartan	antihypertensive	2011	47 (514)
butyl flufenamate	antiinflammatory	1983	19 (316)
halometasone	antiinflammatory	1983	19 (320)
hydrocortisone butyrate	antiinflammatory	1983	19 (320)
isoxicam	antiinflammatory	1983	19 (320)
oxaprozin	antiinflammatory	1983	19 (322)
fisalamine	antiinflammatory	1984	20 (318)
isofezolac	antiinflammatory	1984	20 (319)
piketoprofen	antiinflammatory	1984	20 (322)
pimaprofen	antiinflammatory	1984	20 (322)
alclometasone dipropionate	antiinflammatory	1985	21 (323)
diacerein	antiinflammatory	1985	21 (326)
etodolac	antiinflammatory	1985	21 (327)
nabumetone	antiinflammatory	1985	21 (330)
nimesulide	antiinflammatory	1985	21 (330)
amfenac sodium	antiinflammatory	1986	22 (315)
deflazacort	antiinflammatory	1986	22 (319)
felbinac	antiinflammatory	1986	22 (320)
lobenzarit sodium	antiinflammatory	1986	22 (322)
loxoprofen sodium	antiinflammatory	1986	22 (322)
osalazine sodium	antiinflammatory	1986	22 (324)
prednicarbate	antiinflammatory	1986	22 (325)
AF-2259	antiinflammatory	1987	23 (325)
flunoxaprofen	antiinflammatory	1987	23 (335)
mometasone furoate	antiinflammatory	1987	23 (338)
tenoxicam	antiinflammatory	1987	23 (344)
clobenoside	antiinflammatory	1988	24 (300)
hydrocortisone aceponate	antiinflammatory	1988	24 (304)
piroxicam cinnamate	antiinflammatory	1988	24 (309)
interferon, gamma	antiinflammatory	1989	25 (314)
aminoprofen	antiinflammatory	1990	26 (298)
droxicam	antiinflammatory	1990	26 (302)
fluticasone propionate	antiinflammatory	1990	26 (303)
halobetasol propionate	antiinflammatory	1991	27 (329)
aceclofenac	antiinflammatory	1992	28 (325)
butibufen	antiinflammatory	1992	28 (327)
deprodone propionate	antiinflammatory	1992	28 (329)
amtolmetin guacil	antiinflammatory	1993	29 (332)

GENERIC NAME	INDICATION	YEAR INTRODUCED	ARMC VOL., (PAGE)
zaltoprofen	antiinflammatory	1993	29 (349)
actarit	antiinflammatory	1994	30 (296)
ampiroxicam	antiinflammatory	1994	30 (296)
betamethasone butyrate propionate	antiinflammatory	1994	30 (297)
dexibuprofen	antiinflammatory	1994	30 (298)
rimexolone	antiinflammatory	1995	31 (348)
bromfenac sodium	antiinflammatory	1997	33 (329)
lornoxicam	antiinflammatory	1997	33 (337)
sivelestat	antiinflammatory	2002	38 (366)
lumiracoxib	antiinflammatory	2005	41 (455)
nepafenac	antiinflammatory	2005	41 (456)
canakinumab	antiinflammatory	2009	45 (484)
golimumab	antiinflammatory	2009	45 (503)
mefloquine hydrochloride	antimalarial	1985	21 (329)
artemisinin	antimalarial	1987	23 (327)
halofantrine	antimalarial	1988	24 (304)
arteether	antimalarial	2000	36 (296)
bulaquine	antimalarial	2000	36 (299)
alpiropride	antimigraine	1988	24 (296)
sumatriptan succinate	antimigraine	1991	27 (333)
naratriptan hydrochloride	antimigraine	1997	33 (339)
zomitriptan	antimigraine	1997	33 (345)
rizatriptan benzoate	antimigraine	1998	34 (330)
lomerizine hydrochloride	antimigraine	1999	35 (342)
almotriptan	antimigraine	2000	36 (295)
eletriptan	antimigraine	2001	37 (266)
frovatriptan	antimigraine	2002	38 (357)
dexfenfluramine	antiobesity	1997	33 (332)
orlistat	antiobesity	1998	34 (327)
sibutramine	antiobesity	1998	34 (331)
rimonabant	antiobesity	2006	42 (537)
lorcaserin hydrochloride	antiobesity	2012	48 (506)
quinf amideamebicide	antiparasitic	1984	20 (322)
ivermectin	antiparasitic	1987	23 (336)
atovaquone	antiparasitic	1992	28 (326)
epoprostenol sodium	antiplatelet	1983	19 (318)
beraprost sodium	antiplatelet	1992	28 (326)
iloprost	antiplatelet	1992	28 (332)
prasugrel	antiplatelet	2009	45 (519)
cabergoline	antiprolactin	1993	29 (334)

GENERIC NAME	INDICATION	YEAR INTRODUCED	ARMC VOL., (PAGE)
acitretin	antipsoriasis	1989	25 (309)
calcipotriol	antipsoriasis	1991	27 (323)
tacalcitol	antipsoriasis	1993	29 (346)
tazarotene	antipsoriasis	1997	33 (343)
alefacept	antipsoriasis	2003	39 (267)
efalizumab	antipsoriasis	2003	39 (274)
ustekinumab	antipsoriasis	2009	45 (532)
timiperone	antipsychotic	1984	20 (323)
amisulpride	antipsychotic	1986	22 (316)
zuclopenthixol acetate	antipsychotic	1987	23 (345)
remoxipride hydrochloride	antipsychotic	1990	26 (308)
clospipramine hydrochloride	antipsychotic	1991	27 (325)
nemonapride	antipsychotic	1991	27 (331)
risperidone	antipsychotic	1993	29 (344)
olanzapine	antipsychotic	1996	32 (313)
sertindole	antipsychotic	1996	32 (318)
quetiapine fumarate	antipsychotic	1997	33 (341)
ziprasidone hydrochloride	antipsychotic	2000	36 (312)
perospirone hydrochloride	antipsychotic	2001	37 (270)
aripiprazole	antipsychotic	2002	38 (350)
paliperidone	antipsychotic	2007	43 (482)
blonanserin	antipsychotic	2008	44 (587)
asenapine	antipsychotic	2009	45 (479)
lurasidone hydrochloride	antipsychotic	2010	46 (473)
drotrecogin alfa	antisepsis	2001	37 (265)
tiquizium bromide	antispasmodic	1984	20 (324)
cimetropium bromide	antispasmodic	1985	21 (326)
romiplostim	antithrombocytopenic	2008	44 (619)
eltrombopag	antithrombocytopenic	2009	45 (497)
indobufen	antithrombotic	1984	20 (319)
defibrotide	antithrombotic	1986	22 (319)
alteplase	antithrombotic	1987	23 (326)
APSAC	antithrombotic	1987	23 (326)
picotamide	antithrombotic	1987	23 (340)
cilostazol	antithrombotic	1988	24 (299)
limaprost	antithrombotic	1988	24 (306)
ozagrel sodium	antithrombotic	1988	24 (308)
argatroban	antithrombotic	1990	26 (299)
ethyl icosapentate	antithrombotic	1990	26 (303)
cloricromen	antithrombotic	1991	27 (325)

GENERIC NAME	INDICATION	YEAR INTRODUCED	ARMC VOL., (PAGE)
sarpogrelate hydrochloride	antithrombotic	1993	29 (344)
reteplase	antithrombotic	1996	32 (316)
anagrelide hydrochloride	antithrombotic	1997	33 (328)
clopidogrel hydrogensulfate	antithrombotic	1998	34 (320)
tirofiban hydrochloride	antithrombotic	1998	34 (332)
eptilfibatide	antithrombotic	1999	35 (340)
bivalirudin	antithrombotic	2000	36 (298)
fondaparinux sodium	antithrombotic	2002	38 (356)
ticagrelor	antithrombotic	2010	46 (488)
apixaban	antithrombotic	2011	47 (509)
edoxaban	antithrombotic	2011	47 (527)
levodropropizine	antitussive	1988	24 (305)
nitisinone	antityrosinaemia	2002	38 (361)
sofalcone	antiulcer	1984	20 (323)
teprenone	antiulcer	1984	20 (323)
enprostil	antiulcer	1985	21 (327)
famotidine	antiulcer	1985	21 (327)
misoprostol	antiulcer	1985	21 (329)
rosaprostol	antiulcer	1985	21 (332)
roxatidine acetate hydrochloride	antiulcer	1986	22 (326)
troxipide	antiulcer	1986	22 (327)
benexate hydrochloride	antiulcer	1987	23 (328)
nizatidine	antiulcer	1987	23 (339)
ornoprostil	antiulcer	1987	23 (339)
plaunotol	antiulcer	1987	23 (340)
roxithromycin	antiulcer	1987	23 (342)
spizofurone	antiulcer	1987	23 (343)
omeprazole	antiulcer	1988	24 (308)
irsogladine	antiulcer	1989	25 (315)
rebamipide	antiulcer	1990	26 (308)
lansoprazole	antiulcer	1992	28 (332)
ecabet sodium	antiulcer	1993	29 (336)
tretinoin tocoferil	antiulcer	1993	29 (348)
polaprezinc	antiulcer	1994	30 (307)
pantoprazole sodium	antiulcer	1995	30 (306)
ranitidine bismuth citrate	antiulcer	1995	31 (348)
ebrotidine	antiulcer	1997	33 (333)
rabeprazole sodium	antiulcer	1998	34 (328)
dosmalfate	antiulcer	2000	36 (302)
egualen sodium	antiulcer	2000	36 (303)

GENERIC NAME	INDICATION	YEAR INTRODUCED	ARMC VOL., (PAGE)
esomeprazole magnesium	antiulcer	2000	36 (303)
lafutidine	antiulcer	2000	36 (307)
dexlansoprazole	antiulcer	2009	45 (492)
rimantadine hydrochloride	antiviral	1987	23 (342)
zidovudine	antiviral	1987	23 (345)
ganciclovir	antiviral	1988	24 (303)
foscarnet sodium	antiviral	1989	25 (313)
didanosine	antiviral	1991	27 (326)
zalcitabine	antiviral	1992	28 (338)
sorivudine	antiviral	1993	29 (345)
famciclovir	antiviral	1994	30 (300)
propagermanium	antiviral	1994	30 (308)
stavudine	antiviral	1994	30 (311)
lamivudine	antiviral	1995	31 (345)
saquinavir mesvlate	antiviral	1995	31 (349)
valaciclovir hydrochloride	antiviral	1995	31 (352)
cidofovir	antiviral	1996	32 (306)
indinavir sulfate	antiviral	1996	32 (310)
nevirapine	antiviral	1996	32 (313)
penciclovir	antiviral	1996	32 (314)
ritonavir	antiviral	1996	32 (317)
delavirdine mesylate	antiviral	1997	33 (331)
imiquimod	antiviral	1997	33 (335)
interferon alfacon-1	antiviral	1997	33 (336)
nelfinavir mesylate	antiviral	1997	33 (340)
efavirenz	antiviral	1998	34 (321)
fomivirsen sodium	antiviral	1998	34 (323)
abacavir sulfate	antiviral	1999	35 (333)
amprenavir	antiviral	1999	35 (334)
oseltamivir phosphate	antiviral	1999	35 (346)
zanamivir	antiviral	1999	35 (352)
lopinavir	antiviral	2000	36 (310)
tenofovir disoproxil fumarate	antiviral	2001	37 (271)
vaglancirclovir hydrochloride	antiviral	2001	37 (273)
adefovir dipivoxil	antiviral	2002	38 (348)
atazanavir	antiviral	2003	39 (269)
emtricitabine	antiviral	2003	39 (274)
enfuvirtide	antiviral	2003	39 (275)
fosamprenavir	antiviral	2003	39 (277)
influenza virus (live)	antiviral	2003	39 (277)

GENERIC NAME	INDICATION	YEAR INTRODUCED	ARMC VOL., (PAGE)
entecavir	antiviral	2005	41 (450)
tipranavir	antiviral	2005	41 (470)
darunavir	antiviral	2006	42 (515)
telbivudine	antiviral	2006	42 (546)
clevudine	antiviral	2007	43 (466)
maraviroc	antiviral	2007	43 (478)
raltegravir	antiviral	2007	43 (484)
etravirine	antiviral	2008	44 (602)
laninamivir octanoate	antiviral	2010	46 (470)
peramivir	antiviral	2010	46 (477)
boceprevir	antiviral	2011	47 (521)
rilpivirine	antiviral	2011	47 (546)
telaprevir	antiviral	2011	47 (552)
cevimeline hydrochloride	antixerostomia	2000	36 (299)
etizolam	anxiolytic	1984	20 (318)
flutazolam	anxiolytic	1984	20 (318)
mexazolam	anxiolytic	1984	20 (321)
buspirone hydrochloride	anxiolytic	1985	21 (324)
doxefazepam	anxiolytic	1985	21 (326)
flutoprazepam	anxiolytic	1986	22 (320)
metaclazepam	anxiolytic	1987	23 (338)
alpidem	anxiolytic	1991	27 (322)
cinolazepam	anxiolytic	1993	29 (334)
tandospirone	anxiolytic	1996	32 (319)
dexmethylphenidate hydrochloride	attention deficit hyperactivity disorder	2002	38 (352)
atomoxetine	attention deficit hyperactivity disorder	2003	39 (270)
lisdexamfetamine	attention deficit hyperactivity disorder	2007	43 (477)
finasteride	benign prostatic hyperplasia	1992	28 (331)
tamsulosin hydrochloride	benign prostatic hyperplasia	1993	29 (347)
dutasteride	benign prostatic hyperplasia	2002	38 (353)
fenbuprol	biliary tract dysfunction	1983	19 (318)
clodronate disodium	calcium regulation	1986	22 (319)
gallium nitrate	calcium regulation	1991	27 (328)
neridronic acide	calcium regulation	2002	38 (361)
dexrazoxane	cardioprotective	1992	28 (330)
nimodipine	cerebral vasodilator	1985	21 (330)
brovincamine fumarate	cerebral vasodilator	1986	22 (317)
propentofylline propionate	cerebral vasodilator	1988	24 (310)

GENERIC NAME	INDICATION	YEAR INTRODUCED	ARMC VOL., (PAGE)
indacaterol	chronic obstructive pulmonary disease	2009	45 (505)
formoterol fumarate	chronic obstructive pulmonary disorder	1986	22 (321)
tiotropium bromide	chronic obstructive pulmonary disorder	2002	38 (368)
roflumilast	chronic obstructive pulmonary disorder	2010	46 (480)
aclidinium bromide	chronic obstructive pulmonary disorder	2012	48 (481)
levacecarnine hydrochloride	cognition enhancer	1986	22 (322)
oxiracetam	cognition enhancer	1987	23 (339)
exifone	cognition enhancer	1988	24 (302)
choline alfoscerate	cognition enhancer	1990	26 (300)
aniracetam	cognition enhancer	1993	29 (333)
pramiracetam sulfate	cognition enhancer	1993	29 (343)
amrinone	congestive heart failure	1983	19 (314)
bucladesine sodium	congestive heart failure	1984	20 (316)
ibopamine hydrochloride	congestive heart failure	1984	20 (319)
denopamine	congestive heart failure	1988	24 (300)
enoximone	congestive heart failure	1988	24 (301)
xamoterol fumarate	congestive heart failure	1988	24 (312)
dopexamine	congestive heart failure	1989	25 (312)
milrinone	congestive heart failure	1989	25 (316)
vesnarinone	congestive heart failure	1990	26 (310)
flosequinan	congestive heart failure	1992	28 (331)
docarpamine	congestive heart failure	1994	30 (298)
pimobendan	congestive heart failure	1994	30 (307)
carperitide	congestive heart failure	1995	31 (339)
loprinone hydrochloride	congestive heart failure	1996	32 (312)
colforsin daropate hydrochloride	congestive heart failure	1999	35 (337)
levosimendan	congestive heart failure	2000	36 (308)
nesiritide	congestive heart failure	2001	37 (269)
lulbiprostone	constipation	2006	42 (525)
methylnaltrexone bromide	constipation	2008	44 (612)
promegestrone	contraception	1983	19 (323)
gestrinone	contraception	1986	22 (321)
nomegestrol acetate	contraception	1986	22 (324)
norgestimate	contraception	1986	22 (324)
gestodene	contraception	1987	23 (335)

GENERIC NAME	INDICATION	YEAR INTRODUCED	ARMC VOL., (PAGE)
centchroman	contraception	1991	27 (324)
drospirenone	contraception	2000	36 (302)
trimegestone	contraception	2001	37 (273)
norelgestromin	contraception	2002	38 (362)
ulipristal acetate	contraception	2009	45 (530)
biolimus drug–eluting stent	coronary artery disease, antirestenotic	2008	44 (586)
pasireotide	Cushing's Disease	2012	48 (512)
neltenexine	cystic fibrosis	1993	29 (341)
dornase alfa	cystic fibrosis	1994	30 (298)
ivacaftor	cystic fibrosis	2012	48 (501)
amifostine	cytoprotective	1995	31 (338)
kinetin	dermatologic, skin photodamage	1999	35 (341)
gadoversetamide	diagnostic	2000	36 (304)
ioflupane	diagnostic	2000	36 (306)
muzolimine	diuretic	1983	19 (321)
azosemide	diuretic	1986	22 (316)
torasemide	diuretic	1993	29 (348)
alpha–1 antitrypsin	emphysema	1988	24 (297)
telmesteine	expectorant	1992	28 (337)
erdosteine	expectorant	1995	31 (342)
fudosteine	expectorant	2001	37 (267)
agalsidase alfa	Fabry's disease	2001	37 (259)
chenodiol	gallstones	1983	19 (317)
cisapride	gastroprokinetic	1988	24 (299)
cinitapride	gastroprokinetic	1990	26 (301)
itopride hydrochloride	gastroprokinetic	1995	31 (344)
mosapride citrate	gastroprokinetic	1998	34 (326)
alglucerase	Gaucher's disease	1991	27 (321)
imiglucerase	Gaucher's disease	1994	30 (301)
miglustat	Gaucher's disease	2003	39 (279)
rilonacept	genetic autoinflammatory syndromes	2008	44 (615)
febuxostat	gout	2009	45 (501)
somatropin	growth disorders	1987	23 (343)
octreotide	growth disorders	1988	24 (307)
somatomedin-1	growth disorders	1994	30 (310)
somatotropin	growth disorders	1994	30 (310)
lanreotide acetate	growth disorders	1995	31 (345)
pegvisomant	growth disorders	2003	39 (281)
factor VIIa	haemophilia	1996	32 (307)
erythropoietin	hematopoietic	1988	24 (301)
peginesatide	hematopoietic	2012	48 (514)
eculizumab	hemoglobinuria	2007	43 (468)

GENERIC NAME	INDICATION	YEAR INTRODUCED	ARMC VOL., (PAGE)
factor VIII	hemostatic	1992	28 (330)
thrombin alfa	hemostatic	2008	44 (627)
malotilate	hepatoprotective	1985	21 (329)
mivotilate	hepatoprotective	1999	35 (343)
buserelin acetate	hormone therapy	1984	20 (316)
leuprolide acetate	hormone therapy	1984	20 (319)
goserelin	hormone therapy	1987	23 (336)
tibolone	hormone therapy	1988	24 (312)
nafarelin acetate	hormone therapy	1990	26 (306)
paricalcitol	hyperparathyroidism	1998	34 (327)
doxercalciferol	hyperparathyroidism	1999	35 (339)
maxacalcitol	hyperparathyroidism	2000	36 (310)
falecalcitriol	hyperparathyroidism	2001	37 (266)
cinacalcet	hyperparathyroidism	2004	40 (451)
quinagolide	hyperprolactinemia	1994	30 (309)
conivaptan	hyponatremia	2006	42 (514)
mozavaptan	hyponatremia	2006	42 (527)
tolvaptan	hyponatremia	2009	45 (528)
thymopentin	immunomodulator	1985	21 (333)
bucillamine	immunomodulator	1987	23 (329)
centoxin	immunomodulator	1991	27 (325)
schizophyllan	immunostimulant	1985	22 (326)
lentinan	immunostimulant	1986	22 (322)
ubenimex	immunostimulant	1987	23 (345)
pegademase bovine	immunostimulant	1990	26 (307)
filgrastim	immunostimulant	1991	27 (327)
interferon gamma-1b	immunostimulant	1991	27 (329)
romurtide	immunostimulant	1991	27 (332)
sargramostim	immunostimulant	1991	27 (332)
pidotimod	immunostimulant	1993	29 (343)
GMDP	immunostimulant	1996	32 (308)
cyclosporine	immunosuppressant	1983	19 (317)
mizoribine	immunosuppressant	1984	20 (321)
muromonab-CD3	immunosuppressant	1986	22 (323)
tacrolimus	immunosuppressant	1993	29 (347)
gusperimus	immunosuppressant	1994	30 (300)
mycophenolate mofetil	immunosuppressant	1995	31 (346)
pimecrolimus	immunosuppressant	2002	38 (365)
mycophenolate sodium	immunosuppressant	2003	39 (279)
everolimus	immunosuppressant	2004	40 (455)
belatacept	immunosuppressant	2011	47 (516)
follitropin alfa	infertility	1996	32 (307)
follitropin beta	infertility	1996	32 (308)
cetrorelix	infertility	1999	35 (336)
ganirelix acetate	infertility	2000	36 (305)

GENERIC NAME	INDICATION	YEAR INTRODUCED	ARMC VOL., (PAGE)
corifollitropin alfa	infertility	2010	46 (455)
defeiprone	iron chelation therapy	1995	31 (340)
deferasirox	iron chelation therapy	2005	41 (446)
alosetron hydrochloride	irritable bowel syndrome	2000	36 (295)
tegaserod maleate	irritable bowel syndrome	2001	37 (270)
certolizumab pegol	irritable bowel syndrome	2008	44 (592)
linaclotide	irritable bowel syndrome	2012	48 (502)
nartograstim	leukopenia	1994	30 (304)
tesamorelin acetate	lipodystrophy	2010	46 (486)
belimumab	lupus	2011	47 (519)
Lyme disease vaccine	Lyme disease	1999	35 (342)
sildenafil citrate	male sexual dysfunction	1998	34 (331)
tadalafil	male sexual dysfunction	2003	39 (284)
vardenafil	male sexual dysfunction	2003	39 (286)
udenafil	male sexual dysfunction	2005	41 (472)
avanafil	male sexual dysfunction	2011	47 (512)
laronidase	mucopolysaccharidosis I	2003	39 (278)
idursulfase	mucopolysaccharidosis II (Hunter syndrome)	2006	42 (520)
galsulfase	mucopolysaccharidosis VI	2005	41 (453)
palifermin	mucositis	2005	41 (461)
interferon, b-1b	multiple sclerosis	1993	29 (339)
interferon, b-1a	multiple sclerosis	1996	32 (311)
glatiramer acetate	multiple sclerosis	1997	33 (334)
natalizumab	multiple sclerosis	2004	40 (462)
dalfampridine	multiple sclerosis	2010	46 (458)
fingolimod	multiple sclerosis	2010	46 (468)
teriflunomide	multiple sclerosis	2012	48 (530)
afloqualone	muscle relaxant	1983	19 (313)
eperisone hydrochloride	muscle relaxant	1983	19 (318)
tiropramide hydrochloride	muscle relaxant	1983	19 (324)
tizanidine	muscle relaxant	1984	20 (324)
doxacurium chloride	muscle relaxant	1991	27 (326)
mivacurium chloride	muscle relaxant	1992	28 (334)
rocuronium bromide	muscle relaxant	1994	30 (309)
cisatracurium besilate	muscle relaxant	1995	31 (340)
rapacuronium bromide	muscle relaxant	1999	35 (347)
decitabine	myelodysplastic syndromes	2006	42 (519)
lenalidomide	myelodysplastic syndromes, multiple myeloma	2006	42 (523)
tinazoline	nasal decongestant	1988	24 (312)
taltirelin	neurodegeneration	2000	36 (311)
tafamidis	neurodegeneration	2011	47 (550)

GENERIC NAME	INDICATION	YEAR INTRODUCED	ARMC VOL., (PAGE)
sugammadex	neuromuscular blockade, reversal	2008	44 (623)
edaravone	neuroprotective	2001	37 (265)
idebenone	nootropic	1986	22 (321)
bifemelane hydrochloride	nootropic	1987	23 (329)
indeloxazine hydrochloride	nootropic	1988	24 (304)
nizofenzone	nootropic	1988	24 (307)
alcaftadine	ophthalmologic (allergic conjunctivitis)	2010	46 (444)
diquafosol tetrasodium	ophthalmologic (dry eye)	2010	46 (462)
verteporfin	ophthalmologic (macular degeneration)	2000	36 (312)
pegaptanib	ophthalmologic (macular degeneration)	2005	41 (458)
ranibizumab	ophthalmologic (macular degeneration)	2006	42 (534)
aflibercept	ophthalmologic, macular degeneration	2011	47 (507)
OP-1	osteoinductor	2001	37 (269)
APD	osteoporosis	1987	23 (326)
disodium pamidronate	osteoporosis	1989	25 (312)
ipriflavone	osteoporosis	1989	25 (314)
alendronate sodium	osteoporosis	1993	29 (332)
ibandronic acid	osteoporosis	1996	32 (309)
incadronic acid	osteoporosis	1997	33 (335)
raloxifene hydrochloride	osteoporosis	1998	34 (328)
risedronate sodium	osteoporosis	1998	34 (330)
zoledronate disodium	osteoporosis	2000	36 (314)
strontium ranelate	osteoporosis	2004	40 (466)
minodronic acid	osteoporosis	2009	45 (509)
denosumab	osteoporosis	2010	46 (459)
eldecalcitol	osteoporosis	2011	47 (529)
tiludronate disodium	Paget's disease	1995	31 (350)
nafamostat mesylate	pancreatitis	1986	22 (323)
pergolide mesylate	Parkinson's disease	1988	24 (308)
droxidopa	Parkinson's disease	1989	25 (312)
ropinirole hydrochloride	Parkinson's disease	1996	32 (317)
talipexole	Parkinson's disease	1996	32 (318)
budipine	Parkinson's disease	1997	33 (330)
pramipexole hydrochloride	Parkinson's disease	1997	33 (341)

GENERIC NAME	INDICATION	YEAR INTRODUCED	ARMC VOL., (PAGE)
tolcapone	Parkinson's disease	1997	33 (343)
entacapone	Parkinson's disease	1998	34 (322)
CHF-1301	Parkinson's disease	1999	35 (336)
rasagiline	Parkinson's disease	2005	41 (464)
rotigotine	Parkinson's disease	2006	42 (538)
sapropterin hydrochloride	phenylketouria	1992	28 (336)
alglucosidase alfa	Pompe disease	2006	42 (511)
alvimopan	post-operative ileus	2008	44 (584)
histrelin	precocious puberty	1993	29 (338)
dapoxetine	premature ejaculation	2009	45 (488)
atosiban	premature labor	2000	36 (297)
nalfurafine hydrochloride	pruritus	2009	45 (510)
pirfenidone	pulmonary fibrosis, idiopathic	2008	44 (614)
sitaxsentan	pulmonary hypertension	2006	42 (543)
ambrisentan	pulmonary hypertension	2007	43 (463)
pumactant	respiratory distress syndrome	1994	30 (308)
mepixanox	respiratory stimulant	1984	20 (320)
surfactant TA	respiratory surfactant	1987	23 (344)
gabapentin Enacarbil	restless leg syndrome	2011	47 (533)
teduglutide	short bowel syndrome	2012	48 (526)
binfonazole	sleep disorders	1983	19 (315)
brotizolam	sleep disorders	1983	19 (315)
loprazolam mesylate	sleep disorders	1983	19 (321)
butoctamide	sleep disorders	1984	20 (316)
quazepam	sleep disorders	1985	21 (332)
adrafinil	sleep disorders	1986	22 (315)
zopiclone	sleep disorders	1986	22 (327)
zolpidem hemitartrate	sleep disorders	1988	24 (313)
rilmazafone	sleep disorders	1989	25 (317)
modafinil	sleep disorders	1994	30 (303)
zaleplon	sleep disorders	1999	35 (351)
dexmedetomidine hydrochloride	sleep disorders	2000	36 (301)
eszopiclone	sleep disorders	2005	41 (451)
ramelteon	sleep disorders	2005	41 (462)
armodafinil	sleep disorders	2009	45 (478)
plerixafor hydrochloride	stem cell mobilizer	2009	45 (515)
tirilazad mesylate	subarachnoid hemorrhage	1995	31 (351)
balsalazide disodium	ulcerative colitis	1997	33 (329)
acetohydroxamic acid	urinary tract/bladder disorders	1983	19 (313)

GENERIC NAME	INDICATION	YEAR INTRODUCED	ARMC VOL., (PAGE)
sodium cellulose phosphate	urinary tract/bladder disorders	1983	19 (323)
propiverine hydrochloride	urinary tract/bladder disorders	1992	28 (335)
naftopidil	urinary tract/bladder disorders	1999	35 (344)
solifenacin	urinary tract/bladder disorders	2004	40 (466)
darifenacin	urinary tract/bladder disorders	2005	41 (445)
silodosin	urinary tract/bladder disorders	2006	42 (540)
imidafenacin	urinary tract/bladder disorders	2007	43 (472)
fesoterodine	urinary tract/bladder disorders	2008	44 (604)
mirabegron	urinary tract/bladder disorders	2011	47 (542)
tiopronin	urolithiasis	1989	25 (318)
cadexomer iodine	wound healing agent	1983	19 (316)
epidermal growth factor	wound healing agent	1987	23 (333)
prezatide copper acetate	wound healing agent	1996	32 (314)
acemannan	wound healing agent	2001	37 (259)

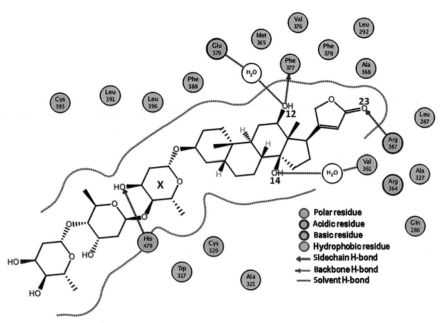

Plate 12.1 Crystal structure of the RORγt ligand binding pocket in complex with digoxin, shown as a two-dimensional ligand interaction diagram.[41]

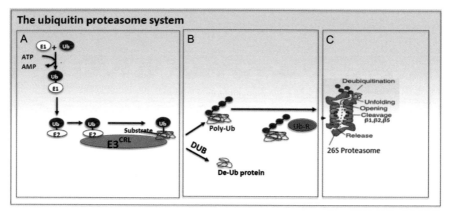

Plate 14.1 The ubiquitin proteasome system (adapted with permission[3]). (A) Unwanted proteins are tagged for degradation through attachment of ubiquitin (Ub) by a cascade of enzymes (E1, E2, and E3s). (B) The ubiquitin label is removed by a deubiquitinating enzyme (DUB) or the protein tagged with a poly-Ub chain is chaperoned to the proteasome. (C) Before the tagged protein can enter the core of the proteasome, the poly-Ub chain must be removed by a DUB enzyme present in the proteasome. The protein is then unfolded and cleaved by beta subunits in the core of the proteasome.

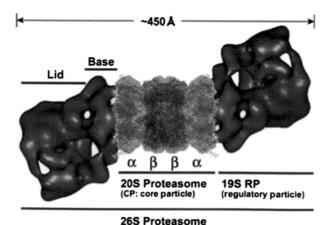

Plate 14.2 Composition of the 26S proteasome (used with permission).[8] The 26S consist of the 20S core particle and two 19S regulatory particles. The numbers refer to protein subunit size as determined by centrifugation.

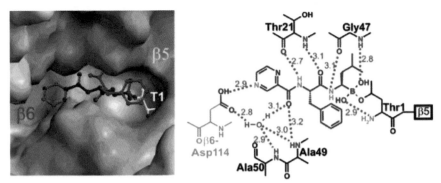

Plate 14.3 Bortezomib bound to the β5 subunit of the yeast 20S proteasome (used with permission).[12] X-ray structure and schematic representation of covalent binding of **1** to the β5 subunit.

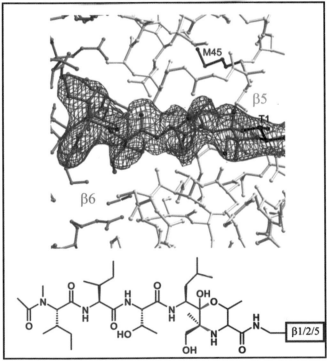

Plate 14.4 Interaction of an epoxyketone with the β5 subunit of the yeast 20S proteasome (used with permission).[21] The X-ray structure of epoxomicin (**7**) bound to the yeast β5 subunit and the schematic representation.

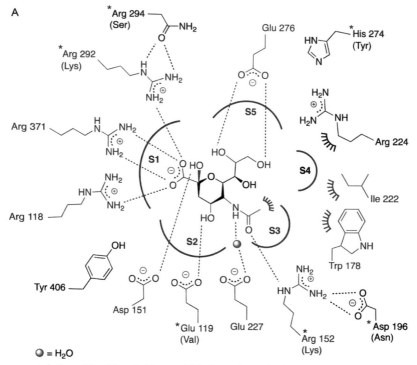

A

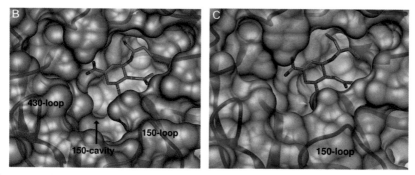

Plate 16.1 (A) A schematic representation of neuraminidase active site in complex with sialic acid (N2 numbering is used). Key binding interactions are shown as dashed lines, while the binding pockets with respective subsites (S1–S5) are highlighted in colors. Observed point mutations in some circulating influenza strains are indicated with an asterisk. (B) Neuraminidase N1-oseltamivir carboxylate complex showing the open 150-loop, 150-cavity, and 430-loop (PDB ID 2HU). (C) Neuraminidase N2-oseltamivir carboxylate complex with a closed 150-loop (PDB ID 2HU4). (B) and (C) were generated using UCSF Chimera package.[19]

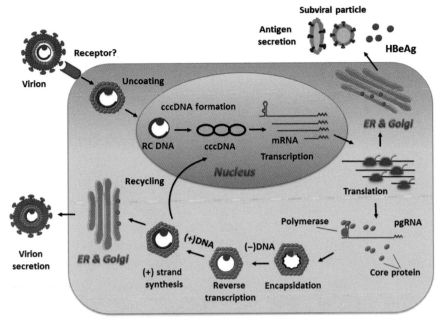

Plate 17.1 Schematic representation of the HBV life cycle.

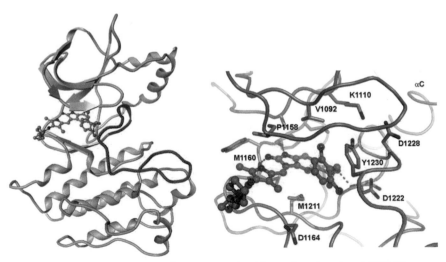

Plate 25.3 Cocrystal structure of **PHA-665752** with unphosphorylated MET KD.

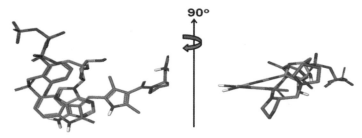

Plate 25.4 Overlap of **PHA-665752** (green) with ATP (cyan) in MET cocrystal structures.

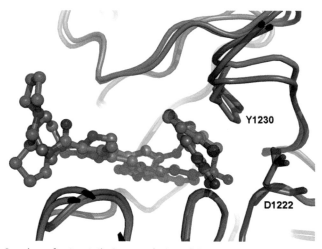

Plate 25.7 Overlay of crizotinib (gray color) and **PHA-665752** (cyan color) bound to unphosphorylated MET kinase domain.

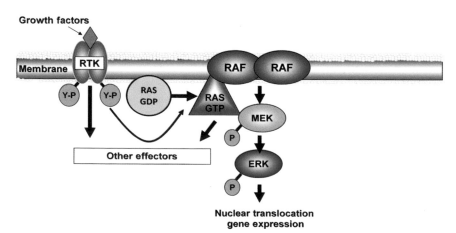

Plate 26.1 RAS–RAF–MEK–ERK signaling pathway.

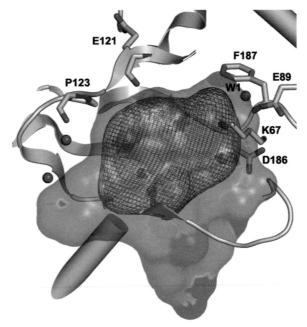

Plate 26.3 Kinase scaffolds bound in the ATP-binding pocket.

Plate 26.4 7-Azaindole bound in PIM-1 kinase.

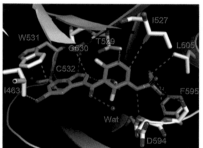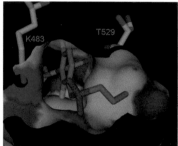

Compound **10**: PLX4720

$IC_{50} = 0.008\ \mu M$

Plate 26.7 Co-crystal structure of compound **10** (PLX4720) in BRAF.

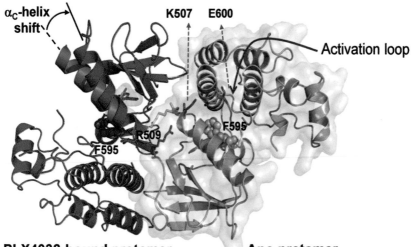

α_C-helix shift

K507   E600

Activation loop

R509

F595

F595

PLX4032-bound protomer   **Apo protomer**

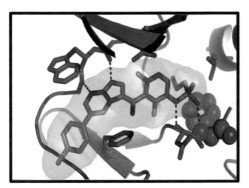

Plate 26.9 Co-crystal structure of vemurafenib.

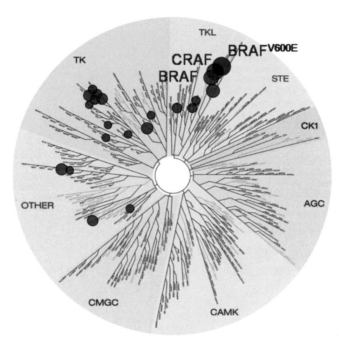

Plate 26.10 Kinome selectivity of vemurafenib.